PROTEIN MODIFICOMICS

PROTEIN MODIFICOMICS

From Modifications to Clinical Perspectives

Edited by

Tanveer Ali Dar
Laishram Rajendrakumar Singh

Academic Press is an imprint of Elsevier
125 London Wall, London EC2Y 5AS, United Kingdom
525 B Street, Suite 1650, San Diego, CA 92101, United States
50 Hampshire Street, 5th Floor, Cambridge, MA 02139, United States
The Boulevard, Langford Lane, Kidlington, Oxford OX5 1GB, United Kingdom

Notices

Knowledge and best practice in this field are constantly changing. As new research and experience broaden our understanding, changes in research methods, professional practices, or medical treatment may become necessary.

Practitioners and researchers must always rely on their own experience and knowledge in evaluating and using any information, methods, compounds, or experiments described herein. In using such information or methods they should be mindful of their own safety and the safety of others, including parties for whom they have a professional responsibility.

To the fullest extent of the law, neither the Publisher nor the authors, contributors, or editors, assume any liability for any injury and/or damage to persons or property as a matter of products liability, negligence or otherwise, or from any use or operation of any methods, products, instructions, or ideas contained in the material herein.

Library of Congress Cataloging-in-Publication Data
A catalog record for this book is available from the Library of Congress

British Library Cataloguing-in-Publication Data
A catalogue record for this book is available from the British Library

ISBN: **978-0-12-811913-6**

For information on all Academic Press publications visit our website at https://www.elsevier.com/books-and-journals

Publisher: Andre G. Wolff
Acquisition Editor: Peter B. Linsley
Editorial Project Manager: Timothy Bennett
Production Project Manager: Maria Bernard
Cover Designer: Matthew Limbert

Typeset by SPi Global, India

Contents

3. Phosphorylation and Acetylation of Proteins as Posttranslational Modification: Implications in Human Health and Associated Diseases

SANA QAUSAIN, HEMALATHA SRINIVASAN, SHAZIA JAMAL, MOHAMMAD NASIRUDDIN, MD. KHURSHID ALAM KHAN

4. Protein Modifications and Lifestyle Disorders

SHIVANI ARORA, ANJU KATYAL

5. Ubiquitin Mediated Posttranslational Modification of Proteins Involved in Various Signaling Diseases

LAVANYA V, SHAZIA JAMAL, NEESAR AHMED

6. Role of Glycosylation in Modulating Therapeutic Efficiency of Protein Pharmaceuticals

PARVAIZ AHMAD DAR, USMA MANZOOR, SNOWBER SHABIR WANI, FASIL ALI, TANVEER ALI DAR

7. Posttranslational Modification of Heterologous Human Therapeutics in Plant Host Expression Systems

AYYAGARI ARCHANA, LAKSHNA MAHAJAN, SAFIKUR RAHMAN, RINKI MINAKSHI

8. Protein Modification in Plants in Response to Abiotic Stress

HILAL AHMAD QAZI, NELOFER JAN, SALIKA RAMAZAN, RIFFAT JOHN

9. Posttranslational Modifications Associated With Cancer and Their Therapeutic Implications

ANIKET KUMAR BANSAL, LAISHRAM RAJENDRAKUMAR SINGH, MAJID RASOOL KAMLI

12. Posttranslational Modifications in Algae: Role in Stress Response and Biopharmaceutical Production

PARVEZ AHMAD, FAREHA BANO

13. Protein Glycosylation: An Important Tool for Diagnosis or Early Detection of Diseases

HUMAYRA BASHIR, BARQUL AFAQ WANI, BASHIR A. GANAI, SHABIR AHMAD MIR

Contributors

Parvez Ahmad Protein Conformation and Enzymology Lab, Department of Biosciences, Jamia Millia Islamia (A Central University), New Delhi, India

Neesar Ahmed School of Life Sciences, B.S. Abdur Rahman Crescent Institute of Science & Technology, Chennai, India

Fasil Ali Department of PG Studies and Research in Biochemistry, Jnana Kaveri PG Centre, Mangalore University, Chikka Aluvara; Clinical Biochemistry, University of Kashmir, Srinagar, India

Shajrul Amin Department of Biochemistry, University of Kashmir, Srinagar, India

Ayyagari Archana Department of Microbiology, Swami Shraddhanand College, University of Delhi, New Delhi, India

Shivani Arora Dr. B.R. Ambedkar Center for Biomedical Research, University of Delhi, Delhi, India

Fareha Bano Department of Biology, College of Science and Arts (Female Branch), Al Ula Campus, Taibah Universty, Al Ula, Madina Province, Saudi Arabia

Aniket Kumar Bansal Dr. B.R. Ambedkar Center for Biomedical Research, University of Delhi, Delhi, India

Humayra Bashir Centre of Research for Development, University of Kashmir Srinagar, Srinagar, India

Ajaz A. Bhat Division of Translational Medicine, Research Branch, Sidra Medicine, Doha, Qatar

Reshmee Bhattacharya Dr. B.R. Ambedkar Center for Biomedical Research, University of Delhi, Delhi, India

M.Z. Chishti Centre of Research for Development, University of Kashmir, Srinagar, India

Jehangir Shafi Dar Centre of Research for Development, University of Kashmir, Srinagar, India

Parvaiz Ahmad Dar Clinical Biochemistry, University of Kashmir, Srinagar, India

Tanveer Ali Dar Clinical Biochemistry, University of Kashmir, Srinagar, India

Bashir A. Ganai Centre of Research for Development, University of Kashmir, Srinagar, India

Shazia Jamal School of Life Sciences, B.S. Abdur Rahman Crescent Institute of Science & Technology, Chennai, India

Nelofer Jan Plant Molecular Biology Lab, Department of Botany, University of Kashmir, Srinagar, India

Riffat John Plant Molecular Biology Lab, Department of Botany, University of Kashmir, Srinagar, India

Majid Rasool Kamli Center of Excellence in Bionanoscience Research; Department of Biological Sciences, Faculty of Science, King Abdulaziz University, Jeddah, Saudi Arabia

Anju Katyal Dr. B.R. Ambedkar Center for Biomedical Research, University of Delhi, Delhi, India

Iram Ashaq Kawa Department of Biochemistry; Department of Clinical Biochemistry, University of Kashmir, Srinagar, India

Md. Khurshid Alam Khan School of Life Sciences, B.S. Abdur Rahman Crescent Institute of Science and Technology, Chennai, India

Sheeza Khan School of Life Science, B. S. Abdur Rahman Crescent Institute of Science and Technology, Chennai, India

Lavanya V School of Life Sciences, B.S. Abdur Rahman Crescent Institute of Science & Technology, Chennai, India

Aqib Rehman Magray Centre of Research for Development, University of Kashmir, Srinagar, India

Lakshna Mahajan Department of Microbiology, Swami Shraddhanand College, University of Delhi, New Delhi, India

Usma Manzoor Clinical Biochemistry, University of Kashmir, Srinagar, India

Akbar Masood Department of Biochemistry, University of Kashmir, Srinagar, India

Rinki Minakshi Department of Microbiology, Swami Shraddhanand College, University of Delhi, New Delhi, India

Shabir Ahmad Mir College of Applied Medical Science, Majmah University, Al-Majmaah, Saudi Arabia

Mir Faisal Mustafa Department of Biochemistry, University of Kashmir, Srinagar, India

Mohammad Nasiruddin Molecular Oncology and Transplant Immunology, MedGenome Labs Pvt. Ltd., Bangalore, India

Sana Qausain School of Life Sciences, B.S. Abdur Rahman Crescent Institute of Science and Technology, Chennai, India

Hilal Ahmad Qazi Plant Molecular Biology Lab, Department of Botany, University of Kashmir, Srinagar, India

Safikur Rahman Department of Medical Biotechnology, Yeungnam University, Gyeongsan, South Korea

Salika Ramazan Plant Molecular Biology Lab, Department of Botany, University of Kashmir, Srinagar, India

Fouzia Rashid Department of Clinical Biochemistry, University of Kashmir, Srinagar, India

Gurumayum Suraj Sharma Dr. B.R. Ambedkar Center for Biomedical Research, University of Delhi, Delhi, India

Laishram Rajendrakumar Singh Dr. B.R. Ambedkar Center for Biomedical Research, University of Delhi, Delhi, India

Hemalatha Srinivasan School of Life Sciences, B.S. Abdur Rahman Crescent Institute of Science and Technology, Chennai, India

Irfan-ur-Rauf Tak Centre of Research for Development, University of Kashmir, Srinagar, India

Snowber Shabir Wani Clinical Biochemistry, University of Kashmir, Srinagar, India

Barqul Afaq Wani Govt. Medical College, Srinagar, India

Preface

The subject of the book, *Protein Modificomics: From modifications to clinical perspectives,* mainly takes into account the large set of modifications that occur to certain amino acid residues of a folded protein after protein translation and also the consequences of any undesired modification. These chemical modifications, collectively called posttranslational modifications (PTMs), mainly include, but are not limited to phosphorylation, glycosylation, ubiquitination, S-nitrosylation, methylation, N-acylation, lipidation, SUMOylation, etc.

The main purpose of a controlled PTM is to produce proteins with functional diversity beyond those encoded by the amino acids present in the polypeptide chains. In fact, PTMs expand nature's gamut by increasing the side chain inventory available to proteins. Knowledge of PTMs is enormously important as they bring about changes in the physical and chemical properties, conformation, folding, and stability of proteins, which ultimately affects the protein function. Importantly, many of such modifications are also involved in almost all of the biological pathways, namely, signal transduction, protein-protein interaction, cell-cycle, cell proliferation, apoptosis, cancer, etc. Nowadays understanding the protein structure-functional relationship therefore requires detailed information not only about its amino acid sequence, but also about the type of modification present on the protein. Furthermore, certain covalent modifications are the emblematic signatures of stressful conditions, for example, hyperglycemia, oxidative stress, hyperosmotic stresses, etc., and therefore are linked with various human patho-physiologies like diabetes, Alzheimer's disease, Huntington's disease, heart disease, age-related disorders, cancer, etc. Thus the allosteric insight into the regulation or deregulation of the protein structure due to modifications is of immense importance for the therapeutic intervention of diseases. PTMs also play a pivotal role in plant physiology and the production of medicinally important primary and secondary metabolites. Understanding of PTMs in plants will help us to enhance the production of these metabolites without greatly altering the genome. This will give us robust eukaryotic systems for the production and isolation of desired products without considerable downstream and isolation processes. Keeping in view the importance of PTMs in humans, plants, and other microbial systems, a thorough insight into, if not all, most of the significant PTMs is important. Therefore the intention of this book was to comprehensively cover all aspects of PTM biology, that is, cell-process regulation, disease etiology, the ability to exploit different bio-systems to yield favorable results with our knowledge and understanding, translational modifications as a stress response, and also the role of PTMs in enhancing the efficiency of protein pharmaceuticals. In fact, we thought of compiling this edited book volume as, despite being so important to protein function, we could not find any comprehensive collection of PTMs of proteins viz-a-viz their importance in various cellular physiological processes, like stress response, signaling, etc., or their clinical perspectives in the literature.

The book comprises 13 chapters that are organized with the intention of providing an introduction to PTMs and then discussing their clinical perspectives and role in stress response. Chapter 1 includes the introduction to PTMs and their emerging role in various biological and disease processes. Chapter 2 describes the mechanism and diseases associated with nonenzymatic PTMs. Chapter 3 introduces the clinical perspectives of PTMs, while Chapters 4–7 describe the most significant PTMs, like phosphorylation, acetylation, ubiquitination, etc., and their role in human health and diseases. The role of these modifications in improving the pharmaceutical properties of the proteins has been described in Chapters 8 and 9. Chapter 10 describes the role of PTMs in cancer and their utilization in cancer therapeutics. Chapters 11 and 12 describe PTMs in other organisms, like plants and algae, in response to the environmental stress as an adaptive measure. The last chapter, Chapter 13, discusses the role of PTMs as markers for early disease diagnosis.

Growing knowledge about the biological role of PTMs on one hand and advancements in sophisticated analytical instrumentation on the other hand, have already led to an increased interest in PTMs. *Protein Modificomics: From modifications to clinical perspectives* intends to serve as an important source of information not only for researchers and scientists working in the field of protein structure-function relationships, but also for people in the pharmaceutical industries.

CHAPTER

1

Posttranslational Modifications of Proteins and Their Role in Biological Processes and Associated Diseases

Irfan-ur-Rauf Tak, Fasil Ali†, Jehangir Shafi Dar*, Aqib Rehman Magray*, Bashir A. Ganai*, M.Z. Chishti**

***Centre of Research for Development, University of Kashmir, Srinagar, India**
†Department of PG Studies and Research in Biochemistry, Jnana Kaveri PG Centre, Mangalore University, Chikka Aluvara, India

1 INTRODUCTION

Generally, a posttranslational modification (PTM) is defined as a chemical modification event resulting from either the covalent addition of some functional groups, or proteolytic cleavage to the premature polypeptide chain after translation so that the protein may attain a structurally and functionally mature form. It depicts an imperative means for diversifying and regulating the cellular proteome. Due to the tremendous scope of these chemical alterations in various biological processes like protein regulation, localization, and synergistic relation with other molecules (nucleic acids, lipids, carbohydrates, cofactors), PTMs do play a significant part in functional proteomics. Their significance in proteome functioning is due to their ability to control protein action, location, and synergy with other cell molecules like nucleic acids, proteins, fatty acids, and cofactors. The primary structure of a protein obtained after the process of translation is just the linear sequence of amino acids, which is insufficient to elucidate

https://doi.org/10.1016/B978-0-12-811913-6.00001-1

the protein's biological activity and their regulatory functions. Posttranslational modifications do play a critical role in determining the native functional structure of proteins.

Research in the proteomics field in the last few decades has shown that the complexity of the human proteome is greater than that of the human genome. The human genome is believed to comprise around 20,000–25,000 genes,[1] while over 1 million proteins are known to be present in the human proteome.[2] This is because a single gene encodes a number of proteins. This enormous diversity of proteins is due to various processes including recombination of genomes, alternative promoter transcription initiation, discrepancies in transcription termination, unusual transcript splicing, and, most importantly, posttranslational modifications.[3] The changes that occur to the mRNA at the level of transcription lead to more diversity of transcriptome than the genome, and the innumerable types of PTMs increase proteome diversity many times over compared to transcriptome and genome.

Posttranslational modifications take place in different amino acids side chains, or at peptide linkages, which are frequently mediated by enzymatic activity. Approximately 5% of the proteome is considered to be comprised of enzymes that are identified to carry out more than 200 types of PTMs.[4] These enzymes include phosphatases, kinases, ligases, transferases, etc. Some of them add various functional groups to the amino acid side chains, while some others remove the functional groups from them. Furthermore, some proteases cleave the peptide bonds of the proteins to remove their specific sequences. These include some enzymes that add or remove the regulatory subunits of the proteins, and hence they play an essential role in regulation. Some proteins even have autocatalytic domains, which have the ability to modify themselves.

A large number of routine cellular processes are regulated by PTMs; for example, phosphorylation of protein, which has been seen as one of the vital control mechanisms that governs the major aspects of cellular life. A majority of the mammalian proteins, which account for nearly 1/3rd of them, are known to contain covalently bound phosphates; the levels of these are said to be controlled by the activities of protein phosphatases and protein kinases, as well as their regulatory subunits. Biological synthesis of the active neuropeptides, which serve as neurotransmitters modulators in both CNS and PNS, is another example of PTM proteolytic processing that involves multiple protease classes. Posttranslational proteolysis also helps in making active enzymes by the conversion of inactive enzyme into its active form, e.g., zymogen (trypsinogen into trypsin).

One of the essential roles of PTM is the macromolecular transport to different cellular spaces by posttranslational glycosylation (e.g., receptor transport by membranes). The intra disulfide bond formed between the two residues of cysteine, which are the backbone for the complex structure and for the purposeful expression of many proteins in enzyme activity, is also well studied by PTM. Many mono-oxygenases (enzymes) which require o_2 for their activity are also linked with PTM; e.g., the amidation of c-terminal peptide transmitters/modulators is catalyzed by peptidyglycine α-amidating monooxygenase and the hydroxylation reactions in hypoxia of the proline residues-inducible factor-1, which is the transcriptional activator catalyzed by prolyl hydroxylases. PTM regulate other enzymes involved in the redox reactions; as a result of this, o_2 availability and redox state alter PTM reactions. Posttranslational modification may take place at any stage during the maturation of the protein. For example, shortly after translation is completed on ribosomes, many proteins are modified to intercede to correct folding/stability of proteins, or direct the nascent protein to specific cellular destinations like membranes, nucleus, lysosomes, etc. Other modifications usually take place after the process of folding and sorting of proteins and are responsible for their catalytic activity.

Some proteins are covalently linked with certain functional groups that are targeted for degradation. Depending upon the nature of modification, posttranslational modification of proteins can be reversible. For example, phosphorylation by protein kinase to the proteins at specific amino acid side chains which are responsible for catalytic activation and inactivation. On the other hand, phosphatases catalyze the hydrolysis of the phosphate group from the protein and, thus, reverse the biological activity. The peptide bond hydrolysis of proteins is a thermodynamically stable reaction and, thus, removes a specific peptide sequence or a regulatory domain permanently. Consequently, analysis of proteins and their PTMs are significant in elucidating the pathological mechanism of diseases like heart disease, cancers, neurodegenerative diseases, arthritis, diabetes, etc. In addition to this, PTMs play a significant part in the functioning of homeostatic proteins, which consequently have a wide range of effects on their capability to interact with other proteins. The characterization of PTMs, although challenging, provides a deep understanding of cellular functions underlying etiological processes. Errors that may occur during posttranslational modifications, either due to hereditary changes or due to environmental effects, may cause a number of human diseases like heart and brain diseases, cancer, diabetes and several other metabolic disorders. Development of specific purification methods are the main challenges that come while going through posttranslationally modified proteins. These challenges are, however, being overcome by using refined proteomic technologies.

For their continued existence, cells should have the ability to interact with other cells and should be able to respond to the external environment. This process of communication between the cells, known as cell signaling, greatly depends upon reversible posttranslational modifications and the quick reprogramming of functions individually. The study of PTMs and their mechanism of regulating various cellular signaling pathways has significant medical implications, in both prevention and cure. In the modern era, understanding the mechanisms of the role key molecules play in signal transduction mechanisms has been thoroughly accumulated. More recently, the emergence of the latest techniques, such as microarray analysis, has provided a detailed account of the changes that occur during downstream transcription level following a variety of stimuli. However, most important processes that are involved in cellular responses are mediated by changes in PTMs rather than the transcriptional changes.

Among other limitations, the major limitation of the current proteomic technology is their limited use in spotting only simple and easy modifications in a large quantity of modified samples but not for complete mapping of all PTMs that occur within a cell. However, with the advancement of new proteomic techniques, there is tremendous scope for understanding the underlying mechanisms in PTMs. Many efforts are also making headway for enriching modified samples and specific detection of modifications.

2 MAJOR POSTTRANSLATIONAL MODIFICATIONS

Nearly all of the PTMs are led by reversible, covalent additions of small, functional groups, such as acyl, phosphate, acetyl, alkyl groups, or the different sugars, to the side-chains of individual protein amino acid residues. Different forms of modifications can occur at a single position of an amino acid, for example, lysine amino acid can undergo methylation (mono, di, or tri) at the N position (also known as the lysine ε-amino site); reversible acetylation may also occur at the same site. Likewise, hydroxyl groups in serines and threonines can be phosphorylated as well as glycosylated. The main purpose of PTMs is to increase the function of the target protein and to provide extra mechanisms for the regulation of modified proteins; i.e., PTMs can regulate the protein's activity by controlling its interactions with other proteins, modifying its enzymatic activity, and changing the stability of the protein. The most frequent PTMs, in the case of mammals, are the phosphorylation of serine amino acid followed by phosphorylation of lysine, representing almost 15% of all known amino acid modifications.[5] More than 200 posttranslational modifications have been observed and most of them have been discovered recently. Surprisingly, a great portion of them are catalyzed by altering enzymes. It is estimated that the human genome encodes for about 518 kinases and more than 150 phosphatases, whereas modifying enzymes are almost all coded by 5% of human genes. Likewise, about 600 and 80 E3 ubiquitinating ligases and deubiquitinases are also coded by the human genome, respectively. Such modifying enzymes are diversely present in all kingdoms of life, especially in eukaryotes. For example, there are 109 kinases and 300 phosphatases coding genes in the genome of Arabidopsis and about 119 kinase genes in the yeast genome. On the other hand, regardless of the growing recognition of their significance, the frequency and full purposeful repertoire of PTMs are still unknown.

Some of the important posttranslational modifications are discussed below.

2.1 Acetylation

Acetylation involves the N-terminal addition of acetyl group to the amino group of the polypeptide chain, affecting 80% of all proteins. Since nonacetylated proteins within the cell are prone to degradation by intracellular proteases, this PTM plays a significant role in the regulation of the life span of intracellular proteins. Acetyl groups are added to the N-terminal end of lysine amino acid, in addition to specific internal residues in proteins, as depicted by the chemical reaction shown in Scheme 1.

Acetylation is an essential modification occurring as the co-translational and posttranslational modification of proteins and is most commonly observed in histone proteins, p53, and tubulins. The process of acetylation and de-acetylation of histone proteins, facilitated by histone acetyl transferases (HATs) and histone deacetyl transferases (HDATs), plays a significant role in gene regulation. However, proteomics studies have identified thousands of acetylated mammalian proteins and all these have remarkable influence on gene expression and metabolism.

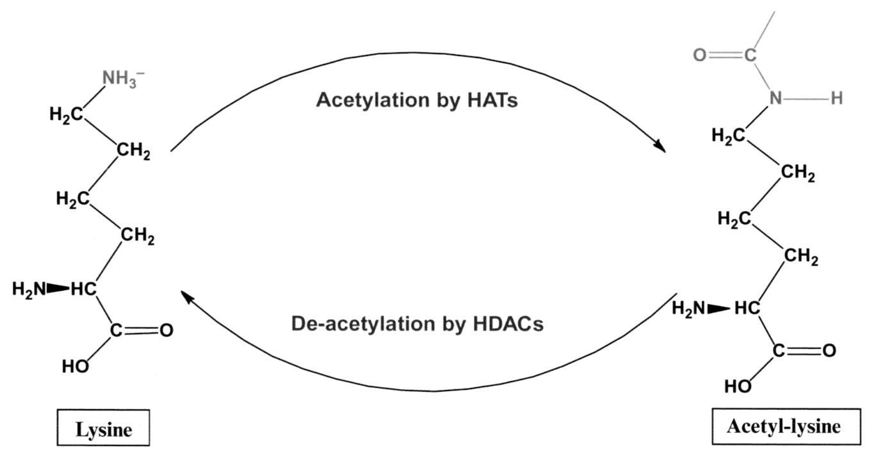

SCHEME 1 Posttranslational modification of proteins by N-terminal addition of acetyl groups.

2.2 Glycosylation and Glycation

Glycosylation involves the addition of a carbohydrate group (glycosyl donor) to a hydroxyl or other functional group of proteins, lipids, and other organic molecules. The majority of proteins are glycosylated after their synthesis on the rough endoplasmic reticulum. It is an enzyme directed, substrate specific, reversible, and tightly regulated process, whereas glycation is a random, nonenzymatic chemical reaction that is involved in the formation of nonfunctional and defective biomolecules (Fig. 1).

Glycosylated molecules (glycans) are classified into five groups:

- N-linked glycans with carbohydrate chains linked to nitrogen of the aspargine or arginine side chain. Their synthesis involves partaking of a particular lipid molecule, known as dolicol phosphate.
- O-linked glycans with carbohydrate chains linked to the –OH of serine, tyrosine, threonine, hydroxyllysine, or hydroxylproline side chains. O-linked glycans participate in a variety of cellular processes.
- C-linked glycans are the forms of glycol sylation products where a sugar residue is linked to the tryptophan side chain carbon atom.
- Phospho-glycans, wherein the sugar groups are locked through the phosphate group of phospho-serine.
- Glypiation, which involves the addition of glycosylphosphitidylinositol anchor linking proteins to lipids through glycan linkage.

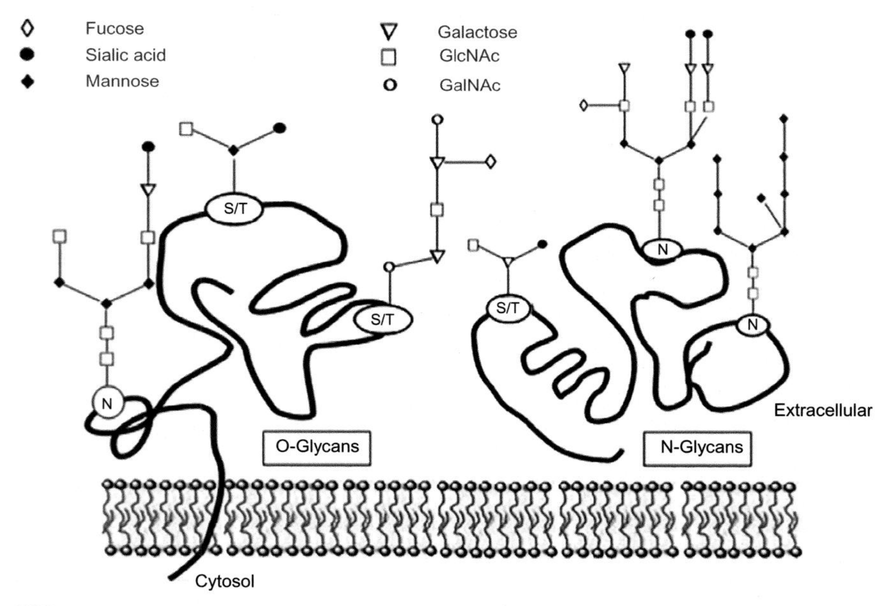

FIG. 1 N-linked and O-linked glycosylation of proteins.

Glycosylation, in general, plays an essential role in protein–protein interactions, immune responses, structural stability of the cell, protein localization, and regulation of cell signaling. Thus, any defective glycosylation can lead to diseases like cirrhosis, diabetes, and exacerbated HIV infection.

2.3 Hydroxylation

Hydroxylation is a reversible PTM carried out by the enzymes known as hydroxylases, and has the utmost importance in cellular physiology. In human proteins, proline is the most frequently hydroxylated residue (Scheme 2).

Proline hydroxylation-mediated collagen modification has been studied broadly because of its crucial role in the structural physiology of the cell. Collagen, an essential part of connective tissue, forms 25%–35% of the proteins in our bodies. The hydroxylation occurs at the γ-Carbon atom of Proline, forming hydroxyproline (Hyp), which is an essential component of collagen occurring at every 3rd residue in its amino acid sequence. In some instances, hydroxylation of Proline at its β-Carbon may also occur. Lysine may be hydroxylated on the δ-C atom, forming hydroxylysine. This reaction is catalyzed by the enzymes; namely, prolyl-4-hydroxylase, prolyl-5, 3-hydroxylase, and lysyl-5-hydroxylase

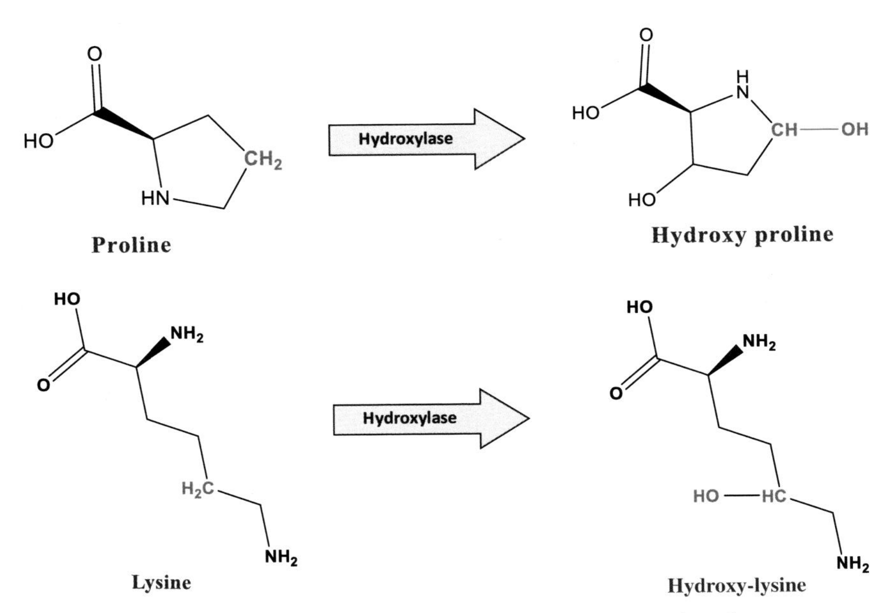

SCHEME 2 Hydroxylation of proline and lysine amino acids by hydroxylase enzyme as a posttranslational modification of proteins.

multi subunit enzyme, in that order. Vitamin C acts as a cofactor in the process. Hydroxylation of proteins imparts tensile strength by allowing fibers to cross-link within the proteins.

2.4 Phosphorylation

Phosphorylation involves the reversible addition of a phosphate group to an amino acid—a critical and most significant PTM that is essential for cell metabolism, enzymatic reactions, protein degradation, and intracellular signaling. This process involves the mediation of numerous protein kinases (PKs) in the cell. The Reversible phosphorylation of proteins brings about a change in the conformation of many enzymes and receptors, which results in their activation and deactivation. On the other hand, de-phosphorylation is involved in the elimination of phosphate groups catalyzed by various phosphatases.

The mitogen-activated central cell proliferation protein kinase pathway—that is, the ERK1/ERK2-MAPK signaling pathway—intercepts the receptor tyrosine kinase (RTKs) pathway. In addition to this, CDKs are some of the protein networks affected by the process of phosphorylation/de-phosphorylation. The most commonly phosphorylated amino acid residues are serine/threonine (Ser/Thr) and tyrosine (Scheme 3), which are often associated with cancer progression. Okadaic acid, which is present in shellfish poisoning, brings about the phosphorylation of Ser/Thr in the cells, which promotes uncontrolled cell proliferation. Phosphorylation plays the foremost function in regulating the light distribution between Photosystems (PS) I and II (state transitions) and in t the repair cycle of PSII. In addition, the Calvin cycle enzymes are known to be regulated through thioredoxin-mediated redox regulation; thereby, determining the efficiency of carbon assimilation. At the time of writing, Lys and N-terminal Lys methylation, acetylation, Tyr nitration and nitrosylation, sumoylation, glutathonylation, and also the glycosylation of chloroplast proteins, have been described.

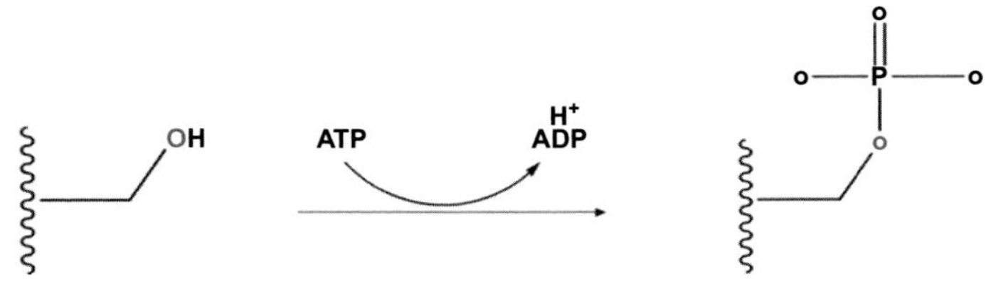

SCHEME 3 Phosphorylation of serine as posttranslational modification.

2.5 Ubiquitination

The process of Ubiquitination mainly involves addition of ubiquitin to the protein substrate. Ubiquitin, found in almost all eukaryotes, is a small regulatory protein with a molecular weight of 8.5 KDa. Ubiquitination of proteins bring about a variety of protein processes. It affects the protein activity, regulates protein–protein interactions, signals them for proteasomic degradation, and alters their cellular locations. The process of ubiquitination mainly involves three steps: firstly, the activation of ubiquitin by ubiquitin-activating enzymes (EIs); secondly, the conjugation by ubiquitin conjugating enzymes (E2s); and thirdly, the ligation by ubiquitin ligases (E3s) as shown in Fig. 2. This sequential cascade of reactions brings about ubiquitin on lysine residues via an isopeptide bond, cystine residues via a thioester bond, and threonine and serine amino acids by an ester bond or via a peptide bond between the amino group of the protein's N-terminus.

2.6 Methylation

Methylation is a type of PTM that is associated with the addition of one or more methyl groups to the nucleophilic side chain of the protein. The process is mediated by the primary methyl group donors like methyltransferases and S-adenosyl methionine (SAM), as shown in Scheme 4. SAM is the second most commonly used substrate after ATP in enzyme reactions. Protein methylation regulates a plethora of cellular events like regulation of transcription, aging, stress response, nuclear transport, T-cell activation, protein repair, ion channel function, neuronal differentiation, and cytokine signaling.

2.7 Amidation

C-terminal alpha-amidation is the most important PTM for various important biological activities like signal transfer and receptor recognition. This process is catalyzed by a single gene encoded protein, Peptidyl-glycine Alpha-amidatingmonooxygenase (PAM). The process of amidation takes place in two steps. In the first step, PAM's peptidylglycine alpha-hydroxyl monooxygenase (PHM) and ascorbate form a transitional product from a glycine-protracted peptide to a hydroxyl in presence of molecular oxygen. Second enzyme, peptidyl-alpha-hydroxyglycine alpha-amidating lyase (PAL) converts this intermediate product into alpha-amide peptide and glycosylate. The significance of amidation lies in the fact that half of the peptides in the case of mammals and more than 80% of insect hormones consist of an alpha-amide C terminal part (Scheme 5).

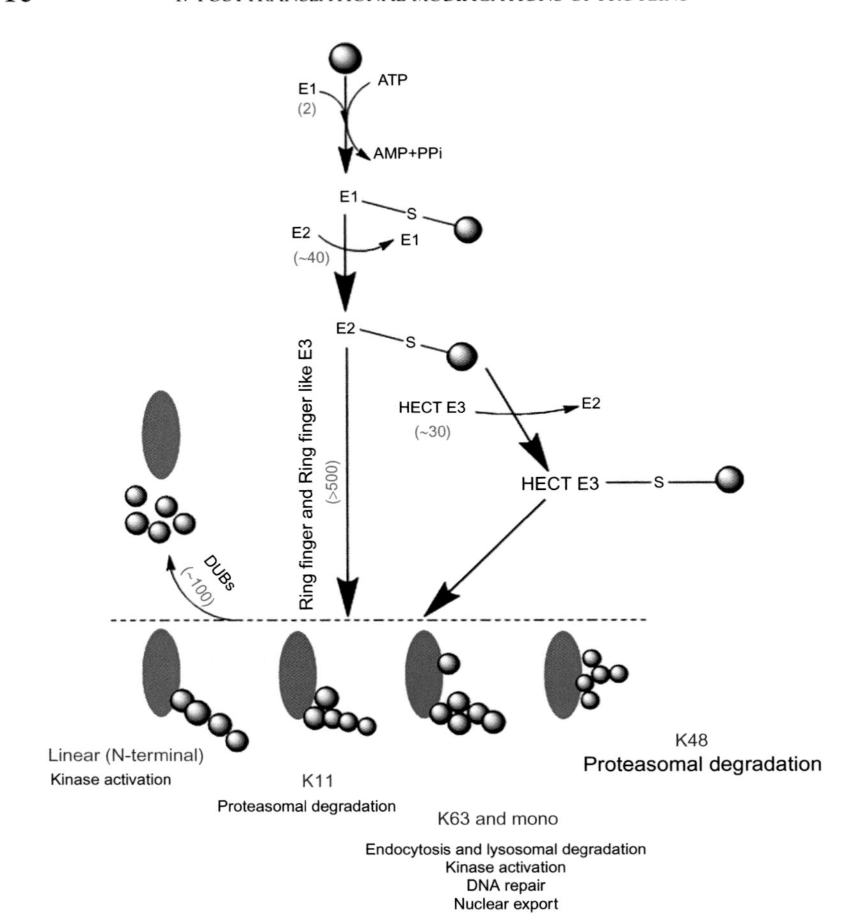

FIG. 2 Ubiquitinylation as a multistep enzymatic PTM of cellular proteins for degradation of unwanted proteins.

2.8 Palmitoylation

Palmitoylation involves the addition of a 16-carbon fatty acid, palmitic acid, to the cysteine residue of proteins by thioester bond. Protein acyl Transferases (PATs) are enzymes responsible for catalyzing the addition of palmitate to the substrate. The process of N-palmitoylation occurs

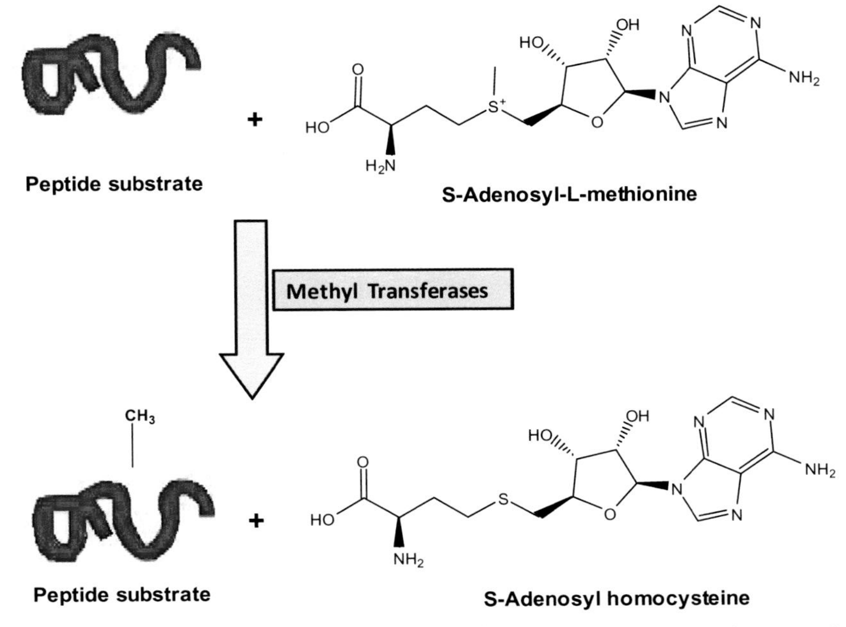

SCHEME 4 Methylation of proteins as posttranslational modification of proteins by methyltransferases with S-adenosyl methionine as a primary methyl group donor.

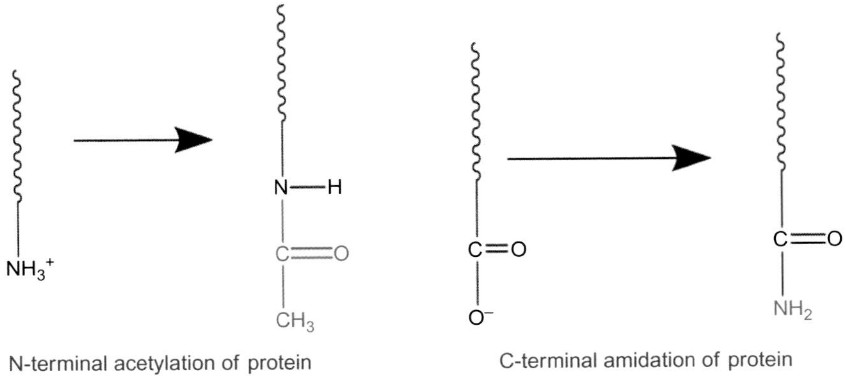

N-terminal acetylation of protein

C-terminal amidation of protein

SCHEME 5 N-terminal acetylation and C-terminal amidation of proteins by peptidylglycine alpha-hydroxyl monooxygenase and peptidyl-alpha-hydroxyglycine alpha-amidating lyase.

through a thioester intermediate using the thiol group of the cysteine amino acid, followed by a spontaneous rearrangement forming amide linkage (Scheme 6). Palmitoylation is known to facilitate various biological activities like interaction of membrane proteins, mediation of protein trafficking and regulation of protein stability and enzyme activity.

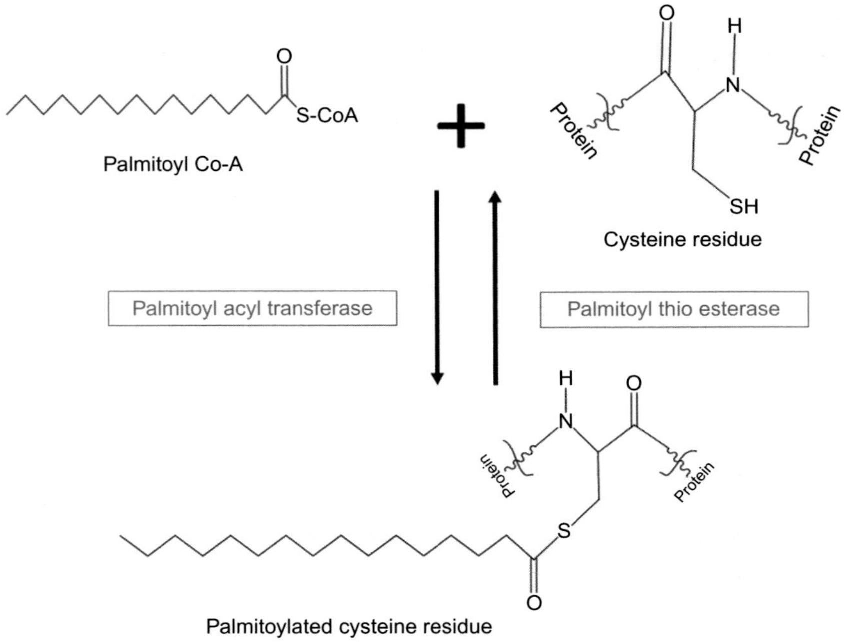

Palmitoyl Co-A

Cysteine residue

Palmitoyl acyl transferase

Palmitoyl thio esterase

Palmitoylated cysteine residue

SCHEME 6 Addition of Palmitoyl group on the cysteine residues of the protein byenzymes Palmitoyl acyl transferases and palmitoyl thio esterase.

2.9 Myristoylation

N-myristoylation is an irreversible covalent modification involving the addition of a 14-carbon, myristic acid to the N-terminal glycine of the target protein and is arbitrated by N-Myristoyl Transferase (NMT), a ubiquitous and crucial enzyme in eukaryotes. Many NMT proteins play a significant role in cellular structures and function regulation (Scheme 7). These proteins facilitate protein-protein interaction and a diversity of cellular processes like signaling pathways, membrane association, subcellular localization, oncogenesis, or viral replication.

2.10 Prenylation

Prenylation, also known as isoprenylation, is also a posttranslational protein modification involving covalent addition of isoprenyl lipid molecules via 15 carbon farnesyl ($C15H25$) or a 20-carbon geranylgeranyl group ($C20H33$) to the carboxy terminal of the protein (Scheme 8). This addition creates a hydrophobic tail and plays a paramount role in membrane association, cell signal transduction, and vesicle trafficking and cell

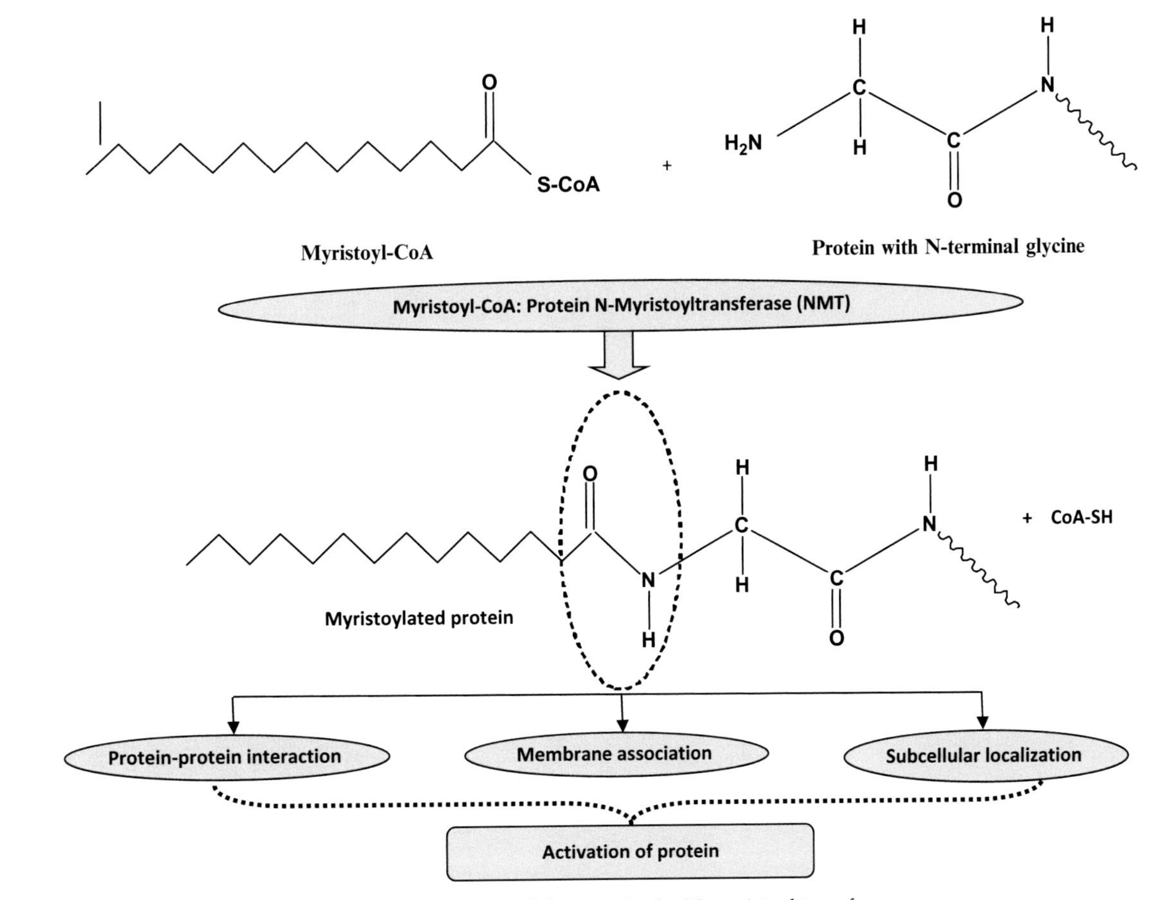

SCHEME 7 Addition of myristic acid to N-terminal glycine of the proteins by N-myristoyl transferase.

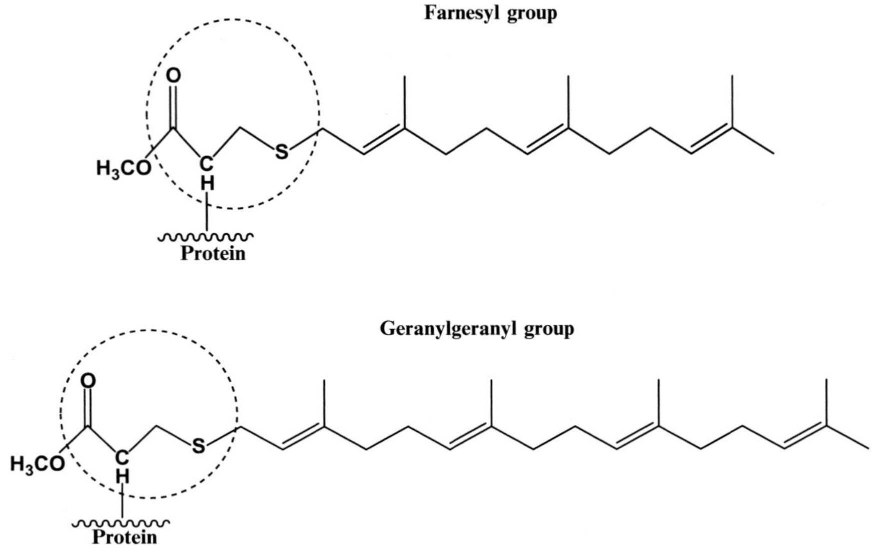

SCHEME 8 Posttranslational protein modification by covalent addition of isoprenyl lipid molecules, i.e., farnesyl and geranylgeranyl group to carboxy terminal of the target protein.

cycle progression. The enzymes involved in the prenylation process are farnesyl transferases (PFT) and protein geranylgeranyl transferases type I and type II.

2.11 Proteolytic Cleavage

This type of posttranslational protein modification involves enzymatic cleavage of the amino acid backbone, which results in the removal of few amino acids from the carboxyl or amine-terminus of a polypeptide chain. This type of PTM is sometimes known as protein processing and is mediated by the enzymes called proteases. Proteolytic cleavage is involved in the action of many activating enzymes that function in blood coagulation and apoptosis. This type of PTM also generates the active peptide hormones like epidermal growth factor (EGF) and insulin from their large precursors. In bacteria and some eukaryotes, an unusual form of protein self-splicing takes place, in which an internal protein chain segment is removed and then its ends are re-joined. An autocatalytic process called protein self-splicing proceeds by itself without partaking of enzymes. In some vertebrate cells, some of the proteins are processed by auto cleavage, but the succeeding ligation step is absent. Insulin is one of the best examples of such modification, as shown in Fig. 3.

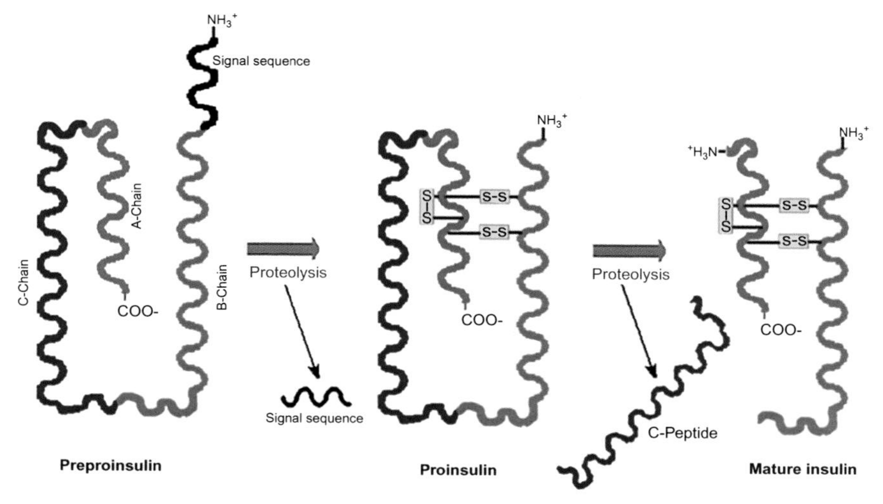

FIG. 3 Proteolytic cleavage of Insulin as a posttranslational protein modification for maturation of target protein.

3 FUNCTIONS OF POSTTRANSLATIONAL MODIFICATIONS

Posttranslational modifications affect the protein structure and dynamics, thereby affecting their function. Alternatively, modified residues serve as the binding sites and are directly recognized by their partners. For example, SH2 domains directly target the phosphotyrosine residues of protein and bromo domains target acetyl lysine residues. Similarly, ubiquitination of proteins target them for proteolytic cleavage by degradation pathways. Biologically, posttranslational modifications are crucial for many activities like regulation of enzymatic activity, regulation of gene expression, regulation of apoptosis and protein stability, destruction, cancer, mediation of protein interaction, etc. The posttranslational modification of proteins is essential for the modulation of proteins in signaling and regulatory mechanisms of cell function.

3.1 Role of PTM in Apoptosis and Cancer

3.1.1 Regulation of Apoptosis by Posttranslational Modifications of Caspases

Apoptosis is the programmed cell death that is responsible for controlling the homeostasis and development of multicellular organisms. Apoptosis is preceded by the assemblage of signaling complexes, which, after their considerable posttranslational regulation via modification by the ubiquitin, brings about the activation of the cell death programme. The process of ubiquitination often destabilizes the proteins and targets them for their proteasomal degradation by the apoptotic pathway.

The key initiators and executors of the progression of apoptosis are the cysteine-dependent aspartate proteases called caspases. There are a number of regulatory circuits that tightly control the processing and activity of caspases. Out of these, the most important, but poorly understood, controlled mechanism of activation of caspases is the posttranslational modification of caspases. The removal or addition of chemical groups drastically affects the catalytic activity of caspases or modulates their enzymatic activity, which in turn controls their nonapoptotic functions. The binding of functional c groups or proteins modulates the caspase enzymatic activity or can also control their nonapoptotic functions. Posttranslational modifications, which include phosphorylation, nitrosylation, glutathionylation, acetylation, SUMOylation, and ubiquitination are essentially associated with the modulation of caspase activity and cell death. They signify a distinctive code, through which the strength of apoptotic response can be regulated and additional possibility can be provided for fine-tuning death signaling downstream of apoptotic stimuli.

It has become obvious now that the PTM of apoptotic proteins by Ubiquitination has a role in the regulation of key components in the apoptotic signaling cascades. For example, ubiquitin E3 ligase such as Mouse Double Minute (MDM2) ubiquitinates p53 and inhibitor of apoptotic proteins (IAP) and deubiquitinases like A20 and ubiquitin-specific proteases 9X (USP9X) regulate the degradation of receptor-interacting protein 1 and myeloid leukaemia cell differentiation 1 respectively. Therapeutic agents targeting the apoptotic regulatory proteins might afford clinical benefits. The highly conservative nature of apoptosis regulation between different organisms can be explained on the bases of evolutionary conservative nature of initiator and affector caspase phosphorylations. Caspase phosphorylation and dephosphorylation presents the additional activation and deactivation networks that regulate cell death in an extremely dynamic manner. The pro-apoptotic and antiapoptotic roles played by caspase ubiquitinations depends on the type of ubiquitin chains and the site of ubiquitination within the structure of caspase.

3.1.2 Role in Cancer

Positioned on the short arm of chromosome number 17 (17 p 13.1), p53 (a tumor suppressor gene) is one of the most important cell cycle regulators, which is translated into a 53 KD protein—an important transcriptional factor that regulates cell cycle by arresting it at G1 phase of interphase. Recognized as the "guardian of genome," p53 exerts its suppressive effect on the cell cycle by significantly organizing a regulatory circuit that checks and responds to a number of stress signals like DNA damage, telomere attrition, hypoxia, and abnormal oncogenic events. P53 is also believed to play a vital role in a transcription independent manner. Human double minute (HDM2), mouse orthologi is mdm2 and Human double minute X (HDMX), mouse ortholog MDMX are the negative regulators that keep the p53 level low by repressing its gene. However, p53 is stabilized under stress conditions and also released from repression and is activated further in a promoter-specific fashion., It has been found that p53 protein or tumor suppressor protein is modified by as many as 50 individual PTMs. Most of these modifications take place because of genotoxic and nongenotoxic stresses; they are also independent of one another in such a way that one, or more than one, modification can nucleate consequent events. This proposes a pathway that is known to function through manifold cooperative events contrary to distinct functions for individual isolated modifications.

MDM2 is vital for maintaining p53 levels both in developing as well as in adult mice, and along with other p53 targeted ubiquitin ligases like ARF-BPI, CHIP, COP1, E6-AP, Pirh2, TOPORS, and TRIM24, it contributes to p53 turnover. The seven carboxy-terminal lysines (K370, K372, K373, K381, K382, and K386) are the chief p53 targets of MDM2-mediated ubiquitination. Acetylation of lysine (K126) residue of p53 is impelled by DNA damage, and this acetylated p53 localizes preferentially to be the promoter of key pro-apoptotic genes without those involved in cell cycle arrest. Induction of p53 involves its uncoupling from the negative regulators, mainly MDM2 and its related protein MDM4, like MDM2 also inhibits p53-mediated transactivation. In the case of DNA damage, response is mediated by PTM of p53, which plays a critical role in this process. MDM2 is itself the main target of stress signaling pathways that interrupt the activity of p53 and thus provides the complementary, but not mutually exclusive, model. It has been proposed that MDM2/MDM4 inhibition and/or degradation generally causes the rapid accumulation of p53 and subsequently, the activation of its transcription functions.

This can be summarized into the following points:

(a) Acetylation does not influence the p53-MDM2 negative feedback regulatory loop but plays an important role in p53-mediated cell fate.
(b) The mechanisms of transactivation by p53 differ depending on the promoter.
(c) There is a degree of redundancy between various acetylation sites so that loss of one can be compensated by the presence of others.

3.2 Role of PTM in Signaling

For the continued existence of cells, their ability to communicate with other cells and act in response to their outside environment is crucial. For this communication to occur, it is important that the external signal must penetrate the lipid bilayer in some way. In most cases, the signal is transmitted through specific proteins that are present on the surface of the cell membrane, rather than the signaling molecule itself entering into the cell. Thus, communication occurs between these proteins and other additional proteins that are associated with the intracellular domain of the membrane; this is known as cellular signaling, which briefly relies on reversible posttranslational protein modifications to reprogram individual protein functions quickly. Eukaryotic cells are known to use posttranslational modifications as the indispensable mechanisms to dynamically coordinate their signaling networks and diversify their protein functions. Numerous human diseases and developmental disorders have been linked to defects in PTMs, thereby highlighting the importance of PTMs in maintenance of normal cellular states.

PTMs are known to impact cell function by the process of modifying histones, enzymes, and their associated activity, assembling the protein complexes as well as in the recognition and phenomenon of targeting in the genome or, for that matter, to other cellular compartments. In the context of single modifications and gene expression, acetylation of certain lysines (i.e., Histone 3 lysine [9-H3K9]) correlates with activation, while tri-methylation of this same residue is most often associated with compaction and gene repression. Lysine can be mono-, di- or tri-methylated in lysine methylation, while in an asymmetric or symmetric fashion, arginine can be mono- or di-methylated. Each degree of methylation for lysines and arginines serves as its own PTM and affects biological output. Most PTMs do not exist alone in the chromatin environment and the combination of these states can reinforce one another. For example, one PTM can serve as a docking site for a binding domain called a "reader" within one protein, while another "reader" within the same protein can recognize another residue. This is the case for the reader protein BPTF, which binds both H3K4me3 and H4K16 acetylation. Therefore, modulating the various types and degrees of modifications will affect output. For these reasons, the cell has developed a series of enzymes that are important for establishing and maintaining these PTMs, which are often referred to as writers or erasers. Many of these enzymes have emerged as critical therapeutic targets and have been identified as key regulators of diseases such as cancer. These observations have also made their associated PTMs candidates for biomarkers in cancer and other diseases.

4 DISEASES ASSOCIATED WITH POSTTRANSLATIONAL MODIFICATIONS

As already discussed, modulation of proteins by the PTMs leads to a diverse range of cellular functions. The identification, characterization, and mapping of posttranslational modifications to specific amino acid residues on proteins are very important for understanding their practical implications in a biological context. The elucidation of the proteome data obtained by newly designed methods is extremely difficult without having proper information about protein modifications. Therefore, an accurate understanding of protein PTMs is very important, not only for gaining an insight into a multitude of cellular functions, but also toward designing drug therapies for most of the life threatening diseases such as neurodegenerative disorders and cancer.

4.1 Relation Between Mutated Posttranslational Modification Sites and Diseases

There are a number of mutations that occur at the posttranslational target sites that have been found to be directly involved in many diseases. One such example is the Spongiform encephalopathy, an N-liked glycosylation loss in prions (autosomal dominant disease) where there occurs the substitution of T183A. This is associated with numerous clinical symptoms like cerebral atrophy early-onset dementia, and hypometabolism. A suitable example is of Kennedys disease (an inherited neurodegenerative disorder) where in androgen receptor losses acetylation sites.

However, in certain cases, nonacetylated mutations misfold the proteins due to proteasomal degradation and form aggregates with other proteins, like ubiquitin ligase E3. One more example is the familial advanced sleep phase syndrome, (an autosomal dominant disorder) involving serine phosphorylation of circadian protein homolog 2 (PER2).

4.2 Acetylation and Related Diseases

Studies have shown that there is a link between protein acetylation with different diseases, and that it also contributes significantly to the pathophysiology of disease. Increased acetylation of the cytoskeleton proteins in response to reactive oxygen species (ROS) in chronic progressive external opthalmoplegia syndrome patients has been shown to increase the mitochondrial dysfunction. In addition to this, acetylated-proteins have been seen to be involved in various cognitive disorders like Alzheimer's disease and dementia. In dementia, lysine acetylation of tau proteins has been found to result in tau tangles, while in Alzheimer's disease, it has been seen that lysine hyperacetylation to β-amyloid peptide results in impaired cognition. Mouse models have also demonstrated that alteration in histone acetylation patterns plays a very important role in age dependent memory impairment.

4.3 Glycosylation and Related Diseases

Excessive oxidative stress in a biological system leads to protein carbonylation. Sometimes, it leads to nonfunctional protein formation, which is a permanent PTM and can be the principal causative agent for many diseases.[6] Increased synthesis of reactive oxygen and nitrogen species, i.e., ROS and RNS, in the cell is key to autoimmune diseases and cancer. However, less is known about their role in oxidation of proteins. Protein carbonylation has demonstrated an essential role as indicator of oxidative or genotoxic stress. In Alzheimer's disease, the role of O-glycosylation and phosphorylation is beneficial and it diminishes the formation of neurofibrillary masses in neurons.[7] Furthermore, glycosylation of prion (PrP), a cell surface protein and a contagious agent, is an element of the ultimate disease outcome in the host. Recently it has been seen that description of glycosylation sites on apolipoprotein E (apoE) has exposed a novel glycosylation site in tally to the sites that were already known, and also at least 8 new complex glycans secreted and cellular apoE were revealed.[8] The involvement of apoE in a number of diseases has been well documented and this novel evidence can give insight toward the empathetic role of glycosylated apoE residues in these diseases. Incomplete glycosylation in the Fc receptor for immunoglobulin A has been shown to impact the IgA-mediated immune response, which in turn may cause many diseases, like HIV, alcoholic lever cirrhosis, and other neuropathies.

4.4 PTM of Proteins During Intermittent Hypoxia

PTM has been shown to be the best mechanism for regulation of protein function by chronic hypoxia. Electrophoretic partition of tissue or cellular proteins followed by immuno-labeling by employing antibodies specific to native and posttranslationally modified forms is usually done during the analysis of protein PTMs.[9] Studies have demonstrated that CIH affects the state of phosphorylation of a subset of proteins associated with activation of transcriptional factors, neurotransmitter synthesis, and signaling pathways that catalyzes specific phosphorylation reactions depending upon the pattern, severity, and duration of hypoxia.[10] Investigations with respect to association of PTMs with CIH are at their preliminary stage and applications of proteomic techniques are important for elucidation of PTMs associated with various hypoxia activated metabolic and signalling ways.[10]

5 ROLE OF POSTTRANSLATIONAL MODIFICATIONS IN PROTEIN-PROTEIN INTERACTION

Protein posttranslational modifications modulate protein functions in a cell by regulating protein–protein interactions. In fact, an increase in structural and biophysical diversity of proteins has been observed by covalent modifications of PTM, thus enhancing the genome information. There are many PTMs that are used by the cell to get a required result—a protein can go through a single PTM or many PTMs that may be involved in several tasks. The different modifications can alter a single position on the protein so that switching between many functions can be regulated by identification of the particular position of PTM.

Many complex and dynamic cellular processes are controlled by PTM through regulation of interactions between key proteins, In order to understand the regulatory mechanisms it is difficult to plot the PTM dependent protein–protein interactions using available approaches. CLASPI (Cross Linking Assisted and Stable isotope labeling in cell culture based Protein Identification) is a recent development in the chemical proteomics approach, which is used to examine methylation-mediated protein–protein interactions in human cell lysates.

Besides other functions, posttranslational modifications are believed to show their function through the modulation of protein–protein interactions. The proteins that undergo PTM are observed to be engaged in more interactions and positioned in more central locations than non-PTM proteins. Phosphorylated proteins are mostly situated in the central network locations[11] and to the broadest interaction spectrum of proteins carrying other PTM types, whereas at the periphery of the network, glycosylated proteins are located.[11] For human interactome, proteins found with the quality network properties undergo sumoylation or proteolytic cleavage. The properties of PTM-type specific protein interaction network (PIN) properties can be rationalized with regard to the function of the respective PTM-carrying proteins. The human proteins involved in disease processes that undergo PTMs are also associated with characteristic PIN properties. The global protein interaction networks and specific PTMs integration involves a new approach to solve the role of PTMs in cellular processes.

6 ROLE OF PTMs IN REPLICATION

In each cell cycle, the genome must be replicated equally. For the completion of genomic DNA synthesis in eukaryotic cells, the replication of DNA starts in S phase at various origins on the entire chromosome throughout the genome. For the prevention of genome instability, a single cell cycle of DNA replication should be done. Gene amplification and DNA damage[12] can result from *Re*-replication, but for this prevention there are a number of mechanisms. DNA replication is mainly regulated through two distinct and temporally stages, i.e., G1 phase origins are "licensed" during the low activity of Cyclin Dependent Kinase (CDKs), and the initiation of replication from these permitted origins occurs in the subsequent S phase as CDK accumulates.[13] The potential sites of replication initiation take place only after certifying in G1 phase, associated with lower CDK activity, that mainly occurs as Mcm2-7 helicase is loaded by the Origin Recognition Complex (ORC, orc 1-6), cdc6 and cdt1 constituting the pre-Replication Complex (pre-RC).[14–16] Low CDK activity during G1 prevents release of Mcm2-7 double hexamer. Sanctioning of origins takes place at G1 phase of interphase and later on is used in the following S phase.[17] This leads to the inactive origins that are not released in an undisturbed cell cycle but has significance for the response in DNA replication pressure.[18–20] The dissembling of dormant origins takes place by reflexive replication, thus preventing their activation and re-activation.[21]

The critical role of PTMs has been revealed by new emerging genetic and biochemical studies in some novel model organisms in the regulation of replication initiation. On the other hand, the best-proven class of PTMs is kinase-mediated phosphorylation, with some essential additional forms of PTMs like ubiquitinylation, methylation, and acetylating, which participate in the control of initiation of replication.[22] The fast modulation of protein function in response to environmental and metabolic changes is enabled by posttranslational modifications.

6.1 Ubiquitinylation by Unperturbed DNA Replication

Ubiquitinylation is one of the fascinating posttranslational modifications of proteins and is one of the most extensive, versatile mechanisms of protein regulation utilized by eukaryotic cells. The process of ubiquitinylation has a significant role in synchronization of the DNA damage response, and is a property of every cellular process. A small (76 residues) polypeptide is linked covalently and linked to the substrate protein during ubiquitinylation. In eukaryotes, there is conserved sequence of ubiquitin and its three dimensional structure, because of which ubiquitin attachment and signaling is much preserved throughout the evolution.

Substrate protein resides in ubiquitin, which contains a conserved C-terminal glycine with an isopeptide bond by a lysine residue next in sequence. A cascade of three enzymes catalyze the reaction: the activated ubiquitin is passed by a ubiquitin-activating enzyme (E1) to a ubiquitin conjugating enzyme (E2), which, with the help of a ubiquitin ligase (E3), can attach to the substrate.[23] In many ways, substrates can be modified with ubiquitin, which can be monoubiquitinylated, multimonoubiquitinylated, or polyubiquitinylated. Formation of ubiquitin chains is carried out by polyubiquitinylation. Further ubiquitin attachment, i.e., attachment of a ubiquitin chain to the substrate, is possible as there are seven lysine residues in the sequence of ubiquitin (K6, K11, K27, K29, K33, K48, and K63). Homogenous polyubiquitin chains can exhibit seven different linkages, which it depends on to form the chain with which lysine within ubiquitin is modified (the same lysine is used for linkage in all ubiquitins) as well as linear chains linked through its N terminal methionine, mixed heterogeneous linkage chains and even branched structure.[23] Ubiquitin attachment to the substrate affects the substrate's activity, localization and fate, and the overall change in three-dimensional structure of the substrate. The unique three-dimensional structure is possessed by each type of chain, thus producing different signals and different substrates. The protein is modified with K11 and K48 linked chains that target it for proteasomal degradation, while as modification with K63, linked chains have a vital role in DNA damage response (DDR). Signaling, at this stage, is the best-studied form of ubiquitinylation so far observed.[24] Removal from the substrate or editing of ubiquitin is carried out by de-ubiquitinylating enzymes (DUBs). It has been strongly advocated that ubiquitinylation, being very plastic and versatile in nature, is one of the major types of protein modification in cellular systems.[23] Ubiquitinylation's role in DDR regulation and other replicative stress processes has been studied for many years,[24] but its essentiality in regulation of undisturbed DNA replication has been tested out recently.

DNA replication initiation requires increased expression of the S phase kinases (CDK and Cdc7) and expression of cell-cycle regulated initiation factors. In order to maintain an increase in CDK level, CDK inhibitors are broken, which leads to decreased activity of APC/C-Cdh1. The ubiquitinylation directed proteosomal degradation lead to the activation of G1-(Cyclin Dependent Kinases) CDKs at the mammalian G1 "restriction point" (START point in yeast) which is only possible when G1 cyclins are not APC/C substrates. One of the fascinating roles of Cdk2 is to initiate S phase and complete G1, whereas mammalian cells in early G1 are in need of Cdk4 and Cdk6. The expression of Cyclin D is stimulated by positive growth in G1, which, when bound with Cdk4 and Cdk6, stimulates a program of gene expression of S phase factors. The complete gene expression program is regulated by the retinoblastoma protein family (Rb), transcription factors E2F1-3, and Myc. The expression of Cyclin *E*-Cdk2 is induced by a transcriptional program that, due to increased levels of CKIs (p27 and p21), remains inactive throughout most of the G1 phase. The ubiquitinylation and degradation of p27 occurs by three ubiquitin ligases in mammalian cells:

- In the G0 and G1 phases, the export of p27 from the nucleus is targeted by KPC ligase to the cell cytoplasm.
- From the G1 to S phase, the expression of Pirh2 gets increased.
- SCFSkp2 targeting nuclear p27 from the early S phase.

Recent studies on replication regulation have shown that ubiquitinylation modulates the process of unperturbed DNA replication, which brought a small explosion of examples of substrates and mechanisms. Now many questions have been raised with the recent knowledge about the regulation, modification, and synchronization of all these ubiquitin dependent mechanisms. A tempting target for the next generation of cancer therapies is the emerging network of ubiquitinylation enzymes, which is essential for faultless execution of DNA replication.[25] The therapies mostly targeting traditional DNA replication mechanism and DNA repair processes are not specific for cancer, so due to its drawbacks, this exhibits a restricted therapeutic window. Many enzymes are deregulated in cancer, which is involved in the ubiquitin regulatory network, thus providing the opportunity for targeted drug therapy.[26] A small molecule inhibitor of the Nedd8 stimulating enzyme—a prospective drug, MLN4924—blocks the activity of the cullin-RING ubiquitin ligases CRLs. Uninhibited DNA synthesis through re-replication foremost to DNA damage and stimulation of apoptosis is the reciprocal treatment effect of numerous human tumor cell lines with MLN4924.[27–29] The ubiquitin regulated DNA replication and its opportunity for cancer treatment will become conventional with improved understanding.

6.2 PTMs of Replication Protein A

In all eukaryotic cells, a conserved single stranded DNA-binding protein called replication protein A (RPA) is conventional in action.[30, 31] ssDNA intermediates in repair, recombination, and replication of DNA is bound by RPA and is involved in damage recognition in the cellular response to DNA damage. RPA is composed of three subunits of roughly 70, 32, and 14-kDa (referred to as RPA1, RPA2, and RPA3, respectively) existing as a stable complex. Multiple posttranslational modifications are carried out by RPA. These include phosphorylation of threonine and serine residues, SUMOylation, and poly-ADP ribosylation.[31–33] Posttranslational modification of RPA primarily occurs after DNA damage, although limited phosphorylation of RPA is observed in S and M phases,[32, 34] wherein both RPA1 and RPA2 subunits have been modified.

7 ROLE OF PTMs IN TRANSCRIPTION

Expression of a gene, known as transcription, is a process in which the enzyme RNA polymerase transcribes a DNA segment into RNA (especially mRNA). Both DNA and RNA are nucleic acids, which use nucleotide base pairs as a complementary language. A diverse family of proteins, called Transcription factors, which work as multisubunit protein complexes, bind directly to specific regions of the DNA "promotor," sometimes to the upstream of the coding region in a gene, or directly to the RNA polymerase molecule. The process of transcribing DNA into RNA is mainly regulated by transcription factors. Excluding RNA polymerase that mainly helps in the start and regulation of transcription of genes, transcription factors involve a number of proteins. Sometimes transcription factors may work individually or in a complex with activator or repressor, which promote and block the organization of RNA polymerase to specific genes.

7.1 Regulation of Activity of Transcription Factor by PTMs

About 7% of the human proteome is composed of the transcription factor, helping in the cellular function by syndicating the external signal information into gene expression programs that, in turn, reconfigure cellular physiology at a basic level. PTMs help surface-initiated cell signaling pathways converge on transcription factors that are dependent on combinational functions, arrangements, time, and space. The activity of transcriptional factors is arranged by the PTMs beautifully throughout entire life span and thus, play a major function in epigenetic regulation of gene expression, which begins from the subcellular localization to protein–protein interactions, sequence-specific DNA binding, transcriptional regulatory activity, and protein stability.[35]

For the development of therapeutic agents to treat disease conditions, the assembly of PTMs of transcription factors also offers numerous impending points of intervention. The information and signaling from the cell surface to the nucleus is provided by site-specific DNA binding transcription factors, which behave as critical targets and effectors of signal transduction pathways. Many such transcriptional regulators associate into large families, which have many homologous DNA-binding domains (DBD) that can bind the DNA, having similar sequences.[36] In practice, transcription factors regulate different sets of target genes, whether they are temporally or spatially patterned, and for the normal development in a controlled way. Now the question is how is specificity and accuracy maintained? The answer to this is complex but is of utmost importance because most of the human diseases including cancer are likely due to the mis-regulation of transcriptional response. Members of the ETS (E twenty six),a transcription factor super family, operating as a dynamic and reversible sensor of upstream signaling events, may provide a way to modulate all facets of transcription factor function.[37] In metazoans transcription factors ETS are conserved and, during development, play a major role, which function as own stream effectors of signal transduction cascades to control the broad spectrum of cell processes. ETS play a vital role in regulation of cell differentiation, proliferation, apoptosis, migration, and epithelial-mesenchymal interactions during normal development; whereas mis-regulated ETS proteins, by a variety of mechanisms, lead to both the initiation and progression of many human cancers.[37] ETS transcription factors are formed of a vastly conserved, 85 amino acid motif called the ETS domain, which belongs to the super family of helix-turn helix (HTH) DNA binding domains and, by a recognition sequence GGAA/T, is referred to as the ETS binding site (EBS). There are variable sequences for EBS, which subsidize to the individual specificity of individual ETS transcription factors, which are approximately 8 in drosophila and 30 in mammals. They function as transcriptional activators or repressors and sometimes may act as both.

Depending on the requirements of multicellular organisms, numerous strategies have been developed for the regulation of activity of transcription factors, providing the sequential and spatial specificity. The coinciding expression patterns of the large number of family members, and their similar or even indistinguishable DNA binding preferences, has given specific importance for ETS transcription factors.[38] This concept of specificity is applied for currently less studied multiprotein transcription factor super families, thus the principles arising from the studies of the ETS family are generally applicable in future. The higher eukaryotes possess small a genome with enormous function coding for protein formation and also posttranslational modifications for increasing protein complements. Due to the multiple spectrum of covalent modifications either individually or in complex combinational patterns, dynamically and reversibly effecting DNA-protein interactions, sub-cellular localizations, protein–protein interactions, stability, activity, and other posttranslational modifications of the target protein, there is a significant increase in the functional complexity of the proteome.

The studies on posttranslation modifications have revealed that phosphorylation plays an important role in modification of the activities of the cellular proteins, which also include transcription factors.[39] By two broad families of kinases, Y protein kinases phosphate and S/T protein kinases groups are added; i.e., the addition of phosphate group to the hydroxyl group of tyrosine (Y) brings phosphorylation, or serine (S)/threonine (T) residues in an ATP-dependent reaction.[40] Phosphorylation is reversible by dephosphorylation mediated by phosphates of either S/T, Y, or by both.[41] The most widespread phosphorylation in S/T is the Y phosphorylation. Besides phosphorylation, glycosylation plays an influencing role in transcription factor activity. The major role played by ETS transcription factor Elf-1 includes mainly nuclear pore proteins, RNA pol II and its associated transcription factors, proteasome components, hormone receptors, kinases and phosphates in nuclear chromatin associated proteins, chromatin structure, transport, signaling, and transcription, and protein turnover is fascinating and tremendous. The nuclear and cytosolic proteins undergo glycosylation by the addition of simple monosaccharide O-linked β-*N*-acetyl glucosamine (O-GlcNAc) to the hydroxyl group of either S or T residues.[42–44] O-GlcNAc levels are dependent on the balance between O-GlcNAc transferase (OGT) and O-GlcNAc-ase; the O-GlcNAc response to the cell cycle, cell stress, metabolism of glucose, and insulin signaling pathways suggest upstream metabolic and signaling events. The glycosylation and phosphorylation play competing and antagonistic roles in recognition by S/T residues as there is no consensus motif for O-GlcNAc attachment, nor any known protein interaction motif that recognizes glycosylated S/T residues. There is an increase in the level of O-GlcNAc and decrease in kinase inhibitors by phosphate inhibitors, by such a reciprocal relationship. With high scoring PEST sequences, O-GlcNAc is found. The effect of PEST sequences is neutralized by O-GlcNAc by inhibition of phosphorylation and subsequent degradation. Thus, in future investigations for frameworks and concerns of O-GlcNAc modification, transcription factor would appear to be an important priority.

8 ROLE OF PTMs IN TRANSLATION

Modifications introduced after protein translation from RNA results in the modulation of their function. Since proteins play a very important role in the cellular functions, such as catalysis, structural building material, transport, and many other functions[45] and modulation of their function by PTMs in turn affects all other cellular processes. Beyond protein diversity, because of mRNA splicing and PTMs of proteins, there occurs further modification of proteins by attaching covalently small chemical moieties to some selected amino acid residues. Many aspects of cell functions like metabolism, signal transduction, and stability of protein have been affected because of identification of more than 200 different proteins.[46, 47] The modifications are mainly because of phosphorylation, methylation, glycosylation, amidation, and acetylation, and many other types, more studied by Uniport.[48]

Many studies on PTMs have specificity for their function and with phosphorylation, which represents the most actively researched PTM-type,[49–53] the focus of attention is the interplay between different PTM-types and has moved most recently[54–57]; e.g., for a genome reduced bacterium *Mycoplasma pneumonia*, an important evidence of an interdependence of phosphorylation and acetylation has been reported.[58] To modulate protein function, integrative PTN spots (PTMi) have been recognized as the site in protein at which diverse PTMs are operated in a combinational manner.[59] On the co-evolution between 13 frequent PTM types in 8 eukaryotic species, a global view of the interplay between PTM types was presented.[60] The most conserved identified is carboxylation whereas phosphorylation found in those PTMs play a key role in protein modulation.

In addition to PTMs, with the help of noncovalent protein–protein interactions, protein functions can be modulated.[61–66] Many posttranslational modifications control the binding affinities between proteins by altering the structural and electrostatic properties of the involved interaction sites. There are more than 60% of PTM sites that are related to functional domains of proteins and preferentially engage in direct protein-protein interactions. Therefore, there is a good chance that proteins possessing the particular PTM-type may possess specific interaction characteristics. It has been seen that in yeast, phosphorylated proteins are involved in more various protein–protein interactions than that of unphosphorylated proteins.[67] Thus, modulation of many different interactions is caused by phosphorylation of a single gene protein, thus altering molecular processes simultaneously.

9 CHEMICAL AND FUNCTIONAL ASPECTS OF PROTEIN PTMs

There are mainly two groups of posttranslational modifications. In the first group, cleavages of peptide bond takes place, resulting in removal of polypeptide fragments leading to unity of proteolytic processes. The second group deals with the modification of amino acid residues but not interfering with the polypeptide backbone.[68] The modifications are diverse in chemical nature and function. The modifications mainly depend on type of amino acid residues.

The posttranslational modification of amino acid residue side chains are mainly involved in four main groups of protein functions. The presence of certain specific prosthetic groups covalently bonded to the polypeptide chain is required by the functional activity of a wide range of proteins. In protein activity, complex organic molecules often take part, e.g., activation of enzymes. Another group of posttranslational modifications is mainly involved in regulation of biochemical process by altering the enzyme activity, whereas the other group involves protein tags, in which intercellular localization of protein and transport of proteins to the proteasome occurs by hydrolyzing and proteolysing reactions. Additionally, some protein posttranslational modifications alter the spatial arrangement of proteins.

10 ROLE OF PTMs IN HISTONE MODIFICATION

Posttranslational covalent modification of histone proteins includes acetylation, methylation, phosphorylation, sumoylation and ubiquitinylation.[69] By changing chromatin's structure or enrolling histone modifiers, gene expression can be altered.[69] The histones are mainly involved in the packing of DNA wrapping the eight histones into chromosomes. The diverse biological processes are acted upon by histone modifications like transcriptional activity switch on/off. In the case of packaging of chromosome and DNA damage/repair mechanism, there is variation of histone activity. For example, in most species, acetylation of H3 histone primarily occurs at lysine, 9, 14, 18, 23, and 56; methylation occurs at arginine 2 and lysine 4, 9, 27, 36, and 79, and phosphorylation occurs at ser 10, ser 28, thr3, and thr11; H4 histone is methylated at position arginine 3 and lysine 20, and primarily acetylated at lysine 5, 8, 12, and 16, and in addition, phosphorylated at serine 1. Thus, beneficial information is obtained from the quantitative detection of various histone modifications for understanding regulation of epigenetics at cellular levels and the expansion and intervention of histone modifying enzyme-targeted drugs.

The majority of the cellular processes like transcription and chromatin dynamics, apoptosis, cell cycle progression, differentiation, DNA repair, DNA replication, gene silencing, nuclear import, and neuronal repression are mostly regulated through histone acetylation.[70] Histone modifying enzymes called histone acetyltransferases (HATs) are involved in histone acetylation, for example, histone H3 and H4 acetylation. There are 20 HATs so far identified and these can be classified as MYST, GNAT1, P300/CBP, TAFII250, and nuclear receptor coactivators such as ACTR. The antagonist effect on inhibition activity of histone deacetylase (HDACs) increases the activity of histone H3 acetylation, which is in contrary to HAT inhibition as its activity is decreased by it. Histone deacetylases play an essential role in hydrolytic removal of acetyl groups from histone lysine residues. The equilibrium of histone acetylation is destabilized in tumorigenesis and cancer progression, and further characterization of acetylation patterns or sites is done by knowing that H3 histone is acetylated at its lysine residues, thus leading to the better understanding of epigenetic regulation of gene activation as well as the development of HAT-targeted drugs. HDACs play an essential role in many cellular processes that involve H3 and H4 histone modifications, which are somehow similar to HATs. There are four classes of HDACs which have been identified; For example; Class I HDACs include 1, 2, 3, and 8; Class II HDACs contains 4, 5, 6, 7, 9, and 10; Sirtuins—also called Class III enzymes—require NAD^+ cofactors for their function, and mostly include SIRTs 1–7. The class IV enzyme, which contains only HDAC11, has some essential features as that of both classes I and II. HDAC inhibition displays important results on apoptosis, cell cycle arrest, and differentiation in cancer cells. However, HDAC inhibitors are presently being developed as one of the novel anticancerous agents.

Enzyme histone methyl transferases (HMTs) carry out the histone methylation, i.e., transfer of variable number of methyl groups, usually 1–3 from S-adenosyl–L-methionine of histone proteins to arginine or lysine residues. Through chromatin dependent transcriptional activation or repression, HMTs control or regulate DNA methylation. The histone genes may be silenced or activated by histone methylation in the cell nucleus. The specificity of lysine or arginine residue is determined by the different histone methyl-transferases that are modified. On histone H3, e.g., histone methyl transferases SET7/9, Ash1, SET1, MLL, ALL-1, Trx, and ALR, and SMYD3catalyzes methylation of histone H3 at lysine 4 (H3-K4) in mammalian cells. ESET, G9a, SUV39-h1, SUV39-h2, dim-5, SETDB1 and Eu-HMTase are histone methyl-transferases catalyzing methylation of histone H3 at lysine 9 (H3-K9) in mammalian cells. G9a and polycomb group enzymes such as EZH2 are histone methyl transferases in mammalian cells, which catalyze methylation of histone H3 at lysine 27(H3-K27). H3-K9 and H3-K27 methylation leads to formation of heterochromatin and gene silencing in gene expression at euchromatin sites. In some compulsive processes such as cancer development, the H3-K27 methylation is involved, whereas transcriptional activity is promoted by methylation of arginine on histones H3 and H4 and is mediated by a family of protein arginine methyl-transferases (PRMTs). There are nine types of PRMTs observed in humans but only nine members are found to methylate histones. They can mediate single arginine residues demethylation. Based on methyl group position, PRMTs can be classified into type I (CARM1, PRMT2, PRMT3, PRMT1, PRMT6, and PRMT8) and type II (PRMT5 and PRMT7). Type II PRMTs are mostly found in disease conditions like cancers, e.g., in certain tumor suppressor gene repression such as RB tumor suppressors PRMT5 plays a major role, whereas its over expression is seen in breast cancer.[71] In order to elucidate out mechanisms of epigenetic regulation of gene activation, as well as benefiting cancer diagnostics and therapeutic intervention, detection of activity and its inhibition of type II PRMTs as well as other HMTs is important.

It has been seen that by removing the methyl groups from modified histone proteins, histone demethylases are used in histone demethylation. The oncogenic functions and other pathological processes have been found to be the property of histone demethylation. Histone demethylation has been seen to be a dynamic process. Lysine specific demethylase 1 and jumonji domain that contain histone demethylases are the two major families that have been discovered. The degree of methylation of a specific amino acid residue in turn dictates the degree of demethylation of enzyme. Histone remethylation at specific residues is done by inhibition of histone demethylases. It would be helpful in determining the epigenetic regulation of gene activation and silencing by studying activity and inhibition of important enzymes, and may therefore be helpful in the treatment of cancer.

11 CONCLUSION AND FUTURE PROSPECTUS

Post translational modifications have provided a boon in the field of protein biology and identifying and characterizing these PTMs has become critical in cell biology, prevention, and treatment of a multitude of diseases. Keeping this in view, in this chapter we have tried to summarize the emerging roles of some of the most significant PTMs like phosphorylation, acetylation, glycosylation, ubiquitination, etc. In addition to this, PTMs maintain functioning of major homeostatic proteins, which in turn regulate various cellular processes, and the ability to interact with proteins is effected by secondary level changes to homeostatic proteins. Furthermore, noncovalent binding of allosteric effectors regulated by posttranslational modifications serve as short-term mechanisms mainly involved in the enzyme activity modulation and various cellular metabolic processes. For example, phosphoprotein enzymes found in *Saccharomyces cerevisiaee* regulate its major metabolic network although having few functional objectives. Enzymes regulated by PTMs, therefore, provide a promising strategy for cells to adjust.

The current advances in the field of proteomics and the determination and quantification of PTMs can provide an improved way of targeting some diseases like heart diseases, cancer, neurodegenerative diseases, and diabetes. Many cellular processes like DNA replication, transcription, cell cycle progression, chromatin dynamics and gene silencing, apoptosis, differentiation, DNA repair, nuclear import, and neuronal repression are mainly controlled through histone acetylation and can therefore play an important role in treatment of the above-mentioned diseases. It will not be out of place to claim that PTMs hold tremendous scope in every field of science and technology, be it proteomics, genomics, and recombinant DNA technology, or even the important biological processes that govern the occurrence of all the deadly diseases. Critical and well-planned research focusing and characterizing the emerging role of PTMs with their clinical perspectives seems highly warranted.

References

1. International Human Genome Sequencing Consortium. Finishing the euchromatic sequence of the human genome. *Nature* 2004;**431**:931–45.
2. Jensen ON. Modification-specific proteomics: characterization of post-translational modifications by mass spectrometry. *Curr Opin Chem Biol* 2004;**8**:33–41.
3. Ayoubi TA, Van De Ven WJ. Regulation of gene expression by alternative promoters. *FASEB J* 1996;**10**:453–60.
4. Walsh C. *Posttranslational modification of proteins: expanding nature's inventory.* Englewood, Colo: Roberts and Co. Publishers; 2006490.
5. Khoury GA, Baliban RC, Floudas CA. Proteome-wide post-translational modification statistics: frequency analysis and curation of the swiss-prot database. *Sci Rep* 2011;**1**:90.
6. Petrov D, Zargovic B. Microscopic analysis of protein oxidative damage: effect of Carbonylation on structure, dynamics, and Aggregability of villin headpiece. *J Am Chem Soc* 2011;**133**:7016–24.
7. Porowska AM, Wasik U, Goras M, Filipek A, Niewiadomska G. Tau protein modifications and interactions: their role in function and dysfunction. *Int J Mol Sci* 2014;**15**:4671–713.
8. Liu K, Paterson AJ, Zhang F, McAndrew J, Fukuchi K, Wyss JM, Peng L, Hu Y, Kudlow JE. Accumulation of protein O-GlcNAc modification inhibits proteasomes in the brain and coincides with neuronal apoptosis in brain areas with high O-GlcNAc metabolism. *J Neurochem* 2004;**89**:1044–55.
9. Kumar V, Calamaras TD, Haeussler D, Colucci WS, Cohen RA, McComb ME, Pimentel D, Bachschmid MM. Cardiovascular redox and ox stress proteomics. *Antioxid Redox Signal* 2012;**17**:1528–59.
10. Kumar GK, Prabhakar NR. Post-translational modification of proteins during intermittent hypoxia. *Respir Physiol Neurobiol* 2008;**164**:272–6.
11. Duan G, Walther D. The roles of post-translational modifications in the context of protein interaction networks. *PLoS Comput Biol* 2015;**11**:.
12. Fragkos M, Ganier O, Coulombe P, Méchali M. DNA replication origin activation in space and time. *Nat Rev Mol Cell Biol* 2015;**16**:360–74.
13. Yeeles JTP, Deegan TD, Janska A, Early A, Diffley JFX. Regulated eukaryotic DNA replication origin firing with purified proteins. *Nature* 2015;**519**:431–5.
14. Cook JG, Park CH, Burke TW, Leone G, DeGregori J, Engel A, Nevins JR. Analysis of Cdc6 function in the assembly of mammalian prereplication complexes. *Proc Natl Acad Sci U S A* 2002;**99**:1347–52.
15. Rialland M, Sola F, Santocanale C. Essential role of human CDT1 in DNA replication and chromatin licensing. *J Cell Sci* 2002;**115**:1435–40.
16. Woo RA, Poon RYC. Cyclin-dependent kinases and S phase control in mammalian cells. *Cell Cycle* 2003;**2**:316–24.
17. Cayrou C, Coulombe P, Vigneron A, Stanojcic S, Ganier O, Peiffer I, Rivals E, Puy A, Laurent-Chabalier S, Desprat R. Genome-scale analysis of metazoan replication origins

reveals their organization in specific but flexible sites defined by conserved features. *Genome Res* 2011;**21**:1438–49.
18. Blow JJ, Ge XQ, Jackson DA. How dormant origins promote complete genome replication. *Trends Biochem Sci* 2011;**36**:405–14.
19. Ge XQ, Jackson DA, Blow JJ. Dormant origins licensed by excess Mcm2–7 are required for human cells to survive replicative stress. *Genes Dev* 2007;**21**:3331–41.
20. Yekezare M, Gómez-González B, Diffley JFX. Controlling DNA replication origins in response to DNA damage—inhibit globally, activate locally. *J Cell Sci* 2013;**126**:1297–306.
21. Santocanale C, Sharma K, Diffley JFX. Activation of dormant origins of DNA replication in budding yeast. *Genes Dev* 1999;**13**:2360–4.
22. Khan SA, Reddy D, Gupta S. Global histone post-translational modifications and cancer: Biomarkers for diagnosis, prognosis and treatment. *World J Biol Chem* 2015;**6**:333–45.
23. Komander D, Rape M. The ubiquitin code. *Annu Rev Biochem* 2012;**81**:203–29.
24. Ulrich HD, Walden H. Ubiquitin signalling in DNA replication and repair. *Nat Rev Mol Cell Biol* 2010;**11**:479–89.
25. Cohen P, Tcherpakov M. Will the ubiquitin system furnish as many drug targets as protein kinases? *Cell* 2010;**143**:686–93.
26. Zhao Y, Sun Y. Cullin-RING ligases as attractive anti-cancer targets. *Curr Pharm Des* 2013;**19**:3215–25.
27. Lin JJ, Milhollen MA, Smith PG, Narayanan U, Dutta A. NEDD8-targeting drug MLN4924 elicits DNA rereplication by stabilizing Cdt1 in S phase, triggering checkpoint activation, apoptosis, and senescence in cancer cells. *Cancer Res* 2010;**70**:10310–20.
28. Soucy TA, Smith PG, Milhollen MA, Berger AJ, Gavin JM, Adhikari S, Brownell JE, Burke KE, Cardin DP, Critchley S. An inhibitor of NEDD8-activating enzyme as a new approach to treat cancer. *Nature* 2009;**458**:732–6.
29. Soucy TA, Smith PG, Rolfe M. Targeting NEDD8-activated cullin-RING ligases for the treatment of cancer. *Clin Cancer Res* 2009;**15**:3912–6.
30. Oakley GG, Patrick SM. Replication protein A: directing traffic at the intersection of replication and repair. *Front Biosci* 2010;**15**:883–900.
31. Wold MS. Replication protein a: a heterotrimeric, single-stranded DNA-binding protein required for eukaryotic DNA metabolism. *Annu Rev Biochem* 1997;**66**:61–92.
32. Dou H, Huang C, Singh M, Carpenter PB, Yeh ET. Regulation of DNA repair through DeSUMOylation and SUMOylation of replication protein a complex. *Mol Cell* 2010;**39**:333–45.
33. Eki T, Hurwitz J. Influence of poly(ADP-ribose) polymerase on the enzymatic synthesis of SV40 DNA. *J Biol Chem* 1991;**266**:3087–100.
34. Binz SK, Sheehan AM, Wold MS. Replication protein a phosphorylation and the cellular response to DNA damage. *DNA Repair* 2004;**3**:1015–24.
35. Berg JM, Tymoczko JL, Stryer L. Section 31.3, Transcriptional activation and repression are mediated by protein-protein interactions. In: *Biochemistry*. 5th ed. New York: W H Freeman; 2002.
36. Alberts B, Johnson A, Lewis J. DNA-binding motifs in gene regulatory proteins. In: *Molecular biology of the cell*. 4th ed. New York: Garland science; 2002.
37. Tina L, Ilaria R. Post-translational modifications influence transcription factor activity: a view from the ETS superfamily. *Bioessays* 2005;**27**:285–98.
38. Graves BJ, Petersen JM. Specificity within the ets family of transcription factors. *Adv Cancer Res* 1998;**75**:1–55.
39. Whitmarsh AJ, Davis RJ. Regulation of transcription factor function by phosphorylation. *Cell Mol Life Sci* 2000;**57**:1172–83.
40. Hunter T. Protein kinases and phosphatases: the yin and yang of protein phosphorylation and signaling. *Cell* 1995;**80**:225–36.

41. Denu JM, Stuckey JA, Saper MA, Dixon JE. Form and function in protein dephosphorylation. *Cell* 1996;**87**:361–4.
42. Comer FI, Hart GW. O-glycosylation of nuclear and cytosolic proteins. Dynamic interplay between O-GlcNAc and O-phosphate. *J Biol Chem* 2000;**275**:29179–82.
43. Hart GW. Dynamic O-linked glycosylation of nuclear and cytoskeletal proteins. *Annu Rev Biochem* 1997;**66**:315–35.
44. Zachara NE, Hart GW. The emerging significance of O-GlcNAc in cellular regulation. *Chem Rev* 2002;**102**:431–8.
45. Lodish H, Berk A, Zipursky SL, Matsudaira P, Baltimore D. *Molecular cell biology*. ISBN-13: 978-0716737063.
46. Deribe YL, Pawson T, Dikic I. Post-translational modifications in signal integration. *Nat Struct Mol Biol* 2010;**17**:666–72.
47. Zhao S, Xu W, Jiang W, Yu W, Lin Y. Regulation of cellular metabolism by protein lysine acetylation. *Science* 2010;**327**:1000–4.
48. The UniProt Consortium. The universal protein resource (UniProt) in 2010. *Nucleic Acids Res* 2010;**38**:D142–8.
49. Choudhary C, Kumar C, Gnad F, Nielsen ML, Rehman M. Lysine acetylation targets protein complexes and co-regulates major cellular functions. *Science* 2009;**325**:834–40.
50. Oliveira AP, Ludwig C, Picotti P, Kogadeeva M, Aebersold R. Regulation of yeast central metabolism by enzyme phosphorylation. *Mol Syst Biol* 2012;**8**:623.
51. Roux PP, Thibault P. The coming of age of phosphoproteomics; from large data sets to inference of protein functions. *Mol Cell Proteomics* 2013;**12**:3453–64.
52. Ubersax JA, Ferrell JE. Mechanisms of specificity in protein phosphorylation. *Nat Rev Mol Cell Biol* 2007;**8**:530–41.
53. Zielinska DF, Gnad F, Wiśniewski JR, Mann M. Precision mapping of an in vivo N-glycoproteome reveals rigid topological and sequence constraints. *Cell* 2010;**141**: 897–907.
54. Brooks CL, Gu W. Ubiquitination, phosphorylation and acetylation: the molecular basis for p53 regulation. *Curr Opin Cell Biol* 2003;**15**:164–71.
55. Danielsen JMR, Sylvestersen KB, Bekker-Jensen S, Szklarczyk D, Poulsen JW. Mass spectrometric analysis of lysine ubiquitylation reveals promiscuity at site level. *Mol Cell Proteomics* 2011;**10**. M110.003590.
56. Hunter T. The age of crosstalk: phosphorylation, ubiquitination, and beyond. *Mol Cell* 2007;**28**:730–8.
57. Latham JA, Dent SYR. Cross-regulation of histone modifications. *Nat Struct Mol Biol* 2007;**14**:1017–24.
58. Van Noort V, Seebacher J, Bader S, Mohammed S, Vonkova I. Cross-talk between phosphorylation and lysine acetylation in a genome-reduced bacterium. *Mol Syst Biol* 2012;**8**:571.
59. Woodsmith J, Kamburov A, Stelzl U. Dual coordination of post translational modifications in human protein networks. *PLoS Comput Biol* 2013;**9**:.
60. Minguez P, Parca L, Diella F, Mende DR, Kumar R. Deciphering a global network of functionally associated post-translational modifications. *Mol Syst Biol* 2012;**8**:599.
61. Arabidopsis Interactome Mapping Consortium. Evidence for network evolution in an *Arabidopsis* Interactome map. *Science* 2011;**333**:601–7.
62. De Las Rivas J, Fontanillo C. Protein-protein interaction networks: unraveling the wiring of molecular machines within the cell. *Brief Funct Genomics* 2012;**11**:489–96.
63. Gavin A-C, Bösche M, Krause R, Grandi P, Marzioch M. Functional organization of the yeast proteome by systematic analysis of protein complexes. *Nature* 2002;**415**:141–7.
64. Nishi H, Hashimoto K, Panchenko AR. Phosphorylation in protein-protein binding: effect on stability and function. *Structure* 2011;**19**:1807–15.
65. Seet BT, Dikic I, Zhou M-M, Pawson T. Reading protein modifications with interaction domains. *Nat Rev Mol Cell Biol* 2006;**7**:473–83.

66. Vinayagam A, Stelzl U, Foulle R, Plassmann S, Zenkner M. A directed protein interaction network for investigating intracellular signal transduction. *Sci Signal* 2011;**4**:rs8.
67. Yachie N, Saito R, Sugiyama N, Tomita M, Ishihama Y. Integrative features of the yeast Phosphoproteome and protein–protein interaction map. *PLoS Comput Biol* 2011;**7**:.
68. Rogers LD, Overall CM. Proteolytic post-translational modification of proteins: proteomic tools and methodology. *Mol Cell Proteomics* 2013;**12**:3532–42.
69. Kurdistani SK, Grunstein M. Histone acetylation and deacetylation in yeast. *Nat Rev Mol Cell Biol* 2003;**4**:276–84.
70. Ma H, Marti-Gutierrez N, Park SW, Wu J, Lee Y, Suzuki K, Koski A, Ji D, hayama T, Ahmed R, darby H, Dyken CV, Li Y, kang E, Park AR, Kim D, Kim ST, Gong J, Gu Y, Xu X, Battaglia D, Krieg SA, Lee DM, Wu DH, Wolf DP, Heitner SB, Carlos J, Belmonte I, Amato P, Kim JS, Kaul S, Mitalipov S. Correction of a pathogenic gene mutation in human embryos. *Nature* 2017;**548**:413–9.
71. Han W, Rasika M, Kevin L, Tao L. Protein arginine methylation of non-histone proteins and its role in diseases. *Cell Cycle* 2014;**13**:32–41.

Further Reading

1. Anantha RW, Borowiec JA. Mitotic crisis: the unmasking of a novel role for RPA. *Cell Cycle* 2009;**8**:12903–8.
2. Anantha RW, Vassin VM, Borowiec JA. Sequential and synergistic modification of human RPA stimulates chromosomal DNA repair. *J Biol Chem* 2007;**282**:35910–23.
3. Block WD, Yu Y, Lees-Miller SP. Phosphatidyl inositol 3-kinase-like serine/threonine protein kinases (PIKKs) are required for DNA damage-induced phosphorylation of the 32 kDa subunit of replication protein a at threonine 21. *Nucleic Acids Res* 2004;**32**:997–1005.
4. Dinkel H, Chica C, Via A, Gould CM, Jensen LJ. Phospho.ELM: a database of phosphorylation sites—update 2011. *Nucleic Acids Res* 2011;**39**:D261–7.
5. Duan G, Walther D, Schulze W. Reconstruction and analysis of nutrient-induced phosphorylation networks in *Arabidopsis thaliana*. *Front Plant Sci* 2013;**4**:540.
6. Gnad F, Gunawardena J, Mann M. PHOSIDA 2011: the post-translational modification database. *Nucleic Acids Res* 2011;**39**:D253–60.
7. Kim SC, Sprung R, Chen Y, Xu Y, Ball H. Substrate and functional diversity of lysine acetylation revealed by a proteomics survey. *Mol Cell* 2006;**23**:607–18.
8. Lu C-T, Huang K-Y, Su M-G, Lee T-Y, Bretaña NA. dbPTM 3.0: an informative resource for investigating substrate site specificity and functional association of protein post-translational modifications. *Nucleic Acids Res* 2013;**41**:D295–305.
9. Minguez P, Letunic I, Parca L, Bork P. PTMcode: a database of known and predicted functional associations between post-translational modifications in proteins. *Nucleic Acids Res* 2013;**41**:D306–11.
10. The UniProt Consortium. Activities at the universal protein resource (UniProt). *Nucleic Acids Res* 2013;**42**:D191–8.
11. Zou Y, Liu Y, Wu X, Shell SM. Functions of human replication protein a (RPA): from DNA replication to DNA damage and stress responses. *J Cell Physiol* 2006;**208**:267–73.
12. Zulawski M, Braginets R, Schulze WX. PhosPhAt goes kinases—searchable protein kinase target information in the plant phosphorylation site database PhosPhAt. *Nucleic Acids Res* 2013;**41**:D1176–84.

CHAPTER

2

Clinical Perspective of Posttranslational Modifications

*Iram Ashaq Kawa**,†, *Akbar Masood**, *Shajrul Amin**, *Mir Faisal Mustafa**, *Fouzia Rashid*†

***Department of Biochemistry, University of Kashmir, Srinagar, India**
†Department of Clinical Biochemistry, University of Kashmir, Srinagar, India

1 INTRODUCTION

Posttranslational modifications (PTMs) are basically a chain of reversible or irreversible chemical alterations of proteins that occur after the completion of translational process. The modification affects both structure and functions of protein and makes it more diverse. This makes coding of eukaryotic genome more complex, enhancing in turn the coding capacity of genes, which helps to regulate a protein's localization, activity, and its interaction with other cellular molecules. Typically, PTMs include functional group additions (e.g., phosphorylation, alkylation, acylation, glycosylation); altering the chemical nature of amino acids (deimidation, oxidation, deamidation, etc.); attachment of other proteins and peptides (e.g., SUMOylation, ubiquitination); and also peptide bond cleavage by enzymes.[1] Proteins under physiological conditions are posttranslationally modified and participate in a range of tasks such as mediation of protein–protein interactions, turnover of protein and localization, enzyme activity, modulation for various signaling cascades, gene expression regulation, cell division, and DNA repair. These reactions may be considered as usually playing a physiological role, and alterations due to inherited or environmental factors, resulting in functional changes and, in some, cause overt disease. PTMs are nowadays greatly studied, owing to their huge

Protein Modificomics
https://doi.org/10.1016/B978-0-12-811913-6.00002-3

significance in understanding the basic pathophysiology of disease processes. In addition to this, we can analyze the prognosis of specific treatment strategies. To date, around 200 posttranslational modifications have been identified[2] and it is impossible to discuss all of them here. Keeping in view the vast diversity and complexity of these modifications, this chapter focuses only on those associated modifications which are not only diseases associated but also very well characterized. Insights for future directions and the challenges faced have also been highlighted.

2 GLYCOSYLATION

Glycosylation is a posttranslational and co-translational mechanism, in which the complex oligosaccharides—glycans—are added to proteins. Being an enzymatic process, glycosylation in comparison to nonenzymatic chemical reaction of glycation, is substrate and site specific. Glycosylation has an important function in localization of protein, protein-protein interactions, immune responses, cell signaling modulation, and cell stability.[3, 4] Eight major glycosylation pathways populate in the endoplasmic reticulum-Golgi network. Among these pathways in mammals, the profound form is N-linked glycosylation (ER), which starts by the stepwise assembly of a universal precursor containing three glucose, nine mannose, and two *N*-acetylglucosamine units (Glc_3 Man_9 $(GlcNAc)_2$) on a lipid carrier, dolichol, to form a lipid-linked oligosaccharide. The transfer of entire glycan then occurs on a nascent polypeptide chain at asparagine in the triplet sequence Asn-X-Ser/Thr (where X is any amino acid except proline). In O-linked glycosylation, glycans are added directly on the hydroxyl groups of threonine or serine and mannose, xylose, *N*-acetylgalactosamine (GalNAc), Fuc, GlcNAc, or Glc. The complex carbohydrates are added as template-independent and likely both environmental and genetic factors together decide glycosylation patterns in various pathological processes. In fact, glycosylation disorders span a magnificent clinical spectrum, and cause abnormality in nearly every organ system[5] as discussed below.

2.1 Congenital Diseases

2.1.1 Congenital Disorders of Glycosylation (CDGs)

Errors in glycosylation lead to several inherited diseases; one example is CDGs, previously termed the carbohydrate-deficient glycoprotein syndromes. CDGs arise due to genetic defects leading to the deficiency or loss of enzyme activity involved in glycan synthesis and processing, or to deficiency of specific transporters.[6] They have been subdivided into N- and

O-glycosylation disorders and the most frequently identified types of CDG are associated with disrupted or defective N-glycosylation pathway. CDGs of N-glycosylation include two groups (CDG-I and CDG-II). CDG-I corresponds to enzymes involved in assembling of the glycoprotein. Enzyme defects that are known are found to be in cytosol (CDG-Ia, CDG-Ib) and ER (CDG-Ic, CDG-Id, and CDG-Ie). CDG-II involves processing either late in the ER (CDG-IIb) or in Golgi (CDG-IIa and CDG-IIc). The effects of this syndrome are mostly observed in the nervous system in addition to other organ systems. Patients typically have structural abnormalities, myopathies, epileptic seizures, intellectual disabilities/developmental delays, hypotonia, and metabolic abnormalities in multiple organ systems. Genetic disease referred to as I-cell disease, inherited in an autonomic recessive manner, is another defect where absence of Golgi GlcNAc phosphotransferase has been reported. This enzyme labels freshly formed proteins by mannose-6-phosphate for targeting them to lysosomes. In the absence of transferase activity, these enzymes are released and thus become a reason for changed turnover of proteins and other molecules inside the cells. Patients usually present with short-trunk dwarfism, mental retardation, and gingival hyperplasia with coarse facial features. These individuals usually die in the first decade of life.[7]

2.1.2 Congenital Muscular Dystrophies (CMDs)

Congenital muscular dystrophies are an O-mannosylation related set of disorders called α dystroglycanopathies and result from the defective O-mannosylation of αDG.[8] αDG is a component of the DGC (dystrophin-glycoprotein complex), which has an essential function in neuromuscular junctions. It relates laminin, extracellular matrix molecule to skeletal muscle cell cytoskeleton and also carries O-linked α- mannose glycans. The other components of DGC besides the αDG are βDG and dystrophin. Dystroglycan, which is a dystrophin-associated glycoprotein and a single polypepetide after proteolytic cleavage produce αDG and βDG.[9] Thus CMDs underlying mechanism involve DGC component mutations, which in turn make whole complex unstable. Taken together for all DGC components, αDG is an extracellular peripheral membrane glycoprotein, which is highly glycosylated. The binding of various extracellular matrix proteins are influenced by its glycans such as agrin that has a role in neuromuscular junction formation and is a synaptic glycoprotein, the neurexins, which are a neuronal-cell-surface protein family, perlecan and laminin via their laminin G domains.[10]

A muscular dystrophy mouse model, myd, showed hypoglycosylation of αDG, indicating that hypoglycosylation is the reason for the disorder.[11, 12] Mutation in enzymes (POMT1 and POMT2, protein *O*-mannosyltransferases in the ER and POMGnT1, protein *O*-mannose β1,2-*N*-acetylglucosaminyltransferase in the Golgi) responsible for mannose residue addition to Ser/Thr in the α-configuration are also known to cause muscular dystrophies.[13, 14] Muscular dystrophies with the αDG hypoglycosylation are also suggested to be caused by few gene mutations, e.g., fukutin related protein (FKRP), fukutin, and LARGE. The functions of these genes are not clear, although they resemble glycosyltransferases.[15–17] However, the gene mutations described above can explain only congenital muscular dystrophy and identification of the functions of the other mutated genes associated with α-dystroglycanopathies will make it possible to diagnose patients with an α-dystroglycanopathy with an assay for glycosyltransferase activity. The clinical spectrum of α dystroglycanopathies is broad and ranges from early-onset muscular dystrophy, to severe brain and eye malformations, to late-onset muscle weakness with normal intelligence, and the disorders include Fukuyama congenital muscular dystrophy, Walker-Warburg syndrome, Muscle-eye-brain disease, and Limb-girdle muscular dystrophies. Clinical findings typically include muscle weakness, contractures, hypotonia, seizures, or mental retardation together with elevated serum creatine kinase activity.

2.2 Neurodegenerative Diseases

Glycosylation defects have also been associated with a neurodegenerative disorder known as Alzheimer's disease (AD). AD patients suffer from a range of difficulties like progressive loss of memory, speech, task performance, and recognition of objects and people. In the brains of AD patients, accumulation of NFT (neurofibrillary tangles) and Aβ amyloid plaques are found. Amyloid β-peptide (Aβ) is generated from a highly glycosylated membrane glycoprotein known as amyloid precursor protein (APP), which has both N- and O-glycans. Neurofibrillary tangles are made of PHFs (paired helical filaments), which comprise pathological filamentous aggregations of abnormally phosphorylated tau protein (phosphorylation at nonphysiological sites). Tau proteins have a role in stability and assembly of microtubules, which is important for the proper functioning of neurons. Glycosylation defects of proteins like APP, tau have been reported in AD.

Three potential N-glycosylation sites are present on tau proteins. The occurrence of N-linked glycans on tau has been confirmed by N-glycosidase F treatments, monosaccharide composition analysis, and lectin staining.[18–20]. In a study by Wang et al., abnormally (or nonphysiologically) glycosylated Tau proteins were found in AD patients' brains and not from the brains of controls. Experimental studies have revealed that if glycosylation of tau tangles is removed, they convert to bundles of straight filaments; this in turn makes them accessible to microtubules.[20] Furthermore, it was observed that nonhyperphosphorylated tau in brains of AD patients are glycosylated in comparison to brains of normal individuals. This glycosylation somehow facilitates further tau protein hyperphosphorylation in the brains of patients with AD.[18] However, the process of N-Glycosylation of tau protein is not clear.

For APP to be functionally active, N-glycans are important. N-glycosylation of APP occurs at two specific sites: Asn467 and Asn496. An experimental setup on Chinese Hamster Ovary Cells (CHO cells) demonstrated that when APP structure was altered by the deletion of Asn467 and Asn496, it resulted in decreased microsomal localization and decreased secretion of APP, thereby highlighting the role of glycans in intracellular sorting of the protein.[21] In another study on mutant CHO cell lines, which are glycosylation-defective and soluble inhibitors of glycosylation, secretion of APP decreased owing to less N-glycan processing or core N-glycosylation.[22] In an experimental setup, a mannosidase inhibitor was used in hamsters, which decreased the movement of glycoproteins including APP toward the synaptic membranes. This inhibitor actually disturbs the formation of hybrid N-glycans (deoxymannojirimycin).[23] This interference of mannosidase inhibitors also results in diminished secretion of APP.[24] APP has a number of O-glycosylation sites in addition to the already discussed N-glycosylation sites, and O-glycans are reported to have an effect on the functions of APP. An experimental study showed that in cells with mutant APP has defective *O*-glycosylation having otherwise normal protein metabolism, the maximum cleavage of APP by secretases α-, β-, and γ occur after *O*-glycosylation. These cells markedly decreased intracellular production of carboxyl-terminal fragment of APP (αAPP COOH), which is α-secretase product plus Aβ42 and Aβ40 in medium, a product of γ- and β-secretases.[25] APP is also O-GlcNAcylated[26] and O-GlcNAcylation, a particular type of O-glycosylation leads to less secretion of Aβ and elevated levels of soluble APPα (sAPPα), owing to a change in APP processing.[27]

There are reports that show that Sialylation of protein is changed in AD. Sialylation is a modification of proteins, in which the addition of sialic acid unit occurs to the end of an oligosaccharide chain in a glycoprotein. The reaction is catalyzed by enzymes called sialyltransferases (ST). These STs are located on the membranes of Golgi apparatus. A study of 12 patients with AD and 12 healthy individuals showed decreased activity of soluble sialyltransferase (ST) in serum.[28] The sialylation differences between the above group of individuals was shown by lectin blotting analysis of cerebrospinal fluid (CSF) proteins.[29] Furthermore, a gene for cell surface immune receptor-sialic acid binding receptor (cluster of differentiation 33) is associated with late-onset AD. This receptor binds to extracellular sialylated glycans.[30–32]

The two major proteases γ-secretase and β-secretase or BACE-1 (β-site APP-cleaving enzyme 1) accountable for producing the toxic Aβ also play an important role in protein glycosylation. The γ, β-secretase alter APP by affecting the extent of sialylation and complex N-glycosylation. BACE-1 cleaves the membrane-bound STs and makes them both soluble and secretable. Not only APP, but several other proteins related to AD pathophysiology like reelin, NCAM (neural cell adhesion molecule), butyrylcholinesterase, v-ATPase, nicastrin, and tyrosine-related kinase B have defective glycosylation as a result of the absence of one of the key components of presenilin and γ-secretase. The inter relation between AD pathophysiology and protein glycosylation needs to be explored further, owing to the benefits it can lead to in terms of diagnosis, management, and treatment of AD.

Several in vitro studies have shown prion diseases are highly reflectant of degree of prion (PrP) glycosylation, which is a protein present on cell surface and main infectious agent of prion disease.[33] The transmission of prion protein (PrP) diseases occurs due to the alteration of normal conformation prion protein (PrP^{C}) to the scrambled form of protein (PrP^{Sc}), which is rich in fibrillar structure. Glycosylation seems to have a role here also, as PrP^{C} has two glycosylation sites involving complex N-glycans, which somehow influence progression to prion disease. Also the association of T183A mutation with the N-linked glycosylation loss in the prion protein has been found in human prion disease.[34] Certain symptoms of prion disease, like early-onset dementia, hypometabolism, and cerebral atrophy may arise owing to this mutation.

2.3 Cancer

In cancer cells one of the profound changes, among many others, is altered glycosylation. Glycans are involved in essential cell biology and molecular processes taking place in cancer, such as hematogenous metastasis and angiogenesis, invasion, tumor proliferation, epithelial-mesenchymal transition, cell to cell contact, and cancer progression and metastasis.[35, 36] Most of the protein biomarkers used nowadays for cancer are glycoproteins and in cancer, their glycosylation is largely affected. This, in turn, provides a set of specific targets for therapeutic intervention, e.g., Alpha-fetoprotein (AFP) in germ cell tumors and Liver cancer, prostate specific antigen (PSA) in Prostate cancer, beta-human chorionic gonadotropin (Beta-hCG) in testicular cancer, choriocarcinoma MUC16 (CA-125) in Ovarian cancer, Carcinoembryonic antigen (CEA) in breast cancer and colorectal cancer, and many more. In addition, glycosylation is changed in several other diseases like acute and chronic inflammatory diseases (sepsis, diabetes, pancreatitis, rheumatoid arthritis,) and infection (HIV/AIDS).[37]

3 ACETYLATION

Protein acetylation is another very common protein modification. Here, the acetyl group is co- or posttranslationally attached either to the lysine residues at ε-amino group or to the N-terminus of protein at α-amino group. Lysine acetylation and deacetylation of proteins was widely studied in histones first. It is a reversible reaction catalyzed by lysine acetyltransferases (KATs, previously known as HATs, histone acetyltransferases) and deacetylases (KDACs, generally termed histone deacetylases or HDACs). HAT enzymes catalyze transfer of acetyl groups to ε-amino group of the lysine residue from acetyl-CoA, whereas HDACs eliminate the acetyl group from acetylated lysine residues, liberating an acetate molecule in the presence of water. The interaction between acetylation and deacetylation is important for many viral cellular processes and any perturbations in the regulation and/or function of the KATs and KDACs could be associated with diseases such as neurodegenerative disorders, cancer, aging, diabetes, cardiovascular diseases and autoimmune disorders.[38, 39]

Histones and transcription factors are the major targets of acetylation. Histone acetylation and deacetylation, which has attained much attention owing to its importance in gene regulation, happens at the N-terminal tail of lysine residue of histone protein.[40, 41] Acetylation neutralizes the positive charge on lysine residues leading to a relaxed chromatin conformation, which in turn facilitates access for transcriptional regulators. The equilibrium between acetylation and deacetylation is significant for gene expression regulation. Therefore, when chromosomal regions that are usually silenced get hyperacetylated, or regions that are usually transcribed get deacetylated, it leads to a number of disorders involving developmental and proliferative diseases. Studies have reported that various cardiovascular diseases (CVDs), such as hypertension,[42, 43] diabetic cardiomyopathy,[44] myocardial infarction,[45] and pulmonary arterial hypertension,[46, 47] and cellular disorders, such as vascular smooth muscle cell proliferation,[48] and apoptosis[49] are associated with the acetylation of core histone. Increased histone acetylation by IL-1β on H4 at K8 and K12 residues is an important inflammation related epigenetic mark.[50]

This increase in acetylation occurs because of the recruitment of HATs like P300/CBP-Associated Factor (PCAF) and CREB-Binding Protein (CBP) by an essential transcription factor, which controls expression of inflammatory genes, e.g., NF-κB (nuclear factor-κB), to the promoter region of inflammatory genes. Sirtuins (SIRTs-from the founding member from budding yeast called silent mating type information regulation two protein) which are NAD^+ dependent protein deacetylases also play a vital role in regulation of mammalian metabolism. Seven different SIRTs are present in cells, three of which are localized in the mitochondria: SIRT3,

SIRT4, and SIRT5.[51] One study has noted that SIRT3, when abolished, results in the hyperacetylation of mitochondrial proteins and shows altered function that ultimately leads to mitochondrial dysfunction.[52] Another study demonstrated an increased acetylation of microtubule proteins in response to surplus ROS production by defective mitochondria and suppression of SIRT2 (which is present in cytoplasm and is known to associate with microtubules deacetylating α tubulin), which in turn exacerbates the mitochondrial malfunction in chronic progressive external ophthalmoplegia (CPEO) syndrome patients.[53] CPEO is actually a disorder in which there is a slowly progressive paralysis of the extra ocular muscles and patients usually experience symmetrical, bilateral, and progressive ptosis, which is followed by ophthalmoparesis months later.

Deregulation of acetylation-deacetylation dynamics cause unusual expression of certain genes that can act as oncogenic roles. Mutations in p300 acetyltransferase have been linked with breast, colorectal, and brain cancer (glioblastoma).[54, 55] Prostate cancer and its reemergence have been associated with the hypoacetylation of several residues like H3K18.[56] Hyperacetylation of histone is observed in hepatocellular carcinoma.[57] Moreover, human CBP (CREB-Binding Protein) gene mutations are reported in developmental haploinsufficiency disorder, known as Rubinstein-Taybi syndrome (RTS)—a congenital malformation, which presents with craniofacial defects, cardiac anomalies, mental retardation, broad thumbs, and broad big toes.[58] The function of CBP in RTS was supported by an experimental study where heterozygous mice were deficient in a single CBP allele and showed an unusual pattern formation with partial resemblance to RTS.[59] Further, the incidence of translocation of human chromosomes highlighted the CBP's vital role in cell transformation, and fusion of CBP with MYST domain containing monocytic leukemia zinc-finger protein (MOZ) or mixed lineage leukemia (MLL) shows gain-of-function features and leads to unusual hyperacetylation and transcriptional activation, which in turn causes leukemia in humans.[60]

Acetoproteins have been associated with various cognitive disorders like Alzheimer's disease and dementia. Lysine residues of tau proteins are acetylated, forming tau tangles, in the case of dementia; whereas, in Alzheimer's disease, impaired cognition is because of hyperacetylation of in β-amyloid peptide.[61] Studies on mouse models have suggested that acetylation of histone is essential for memory consolidation and pattern of acetylation has a role in age-dependent memory impairment.[62] In addition, the disruption of the delicate balance between HAT and HDAC activities have been shown to result in decreased expression of vital genes in Huntington's disease (HD), which is a deadly disease and patients usually present with symptoms like involuntary movements (chorea), and cognitive and personality changes.[63] Furthermore, the loss of androgen receptor (AR) acetylation is involved in the inherited neurodegenerative disorder known as Kennedy's disease, in which patients usually present with altered control of breathing, talking, and swallowing, tremors in the hands when in action, dysphagia, numbness, weakness of muscles, abnormal contraction of foot muscles in response to stimuli, erectile dysfunction and impotence, hormonal imbalance, enlarged breasts, and elevated serum creatinine kinase. Here, a ligand-dependent nuclear translocation was shown to become significantly slower due to substitution of K630A, or both K632A and K633A.[64] Various neuropsychiatric disorders, like schizophrenia, which is characterized by disordered thoughts, presence of hallucinations and delusions, abnormal behaviors, and antisocial behaviors, also show acetylation of histone proteins. A study including schizophrenia patients and age-matched healthy individuals showed a significant hypoacetylation of H3 histones at lysines 9/14 (H3K9K14) in schizophrenia patients in comparison to controls. Furthermore, several candidate's gene expression for schizophrenia showed alteration through treatment with HDAC inhibitor in mouse brain.[65] These studies, therefore, suggest that novel therapies can be developed for various diseases like psychiatric and neurodegenerative disorders by targeting lysine acetylases and deacetylases.

4 PHOSPHORYLATION

Protein phosphorylation is one of the profound PTMs, in which a phosphate group is added to an amino acid, most commonly tyrosine serine and threonine in eukaryotic cells. It is a reversible reaction catalyzed by protein kinases (PKs) and regulates enzyme reactions, interaction between proteins, cellular metabolism, and protein degradation for many proteins that are engaged in intracellular signaling cascades.[66, 67] Conversely, dephosphorylation is the removal of a phosphate group catalyzed by phosphatases (PPs). The mitogen-activated protein kinase-(MAPK-)mediated pathways, pathways involving protein kinase C (PKC) and protein kinase A (PKA), plus phosphoinositide 3-kinase (PI3K)/Akt/mammalian target of rapamycin (mTOR)-dependent signaling act are important components of several signaling pathways. They are regulators of many fundamental intracellular processes. The activation and deactivation of signaling pathways is regulated by both phosphorylation and dephosphorylation, and is a key to maintaining cellular homeostasis. Dysregulation of kinase phosphatase function is now recognized as a cause or consequence of many human diseases, for example, neurological diseases like dementia and Parkinson's that harbor Lewy bodies. Proteolytic Lewy aggregates are formed by phosphorylation at Ser-129 in α-synuclein protein.[68] Phosphorylation is also important for tau function, which is a microtubule associating protein and plays a role in a number of neurodegenerative diseases. Tau protein aggregates are formed as a result of hyperphosphorylation of tau. These aggregates are key constituents of NFT (neurofibrillary tangles), which are the pathological characteristics of tauopathies like Alzheimer's and Parkinson's.[69, 70] In addition, nuclear factor-κB (NF-κB)—a transcription factor that also has a role in immunity, inflammation, cell proliferation, and apoptosis—is regulated by phosphorylation. Aberrant phosphorylation of NF-κB cascade represents a typical characteristic of chronic immune disorders and cancers.[71] Hyperphosphorylation of Akt, a protein kinase, has been found in prostate cancer cellular models, for example LNCaP cells, and also in prostate cancer specimens, mostly in later stages of the disease.[72] Furthermore, the changes in protein phosphorylation are also associated with other diseases such as cancer, heart disease, diabetes, and rheumatoid arthritis.[73–79]

5 CARBONYLATION

Carbonylation of proteins results from an increase in oxidative stress in physiological systems. Carbonylation represents an irreversible PTM that leads to loss of protein function. It is a measure of disease-derived protein dysfunction and severe oxidative damage. A number of oxidative pathways introduce carbonyl groups into proteins. One of these being metal catalyzed oxidation, in which the reactive hydroxyl radicals oxidize side chains of amino acid (Pro, Arg, Lys, and Thr) or cleave the protein backbone resulting in highly reactive carbonyl derivatives, e.g., glutamic semialdehyde from arginyl and prolyl residue, 2-pyrrolidone from prolyl residue, 2-amino-3-ketobutyric acid from threonyl residue, and α-aminoadipic semialdehyde from lysyl residue. Another way in which protein carbonyl groups are generated is when reactive carbonyl species produced via glyoxidation/glycation (a reaction in which the reducing sugars such as fructose or glucose or their oxidative products react with side chains of arginine and lysine residues) reacts with the primary amino group of lysine residues, resulting in the formation of AGE (advanced glycation end products) carrying carbonylated moieties such as carboxymethyllysine and pentosidine. Finally, the protein bound carbonyls can be formed by the reaction of carbonyl containing lipid peroxidation products (LPP), for example 4-hydroxy-nonenal-protein adduct and malondialdehyde-lysine, which are produced by metal catalyzed oxidation of polyunsaturated fatty acids (PUFA) with the side chains of lysine, histidine, and cysteine. Protein carbonylation may generally cause inhibition of enzymatic activity, increased susceptibility to proteolysis and aggregation, and altered cellular uptake.[80]

Protein carbonyl content is a common biomarker of severe oxidative protein damage. It reflects cellular damage because of various ROS species.[80–82] Proteins that are oxidized can have several fates like repair, proteolytic degradation, or some accumulate in the form of damaged or unfolded proteins. The fate also depends on the degree of carbonylation. The proteasomal system takes care of moderately carbonylated proteins, whereas those proteins that are heavily carbonylated form high-molecular-weight aggregates. These clamps do not degrade, thus usually act as unfolded and damaged proteins, which in turn can inhibit the proteasome activity. Increased protein carbonyl content is seen in a number of human diseases such as inflammatory bowel disease (IBD), Alzheimer's disease (AD), arthritis, and diabetes.[83–88] A study on Murphy Roth's large (MRL) mouse model, showed that treatment with a common environmental contaminant, trichloroethene (TCE), results in an increased formation of protein carbonyl(s) and nitrotyrosine (NT), which in turn have shown to be associated with autoimmune response.[89] In skeletal muscle, under several pathological conditions including chronic obstructive pulmonary disease, diabetes, ischemia-reperfusion, and sepsis, an increase in carbonylated protein levels have been detected.[90, 91] In a study, diet-induced obese mice were found to have increased protein carbonylation in the adipose tissue.[92] The same results were found in obese human subjects.[93] In addition, protein carbonyl content has been reported to be significantly increased in acute pancreatitis patients, and was related to disease severity.[94] Thus, taking these studies into consideration, therapeutic strategies can be designed against various diseases where cellular protein carbonylation can be targeted.

6 METHYLATION

Protein methylation is a prevalent PTM, in which the transfer of methyl group occurs from S-adenosyl-L-methionine (SAM) to histone and other proteins, and occurs mainly on lysine and arginine residues. After the identification of first lysine demethylase protein LSD1 /KDM1, methylation is thought to be regulated dynamically similarly to phosphorylation. It was earlier considered as an enzymatically irreversible modification.[95] The reaction is catalyzed by lysine and arginine methyltransferases (KMTs and PRMTs) and is essential for RNA metabolism, gene expression regulation, and protein function.[96–98] Three methyl groups are transferred to the lysine residues at ε-nitrogen by KMTs and the enzyme activity of almost all lysine methyltransferases, with some exemptions (DOT1), is governed by a distinctive SET domain (suppressors of variegation enhancers of zeste and tristae). PRMTs, on the other hand, transfer one or two methyl groups to arginine residue on proteins at the guanidino-nitrogen. Histones have many arginine and lysine methylation sites (specifically H3 and H4). These regulate the chromatin structure and gene transcription. Some other nonhistone proteins like NF-κB, p53, pCAF, ERα, and several transcription factors having a role in tumorigenesis plus other metabolic disorders such as immune and inflammatory responses, are methylated by methyltransferases.[99]

Studies concerning lysine methylation mainly highlight histone methylation, one of the controlling mechanisms of gene expression. It is actually the site and status of methylated lysine residue in histones that govern it to either activate or repress the gene expression, e.g., euchromatin and transcriptional activation is associated with the H3 histone methylation at Lys-4, 36, and 79, whereas heterochromatin and transcriptional repression is correlated with H3 histone methylation at Lys-9 and 27. These changes brought about by methylation influence how protein complexes interact in this scenario and thus affect the overall functioning of chromatin. Because of the crucial function of histone methylation in transcription, defects in functioning of enzymes that mediate these modifications is likely to cause various diseases, including cancer. In an experimental setup mice deficient of SUV39H1 and SUV39H2 methyltransferases or both, developed leukemia, signifying their important role of cell proliferation regulation.[100, 101]

Methyl transferase G9A, concerned with tumor suppressor gene silencing, is up regulated in many cancer cell types and its homolog, GLP, is also over expressed in brain tumors and multiple myeloma.[102] Furthermore, the role of the NSD (Nuclear receptor-binding SET-domain proteins) family of protein lysine methyltransferases, which includes NSD1, NSD2/WHSC1/MMSET, and NSD3/WHSC1L1, have been associated with various cancers, e.g., overexpression of NSD2 causes multiple myeloma, chromosome translocations of NSD1 or NSD3 to NUP98 leads to acute myeloid leukemia, and NSD3 is upregulated in primary breast carcinomas and breast cancer cell lines.[103, 104] Also dysregulation of SMYD methyltransferases (SET and MYND domain-containing proteins) has been involved in many cancers, e.g., overexpression of SMYD3 is reported in hepatocellular carcinoma, breast, and colorectal cancer.[105–107]

SMYD2 inactivates function of tumor suppressor proteins p53 and RB1 by lysine methylation and acts as an oncogenic protein. In addition, arginine methyltransferases like PRMT1, PRMT2, and PRMT4 are known to have a direct role in a number of cancers. PRMT1 (type I protein arginine methyltransferase) catalyzes the methylation of histone H4R3.[108] Its increased expression is reported in a variety of human malignancies.[109] PRMT5 (type II arginine methyltransferase) methylates histone H2AR3, H3R8, and H4R3 and negatively regulates the expression of tumor-suppressor genes and, therefore, acts as an oncogene. Increased protein levels of PRMT5 are present in leukemia and lymphoma cells.[110] Retinoblastoma binding protein 2 (RBP2 or KDM5A) demethylates tri- and di-methylated lysine 4 in histone H3. These are epigenetic marks for transcriptionally active chromatin. H3K4 methylation is promoted by multiple endocrine neoplasia type 1 (MEN1) tumor suppressor. Genetic ablation of Rbp2 in a mouse model was shown to decrease the formation of tumor and extends survival in Rb1+/− mice and Men1-defective mice and thus associate the activity of RBP2 histone demethylase to tumorigenesis. Thus, RBP2 may act as an important target for cancer therapy.[111] Furthermore, anticancer drugs have been developed by targeting protein methyltransferases and demethylases, and clinical trials have already been started.[112]

Apart from cancer, methylation also has an important function in cardiovascular diseases. A study on a rat model of cardiac disease demonstrated a significant change in H3K4 methylation pattern through different stages of disease progression, suggesting that lysine methylation has a promising role in cardiovascular disorders.[113] All the class I PRMTs are known to cause oxidative stress-mediated disease manifestation. This happens through the formation of asymmetric dimethyl arginine (ADMA) intermediate, which is a competitive inhibitor of NOS (nitric oxide synthase).[114] It reduces the availability of NO (nitric oxide) inside cells, which is a hallmark of cardiovascular diseases.[115] Also, the levels of homocysteine have been involved in cardiovascular diseases and neurological disorders such as Parkinson's disease.[116]

Batten disease, or Juvenile ceroidlipofuscinosis-hereditary disorder, shows visual loss, cognitive and psychomotor deterioration, progressive seizures, and early death, between the age of 15 and 35 years. In case of infantile and late infantile forms of ceroidlipofuscinosis, the onset of symptoms is seen between the ages of 4 months and 3 years of age, and in adult cases the onset of symptoms is not seen until the affected individuals are between 25 and 40 years of age. Massive deposits of autofluorescent cytosomes are present in the tissue of individuals with this disease. These storage bodies usually contain large amounts of mitochondrial proteins. The mechanism that leads to the accumulation of such storage bodies is, as yet, not understood. However, the chromatographic analysis indicates a significant amount of modified amino acid *E-N*-trimethyllysine, which may be due to either the excessive methylation of these stored proteins or else the failure to demethylate the intermediate forms of such proteins.[117]

7 HYDROXYLATION

Protein hydroxylation is a reversible PTM in which a hydroxyl group (–OH) is introduced into amino acid residues, mainly proline, forming hydroxyproline, and the less common hydroxylated residue lysine, forming hydroxylysine. The reaction is catalyzed by hydroxylases, which are the members of a large class of enzymes known as2-oxoglutarate-dependent dioxygenases, and plays several critical roles in biological systems. Collagen is one of the most well-known proteins whose function is mediated by hydroxylation state, whereby both proline and lysine hydroxylation enhance the "tightness" of the triple alpha-helix structure. Modification of collagen mediated by proline hydroxylation is important in terms of cell structure and physiology and has been widely studied.[118] The reaction requires 2-oxoglutarate, ascorbate, Fe2+, and O_2 and involves an oxidative decarboxylation of 2-oxoglutarate. Metabolic disorders like scurvy and some cancers are associated with the lack of proline hydroxylation, in which ascorbate deficiency inhibits the collagen prolyl hydroxylase enzyme and results in characteristic signs of the disease.[119] Scurvy is a very rare disease and usually occurs in elderly people, those that live on a diet lacking vegetables and fresh fruits, or alcoholics. Patients develop edema (swelling) in some body parts, debility, anemia, exhaustion, and at times loss of teeth and ulceration of the gums.

Proline hydroxylation is important for activation of antioxidant defense against hypoxia through hypoxia inducible factor (HIF).[120] HIFs are oxygen-dependent transcriptional activators formed of a labile α-subunit and a stable β-subunit and play a major role during hypoxia in cell proliferation,[121] angiogenesis,[122] and embryonic vasculogenesis.[123] In normoxic conditions, prolyl hydroxylase domain proteins (PHDs) lead to the hydroxylation of HIF-α. It is picked by the ubiquitin E3 ligase than sent for degradation to the proteasome. The lack of oxygen under hypoxic conditions brings about the stabilization of HIF-1α and forms HIF heterodimer with HIF-1β. Collectively, these endorse the transcription of hypoxia-response genes,[124] with functional consequences such as angiogenesis, metabolic adaptation, metastasis, apoptosis, and more.[125–128] Furthermore, tumorogenesis was shown to be promoted by HIF proline hydroxylation in a similar way to that under normoxic conditions.[129, 130] In terms of therapeutics, inhibition of normoxic HIF1α is considered as an alternative therapeutic.[129, 131]

Disruption of protein hydroxylation was discovered to have links with many serious human diseases, e.g., osteogenesis imperfecta (brittle bone disease),[132] osteoporosis,[133] nervous system tumors,[134] kidney disease,[135] breast cancer,[136] and other hormone-related or hypoxia-related cancers.[137–139] In addition, aromatic side chain containing amino acids like tyrosine, tryptophan, and phenylalanine undergo hydroxylation. Lack of hydroxylation in such amino acids has been reported to be linked to several genetic diseases like that of hyperphenylalaninemia and phenylketonuria (PKU). This has been attributed to defects in the enzyme known as phenylalanine hydroxylase, whose function is to change phenylalanine to tyrosine.[140] The phenotypic variations attributed to the changes in phenylalanine concentrations can vary from very little to quite high levels of it, which when left untreated can result in profound and irreversible mental disability. Patients starts to develop symptoms like stunted growth, skin conditions, such as eczema, tremors, seizures, trembling and shaking, hyperactivity, and a musty odor of their breath, skin, or urine.

8 NITRATION

Nitration is a stable and a reversible PTM that occurs when amino acids such as tyrosine (Y), tryptophan (W), cysteine (C), and methionine (M) are exposed to nitrating agents or oxidative stress. However, most studies on tyrosine nitration have suggested that tyrosine residues(s) of many proteins are modified by peroxynitrite, which is formed by the combination of nitric oxide with superoxide, resulting in protein nitration and therefore, the function and structure of each target protein is altered.[141, 142] An increased protein-bound nitrotyrosine level is an oxidative damage marker and is associated with many diseases. The basal protein nitration in plasma has been detected in the cardiovascular system[143, 144] as well as in major cell types like fibroblasts, endothelial cells, myocytes, and vascular smooth muscle cells.[145–147] Some proteins (e.g., fibrinogen) undergo gain-of-function and some (e.g., Mn-SOD) are inactivated by nitration that affects the pathophysiology of the cardiovascular system. In cardiovascular disease, nitrotyrosine acts as a potential novel independent marker.[148]

Protein tyr nitration has been reported in correlation with neurodegenerative diseases, e.g., Parkinson's disease, where elevated levels of nitrated tyrosine residues were seen in Lewy bodies, characteristic hallmarks of Parkinson's disease, and were detected by antibodies specific for 3-nitrotyrosine (3NT) and by means of mass spectrometry (MS). Elevated levels of 3-nitrotyrosine (NT) immunoreactivity were found in neurons from AD patients' brains as compared to healthy individuals of the same age[149] and 3-nitrotyrosine and dityrosine levels were constantly high in the neocortical hippocampus and IPL regions in the brains of AD patients and in ventricular cerebrospinal fluid (VF).[150, 151] In addition, nitrotyrosine, along with peroxynitrite, was shown to contribute to tau protein neurofibrillary tangles and amyloid β-peptide-induced toxicity in AD patients.[152] Furthermore, free NT levels have been found to be increased in mouse models and human patients of Amyotrophic lateral sclerosis (ALS), which is a known motor neuron disease, causing spasticity, muscle weakness, progressive paralysis, atrophy, and finally death within the few years of the start of the disease.[153–155]

Additionally, nitrative stress is also reported in age-related macular degeneration (AMD) disease, which has complex pathophysiology involving genetic predispositions, local inflammation, neovascularization, and accumulation of lipofuscin and drusen. It leads to legal blindness and severe visual loss in the elderly population. Identification of commonly inherited variant Y402H from AMD patients has linked the genetics of the disease with inflammation. In inflammation, increased NO may possibly result in nonenzymatic nitration of components of human Bruch's membrane proteins and two nitrative stress biomarkers, namely 3-nitrotyrosine (3-NT), recognized in the preparations of Bruch's membrane, and nitrated A2E from the aged human Bruch's membrane lipid soluble extract, have been identified.[156] Nitrative stress has also been shown to reduce the expression of histone deacetylase 2 (HDAC2) via nitration of distinct tyrosine residues, which marks it for fast proteasomal degradation and finally the gene regulation is affected inside the cell. The study is of greater importance for novel therapies in cancer that explore the choice of HDAC inhibitors.[157] Increased nitroxidative stress is also involved in mediating the pathological effects in acute and chronic liver diseases including alcoholic fatty liver disease (AFLD) and nonalcoholic fatty liver disease (NAFLD). An increase in nitroxidative stress results in dysfunction of many mitochondrial proteins involved in fat and ammonia metabolism, energy supply, antioxidant defense, and so forth, and interrupting many important signaling pathways (e.g., insulin signaling pathway), contributing to initiation of a cascade of deleterious events. Alternatively, other nitrated proteins may stimulate hepatic macrophage Kupffer cells with elevated cytokines and chemokines, which promote infiltration of neutrophils and inflammatory tissue injury.[158]

9 SULFATION

Another important protein PTM is sulfation, which involves the addition of a sulfate group to a variety of proteins and peptides by oxygen or nitrogen respectively (N-sulfation or O-sulfation). The modification occurs at tyrosine residues and is catalyzed by tyrosyl protein sulfotransferase (TPST). It is estimated that about 1% of total tyrosines in eukaryotic proteins undergo sulfation.[159] The posttranslational sulfation is normally examined for secreted and membrane proteins, which has a crucial role in G-protein receptor signaling, protein-protein interactions, immune responses, and chemokine signaling.[160, 161] Tyrosine-sulfated protein and TPST are implicated in several biological processes, including viral entry into cells, leukocyte rolling on endothelial cells, hemostasis, visual functions, and some ligand binding to receptors.[162–166] Tyrosine sulfation has a role in several diseases including lung diseases, HIV infection, multiple sclerosis, autoimmune response, and cellular enzyme regulation.[160, 167–170]

In AIDS, the role of tyrosine sulfation is being considered and evaluated. Tyrosine sulfation of major chemokine receptor CCR5 (a co-receptor for HIV gp120) is important for HIV-1 infection of T cells. It works jointly with CD4 and leads to viral attachment and its successive invasion. On the other hand, it appears to be less important for the coreceptor CXCR4.[163, 171] It remains to be elucidated how the virus enters via sulfated chemokine receptors. This information suggests that HIV-1 cellular entry can be blocked by inhibiting PTS of CCR5, which can be used as a therapeutic measure against HIV, but it is vital to determine that interfering with this modification does not result in any other toxic effect. Further, Epstein-Barr virus (EBV) synthesizes an oncogenic protein known as Latent membrane protein 1 (LMP1), which has been implicated in metastasis and carcinogenesis of nasopharyngeal carcinoma (NPC). NPC patients usually present with symptoms of cranial nerve palsies (generally related to extension of the tumor into the base of the skull), nasal obstruction (e.g., bleeding, congestion, nasal discharge), and changes in hearing (generally related with blockage of the Eustachian tube, but direct extension into the ear is possible). Juan Xu and co-workers in their study showed that tyrosine sulfation of chemokine receptor CXCR4 is induced by LMP1 through tyrosylprotein sulfotransferase-1 (TPST-1), which adds to the extremely metastatic nature of NPC.[172] In addition, the role tyrosine sulfation has been well described in hematopoietic cells during atherosclerosis development.[173]

The association between HMGs (high molecular weight glycoconjugates), which are formed by cystic fibrosis (CF) respiratory epithelia and airway secretions, has been shown to be adversely affected by heavy sulfation of HMG, which possibly creates a breeding environment for harmful bacteria like *Staphylococcus aureus* and *Pseudomonas aeruginosa* in the CF airways, signifying a promising role of sulfation disease pathogenesis.[167] Furthermore, induction of chemokine signaling by the tyrosine sulfation equally aggravates other lung diseases like asthma and chronic obstructive pulmonary disease (COPD), affecting the downstream signaling molecules along with airway inflammation and leukocyte trafficking, which are important in asthma and COPD.[170, 174] Chemokine receptor antagonists are one of the profound research topics in pharmaceutical research, concentrating on inflammation in COPD and asthma.[175, 176]

10 PALMITOYLATION

Palmitoylation is a unique reversible PTM, in which a 16-carbon saturated fatty acid palmitic acid is attached covalently on cysteine residues in proteins through a thioester linkage (S-acylation). This PTM is mediated by a protein known as a palmitoyltransferase (PAT), or an acyltransferase, which has aspartate-histidine-histidine-cysteine (DHHC) motif located within a cysteine-rich domain ($\sim$50 amino acid), and reversed by specific acyl protein thioesterases. The palmitic acid addition increases the protein hydrophobicity and contributes to their membrane association. This modification is important for the regulation of protein stability, subcellular localization, aggregation, translocation to lipid rafts, trafficking, interaction with effectors, and other aspects of protein function. This modification is unlike other lipid-protein modifications such as myristoylation and isoprenylation/prenylation. These are co-translational and irreversible reactions.[177] Proteins that are S-palmitoylated include GPCRs (G-protein coupled receptors), which are palmitoylated near to their transmembrane domains on cysteine residues, Gα subunits, which are palmitoylated near their N terminus and myristoylated at an N-terminal glycine residue, and the Ras family of proteins, farnesylated at a C-terminal CAAX box and also modified near the C-terminus. Peripheral membrane proteins, such as neuronal proteins GAP43 and SNAP25, are modified with palmitate only.

Palmitoylation was shown to modify specific cysteine residues of regulators of G protein signaling (RGS) proteins, which regulate the G-protein linked cell signaling pathways. RGS mediates the deactivation of GPCR signaling pathway by accelerating the intrinsic GTPase activity of heterotrimeric G protein α subunits of the i, q, and 12 classes. This modification of RGS protein regulates not only its activity but also mediates targeting and localization of certain G-proteins, therefore modulating G-protein signaling.[178] There are reports that show that carboxyl tail cysteines of rhodopsin are palmitoylated, which is important for G-protein signaling.[179] on the other hand, in thyrotropin-releasing hormone (TRH) receptor type 1 (TRH-R1), in which unpalmitolyated form is very active and results in the over production of prolactin and thyrotropins, palmitoylation is essential for maintaining the receptor in an inactive form and is not required specifically by the TRH/TRH-R1 complex for G protein coupling.[179]

Many diseases and disorders have been linked to reduced or dysfunctional palmitoylation. Protein palmitoylation is important for several processes in the nervous system involving synaptic activity and neuronal development. Dysregulation of protein palmitoylation has been shown to be associated with neurodegenerative diseases like Alzheimer's disease and Huntington's disease. In AD, the sequential cleavage of β-amyloid precursor protein (APP) by β- and γ-secretases produces β-amyloid peptides (Aβ), which accumulate in the brains Alzheimer's disease patients. Palmitoylation of amyloid precursor protein (APP) at Cys186 and Cys187 targets APP to lipid rafts and, therefore, enhances its amyloidogenic processing and increases its γ-secretase mediated cleavage.[180] Also the γ-secretase complex itself is subjected to palmitoylation and its inefficient palmitoylationin adversely affects the functional potential and proper trafficking of the neurons in AD.[181] In Huntington's disease, huntingtin protein (HTT) is palmitoylated normally at cysteine 214 and is regulated by palmitoyltransferase huntingtin interacting protein 14 (HIP14), which is important for its function and trafficking. Extension of the polyglutamine tract of HTT responsible for Huntington's disease was demonstrated to reduce the affinity of the protein for its specific PATs HIP14 and, therefore, results in a marked decrease in palmitoylation. This further increases the rapid formation of inclusion bodies and neural toxicity, which in turn worsens the patient prognosis.[182]

eNOS (endothelial nitric oxide synthase) palmitoylation in diabetic vascular disease is important for its efficient bioavailability to the plasma membrane. In insulin-resistant or insulin-deficient patients, lack of eNOS palmitoylation was observed to result in chronic inflammatory response in these patients.[183] In the cardiovascular system, eNOS mediates the synthesis of nitric oxide, which acts as an antioxidant, endogenous vasodilator, and platelet inhibitor, and has a role in regulating the vascular endothelium through its anticoagulant and antithrombogenic properties. Also palmitoylation of Hepatitis virus C core protein, which is the viral nucleocapsid of hepatitis C virus (causative agent of chronic hepatitis) at Cysteine 172 residue is essential for virion particles to multiply, therefore maintaining infection in the infected host cells.[184]

Thioacylation of myelin proteolipid protein (PLP), which is the CNS myelin protein and a potential autoantigen in multiple sclerosis (MS), is important for its normal functioning and in myelin stability.[185] PLP is normally thiopalmitoylated at up to six sites, and it has been shown that degree of thiopalmitoylation of PLP increases markedly during the process of demyelination. Thus, during myelin breakdown in MS, there is the potential for thiopalmitoylated PLP peptides to be released, which might play in the expansion of autoimmune responses and enhancement of disease chronicity.[186] With the aim to help patients with autoimmune diseases like multiple sclerosis, a novel approach is being proposed, which uses the stability of palmitoylated proteins to present as antigens to the MHC-class II immune response.[187] Palmitoylated proteins also have a role in the regulation of four characteristics of cancer, which are constant proliferative signaling (Ras, SFKs: lck, lyn, fyn, Yes and Hck, NTSR-1, LRP6, LAT2), resistance to cell death (Fas&FasR, DR4 &DR6, DCC, UNC5H, BAX,c-abl, RhoB), induction of angiogenesis (Tetraspanins: CD9, CD151, Enos, PAFAH1b3), activation of invasion and metastasis (Tetraspanins: CD9, CD151, KAT1/CD82, Integrins $\alpha3$, $\alpha6$, $\alpha7$ and $\beta4$, CD44, CDCP1, Rac1), and the enabling of characteristic tumor-promoting inflammation (CD44, N-RAS, TfR1, TNFα, PAR2).[188] A study has shown that palmitoylation of oncogenic NRAS is important for leukemogenesis and suggests that targeting NRAS palmitoylation is important for treating malignancies related to acute myeloid leukemia (AML) plus other NRAS-amplified leukemias.[189] Owing to the varied role of palmitoylation, novel probes have been developed by researchers, which can be used for the mass spectrometry and fluorescence microscopy analysis of protein palmitoylation.[190]

In light of the above discussion, Table 1 provides a representation of various PTMs, their physiological effects, and their previously reported associated disease conditions where probably, or in certain cases very surely, the modifications have gone wrong.

TABLE 1 Main Posttranslational Modifications of Proteins and Their Associated Diseases

Modifications	Normal physiological effects	Associated diseases
Glycosylation	Addition of complex oligosaccharides-glycans to proteins: has role in protein-protein interactions, protein localization, immune responses, structural cell stability and modulation of cell signaling	• Congenital disorders of Glycosylation (CDG) • Congenital muscular dystrophies (CMDs) • Neurodegenerative diseases • Cancer • Chronic inflammatory diseases (pancreatitis, diabetes, rheumatoid arthritis) • Infections (HIV/AIDS)
Acetylation	Addition of acetyl group to α-amino group of N-terminus or ε-amino of Lys residue: has role in gene regulation and expression and control of various cellular processes	• Cardiovascular diseases (CVDs) • Inflammation • Neurodegenerative diseases • Cancer • Chronic progressive external ophthalmoplegia (CPEO)
Phosphorylation	Addition of phosphate group to amino acids Ser, Thr, and Tyr: has role in cellular metabolism, protein-protein interactions, and intracellular signaling cascades	• Neurodegenerative diseases (PD) • Cancers • Immune disorders • Diabetes, heart diseases, rheumatoid arthritis
Carbonylation	Metal catalyzed oxidation of residues like Lys, Arg, Pro & Thr, or cleavage of protein backbone to form reactive carbonyls: indicates oxidative stress in biological system and leads to loss of protein function by synthesis of protein carbonyl groups	• Alzheimer's disease (AD) • Inflammatory bowel disease (IBD) • Diabetes • Arthritis • Chronic obstructive pulmonary disease (COPD) • Pancreatitis

TABLE 1 Main Posttranslational Modifications of Proteins and Their Associated Diseases—cont'd

Modifications	Normal physiological effects	Associated diseases
Methylation	Transfer of methyl group from s-Adenosyl methionine to lys/arg residue of proteins: regulates gene expression, protein function, and RNA metabolism	• Cancer • Cardiovascular diseases (CVDs) • Juvenile ceroidlipofuscinosis (Batten disease) • Parkinson's disease
Hydroxylation	Addition of hydroxyl group to amino acids mainly Proline/Lysine: this enhances the tightness of the protein structure	• Scurvy • Osteogenesis imperfecta (brittle bone disease) • Osteoporosis • Nervous system tumors • Kidney disease • Phenylketonuria • Cancer
Nitration	Nitric oxide produces peroxy nitrite when combined with superoxide, which modifies Tyr residues of target proteins and alters their structure and functions	• Cardiovascular diseases (CVDs) • Neurodegenerative diseases (AD) • Amytrophic lateral sclerosis (ALS) • Age-related macular degeneration • Alcoholic/nonalcoholic fatty liver disease (AFLD/NAFLD)
Sulfation	Addition of sulfate group in a variety of proteins usually at Tyr residues: happens in both membrane and secretory proteins and is important for protein-protein interactions	• Infections (HIV/AIDS) • Multiple sclerosis • Lung diseases • Nasopharyngeal carcinoma (NPC) • Cystic fibrosis • Asthma • Chronic obstructive pulmonary disease (COPD)
Palmitoylation	Addition of Palmitic acid to proteins usually on cysteine residues: protein hydrophobicity is increased and adds to their membrane association	• Neurodegenerative diseases (AD) • Diabetic vascular disease • Multiple sclerosis • Leukemia

11 CONCLUSION AND FUTURE PERSPECTIVE

Most human proteins are subject to posttranslational modifications, and there are several ways in which these modifications happen inside a cell. Information about these modifications is very important as these not only alter chemical and physical properties of proteins, but also stability, activity, conformation, folding, distribution, and therefore complete functions of the respective proteins. The study and understanding of PTMs and their involvement in various disease processes is a distant target but still a considerable effort has been put in elucidating the importance of these modifications in the cellular context. A reliable in silico tools and new methods like mass spectrometry (MALDI-TOF) are available for characterization of these modifications and their role in physiology and pathology; however, there are essentially no general methods available that can analyze all types of modifications. Since posttranslational modifications have important roles in biological systems, their study and the recognition of new PTMs are being intensely considered and explored in the field of proteomics research. Some posttranslational modifications are associated with diseases that have high mortality rates, like cancer, hypertension, diabetes, and neurodegenerative disorders like AD, PD, and Huntington's disease. Therefore, a comprehensive knowledge of the PTMs of protein is important, not only for understanding its cellular functions, but also to help in the advancement of novel prognostic markers for cancer, neurodegenerative diseases, and other disorders along with their therapeutic targets.

References

1. Walsh C. *Posttranslational modification of proteins: expanding nature's inventory.* Roberts and Company Publishers; 2006.
2. Mann M, Jensen ON. Proteomic analysis of post-translational modifications. *Nat Biotechnol* 2003;**21**(3):255–61.
3. Rakus JF, Mahal LK. New technologies for glycomic analysis: toward a systematic understanding of the glycome. *Annu Rev Anal Chem* 2011;**4**:367–92.
4. Varki A, Lowe JB. *Biological roles of glycans.* Cold Spring Harbor, NY: Cold Spring Harbor Laboratory Press; 2009.
5. Hennet T. Diseases of glycosylation beyond classical congenital disorders of glycosylation. *Biochim Biophys Acta Gen Subj* 2012;**1820**(9):1306–17.
6. Freeze HH, Schachter H. Genetic disorders of glycosylation. In: Varki A, Cummings RD, Esko JD, Freeze HH, Stanley P, Bertozzi CR, Hart GW, Etzler ME, editors. *Essential of glycobiology.* 2nd ed. Cold Spring Harbor, New York: Cold Spring Harbor Laboratory Press; 2009. p. 585–600.
7. Leroy JG, Spranger JRW, Feingold M, Opitz JM, Crocker AC. I-cell disease: a clinical picture. *J Pediatr* 1971;**79**(3):360–5.
8. Martin PT. Congenital muscular dystrophies involving the O-mannose pathway. *Curr Mol Med* 2007;**7**(4):417–25.
9. Slaughter CA, Ervasti JM. Primary structure of dystrophin-associated glycoproteins linking dystrophin to the extracellular matrix. *Nature* 1992;**355**(6362):696.

10. Blake DJ, Weir A, Newey SE, Davies KE. Function and genetics of dystrophin and dystrophin-related proteins in muscle. *Physiol Rev* 2002;**82**(2):291–329.
11. Holzfeind PJ, Grewal PK, Reitsamer HA, Kechvar J, Lassmann H, Hoeger H, et al. Skeletal, cardiac and tongue muscle pathology, defective retinal transmission, and neuronal migration defects in the Largemyd mouse defines a natural model for glycosylation-deficient muscle-eye-brain disorders. *Hum Mol Genet* 2002;**11**(21):2673–87.
12. Michele DE, Barresi R, Kanagawa M, Saito F, Cohn RD, Satz JS, et al. Post-translational disruption of dystroglycan-ligand interactions in congenital muscular dystrophies. *Nature* 2002;**418**(6896):417–21.
13. de Bernabé DBN-V, Currier S, Steinbrecher A, Celli J, van Beusekom E, van der Zwaag B, et al. Mutations in the O-mannosyltransferase gene POMT1 give rise to the severe neuronal migration disorder Walker-Warburg syndrome. *Am J Hum Genet* 2002;**71**(5):1033–43.
14. Yoshida A, Kobayashi K, Manya H, Taniguchi K, Kano H, Mizuno M, et al. Muscular dystrophy and neuronal migration disorder caused by mutations in a glycosyltransferase, POMGnT1. *Dev Cell* 2001;**1**(5):717–24.
15. Brockington M, Blake DJ, Prandini P, Brown SC, Torelli S, Benson MA, et al. Mutations in the fukutin-related protein gene (FKRP) cause a form of congenital muscular dystrophy with secondary laminin alpha2 deficiency and abnormal glycosylation of alpha-dystroglycan. *Am J Hum Genet* 2001;**69**(6):1198–209.
16. Brockington M, Yuva Y, Prandini P, Brown SC, Torelli S, Benson MA, et al. Mutations in the fukutin-related protein gene (FKRP) identify limb girdle muscular dystrophy 2I as a milder allelic variant of congenital muscular dystrophy MDC1C. *Hum Mol Genet* 2001;**10**(25):2851–9.
17. Yoshida-Moriguchi T, Yu L, Stalnaker SH, Davis S, Kunz S, Madson M, et al. O-mannosyl phosphorylation of alpha-dystroglycan is required for laminin binding. *Science* 2010;**327**(5961):88–92.
18. Liu F, Zaidi T, Iqbal K, Grundke-Iqbal I, Merkle RK, Gong C-X. Role of glycosylation in hyperphosphorylation of tau in Alzheimer's disease. *FEBS Lett* 2002;**512**(1–3):101–6.
19. Sato Y, Naito Y, Grundke-Iqbal I, Iqbal K, Endo T. Analysis of N-glycans of pathological tau: possible occurrence of aberrant processing of tau in Alzheimer's disease. *FEBS Lett* 2001;**496**(2–3):152–60.
20. Wang J-Z, Grundke-Iqbal I, Iqbal K. Glycosylation of microtubule-associated protein tau: An abnormal posttranslational modification in Alzheimer's disease. *Nat Med* 1996;**2**(8):871–5.
21. Yazaki M, Tagawa K, Maruyama K, Sorimachi H, Tsuchiya T, Ishiura S, et al. Mutation of potential N-linked glycosylation sites in the Alzheimer's disease amyloid precursor protein (APP). *Neurosci Lett* 1996;**221**(1):57–60.
22. Påhlsson P, Spitalnik SL. The role of glycosylation in synthesis and secretion of beta-amyloid precursor protein by Chinese hamster ovary cells. *Arch Biochem Biophys* 1996;**331**(2):177–86.
23. McFarlane I, Breen KC, Di Giamberardino L, Moya KL. Inhibition of N-glycan processing alters axonal transport of synaptic glycoproteins in vivo. *Neuroreport* 2000;**11**(7):1543–7.
24. McFarlane I, Georgopoulou N, Coughlan CM, Gillian AM, Breen KC. The role of the protein glycosylation state in the control of cellular transport of the amyloid beta precursor protein. *Neuroscience* 1999;**90**(1):15–25.
25. Tomita S, Kirino Y, Suzuki T. Cleavage of Alzheimer's amyloid precursor protein (APP) by secretases occurs after O-glycosylation of APP in the protein secretory pathway Identification of intracellular compartments in which APP cleavage occurs without using toxic agents that interfere with protein metabolism. *J Biol Chem* 1998;**273**(11):6277–84.

26. Griffith LS, Mathes M, Schmitz B. Beta-amyloid precursor protein is modified with O-linked N-acetylglucosamine. *J Neurosci Res* 1995;**41**(2):270–8.
27. Jacobsen KT, Iverfeldt K. O-GlcNAcylation increases non-amyloidogenic processing of the amyloid-β precursor protein (APP). *Biochem Biophys Res Commun* 2011;**404**(3): 882–6.
28. Maguire TM, Gillian AM, O'Mahony D, Coughlan CM, Breen KC. A decrease in serum sialyltransferase levels in Alzheimer's disease. *Neurobiol Aging* 1994;**15**(1):99–102.
29. Fodero LR, Sáez-Valero J, Barquero MS, Marcos A, McLean CA, Small DH. Wheat germ agglutinin-binding glycoproteins are decreased in Alzheimer's disease cerebrospinal fluid. *J Neurochem* 2001;**79**(5):1022–6.
30. Deng Y-L, Liu L-H, Wang Y, Tang H-D, Ren R-J, Xu W, et al. The prevalence of CD33 and MS4A6A variant in Chinese Han population with Alzheimerâ€™s disease. *Hum Genet* 2012;**131**(7):1245–9.
31. Hollingworth P, Harold D, Sims R, Gerrish A, Lambert J-C, Carrasquillo MM, et al. Common variants at ABCA7, MS4A6A/MS4A4E, EPHA1, CD33 and CD2AP are associated with Alzheimer's disease. *Nat Genet* 2011;**43**(5):429–35.
32. Logue MW, Schu M, Vardarajan BN, Buros J, Green RC, Go RCP, et al. A comprehensive genetic association study of Alzheimer disease in African Americans. *Arch Neurol* 2011;**68**(12):1569–79.
33. Salamat MK, Dron M, Chapuis JRM, Langevin C, Laude H. Prion propagation in cells expressing PrP glycosylation mutants. *J Virol* 2011;**85**(7):3077–85.
34. Grasbon-Frodl E, Lorenz H, Mann U, Nitsch RM, Windl O, Kretzschmar HA. Loss of glycosylation associated with the T183A mutation in human prion disease. *Acta Neuropathol* 2004;**108**(6):476–84.
35. Fuster MM, Esko JD. The sweet and sour of cancer: glycans as novel therapeutic targets. *Nat Rev Cancer* 2005;**5**(7):526–42.
36. Taniguchi N, Kizuka Y. Chapter Two-Glycans and Cancer: Role of N-Glycans in Cancer Biomarker, Progression and Metastasis, and Therapeutics. *Adv Cancer Res* 2015; **126**:11–51.
37. Mariño K, Saldova R, Adamczyk B, Rudd PM, Rauter AP. Changes in serum N-glycosylation profiles: functional significance and potential for diagnostics. In: *Carbohydrate Chemistry: Chemical and Biological Approaches*. vol. 37. RSC Publishing; 2012. p. 57–93.
38. Batta K, Das C, Gadad S, Shandilya J, Kundu TK. Reversible acetylation of non histone proteins. In: *Chromatin and disease*. Springer; 2007. p. 193–214.
39. Selvi RB, Kundu TK. Reversible acetylation of chromatin: implication in regulation of gene expression, disease and therapeutics. *Biotechnol J* 2009;**4**(3):375–90.
40. Choi JK, Howe LJ. Histone acetylation: truth of consequences? This paper is one of a selection of papers published in this Special Issue, entitled CSBMCB's 51st annual Meeting-epigenetics and chromatin dynamics, and has undergone the Journal's usual peer review process. *Biochem Cell Biol* 2009;**87**(1):139–50.
41. Wang C, Tian L, Popov VM, Pestell RG. Acetylation and nuclear receptor action. *J Steroid Biochem Mol Biol* 2011;**123**(3):91–100.
42. Lee H-A, Lee D-Y, Cho H-M, Kim S-Y, Iwasaki Y, Kim I. Histone deacetylase inhibition attenuates transcriptional activity of mineralocorticoid receptor through its acetylation and prevents development of hypertension. *Circ Res* 2013; https://doi.org/10.1161/CIRCRESAHA.113.301071.
43. Mu S, Shimosawa T, Ogura S, Wang H, Uetake Y, Kawakami-Mori F, et al. Epigenetic modulation of the renal [beta]-adrenergic-WNK4 pathway in salt-sensitive hypertension. *Nat Med* 2011;**17**(5):573–80.
44. Vadvalkar SS, Baily CN, Matsuzaki S, West M, Tesiram YA, Humphries KM. Metabolic inflexibility and protein lysine acetylation in heart mitochondria of a chronic model of type 1 diabetes. *Biochem J* 2013;**449**(1):253–61.

45. Thal MA, Krishnamurthy P, Mackie AR, Hoxha E, Lambers E, Verma S, et al. Enhanced angiogenic and cardiomyocyte differentiation capacity of epigenetically reprogrammed mouse and human endothelial progenitor cells augments their efficacy for ischemic Myocardial repairnovelty and significance. *Circ Res* 2012;**111**(2):180–90.
46. Machado RD, Eickelberg O, Elliott CG, Geraci MW, Hanaoka M, Loyd JE, et al. Genetics and genomics of pulmonary arterial hypertension. *J Am Coll Cardiol* 2009;**54**(1):S32–42.
47. Yang Q, Lu Z, Ramchandran R, Longo LD, Raj JU. Pulmonary artery smooth muscle cell proliferation and migration in fetal lambs acclimatized to high-altitude long-term hypoxia: role of histone acetylation. *Am J Physiol Lung Cell Mol Physiol* 2012;**303**(11): L1001–10.
48. Tang Y, Boucher JM, Liaw L. Histone deacetylase activity selectively regulates notch-mediated smooth muscle differentiation in human vascular cells. *J Am Heart Assoc* 2012;**1**(3).
49. Liu Y, Wang Z, Wang J, Lam W, Kwong S, Li F, et al. A histone deacetylase inhibitor, largazole, decreases liver fibrosis and angiogenesis by inhibiting transforming growth factor-β and vascular endothelial growth factor signalling. *Liver Int* 2013;**33**(4):504–15.
50. Ito K, Barnes PJ, Adcock IM. Glucocorticoid receptor recruitment of histone deacetylase 2 inhibits interleukin-1beta-induced histone H4 acetylation on lysines 8 and 12. *Mol Cell Biol* 2000;**20**(18):6891–903.
51. Sack MN, Finkel T. Mitochondrial metabolism, sirtuins, and aging. *Cold Spring Harb Perspect Biol* 2012;**4**(12):a013102.
52. Anderson KA, Hirschey MD. Mitochondrial protein acetylation regulates metabolism. *Essays Biochem* 2012;**52**:23–35.
53. Wu YT, Wu SB, Lee WY, Wei YH. Mitochondrial respiratory dysfunction-elicited oxidative stress and posttranslational protein modification in mitochondrial diseases. *Ann N Y Acad Sci* 2010;**1201**(1):147–56.
54. Gayther SA, Batley SJ, Linger L, Bannister A, Thorpe K, Chin S-F, et al. Mutations truncating the EP300 acetylase in human cancers. *Nat Genet* 2000;**24**(3):300–3.
55. Muraoka M, Konishi M, Kikuchi-Yanoshita R, Tanaka K, Shitara N, Chong J-M, et al. p300 gene alterations in colorectal and gastric carcinomas. *Oncogene* 1996;**12**(7):1565–9.
56. Seligson DB, Horvath S, Shi T, Yu H, Tze S, Grunstein M, et al. Global histone modification patterns predict risk of prostate cancer recurrence. *Nature* 2005;**435**(7046):1262–6.
57. Bai X, Wu L, Liang T, Liu Z, Li J, Li D, et al. Overexpression of myocyte enhancer factor 2 and histone hyperacetylation in hepatocellular carcinoma. *J Cancer Res Clin Oncol* 2008;**134**(1):83–91.
58. Petrij F, Giles RH, Dauwerse HG, Saris JJ. Rubinstein-Taybi syndrome caused by mutations in the transcriptional co-activator CBP. *Nature* 1995;**376**(6538):348.
59. Tanaka Y, Naruse I, Maekawa T, Masuya H, Shiroishi T, Ishii S. Abnormal skeletal patterning in embryos lacking a single Cbp allele: a partial similarity with Rubinstein-Taybi syndrome. *Proc Natl Acad Sci* 1997;**94**(19):10215–20.
60. Yang XJ. The diverse superfamily of lysine acetyltransferases and their roles in leukemia and other diseases. *Nucleic Acids Res* 2004;**32**(3):959–76.
61. Mattson MP. Acetylation unleashes protein demons of dementia. *Neuron* 2010;**67** (6):900–2.
62. Peleg S, Sananbenesi F, Zovoilis A, Burkhardt S, Bahari-Javan S, Agis-Balboa RC, et al. Altered histone acetylation is associated with age-dependent memory impairment in mice. *Science* 2010;**328**(5979):753–6.
63. Sadri-Vakili G, Bouzou BRR, Benn CL, Kim M-O, Chawla P, Overland RP, et al. Histones associated with downregulated genes are hypo-acetylated in Huntington's disease models. *Hum Mol Genet* 2007;**16**(11):1293–306.
64. Thomas M, Dadgar N, Aphale A, Harrell JM, Kunkel R, Pratt WB, et al. Androgen receptor acetylation site mutations cause trafficking defects, misfolding, and aggregation similar to expanded glutamine tracts. *J Biol Chem* 2004;**279**(9):8389–95.

65. Tang B, Dean B, Thomas EA. Disease-and age-related changes in histone acetylation at gene promoters in psychiatric disorders. *Transl Psychiatry* 2011;**1**(12):e64.
66. Ghosh G, Adams JA. Phosphorylation mechanism and structure of serine-arginine protein kinases. *FEBS J* 2011;**278**(4):587–97.
67. Hunter T. Protein kinases and phosphatases: the yin and yang of protein phosphorylation and signaling. *Cell* 1995;**80**(2):225–36.
68. Anderson JP, Walker DE, Goldstein JM, de Laat R, Banducci K, Caccavello RJ, et al. Phosphorylation of Ser-129 is the dominant pathological modification of alpha-synuclein in familial and sporadic Lewy body disease. *J Biol Chem* 2006;**281**(40):29739–52.
69. Alonso ADC, Zaidi T, Novak M, Grundke-Iqbal I, Iqbal K. Hyperphosphorylation induces self-assembly of tau, into tangles of paired helical filaments/straight filaments. *Proc Natl Acad Sci* 2001;**98**(12):6923–8.
70. Lu KP, Liou YC, Vincent I. Proline-directed phosphorylation and isomerization in mitotic regulation and in Alzheimer's Disease. *Bioessays* 2003;**25**(2):174–81.
71. Viatour P, Merville M-P, Bours V, Chariot A. Phosphorylation of NF-kappaB and IkappaB proteins: implications in cancer and inflammation. *Trends Biochem Sci* 2005;**30**(1):43–52.
72. Kreisberg JI, Malik SN, Prihoda TJ, Bedolla RG, Troyer DA, Kreisberg S, et al. Phosphorylation of Akt (Ser473) is an excellent predictor of poor clinical outcome in prostate cancer. *Cancer Res* 2004;**64**(15):5232–6.
73. Hao J, Li F, Liu W, Liu Q, Liu S, Li H, et al. Phosphorylation of PRAS40-Thr246 Involved in Renal Lipid Accumulation of Diabetes. *J Cell Physiol* 2014;**229**(8):1069–77.
74. Kooij V, Venkatraman V, Tra J, Kirk JA, Rowell J, Blice-Baum A, et al. Sizing up models of heart failure: proteomics from flies to humans. *Proteomics Clin Appl* 2014;**8**(9–10):653–64.
75. Nie H, Zheng Y, Li R, Guo TB, He D, Fang L, et al. Phosphorylation of FOXP3 controls regulatory T cell function and is inhibited by TNF-[alpha] in rheumatoid arthritis. *Nat Med* 2013;**19**(3):322–8.
76. Radivojac P, Baenziger PH, Kann MG, Mort ME, Hahn MW, Mooney SD. Gain and loss of phosphorylation sites in human cancer. *Bioinformatics* 2008;**24**(16):i241–7.
77. Sciarretta S, Volpe M, Sadoshima J. Mammalian target of rapamycin signaling in cardiac physiology and disease. *Circ Res* 2014;**114**(3):549–64.
78. Streit U, Reuter H, Bloch W, Wahlers T, Schwinger RHG, Brixius K. Phosphorylation of myocardial eNOS is altered in patients suffering from type 2 diabetes. *J Appl Physiol* 2013;**114**(10):1366–74.
79. Watanabe N, Osada H. Phosphorylation-dependent protein-protein interaction modules as potential molecular targets for cancer therapy. *Curr Drug Targets* 2012;**13**(13):1654–8.
80. Dalle-Donne I, Rossi R, Colombo R, Giustarini D, Milzani A. Biomarkers of oxidative damage in human disease. *Clin Chem* 2006;**52**(4):601–23.
81. Dalle-Donne I, Giustarini D, Colombo R, Rossi R, Milzani A. Protein carbonylation in human diseases. *Trends Mol Med* 2003;**9**(4):169–76.
82. Dalle-Donne I, Scaloni A, Giustarini D, Cavarra E, Tell G, Lungarella G, et al. Proteins as biomarkers of oxidative/nitrosative stress in diseases: the contribution of redox proteomics. *Mass Spectrom Rev* 2005;**24**(1):55–99.
83. Chevion M, Berenshtein E, Stadtman ER. Human studies related to protein oxidation: protein carbonyl content as a marker of damage. *Free Radic Res* 2000;**33**:S99–S108.
84. Halliwell B, Gutteridge JMC. *Free radicals in biology and medicine.* Oxford: Oxford University Press; 1999.
85. Lih-Brody L, Powell SR, Collier KP, Reddy GM, Cerchia R, Kahn E, et al. Increased oxidative stress and decreased antioxidant defenses in mucosa of inflammatory bowel disease. *Dig Dis Sci* 1996;**41**(10):2078–86.

86. Mantle D, Falkous G, Walker D. Quantification of protease activities in synovial fluid from rheumatoid and osteoarthritis cases: comparison with antioxidant and free radical damage markers. *Clin Chim Acta* 1999;**284**(1):45–58.
87. Smith CD, Carney JM, Starke-Reed PE, Oliver CN, Stadtman ER, Floyd RA, et al. Excess brain protein oxidation and enzyme dysfunction in normal aging and in Alzheimer disease. *Proc Natl Acad Sci* 1991;**88**(23):10540–3.
88. Telci A, Cakatay U, Kayali R, Erdoğan C, Orhan Y, Sivas A, et al. Oxidative protein damage in plasma of type 2 diabetic patients. *Horm Metab Res* 2000;**32**(01):40–3.
89. Wang G, Wang J, Ma H, Khan MF. Increased nitration and carbonylation of proteins in MRL+/+ mice exposed to trichloroethene: potential role of protein oxidation in autoimmunity. *Toxicol Appl Pharmacol* 2009;**237**(2):188–95.
90. Anderson EJ, Neufer PD. Type II skeletal myofibers possess unique properties that potentiate mitochondrial H2O2 generation. *Am J Physiol Cell Physiol* 2006;**290**(3): C844–51.
91. Barreiro E, Hussain SNA. Protein carbonylation in skeletal muscles: impact on function. *Antioxid Redox Signal* 2010;**12**(3):417–29.
92. Grimsrud PA, Picklo MJ, Griffin TJ, Bernlohr DA. Carbonylation of adipose proteins in obesity and insulin resistance identification of Adipocyte Fatty acid-binding protein as a cellular target of 4-hydroxynonenal. *Mol Cell Proteomics* 2007;**6**(4):624–37.
93. Frohnert BI, Sinaiko AR, Serrot FJ, Foncea RE, Moran A, Ikramuddin S, et al. Increased adipose protein carbonylation in human obesity. *Obesity* 2011;**19**(9):1735–41.
94. Winterbourn CC, Bonham MJD, Buss H, Abu-Zidan FM, Windsor JA. Elevated protein carbonyls as plasma markers of oxidative stress in acute pancreatitis. *Pancreatology* 2003;**3**(5):375–82.
95. Shi Y, Lan F, Matson C, Mulligan P, Whetstine JR, Cole PA, et al. Histone demethylation mediated by the nuclear amine oxidase homolog LSD1. *Cell* 2004;**119**(7):941–53.
96. Chen D, Ma H, Hong H, Koh SS, Huang S-M, Schurter BT, et al. Regulation of transcription by a protein methyltransferase. *Science* 1999;**284**(5423):2174–7.
97. Schubert HL, Blumenthal RM, Cheng X. Many paths to methyltransfer: a chronicle of convergence. *Trends Biochem Sci* 2003;**28**(6):329–35.
98. Shen EC, Henry MF, Weiss VH, Valentini SR, Silver PA, Lee MS. Arginine methylation facilitates the nuclear export of hnRNP proteins. *Genes Dev* 1998;**12**(5):679–91.
99. Yang X-D, Lamb A, Chen L-F. Methylation, a new epigenetic mark for protein stability. *Epigenetics* 2009;**4**(7):429–33.
100. García-Cao M, O'Sullivan R, Peters AHFM, Jenuwein T, Blasco MAA. Epigenetic regulation of telomere length in mammalian cells by the Suv39h1 and Suv39h2 histone methyltransferases. *Nat Genet* 2004;**36**(1):94–9.
101. Peters AHFM, O'Carroll Dn, Scherthan H, Mechtler K, Sauer S, Schöfer C, et al. Loss of the Suv39h histone methyltransferases impairs mammalian heterochromatin and genome stability. *Cell* 2001;**107**(3):323–37.
102. Huang J, Dorsey J, Chuikov S, Zhang X, Jenuwein T, Reinberg D, et al. G9a and Glp methylate lysine 373 in the tumor suppressor p53. *J Biol Chem* 2010;**285**(13): 9636–9641.
103. Barry ER, Corry GN, Rasmussen TP. Targeting DOT1L action and interactions in leukemia: the role of DOT1L in transformation and development. *Expert Opin Ther Targets* 2010;**14**(4):405–18.
104. Rosati R, La Starza R, Veronese A, Aventin A, Schwienbacher C, Vallespi T, et al. NUP98 is fused to the NSD3 gene in acute myeloid leukemia associated with t (8; 11) (p11. 2; p15). *Blood* 2002;**99**(10):3857–60.
105. Hamamoto R, Furukawa Y, Morita M, Iimura Y, Silva FP, Li M, et al. SMYD3 encodes a histone methyltransferase involved in the proliferation of cancer cells. *Nat Cell Biol* 2004;**6**(8):731–40.

106. Hamamoto R, Silva FP, Tsuge M, Nishidate T, Katagiri T, Nakamura Y, et al. Enhanced SMYD3 expression is essential for the growth of breast cancer cells. *Cancer Sci* 2006;**97**(2):113–8.
107. Tsuge M, Hamamoto R, Silva FP, Ohnishi Y, Chayama K, Kamatani N, et al. A variable number of tandem repeats polymorphism in an E2F-1 binding element in the 5′ flanking region of SMYD3 is a risk factor for human cancers. *Nat Genet* 2005;**37**(10):1104–7.
108. Yang Y, Bedford MT. Protein arginine methyltransferases and cancer. *Nat Rev Cancer* 2013;**13**(1):37–50.
109. Yoshimatsu M, Toyokawa G, Hayami S, Unoki M, Tsunoda T, Field HI, et al. Dysregulation of PRMT1 and PRMT6, Type I arginine methyltransferases, is involved in various types of human cancers. *Int J Cancer* 2011;**128**(3):562–73.
110. Pal S, Vishwanath SN, Erdjument-Bromage H, Tempst P, Sif Sd. Human SWI/SNF-associated PRMT5 methylates histone H3 arginine 8 and negatively regulates expression of ST7 and NM23 tumor suppressor genes. *Mol Cell Biol* 2004;**24**(21):9630–45.
111. Lin W, Cao J, Liu J, Beshiri ML, Fujiwara Y, Francis J, et al. Loss of the retinoblastoma binding protein 2 (RBP2) histone demethylase suppresses tumorigenesis in mice lacking Rb1 or Men1. *Proc Natl Acad Sci* 2011;**108**(33):13379–86.
112. Hamamoto R, Saloura V, Nakamura Y. Critical roles of non-histone protein lysine methylation in human tumorigenesis. *Nat Rev Cancer* 2015;**15**(2):110–24.
113. Gluckman PD, Hanson MA, Buklijas T, Low FM, Beedle AS. Epigenetic mechanisms that underpin metabolic and cardiovascular diseases. *Nat Rev Endocrinol* 2009;**5**(7):401–8.
114. Bełtowski J, Kedra A. Asymmetric dimethylarginine (ADMA) as a target for pharmacotherapy. *Pharmacol Rep* 2006;**58**(159):159–78.
115. Böger RH. Asymmetric dimethylarginine, an endogenous inhibitor of nitric oxide synthase, explains the "L-arginine paradox" and acts as a novel cardiovascular risk factor. *J Nutr* 2004;**134**(10):2842S–2847S.
116. Brosnan JT, Jacobs RL, Stead LM, Brosnan ME. Methylation demand: a key determinant of homocysteine metabolism. *Acta Biochim Pol* 2004;**51**:405–14. English Edition.
117. Katz ML, Rodrigues M. Juvenile ceroid lipofuscinosis. Evidence for methylated lysine in neural storage body protein. *Am J Pathol* 1991;**138**(2):323.
118. Prockop DJ, Juva K. Synthesis of hydroxyproline in vitro by the hydroxylation of proline in a precursor of collagen. *Proc Natl Acad Sci* 1965;**53**(3):661–8.
119. Levene CI, Ockleford CD, Barber CL. Scurvy; a comparison between ultrastructural and biochemical changes observed in cultured fibroblasts and the collagen they synthesise. *Virchows Arch B Cell Pathol* 1977;**23**(1):325–38.
120. Jaakkola P, Mole DR, Tian Y-M, Wilson MI, Gielbert J, Gaskell SJ, et al. Targeting of HIF-alpha to the von Hippel-Lindau ubiquitylation complex by O2-regulated prolyl hydroxylation. *Science* 2001;**292**(5516):468–72.
121. Seagroves TN, Ryan HE, Lu H, Wouters BG, Knapp M, Thibault P, et al. Transcription factor HIF-1 is a necessary mediator of the pasteur effect in mammalian cells. *Mol Cell Biol* 2001;**21**(10):3436–44.
122. Frelin C, Ladoux A, D'Angelo G. Vascular endothelial growth factors and angiogenesis. *Ann Endocrinol (Paris)* 2000;**61**(1):70–4.
123. Yang P, Reece EA. Role of HIF-1α in maternal hyperglycemia-induced embryonic vasculopathy. *Am J Obstet Gynecol* 2011;**204**(4):332.e1–7.
124. Ema M, Hirota K, Mimura J, Abe H, Yodoi J, Sogawa K, et al. Molecular mechanisms of transcription activation by HLF and HIF1alpha in response to hypoxia: their stabilization and redox signal-induced interaction with CBP/p300. *EMBO J* 1999;**18**(7):1905–14.
125. Brahimi-Horn C, Mazure N, Pouysségur J. Signalling via the hypoxia-inducible factor-1alpha requires multiple posttranslational modifications. *Cell Signal* 2005;**17**(1):1–9.
126. Liu W, Shen S-M, Zhao X-Y, Chen G-Q. Targeted genes and interacting proteins of hypoxia inducible factor-1. *Int J Biochem Mol Biol* 2012;**3**(2):165.

127. Semenza GL. Targeting HIF-1 for cancer therapy. *Nat Rev Cancer* 2003;**3**(10):721–32.
128. Ziello JE, Jovin IS, Huang Y. Hypoxia-Inducible Factor (HIF)-1 regulatory pathway and its potential for therapeutic intervention in malignancy and ischemia. *Yale J Biol Med* 2007;**80**(2):51.
129. Chan DA, Sutphin PD, Denko NC, Giaccia AJ. Role of prolyl hydroxylation in oncogenically stabilized hypoxia-inducible factor-1alpha. *J Biol Chem* 2002;**277**(42):40112–7.
130. Siddiq A, Aminova LR, Ratan RR. Hypoxia inducible factor prolyl 4-hydroxylase enzymes: center stage in the battle against hypoxia, metabolic compromise and oxidative stress. *Neurochem Res* 2007;**32**(4–5):931–46.
131. Myllyharju J. HIF prolyl 4-hydroxylases and their potential as drug targets. *Curr Pharm Des* 2009;**15**(33):3878–85.
132. Marini JC, Cabral WA, Barnes AM, Chang W. Components of the collagen prolyl 3-hydroxylation complex are crucial for normal bone development. *Cell Cycle* 2007;**6** (14):1675–81.
133. Napoli N, Armamento-Villareal R. Estrogen hydroxylation in osteoporosis. *Adv Clin Chem* 2007;**43**:211–27.
134. Schlisio S. Neuronal apoptosis by prolyl hydroxylation: implication in nervous system tumours and the Warburg conundrum. *J Cell Mol Med* 2009;**13**(10):4104–12.
135. Wenger RH, Hoogewijs D. Regulated oxygen sensing by protein hydroxylation in renal erythropoietin-producing cells. *Am J Physiol Ren Physiol* 2010;**298**(6):F1287–96.
136. Sepkovic DW, Bradlow HL. Estrogen hydroxylation—the good and the bad. *Ann N Y Acad Sci* 2009;**1155**(1):57–67.
137. Gecit I, Aslan M, Gunes M, Pirincci N, Esen R, Demir H, et al. Serum prolidase activity, oxidative stress, and nitric oxide levels in patients with bladder cancer. *J Cancer Res Clin Oncol* 2012;**138**(5):739–43.
138. Palka J, Surazynski A, Karna E, Orlowski K, Puchalski Z, Pruszynski K, et al. Prolidase activity disregulation in chronic pancreatitis and pancreatic cancer. *Hepatogastroenterology* 2001;**49**(48):1699–703.
139. Salnikow K, Kasprzak KS. Ascorbate depletion: a critical step in nickel carcinogenesis? *Environ Health Perspect* 2005;**113**(5):577.
140. Blau N, van Spronsen FJ, Levy HL. Phenylketonuria. *The Lancet* 2010;**376**(9750):1417–27.
141. Koppal T, Drake J, Yatin S, Jordan B, Varadarajan S, Bettenhausen L, et al. Peroxynitrite-induced alterations in synaptosomal membrane proteins. *J Neurochem* 1999;**72**(1):310–7.
142. Yamakura F, Taka H, Fujimura T, Murayama K. Inactivation of human manganese-superoxide dismutase by peroxynitrite is caused by exclusive nitration of tyrosine 34 to 3-nitrotyrosine. *J Biol Chem* 1998;**273**(23):14085–9.
143. Jamshad K, Brennan MD, Bradley N, Beirong GAO, Bruckdorfer R, Jacobs M. 3-Nitrotyrosine in the proteins of human plasma determined by an ELISA method. *Biochem J* 1998;**330**(2):795–801.
144. Marfella R, Quagliaro L, Nappo F, Ceriello A, Giugliano D. Acute hyperglycemia induces an oxidative stress in healthy subjects. *J Clin Invest* 2001;**108**(4):635–6.
145. Davidge ST, Ojimba J, McLaughlin MK. Vascular function in the Vitamin E-deprived rat. *Hypertension* 1998;**31**(3):830–5.
146. Frustaci A, Kajstura J, Chimenti C, Jakoniuk I, Leri A, Maseri A, et al. Myocardial cell death in human diabetes. *Circ Res* 2000;**87**(12):1123–32.
147. Kajstura J, Fiordaliso F, Andreoli AM, Li B, Chimenti S, Medow MS, et al. IGF-1 over-expression inhibits the development of diabetic cardiomyopathy and angiotensin II-mediated oxidative stress. *Diabetes* 2001;**50**(6):1414–24.
148. Peluffo G, Radi R. Biochemistry of protein tyrosine nitration in cardiovascular pathology. *Cardiovasc Res* 2007;**75**(2):291–302.
149. Smith MA, Harris PLR, Sayre LM, Beckman JS, Perry G. Widespread peroxynitrite-mediated damage in Alzheimer's disease. *J Neurosci* 1997;**17**(8):2653–7.

150. Hensley K, Maidt ML, Yu Z, Sang H, Markesbery WR, Floyd RA. Electrochemical analysis of protein nitrotyrosine and dityrosine in the Alzheimer brain indicates region-specific accumulation. *J Neurosci* 1998;**18**(20):8126–32.
151. Su JH, Deng G, Cotman CW. Neuronal DNA damage precedes tangle formation and is associated with up-regulation of nitrotyrosine in Alzheimer's disease brain. *Brain Res* 1997;**774**(1):193–9.
152. Guix FX, Ill-Raga G, Bravo R, Nakaya T, de Fabritiis G, Coma M, et al. Amyloid-dependent triosephosphate isomerase nitrotyrosination induces glycation and tau fibrillation. *Brain* 2009;**132**(5):1335–45.
153. Beal MF, Ferrante RJ, Browne SE, Matthews RT, Kowall NW, Brown RH. Increased 3-nitrotyrosine in both sporadic and familial amyotrophic lateral sclerosis. *Ann Neurol* 1997;**42**(4):644–54.
154. Bruijn LI, Beal MF, Becher MW, Schulz JB, Wong PC, Price DL, et al. Elevated free nitrotyrosine levels, but not protein-bound nitrotyrosine or hydroxyl radicals, throughout amyotrophic lateral sclerosis (ALS)-like disease implicate tyrosine nitration as an aberrant in vivo property of one familial ALS-linked superoxide dismutase 1 mutant. *Proc Natl Acad Sci* 1997;**94**(14):7606–11.
155. Klivenyi P, Ferrante RJ, Matthews RT, Bogdanov MB, Klein AM, Andreassen OA, et al. Neuroprotective effects of creatine in a transgenic animal model of amyotrophic lateral sclerosis. *Nat Med* 1999;**5**(3):347–50.
156. Murdaugh LS, Wang Z, Del Priore LV, Dillon J, Gaillard ER. Age-related accumulation of 3-nitrotyrosine and nitro-A2E in human Bruch's membrane. *Exp Eye Res* 2010;**90** (5):564–71.
157. Osoata GO, Yamamura S, Ito M, Vuppusetty C, Adcock IM, Barnes PJ, et al. Nitration of distinct tyrosine residues causes inactivation of histone deacetylase 2. *Biochem Biophys Res Commun* 2009;**384**(3):366–71.
158. Abdelmegeed MA, Song B-J. Functional roles of protein nitration in acute and chronic liver diseases. *Oxid Med Cell Longev* 2014;.
159. Huttner WB, Baeuerle PA. Protein sulfation on tyrosine. In: *Modern cell biology*. **6**:New York: Alan R. Liss, Inc.; 1988. p. 97–140.
160. Kehoe JW, Bertozzi CR. Tyrosine sulfation: a modulator of extracellular protein-protein interactions. *Chem Biol* 2000;**7**(3):R57–61.
161. Ouyang Y-b, Lane WS, Moore KL. Tyrosylprotein sulfotransferase: purification and molecular cloning of an enzyme that catalyzes tyrosine O-sulfation, a common post-translational modification of eukaryotic proteins. *Proc Natl Acad Sci* 1998;**95**(6):2896–901.
162. Dong J-F, Li CQ, Lopez JA. Tyrosine sulfation of the glycoprotein Ib-IX complex: identification of sulfated residues and effect on ligand binding. *Biochemistry* 1994;**33** (46):13946–53.
163. Farzan M, Mirzabekov T, Kolchinsky P, Wyatt R, Cayabyab M, Gerard NP, et al. Tyrosine sulfation of the amino terminus of CCR5 facilitates HIV-1 entry. *Cell* 1999;**96** (5):667–76.
164. Kanan Y, Siefert JC, Kinter M, Al-Ubaidi MR. Complement factor H, vitronectin, and opticin are tyrosine-sulfated proteins of the retinal pigment epithelium. *PLoS ONE* 2014;**9**(8).
165. Leyte A, Van Schijndel HB, Niehrs C, Huttner WB, Verbeet MP, Mertens K, et al. Sulfation of Tyr1680 of human blood coagulation factor VIII is essential for the interaction of factor VIII with von Willebrand factor. *J Biol Chem* 1991;**266**(2):740–6.
166. Westmuckett AD, Thacker KM, Moore KL. Tyrosine sulfation of native mouse Psgl-1 is required for optimal leukocyte rolling on P-selectin in vivo. *PLoS ONE* 2011;**6**(5).
167. Cheng P-W, Boat TF, Cranfill K, Yankaskas JR, Boucher RC. Increased sulfation of glycoconjugates by cultured nasal epithelial cells from patients with cystic fibrosis. *J Clin Investig* 1989;**84**(1):68.

168. Hsu W, Rosenquist GL, Ansari AA, Gershwin ME. Autoimmunity and tyrosine sulfation. *Autoimmun Rev* 2005;**4**(7):429–35.
169. Huang C-C, Venturi M, Majeed S, Moore MJ, Phogat S, Zhang M-Y, et al. Structural basis of tyrosine sulfation and VH-gene usage in antibodies that recognize the HIV type 1 coreceptor-binding site on gp120. *Proc Natl Acad Sci U S A* 2004;**101**(9):2706–11.
170. Liu J, Louie S, Hsu W, Yu KM, Nicholas Jr. HB, Rosenquist GL. Tyrosine sulfation is prevalent in human chemokine receptors important in lung disease. *Am J Respir Cell Mol Biol* 2008;**38**(6):738–43.
171. Farzan M, Babcock GJ, Vasilieva N, Wright PL, Kiprilov E, Mirzabekov T, et al. The role of post-translational modifications of the CXCR4 amino terminus in stromal-derived factor 1 alpha association and HIV-1 entry. *J Biol Chem* 2002;**277**(33):29484–9.
172. Xu J, Deng X, Tang M, Li L, Xiao L, Yang L, et al. Tyrosylprotein sulfotransferase-1 and tyrosine sulfation of chemokine receptor 4 are induced by Epstein-Barr virus encoded latent membrane protein 1 and associated with the metastatic potential of human nasopharyngeal carcinoma. *PLoS ONE* 2013;**8**(3):e56114.
173. Westmuckett AD, Moore KL. Lack of tyrosylprotein sulfotransferase activity in hematopoietic cells drastically attenuates atherosclerosis in Ldlr-/-mice. *Arterioscler Thromb Vasc Biol* 2009;**29**(11):1730–6.
174. Rossi A, Bonaventure J, Delezoide AL, Superti-Furga A, Cetta G. Undersulfation of cartilage proteoglycans ex vivo and increased contribution of amino acid sulfur to sulfation in vitro in McAlister dysplasia/atelosteogenesis type 2. *FEBS J* 1997;**248**(3):741–7.
175. Charo IF, Ransohoff RM. The many roles of chemokines and chemokine receptors in inflammation. *N Engl J Med* 2006;**354**(6):610–21.
176. Elsner J, Escher SE, Forssmann U. Chemokine receptor antagonists: a novel therapeutic approach in allergic diseases. *Allergy* 2004;**59**(12):1243–58.
177. Aicart-Ramos C, Valero RA, Rodriguez-Crespo I. Protein palmitoylation and subcellular trafficking. *Biochim Biophys Acta Biomembr* 2011;**1808**(12):2981–94.
178. Druey KM, Ugur O, Caron JM, Chen C-K, Backlund PS, Jones TLZ. Amino-terminal cysteine residues of RGS16 are required for palmitoylation and modulation of Gi-and Gq-mediated signaling. *J Biol Chem* 1999;**274**(26):18836–42.
179. Du D, Raaka BM, Grimberg H, Lupu-Meiri M, Oron Y, Gershengorn MC. Carboxyl tail cysteine mutants of the thyrotropin-releasing hormone receptor type 1 exhibit constitutive signaling: role of palmitoylation. *Mol Pharmacol* 2005;**68**(1):204–9.
180. Bhattacharyya R, Barren C, Kovacs DM. Palmitoylation of amyloid precursor protein regulates amyloidogenic processing in lipid rafts. *J Neurosci* 2013;**33**(27):11169–83.
181. Meckler X, Roseman J, Das P, Cheng H, Pei S, Keat M, et al. Reduced Alzheimer's disease β-amyloid deposition in transgenic mice expressing S-palmitoylation-deficient APH1aL and nicastrin. *J Neurosci* 2010;**30**(48):16160–9.
182. Yanai A, Huang K, Kang R, Singaraja RR, Arstikaitis P, Gan L, et al. Palmitoylation of huntingtin by HIP14is essential for its trafficking and function. *Nat Neurosci* 2006;**9**(6):824–31.
183. Wei X, Schneider JG, Shenouda SM, Lee A, Towler DA, Chakravarthy MV, et al. De novo lipogenesis maintains vascular homeostasis through endothelial nitric-oxide synthase (eNOS) palmitoylation. *J Biol Chem* 2011;**286**(4):2933–45.
184. Majeau N, Fromentin Rm, Savard C, Duval M, Tremblay MJ, Leclerc D. Palmitoylation of hepatitis C virus core protein is important for virion production. *J Biol Chem* 2009;**284**(49):33915–25.
185. Bizzozero OA, Tetzloff SU, Bharadwaj M. Overview: protein palmitoylation in the nervous system: current views and unsolved problems. *Neurochem Res* 1994;**19**(8):923–33.
186. Barrese N, Mak B, Fisher L, Moscarello MA. Mechanism of demyelination in DM20 transgenic mice involves increased fatty acylation. *J Neurosci Res* 1998;**53**(2):143–52.

187. Pfender NgA, Grosch S, Roussel G, Koch M, Trifilieff E, Greer JM. Route of uptake of palmitoylated encephalitogenic peptides of myelin proteolipid protein by antigen-presenting cells: importance of the type of bond between lipid chain and peptide and relevance to autoimmunity. *J Immunol* 2008;**180**(3):1398–404.
188. Yeste-Velasco M, Linder ME, Lu Y-J. Protein S-palmitoylation and cancer. *Biochim Biophys Acta Rev Cancer* 2015;**1856**(1):107–20.
189. Cuiffo B, Ren R. Palmitoylation of oncogenic NRAS is essential for leukemogenesis. *Blood* 2010;**115**(17):3598–605.
190. Li L, Dong L, Xia L, Li T, Zhong H. Chemical and genetic probes for analysis of protein palmitoylation. *J Chromatogr B* 2011;**879**(17):1316–24.

CHAPTER

3

Phosphorylation and Acetylation of Proteins as Posttranslational Modification: Implications in Human Health and Associated Diseases

Sana Qausain, Hemalatha Srinivasan*, Shazia Jamal*, Mohammad Nasiruddin*[†]*, Md. Khurshid Alam Khan**

***School of Life Sciences, B.S. Abdur Rahman Crescent Institute of Science & Technology, Chennai, India**

[†]Molecular Oncology and Transplant Immunology, MedGenome Labs Pvt. Ltd., Bangalore, India

1 INTRODUCTION

Posttranslational modifications (PTMs) occur mainly in eukaryotic proteins and regulate its functional activities in cells and interaction with other cellular molecules.[1] Any dysregulation of PTMs' homeostasis can potentially lead to a pathological state. Identifying and understanding the PTM-induced changes in proteins has been suggested to be important in prevention and treatment of some diseases. Different types of modifications in proteins controlled by the cell determine the final outcome of the modified protein in terms of structure, functionality, and interactions.[2] With new techniques emerging such as immunoprecipitation, proximity ligation assay, mass spectrometry, fluorescent staining, etc. Hundreds of different PTMs have been identified.[3, 4] These modifications include

Protein Modificomics
https://doi.org/10.1016/B978-0-12-811913-6.00003-5

phosphorylation, acetylation, glycosylation, ubiquitination, nitrosylation, methylation, and others.[5–7] Introduction of functional groups through PTMs leads to enhancement and extension of properties of amino acid residues. Most modifications help the cells to respond to internal and external changes specifically and rapidly, such as carbonylation and oxidation that occur during specific environmental stress to the cell.[8] Among the different modifications, phosphorylation and acetylation, which are reversible in nature, regulate a large number of eukaryotic signaling and other cellular approaches.[9] These two modifications are tightly regulated and control almost all cellular process through changes in protein stability, localization, and interactions. Aberrant protein phosphorylation and acetylation can lead to the onset of several diseases.[10] An in-depth understanding of effect of PTMs is important not only for understanding the cellular functions important for drug development but also to understand the mechanism of diseases like cancer, Alzheimer's, Parkinson's, etc. In this chapter the discussion covers the role of two important reversible PTMs, namely phosphorylation and acetylation, with their significance in human health and disease.

2 PHOSPHORYLATION

Protein phosphorylation is defined as the process of attachment of a phosphoryl group to side chains of serine, threonine, tyrosine, and histidine amino acids. In fact, Phosphorylation is an esterification reaction in which a phosphate group reacts with the hydroxyl group of side chains of tyrosine, threonine, or serine. Attachment of the phosphate group to the side chains is carried out by the enzyme protein kinase while as its detachment, i.e., removal of the phosphate group is carried out by phosphatases. With the advancement of technology, thousands of in vivo phosphorylation events have been recognized by quantitative phospho-proteomics experiments, suggesting that a large number of proteins might be regulated by phosphorylation events. With the development of computational and experimental approaches, data on phosphorylation is significantly increasing. Studies are now more focused to prioritize phosphosites and assess their functional relevance. A need for a better understanding of the interplay between phosphorylation and allosteric regulation, abnormal phosphorylation in diseases, and identification of novel phosphorylation inhibitors[11] has arisen. Experiments have suggested that there are larger roles of phosphorylation in maintaining the normal cellular conditions and metabolic events.[2]

Phosphorylation is a covalent modification and it carries two negative charges at physiological pH. The phosphate groups can form salt bridges and hydrogen bonds between oxygen of the phosphate group and side chains of positively charged amino acid residues.[12] It has been shown that, as compared to lysine, the side chain of arginine can form much stronger salt bridges with other phosphorylated side chains. On the other hand, interaction of phosphoserine is more stable than phospho-aspartate.[13] Addition or removal of a phosphate group of a protein considerably alters the physio-chemical properties of the protein, which in turn can affect its stability, kinetics, and dynamics.[12] So far, it has been found that around 30% of cellular proteins are phosphorylated at distinct phosphorylation sites in response to external and internal stimuli.[14] The factors that determine the kinase and phosphatase recognition and binding specificity with proteins for phosphorylation and de-phosphorylation, respectively, are still unknown in several cases. The impact of disease-causing mutations affects the normal phosphorylation-mediated signaling.[15] Keeping in view the frequency and importance of phosphorylation as a PTM, this chapter has been designed to provide phosphorylation status of different proteins in diseased states.

2.1 Heat Shock Protein (Hsp27)

Human HSP27 functions as a molecular chaperone and is induced by various heat shock factors expressed under chemical and physical stresses, oxidative stress, and chemotherapies.[16–18] The protein binds to the accumulated misfolded proteins of the cell and translocates them to ATP-dependent chaperones or to proteasome machinery for its degradation. HSP27 phosphorylation is a dynamic process, which is controlled by cellular environmental conditions. The functions of HSP27 are regulated phosphorylation events.[19, 20] The phosphorylation process changes the conformation, which affects the assembly of the protein. The protein undergoes phosphorylation at serine-17, 82, and 78 by MAPKAP 2/3 kinase through activation of the P38 MAPK pathway.[21, 22] After phosphorylation, the HSP27 exist as dimers and tetramers which help in interaction with other proteins.[23–25]

As compared to normal cells, the status of Hsp27 phosphorylation in cancer cells is poorly understood. In the case of prostate cancer, Hsp27 phosphorylation occurs upon activation of the androgen receptor (AR), which leads to displacement of Hsp90 from the AR-Hsp90 complex, affecting AR translocation to the nucleus. In fact, it has been found that on inhibition of Hsp27 phosphorylation, AR becomes associated with MDM2, an E3 ubiquitin ligase, which in turn leads to enhanced AR degradation. Enhanced degradation of AR leads to an increase in the apoptotic rate of prostate cancer cells.[26] In prostate and pancreatic cancer cells, increased HSp27 phosphorylation leads to an increase in HSp27-induced cyto-protection to eukaryotic translational initiation factor 4E (eIF4E).

In case of castration and chemotherapy, the phosphorylated form of Hsp27 protects eIF4E from degradation leading to resistance to apoptosis.[27, 28] An analogous mechanism was found in prostate cancer, in which the phosphorylated form of Hsp27 increases the survival of the cell. This interaction improves the transcriptional activity, shuttling, and stability of AR, which increases survival of prostate cancer cells.[29] The phosphorylated form of Hsp27 is considered as one of the most important signature markers of prostate cancer. Additionally, Hsp27 phosphorylation via regulated epithelial–mesenchymal transition process has been reported to assist in maintaining breast cancer stem cells and NF-B activity.[30]

A number of studies have reported that phosphorylated HSP27 is involved in suppressing cell growth and chemo-sensitivity.[31–35] Interestingly, unusual Hsp27 expression in cancerous tissue plays a noteworthy role in aggressive tumor phenotype, poor patient prognosis, and enhanced therapy resistance. A number of studies in advanced tumors suggest that increase in Hsp27 phosphorylation is interrelated to resistance in treatment.[34, 36–38] Targeting Hsp27 over expression has shown promising results in pancreatic cancer and the phosphorylated HSP27 state is considered to play a critical role in resistance against gemcitabine.[33, 34, 39] The subcellular localization of phosphorylated Hsp27 has also shown its significant role in mRNA processing.[39, 40] Phosphorylated HSP27 has been shown to help in suppressing the growth of hepatocellular carcinomas, whereas its inhibition leads to increased sensitivity of colorectal cancer cells to chemotherapy.[35] Despite some developments, the role of phosphorylated HSP27 is not fully explored and needs to be further investigated in various types of cancers. Since phosphorylated Hsp27 has shown a role in cancer, interfering with its interaction with other proteins could be a potential therapeutic approach to target diseased cells.

2.2 KRAS

RAS proteins belong to the family of small GTPases, which are activated in response to external stimuli and regulate some important signal transduction pathways. One of the causative agents for cancer development is the insensitivity of RAS proteins to external signals due to point mutations. Among these, the most frequent oncogenic RAS mutations are at codons 12, 13, and 61, which lead to inhibition of GTPase activity or interference with the GAP functioning. The presence of a protein kinase C substrate at the C-terminal of K-RAS has been observed. It has been shown that three potent phosphate group acceptors, i.e., S181, T183, and S171 are present at the C-terminal of K-Ras. However, phosphoamino acid analysis of K-Ras from PKC agonists exposed cells have shown that only serine181 acts as a conserved phosphorylation site.[41] It has been found that for binding to HNRNPA2B1, phosphorylation of K-Ras at serine 181 is vital and unavoidable.[42, 43] In fact, serine 181 phosphorylation in turn leads to detachment of KRAS from the cell membrane and is thus a vital determinant for its subcellular location. All this potentially increases the availability of GTP-bound states for extended periods. In addition to this, in vivo studies have shown that phosphorylation of oncogenic KRAS is vital for tumor growth. Experimental analysis including λ phosphatase treatment and 2D-electrophoresis has observed that the presence of phosphorylated KRAS in human cancers highlights the alleged involvement of phosphorylated serine 181 in KRAS-driven human tumors.[44]

2.3 NF-κB and IκB

Nuclear factor-κB (NF-κB), one of the most studied transcription factors, has been shown to control expression of a number of proteins involved in apoptosis, cell proliferation, and inflammation and cell differentiation. The NF-κB is bound to IκB, which prevents its nuclear translocation from cytoplasm. There are a number of factors responsible for its activation, mainly by IKK mediated phosphorylation of IκBα. In fact, unusual activity of NF-κB has been observed as a hallmark of chronic inflammatory diseases and cancers.[45] The phosphorylation of IκBα results in ubiquitin-mediated degradation of IκBα and thus releases NF-κB to be translocated to the nucleus. The degradation of IκBα via phosphorylation also activates a catalytic subunit of protein kinase A. These modifications can affect various signaling pathways.[46] It has been observed that Ser276 phosphorylation of the p65 subunit of NF-κB affects its interaction with RHD (Rel Homology Domain) to facilitate DNA binding by favoring the interaction with p300 and CBP transcriptional coactivators.[47, 48] Phosphorylation of the NF-κB subunit has a significant effect on its transcriptional activity. Transcriptional activity mainly depends upon the site of the phosphorylation on NF-κB site and various kinases involved in its phosphorylation. Phosphorylation of the NF-κB subunit regulates transcription by regulating transcription factors. Significant evidence has shown that phosphorylation at different sites leads to generation of a heterogeneous pool of modified NF-κB.[49] Due to the vital role of NF-κB and IκB phosphorylation in inflammation and cancer, these proteins can serve as ideal targets for designing therapeutic agents.

2.4 Cytochrome c (Cyt-c)

Cyt-*c* is a heme containing protein that transports electrons from bc_1-complex to cytochrome oxidase. Iron moiety of the heme group is coordinated with the polypeptide chain through His18 and Met80.

After ischemia-repurfusion in isolated hearts, the electron flows to cyt-*c*, along with ROS generation, which in turn carries out oxidization of Met80 of cyt-*c* to Met80 sulfoxide (Met-O). All this eventually leads to conversion of cyt-*c* to a peroxidase enzyme. In ischemia, disruption of Met80-Fe ligation has been shown to convert cyt-*c* from an electron carrier to a peroxidase, which in turn inhibits oxidative phosphorylation, resulting in cytochrome *c* Met80 sulfoxide, breakdown of cristae membranes, and cytochrome *c* release. In addition to this, cyt-*c* also functions as a cardiolipin peroxidase.[50, 51] There are number of residues that undergo phosphorylation in cytochrome *c*, playing a role in controlling peroxidase activity of cyt—C.

Throughout the cytochrome *c* sequence different residues have been shown to undergo phosphorylation, as reported from different species. Threonine 29, threonine 41, Serine 48, Tyrosine 47, 49, Threonine 50, Tyrosine 68, 65, Threonine 79, and Tyrosine 98 residues all undergo phosphorylation. Particularly, phosphorylation of Thr28 leads to partial inhibition of respiration at the cytochrome *c* oxidase reaction step. Presence of phosphorylation sites on cyt-*c* in turn confirms that various functions of cyt-c are controlled by cellular signaling pathways.[52, 53] As compared to unphosphorylated wile-type cyt c, the T28E phosphomimetic variant is more stable and also able to degrade reactive oxygen species. Experiments in knockout cells have shown intact mitochondrial membrane potential, cell respiration, and decreased ROS levels in the T28E phosphomimetic mutants compared with wild type.[54] Replacement of phosphomimetic variant of Tyr48 by p-carboxy-methyl-L-phenylalanine (pCMF), a noncanonical amino acid, induces local perturbations leading to increased internal dynamics of the surrounding mutations.[55] Biochemical assays have shown that the Y48pCMFmutation impairs Cc channeling between cytochrome *bc*1 (Cbc1) and CcO, which in turn augments peroxidase activity, thus inducing anti apoptotic function relating to the ischemic hearts.[53]

2.5 Epidermal Growth Factor Receptor

EGFR, belonging to the tyrosine kinase receptor family, is identified as an oncogene in several solid tumors. EGFR expression along with c-Src (non-receptor tyrosine kinase) resulted in increased proliferation, transformation, and tumorigenesis, synergistically. However, the exact mechanism needs to be investigated. It has been observed that phosphorylation of tyrosine at residue 845 and 1101 plays an important role, where tyrosine 845 on EGFR is responsible for cellular proliferation and transformation in various breast cancer cell models.[56] Preventing tyrosine phosphorylation at position 845, either by treating the cells with SFK inhibitor dasatinib or over expression of EGFR Y845F mutant, does not affect the phosphorylation of other tyrosine residues. Therefore, it can be concluded that Y845 phosphorylation is vital for proliferation and transformation of breast cancer cell lines.[57] Importantly the auto-phosphorylated tyrosine 845 residue on EGFR remains phosphorylated with Src family kinase (SFK) inhibitor dasatinib or by over expression of phospho-defective mutant.[58] EGF binding to EGFR results in the dimerization of EGFR. It is observed that sialylation prevents this dimerization and prevents auto-phosphorylation of tyrosine residue of EGFR.[59] Several studies have proved that inhibition of sialylation or mutant L858R/T790M resulted in enhanced phosphorylation of EGFR and showed resistance to gefitinib in TKI-resistant lung cancer cell lines.[60]

2.6 P53

The p53 protein is a short-term transcription factor and tumor suppressor stabilized and activated in response to cellular stresses like DNA damage and hyperproliferation. P53 experiences posttranslational modification on the transactivation domain of the protein by DNA damage inducing agents and glucose deprivation and is the first nonhistone protein acetylated by p300/CBP.[61] In response to DNA damage, phosphorylation of ser15 of p53 protein in response to metabolic stress is carried out by both the ATM and ATR protein kinases.[62, 63] More than 10 kinases are reported to be involved in p53 phosphorylation. Additionally, immunological studies have shown a noticeable increase in p53 phosphorylation in newly processed tumor tissues. Moreover, level of p53 phosphorylation has also been observed at high frequency in tumors with wild type p53. A very weak level of p53 phosphorylation was observed in nontransformed cells as well as tissues distant to tumor cells. The residues undergoing phosphorylation are Ser-6, 9, 20, 37, 46, 81, 389, and 392. After DNA damage, p53 has been found to be phosphorylated on Ser15 and Ser372.[64] Phosphorylation of p53 residues at serine-18, 20, 46, 81, and 389 alters its transcriptional activity and stabilization.[65–67] After extensive studies on the phosphorylation of p53 during cell cycle, a hypothesis has been proposed that tumors may show different phosphorylation patterns either due to the presence of a mutation or dysregulated phosphatases and kinases.[68] Recent studies have shown that, after phosphorylation at Ser-46, p53 targets palmdelphin (isoform of the paralemmin families) to the nucleus for DNA damage induced apoptosis.[69]

2.7 Eukaryotic Translation Initiation Factor 2α (elf2α)

The eIF2α protein is a G protein hetero-trimer required for GTP dependent delivery of tRNA to ribosome and has been shown to play an important role in cellular redox homeostasis.[70] Phosphorylation of eIF2α protein takes place in response to stresses with inferences in antitumor treatments due to chemotherapeutic drugs. The alpha subunit of eIF2α, having serine 51, acts as a principal regulator for adaptation of cells to the genetic loss of the kinases PERK and GCN2. Phosphorylation of the protein leads to translational attenuation, an early event activated under different forms of cell stress. Genetic evidence suggests that the inability to catalyze eIF2a phosphorylation may contribute to diseases such as diabetes. In fact, impaired eIF2α phosphorylation by genetic means makes tumor cells susceptible to oxidative stress induced death Studies on mouse models in which serine-51was substituted by alanine to abolish eIF2a phosphorylation showed that the mutant mice died within the first day of life, as a result of a severe loss in pancreatic beta cells and hyperglycemia. This result indicates the critical role for eIF2a phosphorylation in the development and functioning of pancreatic b-cells.[71] It has also been shown that eIF2αP (phosphorylated eIF2α) facilitates Akt activation in cells under stress. In eIF2αP proficient cells, Akt activation has a significant role and the reduced concentration of activated Akt in cells deficient in eIF2αP encourages cellular death. In fact, it has been shown that, on treatment with H_2O_2, the eIF2αP deficient cells were found to be more prone to death as compared to eIF2αP proficient cells.[72] Enhanced concentration of eIF2αP leads to decreased activity of mTORC1 and thus reduces negative regulation of the PI3K-Akt pathway.[73] Pharmacological inhibition of eIF2αP is quite promising in devising strategies for enhancing anticancerous activity of pro-oxidant drugs on hyperactivated Akt tumors.[74]

2.8 Tau Protein

Tau proteins are associated with microtubules and involved in formation of microtubules. They are important cytoskeletal proteins, mainly expressed in neurons for axon and dendritic development, which in turn are required for neuron transmission. Due to alternative splicing of Tau mRNA, six Tau isoforms are expressed in an adult human brain. A number of in vitro studies have shown involvement of phosphatases and protein kinases in anomalous phosphorylation of this protein. However, the protein kinase responsible for phosphorylation of tau protein in the brain is still not clear. It has been reported that the activity of tau is due to its phosphorylated state. If the level of phosphorylation is normal then the biological function of tau is normal, whereas it loses its biological activity as soon as it is hyperphosphorylated.[75] This difference in biological activity is in relation with the change in structure of the protein, which in turn depends on the extent of phosphorylation. As reported from its longest variant of 441 amino acids, Tau protein contains around 80 potential serine or threonine phosphorylation sites. In vitro studies have found that the phosphorylated recombinant tau is not able to assemble into its paired helical-like filaments, whereas its incubation with sulfated glycosaminoglycans (such as heparin or heparin sulfate) results in the formation of Alzheimer-like filaments.[76–78] Most of the phosphorylation sites reported in Tau are Ser-Pro and Thr-Pro motifs.

Hyperphosphorylated tau in affected neurons results due to the imbalance in the activities of phosphatase and tau kinase. The self-assembly of tau starts by assembling of microtubule binding domains, and abnormal hyperphosphorylation in turn promotes tangling of paired helical and straight fragments by neutralization of charges of basic amino acids of the flanking regions, which otherwise are inhibitory in nature. In fact, abnormal hyperphosphorylation of tau is responsible for neurofibrillary degeneration in Alzheimer's and other tau involved diseases. Currently, most of the tau-based therapeutic strategies for neurodegeneration disorders are based on modifying tau phosphorylation/hyperphosphorylation of neurofibrillary tangles.

2.9 Insulin Resistant Substrate Protein (IRS-1)

IRS-1 belongs to the insulin receptor substrate (IRS) protein family that acts as a signal adapter protein. IRS-1 consists mainly of phosphotyrosine binding (PTB) and pleckstrin homology (pH) domains. One of the main causes of insulin resistance is defective IRS-1 signaling, which undergoes degradation via ubiquitination and phosphorylation on serine and threonine residue.[38] The mTOR/S6K pathway is involved in the hyperphosphorylation of IRS-1 in response to insulin.[79] There are number of reports that correlate dysfunctional phosphorylation of IRS-1 with brain insulin receptor, as seen in many other tissues. Phosphorylation of this protein plays an important role in the pathogenesis of IR and AD. The phosphorylated active level of IRS-1 differs in AD brains and cognitively normal (CN) controls, which suggests that these are more relevant to IR and AD pathogenesis. mTOR and GSK-3β mediated phosphorylation in human hippocampus tissue in combination with JNK pathway inhibition increases phosphorylation on IRS-1 at multiple serine residues, more importantly of serine 312, 616, and 636.[80,81] Rapamycin, a well-known drug for treating various types of cancer, promotes proteasome mediated degradation of IRS-1 through phosphorylation of serine/threonine residue.[82]

3 PROTEIN ACETYLATION

In addition to phosphorylation, protein acetylation is another PTM in eukaryotes that regulates the functional state of various proteins. The protein acetylation takes place in two different ways; either by Nt-acetyltransferase

(NATs) activity, where 80%–90% of proteins became acetylated at N-terminus, or acetylation at the ε-amino group of lysine. NATs is reversible and well regulated. The acetylation process eliminates positive charge on the lysine side chain and thus the functional state of protein changes in terms of its interaction with other proteins, DNA, its turnover number, and cellular localization. It has been found that acetylation enhances DNA interaction of nonhistone proteins including P53, E2F, etc., whereas it is also observed that an opposite effect occurs where acetylation effects DNA binding to FoxO1, HMGI, p65, etc.[83] Lysine acetylation of modification was first discovered in histones by histone acetyltransferase (HATs) and was found to be involved in regulating gene transcription.

3.1 Histone Acetylation

Histone acetylation was first observed on lysine by enzyme HATs. The histone acetylation and de-acetylation plays a vital role in several cellular functions, transcription factors, cellular metabolism, and various interactions critical in many cellular functions. Different groups of HDACs, based on their homology to yeast histone deacetylases, include Class I (HDAC 1–3 and 8), Class II (HDAC 4–7, 9 and 10), Class III (SIRT1–7), and Class IV (HDAC11). Similarly, the three major categories of acetylases are GNAT (Gcn5-related *N*-acetyltransferase), CPB/p300, and the MYST family. Any defect during the acetylation process results in severe diseases including cardiovascular disorders and cancer. Therefore, it is important to consider the physiological outcome of protein acetylation that can be targeted in various diseases.

3.1.1 Signal Transducer and Activator of Transcription (STAT) Protein

In mammalian cells, STAT represents seven classes of proteins, namely STAT 1, STAT 2, STAT 3, STAT 4, STAT 5a/b, and STAT 6. All classes of STATs consist of structurally and functionally conserved amino-terminal domain (NH2), DNA binding domain (DBD), coiled-coiled domain (CCD), and SH2 domain. However, STAT is specific, due to its carboxyl terminal transcriptional activation domain. STATS mainly act as transcription factors and play a significant role in pathogenesis of different human diseases effecting cell growth, immune response, cellular differentiation, apoptosis, and cell survival. Acetylation of STATs depends on the activity of MDACs and Histone acetyltransferase. Generally, transcriptional activity of STAT increases due to acetylation that depends on various acetyl transferases. STAT6 was reported to be the first STAT protein to undergo acetylation, However STAT1, STAT2, STAT3, and STAT 5b have now been very well characterized. K410 and K413 were identified as acetyl on STAT1. STAT1 has dual (positive and negative) effect on NF-kB activity. STAT3 acetylation increases its transcription activity, as seen with increase in nuclear location and DNA binding.

3.1.2 Interferon Regulatory Factor (IRF)

Histone acetyl transferases acetylate histones as well as a few transcription proteins, i.e., nonhistone proteins. IRFs are groups of transcription factors that are known to regulate cell growth, differentiation, the immune system, and hematopoietic system by interaction with interferon regulatory elements (IRF-E). IRFs are modified by HATs, which act as a co-activator in recognizing DNA elements. To date, around nine IRF family proteins (IRF-1–9) have been identified. It has been shown that IRF-1 and IRF-2 both interact with PCAF, which is histone acetyl transferase and regulate the transcription. IRF-2 acetylation is observed during active nonconfluent growth of NIH3T3. IRF-2 acetylation results in the interaction with H4 that regulates cell growth. In addition to this, acetylated IRF-2 also interacts with several other proteins in regulating the gene transcription. For example, nucleolin prefers to associate with acetylated IRF-2, which regulates the expression of H4 protein. IRF-3/HATp300 interaction is also essential for transcription activation of IFN-β. Acetylation of IRF-7 is negatively regulated by its acetylation of Lys-92. It can be concluded that acetylation within DNA binding domain represses its binding activity while acetylation enhances its DNA binding activity.

3.1.3 Hsp 90

Acetylation of molecular chaperones plays a significant role in oncogenic stress, stress induced by increased demand of signaling for uncontrolled growth, and also leads to dynamic reorganization of chaperone complexes.[84] Among these molecular chaperones, Hsp90 acts as an essential molecular chaperone for maintaining cellular protein homeostasis by assisting folding as well as its intracellular trafficking.[85, 86] It is a 90 kDa nuclear and cytosolic protein, which comprises of two subunits Hsp90α of 732 amino acids and Hsp90β of 724 amino acids). In most of the cellular functions of HSp90, N-terminal ATP-binding domain plays a dominant role.[87] Mass spectrometry studies have revealed that the Hsp90α unit of the protein undergoes acetylation. Hyperacetylated Hsp90 has been found in extracellular fluids of in vitro breast cancer cell invasion. However, Hsp90 has been found to be present on the metastatic melanoma cell surfaces.[88] In fact, it is inducible isoform hsp90α, not hsp90β, which is secreted and present on the cancer cell surfaces.[89] It has been observed that on starvation of breast cancerous cells, the secretion and extra-cellular localization of endogenously expressed hsp90α are promoted and the molecular chaperone is hyperacetylated under these conditions.

Hsp90α's significance in both normal and oncogenic signaling prompts us to understand its regulatory mechanism. The stable GR-Hsp90 stable complex becomes a dynamic one upon the acetylation of Hsp90, so as to enable the entry of GR for transcriptional activation in nucleus. Subsequent HDAC6-induced de-acetylation allows Hsp90s to gain entry into productive HDAC6 chaperone complexes, which might then be obligatory for strong Hsp90 activity by stimulating de-acetylated Hsp90.

3.1.4 Pyruvate Kinase

Pyruvate kinase (PK) catalyzes the addition of a phosphate group removed from phosphoenol pyruvate (PEP) to Adenosine diphosphate (ADP), which in turn results in generation of one molecule of pyruvate and ATP. In mammals, genes PKM (pyruvate kinase muscle) and PKL/R (i.e., pyruvate kinase, liver, and RBCs) code for four isoforms (represented as L, R, M1, and M2) of PK in mammals. However, the PKM2 isoform has been found to be the dominant isoform in various diseases in humans. Acetylation of PKM2 at K305 leads to a decrease in its activity. At the same time, acetylation of MK305 on PKM2 increases its interaction with HSC70, which results in its degradation mediated by CMA (Chaperone mediated autophagy). Due to decreased PKM2 enzymatic activity, there is accumulation of fructose-1, 6-bisphophate (FBP), and glucose-6-phosphate (G6P). Furthermore, acetylation mimic PKM2K30Q was found to enhance cell proliferation and promote tumor growth. Therefore, targeting K305 could be of potential therapeutic value in some cancers.

Acetylation of lysine at K433 is also reported to stimulate cell proliferation and tumorigenesis. It has been experimentally observed that the acetylation-mimetic of PKM2 (K433Q) develops tumors very fast in mice. As compared to normal tissue, acetylated PKM2 at K433 expression is significantly higher in breast cancer. As a result of increased PKM2 acetylation, tyrosine and threonine kinase activity is enhanced with the accumulation of PKM2 within the nucleus. It is reported that PKM2 acetylation at K433 is mediated by P300 acetyltransferase that inhibits FBP binding and affects the conversion of PKM2 to active glycolytic enzyme in its tetrameric form in the cytoplasm; furthermore, the dimeric form of PKM2 is translocated to the nucleus and thus regulates gene expression.

3.1.5 Huntingtin Protein (Htt Protein)

Huntington's disease, which affects the brain, is mainly caused by mutant huntingtin protein (3000 amino acids). This protein is expressed in every part of the human body, but the brain is the target organ. The mutant Htt has polyglutamine, which plays a crucial role in its aggregation. Polyglutamine formation is due to CAG trinucleotide repeat in the Htt gene, with more than 36 repeats. The Htt protein is controlled by UPS or autophagic-lysosomal pathway. Many people have shown the role of acetylation in clearance of mutant huntingtin protein, which in turn can serve as an ideal and a potential therapeutic approach. Histone acetyltransferase (HAT) domain of cAMP response element binding protein (CREB binding protein) directly interacts with the Htt mutant protein. Acetylation takes place on lysine 444 (K444), which increases the autophagosomes of mutant protein targeting the mutant protein through acetylation.[90] CREB induced acetylation of the mutant protein improves mutant protein clearance and thus serves as a neuroprotective. Although CREB alone cannot serve as a specific therapeutic target, acetylation of mutant protein also represents a novel pathway that can be targeted to develop selective compounds, which can specifically promote acetylation and hence can increase the clearance rate of mutant Htt in HD. Studies showing the enrichment of autophagosomes with acetylated Htt fragments support the trafficking of soluble and longer Htt fragments to autophagosomes. Lysine at position 8 has also been shown to be acetylated for shorter fragments, suggesting that shorter fragments may also be regulated by acetylation.[91]

4 CONCLUSION AND FUTURE PROSPECTS

PTMs in proteins bring variations in the proteins by adding/removing different groups, thereby modulating the structure and function of proteins. New methodologies for detecting PTMs have helped in knowing the identification of different modifications, which can be used to know the existence of the different forms of proteins in different diseases. More advanced techniques are required to detect the PTMs in the proteins and their localization in the cell. The detection of accurate PTMs in diseased states will help to design drugs against the particular modified state of the protein. Moreover, there is need to know the effect of PTMs on functional changes in the proteins.

References

1. Prabakaran S, Lippens G, Steen H, Gunawardena J. Post-translational modification: nature's escape from genetic imprisonment and the basis for dynamic information encoding. *Wiley Interdiscip Rev Syst Biol Med* 2012;**4**(6):565–83.
2. Karve TM, Cheema AK. Small changes huge impact: the role of protein posttranslational modifications in cellular homeostasis and disease. *J Amino Acids* 2011;**2011**:207691.
3. Larsen MR, Trelle MB, Thingholm TE, Jensen ON. Analysis of posttranslational modifications of proteins by tandem mass spectrometry. *Biotechniques* 2006;**40**(6):790–8.
4. Slade DJ, Subramanian V, Fuhrmann J, Thompson PR. Chemical and biological methods to detect post-translational modifications of arginine. *Biopolymers* 2014;**101**(2):133–43.

5. Brooks CL, Gu W. Ubiquitination, phosphorylation and acetylation: the molecular basis for p53 regulation. *Curr Opin Cell Biol* 2003;**15**(2):164–71.
6. Deribe YL, Pawson T, Dikic I. Post-translational modifications in signal integration. *Nat Struct Mol Biol* 2010;**17**(6):666–72.
7. Zhao S, Xu W, Jiang W, Yu W, Lin Y, Zhang T, et al. Regulation of cellular metabolism by protein lysine acetylation. *Science* 2010;**327**(5968):1000–4.
8. Drazic A, Winter J. The physiological role of reversible methionine oxidation. *Biochim Biophys Acta* 2014;**1844**(8):1367–82.
9. Hunter T. The age of crosstalk: phosphorylation, ubiquitination, and beyond. *Mol Cell* 2007;**28**(5):730–8.
10. Singh BN, Zhang G, Hwa YL, Li J, Dowdy SC, Jiang SW. Nonhistone protein acetylation as cancer therapy targets. *Expert Rev Anticancer Ther* 2010;**10**(6):935–54.
11. Olsen JV, Vermeulen M, Santamaria A, Kumar C, Miller ML, Jensen LJ, et al. Quantitative phosphoproteomics reveals widespread full phosphorylation site occupancy during mitosis. *Sci Signal* 2010;**3**(104):ra3.
12. Johnson LN, Lewis RJ. Structural basis for control by phosphorylation. *Chem Rev* 2001; **101**(8):2209–42.
13. Mandell DJ, Chorny I, Groban ES, Wong SE, Levine E, Rapp CS, Jacobson MP. Strengths of hydrogen bonds involving phosphorylated amino acid side chains. *J Am Chem Soc* 2007;**129**(4):820–7.
14. Cheng HC, Qi RZ, Paudel H, Zhu HJ. Regulation and function of protein kinases and phosphatases. *Enzyme Res* 2011;**2011**. 794089.
15. Ubersax JA, Ferrell Jr JE. Mechanisms of specificity in protein phosphorylation. *Nat Rev Mol Cell Biol* 2007;**8**(7):530–41.
16. Freeman BC, Yamamoto KR. Disassembly of transcriptional regulatory complexes by molecular chaperones. *Science* 2002;**296**(5576):2232–5.
17. Garrido C, Brunet M, Didelot C, Zermati Y, Schmitt E, Kroemer G. Heat shock proteins 27 and 70: anti-apoptotic proteins with tumorigenic properties. *Cell Cycle* 2006;**5**(22): 2592–601.
18. Lindquist S, Craig EA. The heat-shock proteins. *Annu Rev Genet* 1988;**22**:631–77.
19. Benjamin IJ, McMillan DR. Stress (heat shock) proteins: molecular chaperones in cardiovascular biology and disease. *Circ Res* 1998;**83**(2):117–32.
20. Lambert H, Charette SJ, Bernier AF, Guimond A, Landry J. HSP27 multimerization mediated by phosphorylation-sensitive intermolecular interactions at the amino terminus. *J Biol Chem* 1999;**274**(14):9378–85.
21. Guay J, Lambert H, Gingras-Breton G, Lavoie JN, Huot J, Landry J. Regulation of actin filament dynamics by p38 map kinase-mediated phosphorylation of heat shock protein 27. *J Cell Sci* 1997;**110**(Pt 3):357–68.
22. Rouse J, Cohen P, Trigon S, Morange M, Alonso-Llamazares A, Zamanillo D, et al. A novel kinase cascade triggered by stress and heat shock that stimulates MAPKAP kinase-2 and phosphorylation of the small heat shock proteins. *Cell* 1994;**78**(6): 1027–37.
23. Arrigo AP, Virot S, Chaufour S, Firdaus W, Kretz-Remy C, Diaz-Latoud C. Hsp27 consolidates intracellular redox homeostasis by upholding glutathione in its reduced form and by decreasing iron intracellular levels. *Antioxid Redox Signal* 2005;**7**(3–4):414–22.
24. Mehlen P, Hickey E, Weber LA, Arrigo AP. Large unphosphorylated aggregates as the active form of hsp27 which controls intracellular reactive oxygen species and glutathione levels and generates a protection against TNFalpha in NIH-3T3-ras cells. *Biochem Biophys Res Commun* 1997;**241**(1):187–92.
25. Rogalla T, Ehrnsperger M, Preville X, Kotlyarov A, Lutsch G, Ducasse C, et al. Regulation of Hsp27 oligomerization, chaperone function, and protective activity against oxidative stress/tumor necrosis factor alpha by phosphorylation. *J Biol Chem* 1999;**274**(27): 18947–56.

26. Chen L, Fang B, Giorgianni F, Gingrich JR, Beranova-Giorgianni S. Investigation of phosphoprotein signatures of archived prostate cancer tissue specimens via proteomic analysis. *Electrophoresis* 2011;**32**(15):1984–91.
27. Andrieu C, Taieb D, Baylot V, Ettinger S, Soubeyran P, De-Thonel A, et al. Heat shock protein 27 confers resistance to androgen ablation and chemotherapy in prostate cancer cells through eIF4E. *Oncogene* 2010;**29**(13):1883–96.
28. Baylot V, Andrieu C, Katsogiannou M, Taieb D, Garcia S, Giusiano S, et al. OGX-427 inhibits tumor progression and enhances gemcitabine chemotherapy in pancreatic cancer. *Cell Death Dis* 2011;**2**:e221.
29. Zoubeidi A, Zardan A, Beraldi E, Fazli L, Sowery R, Rennie P, et al. Cooperative interactions between androgen receptor (AR) and heat-shock protein 27 facilitate AR transcriptional activity. *Cancer Res* 2007;**67**(21):10455–65.
30. Wei L, Liu TT, Wang HH, Hong HM, Yu AL, Feng HP, Chang WW. Hsp27 participates in the maintenance of breast cancer stem cells through regulation of epithelial-mesenchymal transition and nuclear factor-kappaB. *Breast Cancer Res* 2011;**13**(5):R101.
31. Matsunaga A, Ishii Y, Tsuruta M, Okabayashi K, Hasegawa H, Kitagawa Y. Inhibition of heat shock protein 27 phosphorylation promotes sensitivity to 5-fluorouracil in colorectal cancer cells. *Oncol Lett* 2014;**8**(6):2496–500.
32. Matsushima-Nishiwaki R, Takai S, Adachi S, Minamitani C, Yasuda E, Noda T, et al. Phosphorylated heat shock protein 27 represses growth of hepatocellular carcinoma via inhibition of extracellular signal-regulated kinase. *J Biol Chem* 2008;**283**(27):18852–60.
33. Nakashima M, Adachi S, Yasuda I, Yamauchi T, Kawaguchi J, Itani M, et al. Phosphorylation status of heat shock protein 27 plays a key role in gemcitabine-induced apoptosis of pancreatic cancer cells. *Cancer Lett* 2011;**313**(2):218–25.
34. Taba K, Kuramitsu Y, Ryozawa S, Yoshida K, Tanaka T, Maehara S, et al. Heat-shock protein 27 is phosphorylated in gemcitabine-resistant pancreatic cancer cells. *Anticancer Res* 2010;**30**(7):2539–43.
35. Yasuda E, Kumada T, Takai S, Ishisaki A, Noda T, Matsushima-Nishiwaki R, et al. Attenuated phosphorylation of heat shock protein 27 correlates with tumor progression in patients with hepatocellular carcinoma. *Biochem Biophys Res Commun* 2005;**337**(1):337–42.
36. Sakai A, Otani M, Miyamoto A, Yoshida H, Furuya E, Tanigawa N. Identification of phosphorylated serine-15 and -82 residues of HSPB1 in 5-fluorouracil-resistant colorectal cancer cells by proteomics. *J Proteome* 2012;**75**(3):806–18.
37. Wang HQ, Yang B, Xu CL, Wang LH, Zhang YX, Xu B, et al. Differential phosphoprotein levels and pathway analysis identify the transition mechanism of LNCaP cells into androgen-independent cells. *Prostate* 2010;**70**(5):508–17.
38. Xu Y, Diao Y, Qi S, Pan X, Wang Q, Xin Y, et al. Phosphorylated Hsp27 activates ATM-dependent p53 signaling and mediates the resistance of MCF-7 cells to doxorubicin-induced apoptosis. *Cell Signal* 2013;**25**(5):1176–85.
39. Guo Y, Ziesch A, Hocke S, Kampmann E, Ochs S, De Toni EN, et al. Overexpression of heat shock protein 27 (HSP27) increases gemcitabine sensitivity in pancreatic cancer cells through S-phase arrest and apoptosis. *J Cell Mol Med* 2015;**19**(2):340–50.
40. Bryantsev AL, Kurchashova SY, Golyshev SA, Polyakov VY, Wunderink HF, Kanon B, et al. Regulation of stress-induced intracellular sorting and chaperone function of Hsp27 (HspB1) in mammalian cells. *Biochem J* 2007;**407**(3):407–17.
41. Ballester R, Furth ME, Rosen OM. Phorbol ester- and protein kinase C-mediated phosphorylation of the cellular kirsten ras gene product. *J Biol Chem* 1987;**262**(6):2688–95.
42. Barcelo C, Paco N, Morell M, Alvarez-Moya B, Bota-Rabassedas N, Jaumot M, et al. Phosphorylation at Ser-181 of oncogenic KRAS is required for tumor growth. *Cancer Res* 2014;**74**(4):1190–9.
43. Bivona TG, Quatela SE, Bodemann BO, Ahearn IM, Soskis MJ, Mor A, et al. PKC regulates a farnesyl-electrostatic switch on K-Ras that promotes its association with Bcl-XL on mitochondria and induces apoptosis. *Mol Cell* 2006;**21**(4):481–93.

44. Jasinski P, Zwolak P, Terai K, Borja-Cacho D, Dudek AZ. PKC-alpha inhibitor MT477 slows tumor growth with minimal toxicity in in vivo model of non-Ras-mutated cancer via induction of apoptosis. *Investig New Drugs* 2011;**29**(1):33–40.
45. Perkins ND. Post-translational modifications regulating the activity and function of the nuclear factor kappa B pathway. *Oncogene* 2006;**25**(51):6717–30.
46. Perkins ND. Integrating cell-signalling pathways with NF-kappaB and IKK function. *Nat Rev Mol Cell Biol* 2007;**8**(1):49–62.
47. Zhong H, May MJ, Jimi E, Ghosh S. The phosphorylation status of nuclear NF-kappa B determines its association with CBP/p300 or HDAC-1. *Mol Cell* 2002;**9**(3):625–36.
48. Zhong H, Voll RE, Ghosh S. Phosphorylation of NF-kappa B p65 by PKA stimulates transcriptional activity by promoting a novel bivalent interaction with the coactivator CBP/p300. *Mol Cell* 1998;**1**(5):661–71.
49. Moreno R, Sobotzik JM, Schultz C, Schmitz ML. Specification of the NF-kappaB transcriptional response by p65 phosphorylation and TNF-induced nuclear translocation of IKK epsilon. *Nucleic Acids Res* 2010;**38**(18):6029–44.
50. Kagan VE, Borisenko GG, Tyurina YY, Tyurin VA, Jiang J, Potapovich AI, et al. Oxidative lipidomics of apoptosis: redox catalytic interactions of cytochrome c with cardiolipin and phosphatidylserine. *Free Radic Biol Med* 2004;**37**(12):1963–85.
51. Kagan VE, Tyurin VA, Jiang J, Tyurina YY, Ritov VB, Amoscato AA, et al. Cytochrome c acts as a cardiolipin oxygenase required for release of proapoptotic factors. *Nat Chem Biol* 2005;**1**(4):223–32.
52. Lee I, Salomon AR, Yu K, Doan JW, Grossman LI, Huttemann M. New prospects for an old enzyme: mammalian cytochrome c is tyrosine-phosphorylated in vivo. *Biochemistry* 2006;**45**(30):9121–8.
53. Yu H, Lee I, Salomon AR, Yu K, Huttemann M. Mammalian liver cytochrome c is tyrosine-48 phosphorylated in vivo, inhibiting mitochondrial respiration. *Biochim Biophys Acta* 2008;**1777**(7–8):1066–71.
54. Mahapatra G, Varughese A, Ji Q, Lee I, Liu J, Vaishnav A, et al. Phosphorylation of cytochrome c threonine 28 regulates electron transport chain activity in kidney: Implications for AMP kinase. *J Biol Chem* 2017;**292**(1):64–79.
55. Pecina P, Borisenko GG, Belikova NA, Tyurina YY, Pecinova A, Lee I, et al. Phosphomimetic substitution of cytochrome C tyrosine 48 decreases respiration and binding to cardiolipin and abolishes ability to trigger downstream caspase activation. *Biochemistry* 2010;**49**(31):6705–14.
56. Tice DA, Biscardi JS, Nickles AL, Parsons SJ. Mechanism of biological synergy between cellular Src and epidermal growth factor receptor. *Proc Natl Acad Sci U S A* 1999; **96**(4):1415–20.
57. Mueller KL, Hunter LA, Ethier SP, Boerner JL. Met and c-Src cooperate to compensate for loss of epidermal growth factor receptor kinase activity in breast cancer cells. *Cancer Res* 2008;**68**(9):3314–22.
58. Buettner R, Mesa T, Vultur A, Lee F, Jove R. Inhibition of Src family kinases with dasatinib blocks migration and invasion of human melanoma cells. *Mol Cancer Res* 2008; **6**(11):1766–74.
59. Yen HY, Liu YC, Chen NY, Tsai CF, Wang YT, Chen YJ, et al. Effect of sialylation on EGFR phosphorylation and resistance to tyrosine kinase inhibition. *Proc Natl Acad Sci U S A* 2015;**112**(22):6955–60.
60. Wu M, Yuan Y, Pan YY, Zhang Y. Antitumor activity of combination treatment with gefitinib and docetaxel in EGFR-TKI-sensitive, primary resistant and acquired resistant human non-small cell lung cancer cells. *Mol Med Rep* 2014;**9**(6):2417–22.
61. Ionov Y, Matsui S, Cowell JK. A role for p300/CREB binding protein genes in promoting cancer progression in colon cancer cell lines with microsatellite instability. *Proc Natl Acad Sci U S A* 2004;**101**(5):1273–8.

62. Jones RG, Plas DR, Kubek S, Buzzai M, Mu J, Xu Y, et al. AMP-activated protein kinase induces a p53-dependent metabolic checkpoint. *Mol Cell* 2005;**18**(3):283–93.
63. Meek DW. Tumour suppression by p53: a role for the DNA damage response? *Nat Rev Cancer* 2009;**9**(10):714–23.
64. Buschmann T, Adler V, Matusevich E, Fuchs SY, Ronai Z. p53 phosphorylation and association with murine double minute 2, c-Jun NH2-terminal kinase, p14ARF, and p300/CBP during the cell cycle and after exposure to ultraviolet irradiation. *Cancer Res* 2000;**60**(4):896–900.
65. Banin S, Moyal L, Shieh S, Taya Y, Anderson CW, Chessa L, et al. Enhanced phosphorylation of p53 by ATM in response to DNA damage. *Science* 1998;**281**(5383):1674–7.
66. Bulavin DV, Saito S, Hollander MC, Sakaguchi K, Anderson CW, Appella E, Fornace Jr AJ. Phosphorylation of human p53 by p38 kinase coordinates N-terminal phosphorylation and apoptosis in response to UV radiation. *EMBO J* 1999;**18**(23):6845–54.
67. Hirao A, Kong YY, Matsuoka S, Wakeham A, Ruland J, Yoshida H, et al. DNA damage-induced activation of p53 by the checkpoint kinase Chk2. *Science* 2000;**287**(5459):1824–7.
68. Minamoto T, Buschmann T, Habelhah H, Matusevich E, Tahara H, Boerresen-Dale AL, et al. Distinct pattern of p53 phosphorylation in human tumors. *Oncogene* 2001;**20**(26):3341–7.
69. Dashzeveg N, Taira N, Lu ZG, Kimura J, Yoshida K. Palmdelphin, a novel target of p53 with Ser46 phosphorylation, controls cell death in response to DNA damage. *Cell Death Dis* 2014;**5**.
70. Harding HP, Zhang Y, Zeng H, Novoa I, Lu PD, Calfon M, et al. An integrated stress response regulates amino acid metabolism and resistance to oxidative stress. *Mol Cell* 2003;**11**(3):619–33.
71. Scheuner D, Vander Mierde D, Song B, Flamez D, Creemers JW, Tsukamoto K, et al. Control of mRNA translation preserves endoplasmic reticulum function in beta cells and maintains glucose homeostasis. *Nat Med* 2005;**11**(7):757–64.
72. Sarbassov DD, Guertin DA, Ali SM, Sabatini DM. Phosphorylation and regulation of Akt/PKB by the rictor-mTOR complex. *Science* 2005;**307**(5712):1098–101.
73. Uranga RM, Katz S, Salvador GA. Enhanced phosphatidylinositol 3-kinase (PI3K)/Akt signaling has pleiotropic targets in hippocampal neurons exposed to iron-induced oxidative stress. *J Biol Chem* 2013;**288**(27):19773–84.
74. Krishnamoorthy J, Rajesh K, Mirzajani F, Kesoglidou P, Papadakis AI, Koromilas AE. Evidence for eIF2alpha phosphorylation-independent effects of GSK2656157, a novel catalytic inhibitor of PERK with clinical implications. *Cell Cycle* 2014;**13**(5):801–6.
75. Alonso AC, Zaidi T, Grundke-Iqbal I, Iqbal K. Role of abnormally phosphorylated tau in the breakdown of microtubules in alzheimer disease. *Proc Natl Acad Sci U S A* 1994;**91**(12):5562–6.
76. Goedert M, Jakes R, Spillantini MG, Hasegawa M, Smith MJ, Crowther RA. Assembly of microtubule-associated protein tau into Alzheimer-like filaments induced by sulphated glycosaminoglycans. *Nature* 1996;**383**(6600):550–3.
77. Perez M, Valpuesta JM, Medina M, Montejo de Garcini E, Avila J. Polymerization of tau into filaments in the presence of heparin: the minimal sequence required for tau-tau interaction. *J Neurochem* 1996;**67**(3):1183–90.
78. Hasegawa M, Crowther RA, Jakes R, Goedert M, Alzheimer-like changes in microtubule-associated protein Tau induced by sulfated glycosaminoglycans. Inhibition of microtubule binding, stimulation of phosphorylation, and filament assembly depend on the degree of sulfation. *J Biol Chem* 1997;**272**(52):33118–24.
79. Haruta T, Uno T, Kawahara J, Takano A, Egawa K, Sharma PM, Olefsky JM, Kobayashi M. A rapamycin-sensitive pathway down-regulates insulin signaling via phosphorylation and proteasomal degradation of insulin receptor substrate-1. *Mol Endocrinol* 2000;**14**:783–94.

80. Boura-Halfon S, Zick Y. Phosphorylation of IRS proteins, insulin action, and insulin resistance. *Am J Physiol Endocrinol Metab* 2009;**296**(4):E581–91.
81. Boura-Halfon S, Zick Y. Serine kinases of insulin receptor substrate proteins. *Vitam Horm* 2009;**80**:313–49.
82. Destefano MA, Jacinto E. Regulation of insulin receptor substrate-1 by mTORC2 (mammalian target of rapamycin complex 2). *Biochem Soc Trans* 2013;**41**(4):896–901.
83. Liszczak G, Goldberg JM, Foyn H, Petersson EJ, Arnesen T, Marmorstein R. Molecular basis for N-terminal acetylation by the heterodimeric NatA complex. *Nat Struct Mol Biol* 2013;**20**(9):1098–105.
84. Kovacs JJ, Cohen TJ, Yao TP. Chaperoning steroid hormone signaling via reversible acetylation. *Nucl Recept Signal* 2005;**3**:e004.
85. Pearl LH, Prodromou C. Structure and mechanism of the Hsp90 molecular chaperone machinery. *Annu Rev Biochem* 2006;**75**:271–94.
86. Wiech H, Buchner J, Zimmermann R, Jakob U. Hsp90 chaperones protein folding in vitro. *Nature* 1992;**358**(6382):169–70.
87. Ali MM, Roe SM, Vaughan CK, Meyer P, Panaretou B, Piper PW, et al. Crystal structure of an Hsp90-nucleotide-p23/Sba1 closed chaperone complex. *Nature* 2006;**440**(7087):1013–7.
88. Becker B, Multhoff G, Farkas B, Wild PJ, Landthaler M, Stolz W, Vogt T. Induction of Hsp90 protein expression in malignant melanomas and melanoma metastases. *Exp Dermatol* 2004;**13**(1):27–32.
89. Eustace BK, Sakurai T, Stewart JK, Yimlamai D, Unger C, Zehetmeier C, et al. Functional proteomic screens reveal an essential extracellular role for hsp90 alpha in cancer cell invasiveness. *Nat Cell Biol* 2004;**6**(6):507–14.
90. Jeong H, Then F, Melia Jr TJ, Mazzulli JR, Cui L, Savas JN, et al. Acetylation targets mutant huntingtin to autophagosomes for degradation. *Cell* 2009;**137**(1):60–72.
91. Chesser AS, Pritchard SM, Johnson GV. Tau clearance mechanisms and their possible role in the pathogenesis of alzheimer disease. *Front Neurol* 2013;**4**:122.

CHAPTER

4

Protein Modifications and Lifestyle Disorders

Shivani Arora, Anju Katyal

Dr. B.R. Ambedkar Center for Biomedical Research, University of Delhi, Delhi, India

1 INTRODUCTION

The incidences of lifestyle diseases, viz. obesity, cardiovascular diseases, diabetes, and arthritis are on rise; however, the mechanisms that trigger their development and maintain the progression are under investigation. Clinical and experimental research over the past two decades has placed inflammation, cross talk between modified cellular proteome, and peripheral immune responses at the central position in pathogenesis of these disorders. However, the mechanisms by which immune tolerance is disturbed are complex and poorly understood. Cellular machinery under stress is prone to alternative splicing, translational errors, protein misfolding, and protein modifications. Decreased ATP production during disease conditions leads to accumulation of misfolded proteins. Disruption of normal cellular energetics warrants an adaptive response from endoplasmic reticulum (ER) and mitochondria; typically known as "unfolded protein response" (UPRer and UPRmt).[1, 2] UPRs warrant an increased expression of stress related proteins such as heat shock proteins and glucose regulatory proteins, chaperones as well for protein degradation to re-establish cellular homeostasis via altered gene expression. Under continued stress, this ER-mediated mechanism fails and eventually, apoptosis ensues. Misfolded proteins, protein aggregates, and apoptotic debris jointly initiate a dialogue between stressed cellular entities and the immune system that triggers an inflammatory cascade and local inflammatory reactions. Further, the alterations in the basic tertiary structure of proteins imparts them an entirely new identity and these are no

Protein Modificomics
https://doi.org/10.1016/B978-0-12-811913-6.00004-7

longer identified as self-proteins; instead, they are recognized as neoantigens by the immune system.[3–5]

The breach of tolerance to self-proteins and the resultant development of temporal and innate immune responses is a complex phenomenon. The modifications of amino acid residues following protein translation can result in recognition of a self-protein as a neoantigen. Recognition of these modified proteins as "non-self" or "dangerous", exposure of cryptic epitope (changes in tertiary structure of protein due to modifications can cause surfacing of the otherwise embedded sequence), epitope spreading, and coupling of modified proteins to an exogenous antigen contribute to breakdown of immune tolerance. B and T lymphocytes are able to recognize new antigens.[6] The antibodies specifically targeted towards these neo-antigens/novel epitopes are reported to be generated, imparting the disease its autoimmune character.[7, 8] The modified epitopes in the proteins are probably perceived by the T cells as novel entities that help autoreactive B cells to initiate an intra-molecular dispersion, and develop an obnoxious immune response. The native or adducted proteins can also be engulfed as antigen-antibody complexes by macrophages through IgG receptors, and thereafter an inflammatory cascade ensues. This marks the beginning of autoimmunity, by precipitating cell-mediated, as well as humoral, responses against unmodified proteins as well. In the upcoming sections, we briefly review some of the advances in our understanding of pathogenesis of common lifestyle disorders brought about by proteome disclosure.

2 DEREGULATED ADIPOCYTE PROTEOME IN OBESITY

Obesity is a complex disorder, multi-factorial in origin, spreading like an epidemic pan-globally. It is the disparity in cellular energy homeostasis, characterized not only by miscommunication between the peripheral fat stores and CNS, but also by the emergence of chronic "sterile" inflammation. Incidences of obesity parallel the frequency of fatty liver disease.[9, 10] Nevertheless, obesity is well known to drive the pathogenesis of nonalcoholic fatty liver disease, a major contributor to the pathophysiology of metabolic disorders. Abdominal fat deposition increases the risk for atheromatous plaque formation. Recently, obesity has been recognized as a major co-factor for developing atherosclerosis.[11] Decreased basal metabolic rate (BMR) and, consequently, increased visceral fat, along with obesity, are the predisposing factors for Type 2 diabetes as well.[12] Furthermore, meta-analysis from a prospective study shows positive correlation between obesity and development of dementia later in life.[13] Thus, it would not be an over exaggeration at all to regard obesity as a protagonist of all lifestyle-related diseases.

Obesity is known to impair immune function; it alters the total leukocyte count and cellular immune responses. Various plausible explanations have been put forward to explain the immune responses during obesity. The increase in leptin, the pro-inflammatory adipokine, and a decrease in adiponectin—the antiinflammatory adipokine—observed in obesity seems to influence the activation of immune cells. Free fatty acids can invoke inflammatory responses via multiple routes, alterations in adipokine secretome and the activation of the TLR2 signaling cascade are the major ones to be affected. Furthermore, mitochondrial dysfunction due to substrate overload and adipocyte proliferation can trigger UPRer, and hypoxia in hypertrophied visceral adipose tissue, which upregulates the proflammatory genes and activates the immune cells.[2]

The mitochondrial activity and its biogenesis in adipocytes is dramatically dysregulated during obesity. Malfunctioning of mitochondria has been found to be directly proportional to hypertrophy of adipocytes and substrate overload observed in obese people. A proportionate increase in ROS/RNS generation is accompanied with the increasing loss of mitochondrial functionality. The increased oxidative stress in adipocytes due to dysfunctional mitochondria is a causal factor for abnormal adipokine production and contributes to ER stress.[1]

Dyslipidemia, hypercholesterolemia, and high triglyceride levels are typical features of obesity syndrome. In order to counterweigh high plasma free fatty acid levels, the liver increases the production of apo-B, which carries these triglycerols to visceral adipocytes in order to generate LDL and VLDL.[12] The LDLs and VLDLs thus generated can undergo various types of modifications including glycation, oxidation, and nitrosylation. These modified LDLs are pro-inflammatory in nature and have been documented to play a potential role in atherosclerosis. Nitrosylation of apoA-1 (an HDL) in cholesterol-laden macrophages makes it functionally redundant and nitrosylated HDL is inefficient in stimulating ABCA-1-dependent cholesterol efflux, which consequently results in foam cell formation. Furthermore, chlorination of apoA-1 via MPO-derived HOCl makes it pro-inflammatory; chlorinated apoA-1 activates NF-κB and increases the expression of VCAM on endothelial cell surfaces.[14–16]

Adipose tissue secretes numerous proteins that are collectively termed adipokines, and which constitute the adipose tissue secretome. These adipokines work in paracrine and autocrine manners and are known to regulate various metabolic processes. Adipokine Q/adiponectin is an important adipokine involved in regulation of insulin sensitivity in the liver and muscles; it also regulates the rate of FFA oxidation in these tissues.[17–19] Decreased adiponectin levels are responsible for insulin resistance; induction of hyperinsulinemia[20, 21] can promote atherogenesis[22, 23] as well as increase the risk to CAD.[21] The transcription of adhesion factors and foam cell generation via TNF alpha is inhibited by adiponectin via regulation of TNFα. Leptin is another adipokine secreted from adipocytes. A decrease in its level is also involved in regulation of insulin resistance, hyperinsulinemia, and predisposition to fatty liver syndrome.[24]

Visceral and central adipose tissues, rather than total body fat, play a prominent role in stimulating inflammatory responses. Visceral adipose tissues provide a steady flow of fatty acids and encourage hepatic fat deposition. They are the immediate suppliers of triglycerides to the liver and thereby increase the transcription of critical regulatory proteins responsible for fatty acid synthesis and oxidation, stress responses in peroxisomes, and defects in Acyl CoA oxidation, contributing to the progress of steatohepatitis by shutting down β-oxidation in mitochondria. Furthermore, a direct release of adipokines in portal circulation supports the generation of proinflammatory cytokines in the liver.[25]

High intake of poly-unsaturated fatty acids (PUFAs) has also been reported to alter gut microbiota. It has been reported that their breakdown products change the chemical blueprint of cecal bacteria[26] in a pattern, dependent on cholic acid content.[27] PUFAs can damage the intestinal barrier by downregulating the intestinal tight junction proteins, and changes intestinal permeability, which increases the systemic endotoxin levels. Moreover, prolonged PUFA treatment increases TLR expression on the hepatic macrophages.[28] The lipopolysaccharides in systemic circulation through lipid A portion bind to lipopolysaccharides binding protein (LBP),[29] which further complexes with CD14 on the macrophage membrane and results in its activation.[30] Gut-derived endotoxins also bind to Myeloid Differentiation factor-2 (MD-2) receptor present on macrophages.[31] Tyrosine kinase pathways are activated by both CD-14 and MD-2 in a TLR4-dependent manner.[32, 33] Consequentially activation of MyD88 and/or TRIF-dependent NF-κB, and type-I IFNs pathways result in generation of proinflammatory cytokines.[34, 35] In a nutshell, focal immune responses in the adipose deposits and the innate immune responses during obesity form a vicious loop, wherein both consumate to the pathology of a related metabolic complication.

The above literature corroborates the importance of characterizing the adipocyte proteome and secretome, not only to unravel the mechanisms that underlie pathogenesis of obesity, but also for exposing the missing links between obesity and other complex disorders (Fig. 1). In one such study, subcutaneous and visceral fat depots from 20 diabetic as well as obese and 22 nondiabetic but obese individuals were subjected to proteome analysis. The results from the study revealed that, although the number of proteins being differentially regulated was approximately the same (~600), there were 19 and 41 proteins, respectively, in the subcutaneous and visceral adipocyte proteome of diabetic subjects that were

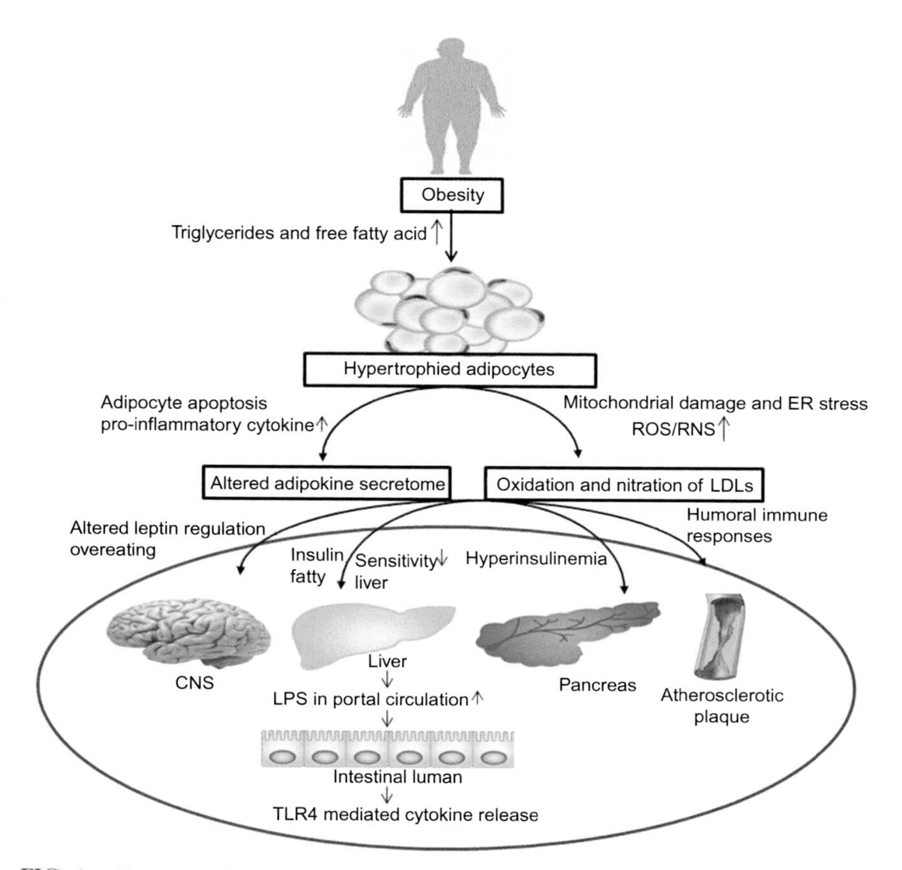

FIG. 1 Obesity mediated stress in mitochondrial machinery results in increased adipocyte apoptosis and recruitment of macrophages, thereby promoting pro-inflammatory cytokine release. Additionally, increased free fatty acids and altered adipocyte secretome combine to affect multiple organs including brain, liver, pancreas, and circulatory system, as well as alter the chemical fingerprint of gut microbes surmounting both the adaptive and humoral immune responses. The reactive species generated due to lipid peroxidation are capable of modifying proteins and adipokines, which in turn provoke humoral and adaptive immune responses and promote the pathogenesis of type 2 diabetes, fatty liver syndrome, and atherosclerosis.

differentially expressed. Furthermore, the proteome of both the tissues in diabetic subjects shared five proteins. Five proteins identified as myosin, 78 kDa glucose-regulated protein, protein cordon-blue, zinc finger protein 611, and cytochrome *c* oxidase subunit 6B1 were differentially expressed in both the tissues. Further investigation of the reported proteins can potentially enhance our existing understanding of adipose tissue-driven pathogenesis of diabetes and obesity, and can help in identifying novel clinical targets in mitigating type 2 diabetes and obesity.[36]

3 PROTEOSTASIS IN DIABETES

The immune responses in type 1 diabetes are easily dodged by modified proteins, thereby masking a mild difference between the self and nonself. Pancreatic β-cells being vulnerable to oxidative stress are potential loci for protein alterations. Chemically transformed proteins are perceived as novel peptides, capable of eliciting the production of cytokines/chemokines by the immune system and by the β-cells.

Any form of stress in the pancreatic β-cells can imbalance its protein homeostasis. New epitopes for autoantibody binding or the ligands for receptors on immune cells can be created by different types of modifications such as phosphorylation, methylation, or glycosylation; alternatively, neoepitopes, which influence antigen processing, can also be generated by modifications, which may or may not be aided by specific enzymes, such as deamidation and citrullination.

In certain cases, posttranslation modifications drive the process of neoantigen formation. In such cases, the abnormal protein serves as the neoepitope that is recognized by T-cells. Alternatively, PTM itself enhances or limits the generation of peptide fragments due to extracellular proteolysis of abnormal proteins by interfering with the cleavage site.

Alternative splicing is the major mechanism for modification of β-cell antigens. About 90% of human genes undergo alternative splicing, which means that several isoforms of a protein can be upshot from the same gene. Prominent tissue specific factors and local influencers during diseased state control this process and create a diverse proteome, which quite frequently leads to potential antigenic peptide generation. Eight different spliced variants have been found for islet-specific glucose-6-phosphatase catalytic subunit-related protein (IGRP), an autoantigen described for type 1 diabetes, out of which, six are unique β-cells antigens. The frame-shifts and neosequences of these isoforms of IGRP generate highly immunogenic T-cell ligands.[6]

Since the list of autoantigens for type 1 diabetes is expanding at a fast pace, it is believed that thousands of spliced variants may exist that are β-cell specific. RNA sequencing is being used, to spot transcripts that are the products of alternative splicing. Islet-specific variants can be spotted by comparing the RNA sequence results from cytokine stimulated β-cells with similar data obtained from other human tissues. The clinical impact of these remain to be elucidated, but it can be hypothesized that the PTMs continually drive the production of alternatively spliced variants locally in the β-cells, which keeps charging the cytokine production unless stopped externally.

Peripheral tissue antigens (PTAs) play a crucial role in sustaining tolerance to self-proteins. Deaf1 is a transcriptional regulator that controls their expression in the pancreatic lymph nodes. The role of a spliced variant of Deaf1 has been put forward as steering autoimmune responses.[6] This spliced variant of Deaf1 results in downregulation of PTAs. It leads to reduced expression of adenosine A1 receptors on stromal cells present in the lymph nodes of the pancreas. As a result, the T cells, which are reactive for the A-1 receptor, are not destroyed because they perceive these A1 receptors as novel. These escaped T cells can now mediate the invasion of glucagon cells that express A1 receptors, which at the very beginning of insulitis are responsible for apoptosis of insulin secreting β-cells.

Posttranslational modifications in insulin A chain also evokes T-cell responses in type 1 diabetics. Modification of cysteine residues and disulfide bond formation between A6 and A7 cystine residues, in the insulin A chain makes it recognizable by T-cells and thereby initiating immune responses.[3–5]

Tissue transglutaminase (tTG), involved in deamidation of gluten peptides, regulates T cell stimulation by improving their binding to differentially amidated glutens, and facilitates their recognition by the HLAs. The extent of deamidation on glutens directly determines the T cell responses in patients with celiac disease. Since tTGs are constitutively expressed in many tissues, including the islet cells, they may play an important role in imparting autoimmune character to type 1 diabetes. Thus, by screening the β-cell proteome for its amidation status, it is possible to narrow down to antigens that facilitate disease associated HLA DQ responses. No circulating antibodies against tTG have been recorded in type 1 diabetic patients, suggesting distinct pathological processes.[37–41] However, alteration in amidation status of WE14—a peptide from chromogranin A—by tissue transglutaminase converts it to a highly stimulatory peptide. This deaminated WE14 is more antigenic for autoreactive T cells and could serve as a potential antigen in type 1 diabetics.[42]

Antibodies against, prolyl 4 hydroxylase, have been detected in the sera of patients suffering from type 1 diabetes. This is an ER protein involved in protein folding. Detection of antibodies against this enzyme suggests that a modification leads to conformational changes making it potentially immunogenic. Malfunctioning of this enzyme lowers the release of insulin from the β-cells in response to glucose. This enzyme is currently being screened for its potential to be a biomarker for early detection of diabetes.

Insulin secreting granules of the β-cell have been under constant scrutiny as a potential source of autoantigen. These were amongst the first targets of β-cell-specific T lymphocytes to be identified by the researchers. Their upregulation is a primary posttranscriptional event in, type 1 diabetes. Nuclear retention of polypyrimidine-tract binding protein 1 (PTBP1), a protein associated with posttranscriptional upregulation of insulin secretary granule protein due to defective phosphorylation, may contribute to impaired insulin biosynthesis.[7, 8]

Integration of genomics, proteomics, bioinformatics, and clinical data is definitely required to improve our understanding of the protein modifications that are of critical importance in type 1 diabetes. A preferred approach to reveal differential modifications that play a role in development of antigenicity would be to screen the existing autoantibodies from type 1 diabetics for their targets. The identity of modifications at the epitopes recognized by disease specific antibodies or T cells will be crucial in deciphering the drivers of stress for the disease.

Since the intrinsic propensity of β-cells to succumb to stress in an organ as well as across a population is different, understanding this difference certainly could yield the in-depth mechanism of disease propagation. Despite the fact that the heterogeneity of type 1 diabetes is a rate-limiting factor, bio-banked samples and longitudinal disease tracking studies can help in pin-pointing biomarkers and/or in devising therapeutic strategies.

4 PROTEIN MODIFICATION IN CARDIOVASCULAR DISEASES

Cardiovascular disorders (CVDs) are a leading cause of death globally in both women and men. In the United States of America alone, it accounts for 37% of all deaths. Proteome homeostasis in cardiac myocyte is necessary for proper functioning of the myocardium. Proteome modification due to external stressors affects their folding and trafficking to subcellular destinations, thereby influencing physiological aspects of their function, half-life, and communication with other biomolecules including other proteins, RNA, and DNA. Obesity, diabetes, hypercholesterolemia, chronic smoking, hypertension, and age are the major risk factors associated with CVD, which act as prominent stressors promoting protein modifications. Lipid lowering, smoking cessation, weight loss, and improved glucose control are the hallmarks for containing progression of CVDs. All the aforementioned risk factors induce oxidative stress that favors AGEs, ROS, and RNS generation capable of efficiently forming stable protein adducts.

The sirtuins (Class III histone deacetylase) have been closely implicated in pathophysiology of cardiac disease.[43–46] Protein regulation and signaling in the myocardium goes haywire due to altered SIRTs activity. Lysine acetylation, an important reversible regulatory PTM, mediated by histone acetyltransferases (HATs) and reversed by histone deacetylases (HDACs), strongly influences gene expression, metabolic processes, and chromatin remodeling,[47–49] and is brought about by SIRTs.[50–52] SIRT1 and SIRT7 play a cardioprotective role by preventing oxidative, stress-mediated, and age induced cellular damage.[53] Inhibition of SIRT1 activity ameliorates their cardioprotective potential.[44, 45] SIRT2, on the other hand, promotes cardiac injury. It binds to one of the components of necrosis-promoting complex—the receptor interacting protein-3 (RIP3)—and, therefore, inhibition of SIRT2 is protective against ischemia reperfusion injury.[54] SIRTs definitely play a crucial role in progression of cardiac diseases, these are the molecules which potentially regulate de novo fatty acid synthesis; however, its mechanism of action, as well as its potential targets, which precipitate particular clinical phenotypes, remains ambiguous.

In diabetics, prolonged hyperglycemia is known to alter vascular tissues at cellular level and potentially to accelerate the pathogenesis of CVD. Nonenzymatic protein and lipid glycosylation, induction of oxidative stress, and activation of protein kinase C (PKC) are the three fundamental mechanisms in the arterial walls that underlie pathological alterations seen in diabetics and set the stage for precipitation of CVDs.[55–58] Glucose is capable of forming Amadori-type early glycosylation products by interacting with the reactive amino groups of the serum and vascular proteins within a few days; hemoglobinA1c is a classic example of an Amadori-type early glycosylation product, which is a widely used clinical biomarker.[59] Some of the long lived early glycosylation products—like collagen—can undergo further rearrangements and yield advanced glycosylation end products (AGEs) that are quite stable and irreversible protein conjugates. AGEs consist of several chemical entities but the most prominent amongst them are N (epsilon)-(carboxymethyl) lysine protein adducts that are majorly recognized as immunological epitopes in AGE modified proteins.[60] Apolipoprotein B and phospholipids are the most researched AGEs adducted proteins that contribute to pathogenesis of CVDs.[61, 62] Glycosylation of apolipoprotein B occurs at the lysine residue located in its receptor binding domain; this conjugation generates a carboxymethyl lysine protein adduct.[63] The modified adduct is no longer recognized by the LDL-receptor and there occurs a significant decrease in clearance of modified ApoB. In contrast, these glycated LDLs are engulfed by the scavenger receptors present on macrophages (aortic intimal cells and monocyte-derived macrophages) and stimulate foam cell formation.[64–66] Another target of glycosylation relevant to CVDs pathophysiology is Regulatory membrane protein CD59—a regulatory protein that restricts the expression of membrane attack complex of complement (MAC). As a result of glycation, Regulatory membrane protein CD59 protein loses its activity, leading to increased production of MAC in blood vessels. Unchecked MAC induces tissue remodeling by increasing the release of fibroblast growth factors, platelet derived growth factors, and cytokines.[67, 68] Amongst other targets of glycosylation are collagen IV, laminin, vitronectin, and transmembrane integrin receptors[59] (Fig. 2).

Specific cell surface receptors are responsible for mediating cellular interactions of AGE.[69] The receptors for AGE (RAGE) belong to the superfamily of immunoglobulin receptors and are present on macrophages,

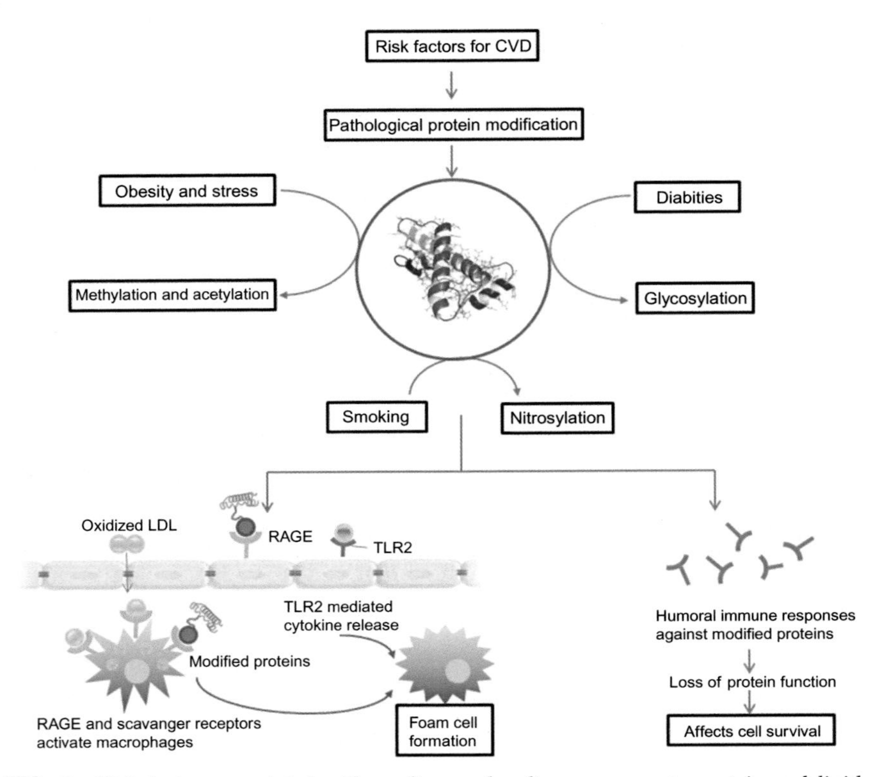

FIG. 2 Risk factors associated with cardiovascular diseases promote protein and lipid modifications. These modified biomolecules stimulate cellular/humoral immune responses and promote foam cell formation, thus affecting endothelial cell survival and consequent atherosclerosis.

endothelial cells, and smooth muscle cells.[69–71] RAGE are involved in regulating AGE turnover rate and are upregulated in the event of increased AGE levels, as well as with aging, as they play a role in degrading senescent proteins too.[72] Interaction of AGEs with RAGE expressed on endothelial cells induces oxidative stress. This, consequently, results in increased expression of transcription factor NF-κB[73, 74] and VCAM-1[75] as well as increases the permeability of endothelial cell monolayers.[76, 77] Ligand binding to RAGE bearing macrophages results in chemotaxis, and production of cytokines, including interleukin-1, tumor necrosis factor-α, platelet-derived growth factor, and insulin growth factor-I.[78–80]

The potential correlation between diabetes and the role of AGE-RAGE interaction in atherogenic plaque formation was confirmed by Park et al.[81] In their study, they used apo$E^{-/-}$ mice that were vulnerable to atherosclerosis, and induced diabetes in them using streptozotocin. An augmentation in the severity of atherosclerosis was observed in hyperglycemic mice in comparison to euglycemic apoE controls. The lesions formed were more complex in terms of fibrous caps, and monocyte infiltration, that extended in the major arteries including the aorta. Increased expression of RAGE and the AGEs were detected in the wall of the vessel and at the lesion site. Furthermore, they used a truncated soluble extracellular domain of RAGE to block AGE-RAGE interaction; this resulted in a marked decrease in the rate and complexity of lesion formation in diabetic mice. Furthermore, the observations were not linearly equivalent to the serum glucose and lipid levels, implicating the direct role of AGE-RAGE interaction in promoting disease physiology.[81] Although it is clear that RAGE-AGEs interactions and glycosylation of proteins are critical in vascular remodeling during atherosclerosis, its precise mechanism, targets, and the relationship between the course of time during which protein modification occurs in diabetics and the phenotypic manifestation of CVDs, are still a point of investigation.

Considerable literature also focuses on the role of nitrosylated plasma and tissue proteins in mediating CVD prognosis. ApolipoproteinA-1, apolipoprotein B, fibrinogen β chain, Igγ-1 chain C region, Igκ-1 chain C region, Igμ chain C region, Igλ chain C region, Ig heavy chain V-VIII, Zinc finger protein, BTB domain-containing protein 1, and protein EFR3 homologue B have been confirmed for the presence and functional consequences of modification by 3-nitrotyrosine.[82, 83] Nitration of apoB-100 and fibrinogen precipitates new proatherogenic and prothrombotic functions, as well as triggering an adaptive immune response via the nitrated epitope. Increased levels of Anti-3-nitrotyrosine antibodies are consistently observed in coronary artery disease patients and atherosclerosis prone mice.[84] The pathological repercussions of this immune response remain unclear at the present time. A strong correlation between plasma 3-nitrotyrosine levels and higher risk for CVDs supports the usefulness of tracking this PTM as a prognostic marker.

5 MODIFIED PROTEOME IN RHEUMATOID ARTHRITIS

Rheumatoid arthritis (RA) is a debilitating disorder, exhibiting sequential involvement of synovial joints with several extra-articular manifestations such as rheumatoid nodes, accelerated atherosclerosis, and pulmonary fibrosis. It is an autoimmune disorder; a condition in predisposed individuals where self-proteins are targeted as antigens due to "breached self-tolerance." Oxidation, citrullination, and carbomylation are the most important PTMs, which have been closely associated with the pathogenesis of RA in poor prognosis as well as progression of rheumatoid arthritis by rendering self-proteins immunogenic.

Several reports by researchers provide strong evidence to support the theory that "Posttranslational modifications (either enzymatic or nonenzymatic) play a fundamental role in its pathophysiology." Most critical amongst them is the finding by Nissam et al.[85] which elaborates on the alterations in immunogenicity of Collagen II (CII) brought about after its citrullination and how this modified CII becomes potentially arthritogenic. CII is the most abundant cartilage collagen protein and a well-acknowledged neoantigen in RA[86–88] and, therefore, detection of antibodies against it is of utmost importance.[87] These antibodies against citrullinated proteins (ACAP), rheumatoid factor (RF), and shared epitopes that have long been established are three independent additive factors that determine prognosis of RA.

Citrullination is the process of deamination of arginine by the enzyme peptidylarginine deiminase (PAD) to citrulline that results in a molecular mass alteration (less than a Da) and loss of a positive charge, which may alter the protein's ability to interact with other proteins. PAD is expressed constitutively in various tissues including epidermis, sweat glands, hair follicles, testes, and ovaries, where it plays several important physiological roles and regulates the critical processes of differentiation, development, and apoptosis. Inducible forms of PAD, which are expressed during high metabolic stress carry out citrullination, which has an important role to play in development of cancer,[89, 90] multiple sclerosis,[91, 92] Alzheimer's disease,[93, 94] polymayocistis, chronic tonsillitis, and inflammatory bowl disease. In the synovium, however, only PAD 2 and 4 are expressed during many inflammatory conditions that are not associated with RA, and therefore, exactly how the rate of citrullination of both intra and extrareticular proteins goes haywire in RA is still a central point of research. PADs are also induced during stress conditions in response to high intracellular calcium concentration, viz. autophagy, NETosis, and apoptosis—the processes closely associated with autoimmune disorders—suggesting PAD mediated citrullination being an inflammation mediated response in AIDs.

Anticitrulline antibodies directed against citrullinated proteins (ACPA), are quite specific for rheumatoid arthritis (RA) and are detectable in about 80% of patients. Collagen, fibrinogen, vimentin, enolase, keratin, filaggrin, and perinuclear factor are some of the citrullinated proteins that show positive reactivity toward ACAP in RA patients.[95–99] ACAPs represent a group of antibodies and only its IgG-isotype is associated with RA. The extra-articular manifestations that are primarily responsible for the severity and poor prognosis of RA also positively correlate with the ACAP titers in the sera,[100] implying that, although the disease follows a similar progressive pattern, ACPA+RA has a differential etiology.[101] The importance of ACAP as a prognostic biomarker is highlighted by independent clinical studies performed by Van Gaalen et al.[102], De Rycke et al.[103], and Berglin et al.[104] The results of these studies revealed that radiological damage and its progression in the following 2 years was more prominent in patients who were ACPA+, whereas such a correlation was missing in patients who were only RF+. Furthermore, the studies showed that ACPA titers in the sera reduced substantially in those patients who responded positively to immunotherapy. Owing to its role in pathogenesis and progression of RA, "ACPA titers," has been included under the RA classification list by the European Union League Against Rheumatism in 2010 for early diagnosis and improving prognosis of RA. Silica exposure and chronic smoking have been linked to RA[105–107] and other autoimmune diseases.[108] Smoking is a risk factor for the development of ACPA-positive RA, especially in people who present HLA-DRB*0104 allele.[109] Chronic smoking increases citrullination of proteins, activates PAD2 and PAD4, and stimulates immune responses.[110] Lung biopsies from chronic smokers revealed an increase in protein citrullination in RA patients, whereas those smokers who had developed lung cancer did not show any protein citrullination.[101]

Unlike citrullination, carbamylation is a nonenzymatic PTM that occurs ubiquitously in the presence of a highly reactive metabolite—cynate. Carbamylation is the process of addition of cynate group on the alpha amino group present in the N-terminal of amino acids like lysine (forming homocitrulline), and arginine. In humans, cynate is generated from urea and exits in delicate equilibrium with it. Under normal physiological conditions, the cynate concentration (50 nmol/L) in blood is too low to source any significant protein modification. External and environmental factors including chronic exposure to tobacco, some biomass smokes, and herbicides containing potassium and sodium cyanate are the major factors that outsource pathological protein carbamylation.

Carbamylation of a protein changes its charge and its conformation, resulting in an entity that could be potentially antigenic, as well as biologically redundant. The number of lysines and arginines in a protein, as well as its turnover rate, are the major determinants that contribute toward the susceptibility of a protein to carbamylation. Carbamylation is a time-dependent process and is more likely to affect proteins that are long-lived or have low turnover rates, as they are more likely to acquire homocitrulline residues over time.

Carbamylation of amino acids changes its charge as well as induces conformational changes in protein, resulting in an entity that could be potentially antigenic and either partially or completely nonfunctional. Carbamylation results in a complete loss of activity of matrix and tissue metalloproteinase-2, and in partial loss of insulin, glucagon, adrenocorticotropic hormone, and erythropoietin activity.

Presence of anticarbamylated protein antibody (anti-CarPA) was first evidenced in 2010 by Mydel and his co-workers[111] in the sera of humans and of experimental RA animals. They revealed that carbamylated-Lys residues trigger immune responses, chemotaxis of CD4+ T cells, IL-10, and production of INF-γ and IL-17 cytokines. Antibodies have been detected against carbamylated fibrinogen, vimentin, albumin, hemoglobin, and enolase in RA patients. Anti-CarPA IgG and IgA are detectable in about 45% of RA patients and Anti-CarPA IgA is documented in 30% of ACPA- RA patients. Carbamylated proteins, with their ability to modulate immune responses, could play an important role in pathology of RA. Carbamylated albumin exerts an inhibitory effect on the polymorphonuclear leukocyte respiratory burst. Carbamylated-LDL plays a central role in inflammatory responses during atherosclerosis, and carbamylated collagen has been documented to activate the production of matrix metalloproteinase-9, which potentially increases the turnover of extracellular matrix. Anti-CarPA generation is independent of ACPA and shows no cross reactivity toward ACAP. This is evidenced by the presence of anti-CarPA and inconsistency in ACPA detectable in the sera of patients suffering from arthralgia[112] and positively correlates with the progression of artheralgia to RA. Despite these encouraging findings, further research is needed for clearly establishing the exact nature of the epitope identified by Anti-CarPA antibodies, and of its individual contribution for effective clinical management of RA. Furthermore, it will be interesting to expose the list of its target proteins, particularly amongst ACPA-patients to determine whether they play any pathological role in progression of RA and to evaluate the clinical relevance of Anti-CarPA as a promising clinical and prognostic marker.

An inverse relationship between antioxidant intake and decrease in inflammation levels in RA patients suggests the role of ROS in pathogenesis of RA. Levels of ROS are considerably higher in autoimmune disorders.[113] Exposure of biomolecules to free radicals invites their biotransformation.[114, 115] Superoxide radical, hydrogen peroxide (H_2O_2), hydroxyl radical, hypochlorous acid (HOCl), nitric oxide, and peroxynitrite ($ONOO^-$), are the primary oxidants found in the joints of 90% of RA patients that promote lipid peroxidation, and formation of foam cells in synovial fluid,[113, 116–118] and consequently tissue damage observed in joints.[118, 119] AGEs also cause damage to collagen; permanent cross-links are formed between the fibers.[120, 121] In RA, immunoglobulins themselves have been reported to undergo glycation that generates AGE-IgG and autoantibodies to AGE modified-IgG. The levels of AGE-IgG correspond to the severity of inflammation; however, it is not associated with RA.[122–124] Tittering autoantibody showing reactivity for AGE-IgG could be a potential diagnostic tool to track RA progression.[122, 125] Chlorinated or nitrosylated IgG can stimulate T-cell responses.[126] Matrix proteins including hyaluronan, collagens, as well as the proteoglycans, are amongst the identified oxidized proteins. There is no specific recognized pattern that can presuppose the targets of oxidation. Both steric and stochastic factors determine the extent as well as the propensity; however, enrichment of YXXK motif in the vicinity of chlorination has been observed.[126] Tyrosine residues are particularly vulnerable to chlorination and nitrosylation. Their modification leads to formation of 3-chlorotyrosine, and 3-nitrotyrosine (3-NT) within the tertiary protein structure.[127–129] Antibodies against 3-NT are documented in RA patients and their titer corresponds to its severity.[127] Tyrosine residues in collagen reportedly form adducts with peroxynitrites that disrupts collagen structure. Treatment with Anti-TNF reduces oxidative stress with proportionate improvement in disease severity,[130–133] however changes in oxidative stress-related protein modifications and antibody against these proteins still need to be tracked after anti-TNF therapy, before confirming their value as prognostic markers.

6 CONCLUSION AND FUTURE ASPECTS

Modification of self-proteins, due to inflammation in general, is the leading cause of the formation of neoantigens, stimulating the release of antibodies. The breach in self-tolerance occurs due to release of nonspecific antibodies, released against modified self-protein, capable of binding with both the native proteins and the neo-antigens. Protein modifications accelerate the pathological progression of disease by altering several physiological pathways, in a receptor-dependent or independent manner. These processes promote disease progression by continually providing novel immune targets, (Fig. 3). What leads to an initial breach in

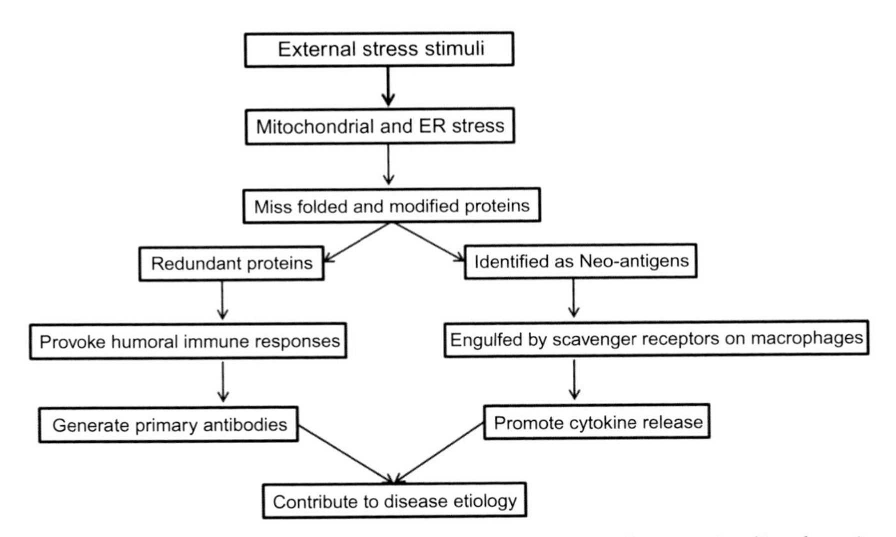

FIG. 3 Mitochondrial and ER stress responses under external stress stimuli go haywire and an overload of misfolded and modified proteins not only drives the cell to apoptosis but also generates adaptive and humoral immune responses which add up to the disease progression.

recognition between self and nonself is still a matter of research. Current research is focused on identifying inflammatory epitope and on developing novel proteins, and/or protein-antibody conjugates that could target the specific receptors on the dendritic cells to prevent cytokine release and modulate immune responses. Furthermore, studies are being conducted to explore the potential of modified proteins in influencing the transcriptional and translational machinery that ultimately leads to the disease pathology. Taken together, an enhanced understanding of these modifications could help clinicians and researchers in identifying potential biomarkers for improving disease prognosis, and could possibly aid in designing targeted therapies for disease mitigation and cure.

References

1. Bournat JC, Brown CW. Mitochondrial dysfunction in obesity. *Curr Opin Endocrinol Diabetes Obes* 2010;**17**:446–52.
2. Monteiro R, Azevedo I. Chronic inflammation in obesity and the metabolic syndrome. *Mediat Inflamm* 2010;**2010**:289645.
3. Mannering SI, Harrison LC, Williamson NA, et al. The insulin A-chain epitope recognized by human T cells is posttranslationally modified. *J Exp Med* 2005;**202**:1191–7.
4. Skowera A, Ellis RJ, Varela-Calviño R, et al. CTLs are targeted to kill beta cells in patients with type 1 diabetes through recognition of a glucose regulated preproinsulin epitope. *J Clin Investig* 2008;**118**:3390–402.
5. Yip L, Su L, Sheng D, et al. Deaf1 isoforms control the expression of genes encoding peripheral tissue antigens in the pancreatic lymph nodes during type 1 diabetes. *Nat Immunol* 2009;**10**:1026–33.
6. Dogra RS, Vaidyanathan P, Prabakar KR, et al. Alternative splicing of G6PC2, the gene coding for the islet-specific glucose-6-phosphatase catalytic subunit-related protein (IGRP), results in differential expression in human thymus and spleen compared with pancreas. *Diabetologia* 2006;**49**:953–7.
7. Knoch KP, Meisterfeld R, Kersting S, et al. cAMP-dependent phosphorylation of PTB1 promotes the expression of insulin secretory granule proteins in beta cells. *Cell Metab* 2006;**3**:123–34.
8. Roep BO, Arden SD, de Vries RR, et al. T-cell clones from a type-1 diabetes patient respond to insulin secretory granule proteins. *Nature* 1990;**345**:632–4.
9. Ishii H. Alcoholic liver disease: the etiologic mechanism and clinical features. *J Jpn Soc Intern Med* 2003;**92**:1623–37.
10. Marchesini G, Brizi M, Bianchi G, et al. Nonalcoholic fatty liver disease a feature of the metabolic syndrome. *Diabetes* 2001;**50**:1844–50.
11. Nelson RH. Hyperlipidemia as a risk factor for cardio vascular disease. *Prim Care* 2013;**40**:195–211.
12. Gómez-Serrano M, Camafeita E, García-Santos E, et al. Proteome-wide alterations on adipose tissue from obese patients as age-, diabetes- and gender specific hallmarks. *Sci Rep* 2016;**6**:25756.
13. Luchsinger JA, Gustafson DR. Adiposity, type 2 diabetes, and alzheimer's disease. *J Alzheimers Dis* 2009;**16**:693–704.
14. Nicholls SJ, Hazen SL. Myeloperoxidase, modified lipoproteins, and atherogenesis. *J Lipid Res* 2009;**50**:S346–51.

15. Shao B, Tang C, Sinha A, et al. Humans with atherosclerosis have impaired ABCA1 cholesterol efflux and enhanced high density lipoprotein oxidation by myeloperoxidase. *Circ Res* 2014;**114**:1733–42.
16. Shishehbor MH, Aviles RJ, Brennan ML, et al. Association of nitrotyrosine levels with cardiovascular disease and modulation by statin therapy. *J Am Med Assoc* 2003;**289**:1675–80.
17. Fruebis J, Tsao TS, Javorschi S, et al. Proteolytic cleavage product of 30-kDa adipocyte complement-related protein increases fatty acid oxidation in muscle and causes weight loss in mice. *Proc Natl Acad Sci U S A* 2001;**98**:2005–10.
18. Scherer PE, Williams S, Fogliano M, et al. A novel serum protein similar to C1q, produced exclusively in adipocytes. *J Biol Chem* 1995;**270**:26746–9.
19. Yamauchi T, Kamon J, Waki H, et al. The fat-derived hormone adiponectin reverses insulin resistance associated with both lipoatrophy and obesity. *Nat Med* 2001;**7**:941–6.
20. Arita Y, Kihara S, Ouchi N, et al. Paradoxical decrease of an adipose-specific protein, adiponectin, in obesity. *Biochem Biophys Res Commun* 1999;**257**:79–83.
21. Hotta K, Funahashi T, Arita Y, et al. Plasma concentrations of a novel, adipose-specific protein, adiponectin, in type 2 diabetic patients. *Arterioscler Thromb Vasc Biol* 2000;**20**: 1595–9.
22. Ouchi N, Kihara S, Arita Y, et al. Novel modulator for endothelial adhesion molecules: adipocyte-derived plasma protein adiponectin. *Circulation* 1999;**100**:2473–6.
23. Ouchi N, Kihara S, Arita Y, et al. Adipocyte-derived plasma protein, adiponectin, suppresses lipid accumulation and class a scavenger receptor expression in human monocyte-derived macrophages. *Circulation* 2001;**103**:1057–63.
24. Shimomura I, Hammer RE, Ikemoto S, et al. Leptin reverses insulin resistance and diabetes mellitus in mice with congenital lipodystrophy. *Nature* 1999;**401**:73–6.
25. Parola M, Marra F. Adepokines and redox signaling: impact on fatty liver disease. *Antioxid Redox Signal* 2011;**15**:461–83.
26. Daniel H, Moghaddas Gholami A, Berry D, et al. High-fat diet alters gut microbiota physiology in mice. *ISME J* 2014;**8**:295–308.
27. Yokota A, Fukiya S, Islam KB, et al. Is bile acid a determinant of the gut microbiota on a high-fat diet? *Gut Microbes* 2012;**3**:455–9.
28. Kirpich IA, Feng W, Wang Y, et al. The type of dietary fat modulates intestinal tight junction integrity, gut permeability, and hepatic toll-like receptor expression in a mouse model of alcoholic liver disease. *Alcohol Clin Exp Res* 2012;**36**:835–46.
29. Tobias PS, Soldau K, Ulevitch RJ. Identification of a lipid a binding site in the acute phase reactant lipopolysaccharide binding protein. *J Biol Chem* 1989;**264**:10867–71.
30. Heumann D, Gallay P, Barras C, et al. Control of lipopolysaccharide (LPS) binding and LPS-induced tumor necrosis factor secretion in human peripheral blood monocytes. *J Immunol* 1992;**148**:3505–12.
31. Shimazu R, Akashi S, Ogata H, et al. MD-2, a molecule that confers lipopolysaccharide responsiveness on toll-like receptor 4. *J Exp Med* 1999;**189**:1777–82.
32. Hoshino K, Takeuchi O, Kawai T, et al. Cutting edge: toll-like receptor 4 (TLR4)-deficient mice are hyporesponsive to lipopolysaccharide: evidence for TLR4 as the Lps gene product. *J Immunol* 1999;**162**:3749–52.
33. Poltorak A, He X, Smirnova I, et al. Defective LPS signaling in C3H/HeJ and C57BL/10ScCr mice: mutations in Tlr4 gene. *Science* 1998;**282**:2085–8.
34. Kawai T, Takeuchi O, Fujita T, et al. Lipopolysaccharide stimulates the MyD88-independent pathway and results in activation of IFN-regulatory factor 3 and the expression of a subset of lipopolysaccharide-inducible genes. *J Immunol* 2001;**167**:5887–94.
35. Takeda K, Akira S. Toll-like receptors in innate immunity. *Int Immunol* 2005;**17**:1–14.
36. Fang L, Kojima K, Zhou L, et al. Analysis of the human proteome in subcutaneous and visceral fat depots in diabetic and non-diabetic patients with morbid obesity. *J Proteomics Bioinform* 2015;**8**:133–41.

37. Dørum S, Arntzen MO, Qiao SW, et al. The preferred substrates for transglutaminase 2 in a complex wheat gluten digest are peptide fragments harboring celiac disease T-cell epitopes. *PLoS ONE* 2010;**5**.
38. Hovhannisyan Z, Weiss A, Martin A, et al. The role of HLA-DQ8 beta57 polymorphism in the anti-gluten T-cell response in coeliac disease. *Nature* 2008;**456**:534–8.
39. Mannering SI, Pang SH, Williamson NA, et al. The A-chain of insulin is a hot-spot for CD4+ T cell epitopes in human type 1 diabetes. *Clin Exp Immunol* 2009;**156**:226–31.
40. Tollefsen S, Arentz-Hansen H, Fleckenstein B, et al. HLA-DQ2 and -DQ8 signatures of gluten T cell epitopes in celiac disease. *J Clin Investig* 2006;**116**:2226–36.
41. Yoshida K, Corper AL, Herro R, et al. The diabetogenic mouse MHC class II molecule I-Ag7 is endowed with a switch that modulates TCR affinity. *J Clin Investig* 2010;**120**:1578–90.
42. Stadinski BD, Delong T, Reisdorph N, et al. Chromogranin A is an autoantigen in type 1 diabetes. *Nat Immunol* 2010;**11**:225–31.
43. Kawashima T, Inuzuka Y, Okuda J, et al. Constitutive SIRT1 over expression impairs mitochondria and reduces cardiac function in mice. *J Mol Cell Cardiol* 2011;**51**:1026–36.
44. Nadtochiy SM, Redman E, Rahman I, et al. Lysine deacetylation in ischaemic preconditioning: the role of SIRT1. *Cardiovasc Res* 2011;**89**:643–9.
45. Nadtochiy SM, Yao H, Mc Burney MW, et al. SIRT1-mediated acute cardio protection. *Am J Phys Heart Circ Phys* 2011;**301**:H1506–12.
46. Tanno M, Kuno A, Horio Y, et al. Emerging beneficial roles of sirtuins in heart failure. *Basic Res Cardiol* 2012;**107**:273.
47. Guan KL, Xiong Y. Regulation of intermediary metabolism by protein acetylation. *Trends Biochem Sci* 2011;**36**:108–16.
48. Norris KL, Lee JY, Yao TP. Acetylation goes global: the emergence of acetylation biology. *Sci Signal* 2009;**2**:76.
49. You L, Nie J, Sun WJ, et al. Lysine acetylation: enzymes, bromodomains and links to different diseases. *Essays Biochem* 2012;**52**:1–12.
50. Morris BJ. Seven sirtuins for seven deadly diseases of aging. *Free Radic Biol Med* 2013;**56**:133–71.
51. Newman JC, He W, Verdin E. Mitochondrial protein acylation and intermediary metabolism: regulation by sirtuins and implications for metabolic disease. *J Biol Chem* 2012;**287**:42436–43.
52. Sack MN. The role of SIRT3 in mitochondrial homeostasis and cardiac adaptation to hypertrophy and aging. *J Mol Cell Cardiol* 2012;**52**:520–5.
53. Chong ZZ, Wang S, Shang YC, et al. Targeting cardiovascular disease with novel SIRT1 pathways. *Futur Cardiol* 2012;**8**:89–100.
54. Narayan N, Lee IH, Borenstein R, et al. The NAD-dependent deacetylase SIRT2 is required for programmed necrosis. *Nature* 2012;**492**:199–204.
55. Grundy SM, Benjamin IJ, Burke GL, et al. Diabetes and cardiovascular disease: a statement for healthcare professionals from the American heart association. *Circulation* 1999;**100**:1134–46.
56. Laakso M. Hyperglycemia and cardiovascular disease in type 2 diabetes. *Diabetes* 1999;**48**:937–42.
57. Nathen DM, Genuth S, Lachin J, et al. The effect of intensive treatment of diabetes on the development and progression of long-term complications in insulin-dependent diabetes mellitus. *N Engl J Med* 1993;**329**:977–86.
58. Nishikawa T, Edelstein D, Du XL, et al. Normalizing mitochondrial superoxide production blocks three pathways of hyperglycaemic damage. *Nature* 2000;**404**:787–90.
59. Brownlee M, Cerami A, Vlassara H. Advanced glycosylation end products in tissue and the biochemical basis of diabetic complications. *N Engl J Med* 1988;**318**:1315–21.

60. Ikeda K, Higashi T, Sano H, et al. N (epsilon)-(carboxymethyl) lysine protein adduct is a major immunological epitope in proteins modified with advanced gly cation end products of the maillard reaction. *Biochemistry* 1996;**35**:8075–83.
61. Bucala R, Makita Z, Koschinsky T, et al. Lipid advanced glycosylation: pathway for lipid oxidation in vivo. *Proc Natl Acad Sci U S A* 1993;**90**:6434–8.
62. Bucala R, Mitchell R, Arnold K, et al. Identification of the major site of apolipoprotein B modification by advanced glycosylation end products blocking uptake by the low density lipoprotein receptor. *J Biol Chem* 1995;**270**:10828–32.
63. Bucala R, Makita Z, Vega G, et al. Modification of low density lipoprotein by advanced glycation end products contributes to the dyslipidemia of diabetes and renal insufficiency. *Proc Natl Acad Sci U S A* 1994;**91**:9441–5.
64. Klein RL, Laimins M, Lopes-Virella MF. Isolation, characterization, and metabolism of the glycated and nonglycated subfractions of low-density lipoproteins isolated from type I diabetic patients and nondiabetic subjects. *Diabetes* 1995;**44**:1093–8.
65. Sobenin IA, Tertov VV, Koschinsky T, et al. Modified low density lipoprotein from diabetic patients causes cholesterol accumulation in human intimal aortic cells. *Atherosclerosis* 1993;**100**:41–54.
66. Steinbrecher UP, Witztum JL. Glucosylation of low-density lipoproteins to an extent comparable to that seen in diabetes slows their catabolism. *Diabetes* 1984;**33**:130–4.
67. Acosta J, Hettinga J, Fluckiger R, et al. Molecular basis for a link between complement and the vascular complications of diabetes. *Proc Natl Acad Sci U S A* 2000;**97**:5450–5.
68. Benzaquen LR, Nicholson-Weller A, Halperin JA. Terminal complement proteins C5b-9 release basic fibroblast growth factor and platelet-derived growth factor from endothelial cells. *J Exp Med* 1994;**179**:985–92.
69. Schmidt AM, Hori O, Brett J, et al. Cellular receptors for advanced glycation end products. Implications for induction of oxidant stress and cellular dysfunction in the pathogenesis of vascular lesions. *Arterioscler Thromb* 1994;**14**:1521–8.
70. Brett J, Schimidt AM, Yann SD, et al. Survey of the distribution of a newly characterized receptor for advanced glycation end products in tissues. *Am J Pathol* 1993;**143**:1699–712.
71. Neeper M, Schmidt AM, Brett J, et al. Cloning and expression of a cell surface receptor for advanced glycosylation end products of proteins. *J Biol Chem* 1992;**267**:14998–5004.
72. Vlassara H. Receptor-mediated interactions of advanced glycosylation end products with cellular components within diabetic tissues. *Diabetes* 1992;**41**(Suppl 2):52–6.
73. Wautier JL, Wautier MP, Schmidt AM, et al. Advanced glycation end products (AGEs) on the surface of diabetic erythrocytes bind to the vessel wall via a specific receptor inducing oxidant stress in the vasculature: a link between surface associated AGEs and diabetic complications. *Proc Natl Acad Sci U S A* 1994;**1994**:7742–6.
74. Yan SD, Schmidt AM, Anderson GM, et al. Enhanced cellular oxidant stress by the interaction of advanced glycation end products with their receptors/binding proteins. *J Biol Chem* 1994;**269**:9889–97.
75. Schmidt AM, Hori O, Chen JX, et al. Advanced glycation end products interacting with their endothelial receptor induce expression of vascular cell adhesion molecule-1 (VCAM-1) in cultured human endothelial cells and in mice. A potential mechanism for the accelerated vasculopathy of diabetes. *J Clin Investig* 1995;**96**:1395–403.
76. Esposito C, Gerlach H, Brett J, et al. Endothelial receptor-mediated binding of glucose-modified albumin is associated with increased monolayer permeability and modulation of cell surface coagulant properties. *J Exp Med* 1989;**170**:1387–407.
77. Wautier JL, Zoukourian C, Chappey O, et al. Receptor-mediated endothelial cell dysfunction in diabetic vasculopathy. Soluble receptor for advanced glycation end products blocks hyperpermeability in diabetic rats. *J Clin Investig* 1996;**97**:238–43.
78. Kirstein M, Aston C, Hintz R, et al. Receptor-specific induction of insulin-like growth factor I in human monocytes by advanced glycosylation end product-modified proteins. *J Clin Investig* 1992;**90**:439–46.

79. Kirstein M, Brett J, Radoff S, et al. Advanced protein glycosylation induces trans endothelial human monocyte chemotaxis and secretion of platelet-derived growth factor: role in vascular disease of diabetes and aging. *Proc Natl Acad Sci U S A* 1990;**87**:9010–4.
80. Vlassara H, Brownlee M, Manogue KR, et al. Cachectin/TNF and IL-1 induced by glucose-modified proteins: role in normal tissue remodeling. *Science* 1988;**240**:1546–8.
81. Park L, Raman KG, Lee KJ, et al. Suppression of accelerated diabetic atherosclerosis by the soluble receptor for advanced glycation end products. *Nat Med* 1998;**4**:1025–31.
82. Thomson L. 3-Nitrotyrosine modified proteins in atherosclerosis. *Dis Markers* 2015;**2015**:1–8.
83. Trujillo M, Alvarez B, Souza JM, et al. Mechanisms and biological consequences of peroxynitrite-dependent protein oxidation and nitration. In: Ignarro LJ, editor. *Nitric oxide biology and pathobiology*. 2nd ed. San Diego, CA: Academic Press; 2010. p. 1010–50.
84. Thomson L, Tenopoulou M, Lightfoot R, et al. Immunoglobulins against tyrosine-nitrated epitopes in coronary artery disease. *Circulation* 2012;**126**:2392–401.
85. Nissim A, Winyard PG, Corrigall V, et al. Generation of neoantigenic epitopes after post-translational modification of type II collagen by factors present within the inflamed joint. *Arthritis Rheum* 2005;**52**:3829–38.
86. Holmdahl R, Bockermann R, Bäcklund J, et al. The molecular pathogenesis of collagen-induced arthritis in mice—a model for rheumatoid arthritis. *Ageing Res Rev* 2002;**1**:135–47.
87. Rowley MJ, Nandakumar K, Holmdahl R. The role of collagen antibodies inmediating arthritis. *Mod Rheumatol* 2008;**18**:429–41.
88. Snir O, Widhe M, Von Spee C, et al. Multiple antibody reactivities to citrullinated antigens in sera from patients with rheumatoid arthritis: association with HLA-DRB1 alleles. *Ann Rheum Dis* 2009;**68**:736–43.
89. Baka Z, Barta P, Losonczy G, et al. Specific expression of PAD4 and citrullinated proteins in lung cancer is not associated with anti-CCP antibody production. *Int Immunol* 2011;**23**:405–14.
90. Stadlera SC, Vincentc CT, Fedorove VD, et al. Dysregulation of PAD4-mediated citrullination of nuclear GSK3β activates TGF-β signaling and induces epithelial-to-mesenchymal transition in breast cancer cells. *Proc Natl Acad Sci U S A* 2013;**110**:11851–6.
91. Moscarello MA, Mastronardi FG, Wood DD. The role of citrullinated proteins suggests a novel mechanism in the pathogenesis of multiple sclerosis. *Neurochem Res* 2007;**32**:251–6.
92. Roth EB, Theander E, Londos E, et al. Pathogenesis of autoimmune diseases: antibodies against transglutaminase, peptidylarginine deiminase and protein-bound citrulline in primary sjögren's syndrome, multiple sclerosis and alzheimer's disease. *Scand J Immunol* 2008;**67**:626–31.
93. Acharya NK, Nagele EP, Han M, et al. Neuronal PAD4 expression and protein citrullination: possible role in production of autoantibodies associated with neurodegenerative disease. *J Autoimmun* 2012;**38**:369–80.
94. Ishigami A, Ohsawa T, Hiratsuka M, et al. Abnormal accumulation of citrullinated proteins catalyzed by peptidylarginine deiminase in hippocampal extracts from patients with alzheimer's disease. *J Neurosci Res* 2005;**80**:120–8.
95. Bang H, Egerer K, Gauliard A, et al. Mutation and citrullination modifies vimentin to a novel autoantigen for rheumatoid arthritis. *Arthritis Rheum* 2007;**56**:2503–11.
96. Kinloch A, Tatzer V, Wait R, et al. Identification of citrullinated alpha-enolase as a candidate autoantigen in rheumatoid arthritis. *Arthritis Res Ther* 2005;**7**:R1421–9.
97. Masson-Bessière C, Sebbag M, Girbal-Neuhauser E, et al. The major synovial targets of the rheumatoid arthritis-specific anti filaggrin autoantibodies are deiminated forms of the α-and β-chains of fibrin. *J Immunol* 2001;**166**:4177–84.
98. Sebbag M, Chapuy-Regaud S, Auger I, et al. Clinical and pathophysiological significance of the autoimmune response to citrullinated proteins in rheumatoid arthritis. *Joint Bone Spine* 2004;**71**:493–502.

99. Sebbag M, Moinard N, Auger I, et al. Epitopes of human fibrin recognized by the rheumatoid arthritis-specific autoantibodies to citrullinated proteins. *Eur J Immunol* 2006;**36**:2250–63.
100. Klareskog L, Stolt P, Lundberg K, et al. A new model for an etiology of rheumatoid arthritis: smoking may trigger HLA-DR (shared epitope)-restricted immune reactions to autoantigens modified by citrullination. *Arthritis Rheum* 2006;**54**:38–46.
101. Aubart F, Crestani B, Nicaise-Roland P, et al. High levels of anti-cyclic citrullinated peptide autoantibodies are associated with co-occurrence of pulmonary diseases with rheumatoid arthritis. *J Rheumatol* 2011;**38**:979–82.
102. De Rycke L, Peene I, Hoffman IEA, et al. Rheumatoid factor and anti-citrullinated protein antibodies in rheumatoid arthritis: diagnosis value, associations with radiological progression rate, and extra-articular manifestations. *Ann Rheum Dis* 2004;**63**:1587–93.
103. Van Gaalen FA, Linn-Rasker SP, van Venrooij WJ, et al. Autoantibodies to cyclic citrullinated peptides predict progression to rheumatoid arthritis in patients with undifferentiated arthritis: a prospective cohort study. *Arthritis Rheum* 2004;**50**:709–15.
104. Berglin E, Johansson T, Sundin U, et al. Radiological outcome in rheumatoid arthritis is predicted by presence of antibodies against cyclic citrullinated peptide before and at disease onset, and by IgA-RF at disease onset. *Ann Rheum Dis* 2006;**65**:453–8.
105. Stolt P, Kallberg H, Lundberg I, et al. Silica exposure is associated with increased risk of developing rheumatoid arthritis: results from the Swedish EIRA study. *Ann Rheum Dis* 2005;**64**:582–6.
106. Stolt P, Yahya A, Bengtsson C, et al. Silica exposure among male current smokers is associated with a high risk of developing ACPA-positive rheumatoid arthritis. *Ann Rheum Dis* 2010;**69**:1072–6.
107. Turner S, Cherry N. Rheumatoid arthritis in workers exposed to silica in the pottery industry. *Occup Environ Med* 2000;**57**:443–77.
108. Speck-Hernandez CA, Montoya-Ortiz G. Silicon, a possible link between environmental exposure and autoimmune diseases: the case of rheumatoid arthritis. *Arthritis* 2012;**2012**:1–11.
109. Machold KP, Stamm TA, Nell VPK, et al. Very recent onset rheumatoid arthritis: clinical and serological patient characteristics associated with radiographic progression over the first years of disease. *Rheumatology* 2007;**46**:342–9.
110. Syversen SW, Gaarder PI, Goll GL, et al. High anti-cyclic citrullinated peptide levels and an algorithm of four variables predict radiographic progression in patients with rheumatoid arthritis: results from a 10-year longitudinal study. *Ann Rheum Dis* 2008;**67**:212–7.
111. Mydel P, Wang Z, Brisslert M, et al. Carbamylation-dependent activation of T cells: a novel mechanism in the pathogenesis of autoimmune arthritis. *J Immunol* 2010;**184**:6882–90.
112. Adams LE, Roberts SM, Donovan-Brand R, et al. Study of procainamide hapten-specific antibodies in rabbits and humans. *Int J Immunopharmacol* 1993;**15**:887–97.
113. Jaswal S, Mehta HC, Sood AK, et al. Antioxidant status in rheumatoid arthritis and role of antioxidant therapy. *Clin Chim Acta* 2003;**338**:123–9.
114. Griffiths HR. Is the generation of neo-antigenic determinants by free radicals central to the development of autoimmune rheumatoid disease? *Autoimmun Rev* 2008;**7**:544–9.
115. Yang ML, Doyle HA, Gee RJ, et al. Intracellular protein modification associated with altered T cell functions in autoimmunity. *J Immunol* 2006;**177**:4541–9.
116. Dai L, Lamb DJ, Leake DS, et al. Evidence for oxidized low density lipoprotein in synovial fluid from rheumatoid arthritis patients. *Free Radic Res* 2000;**32**:479–86.
117. Dai L, Zhang Z, Winyard PG, et al. A modified form of low-density lipoprotein with increased electronegative charges present in rheumatoid arthritis synovial fluid. *Free Radic Biol Med* 1996;**22**:705–10.

118. Kurien BT, Scofield RH, R. H. Autoimmunity and oxidatively modified autoantigens. *Autoimmun Rev* 2008;**7**:567–73.
119. Viguet-Carrin S, Roux JP, Arlot ME, et al. Contribution of the advanced glycation end product pentosidine and of maturation of type I collagen to compressive biomechanical properties of human lumbar vertebrae. *Bone* 2006;**39**:1073–9.
120. Drinda S, Franke S, Canet CC, et al. Identification of the advanced glycation end products Nε-carboxymethyllysine in the synovial tissue of patients with rheumatoid arthritis. *Ann Rheum Dis* 2002;**61**:488–92.
121. Drinda S, Franke S, Rüster M, et al. Identification of the receptor for advanced glycation end products in synovial tissue of patients with rheumatoid arthritis. *Rheumatol Int* 2005;**25**:411–3.
122. Grinnell S, Yoshida K, Jasin HE. Responses of lymphocytes of patients with rheumatoid arthritis to IgG modified by oxygen radicals or peroxynitrite. *Arthritis Rheum* 2005;**52**:80–3.
123. Newkirk MM, Le Page K, Niwa T, et al. Advanced glycation endproducts (AGE) on IgG, a target for circulating antibodies in north American Indians with rheumatoid arthritis (RA). *Cell Mol Biol* 1998;**44**:1129–38.
124. Newkirk MM, Goldbach-Mansky R, Lee J, et al. Advanced glycation end-product (AGE)-damaged IgG and IgM autoantibodies to IgG-AGE in patients with early synovitis. *Arthritis Res Ther* 2003;**5**:R82–90.
125. Ahmad S, Habib S, Moinuddin, et al. Preferential recognition of epitopes on AGE-IgG by the autoantibodies in rheumatoid arthritis patients. *Hum Immunol* 2013;**74**:23–7.
126. Bergt C, Fu X, Huq NP, et al. Lysine residues direct the chlorination of tyrosines in YXXK motifs of apolipoprotein A-I when hypochlorous acid oxidizes high density lipoprotein. *J Biol Chem* 2004;**279**:7856–66.
127. Khan F, Siddiqui AA. Prevalence of anti-3-nitrotyrosine antibodies in the joint synovial fluid of patients with rheumatoid arthritis, osteoarthritis and systemic lupus erythematosus. *Clin Chim Acta* 2006;**370**:100–7.
128. Kaur H, Halliwell B. Evidence for nitric oxide-mediated oxidative damage in chronic inflammation nitrotyrosine in serum and synovial fluid from rheumatoid patients. *FEBS Lett* 1994;**350**:9–12.
129. Sandhu JK, Robertson S, Birnboim HC, et al. Distribution of protein nitrotyrosine in synovial tissues of patients with rheumatoid arthritis and osteoarthritis. *J Rheumatol* 2003;**30**:1173–81.
130. Kageyama Y, Takahashi M, Nagafusa T, et al. Etanercept reduces the oxidative stress marker levels in patients with rheumatoid arthritis. *Rheumatol Int* 2008;**28**:245–51.
131. Lemarechal H, Allanore Y, Chenevier-Gobeaux C, et al. Serum protein oxidation in patients with rheumatoid arthritis and effects of infliximab therapy. *Clin Chim Acta* 2006;**372**:147–53.
132. Túnez I, Feijóo M, Huerta G. The effect of infliximab on oxidative stress in chronic inflammatory joint disease. *Curr Med Res Opin* 2007;**23**:1259–67.
133. Kageyama Y, Takahashi M, Ichikawa T, et al. Reduction of oxidative stress marker levels by anti-TNF-α antibody, infliximab, in patients with rheumatoid arthritis. *Clin Exp Rheumatol* 2008;**26**:73–80.

Further Reading

134. Hitchon CA, El-Gabalawy HS. Oxidation in rheumatoid arthritis. *Arthritis Res Ther* 2004;**6**:265–78.

CHAPTER

5

Ubiquitin Mediated Posttranslational Modification of Proteins Involved in Various Signaling Diseases

Lavanya V, Shazia Jamal, Neesar Ahmed

School of Life Sciences, B.S. Abdur Rahman Crescent Institute of Science & Technology, Chennai, India

1 INTRODUCTION

Ubiquitination, a multifaceted protein posttranslational modification, plays a key role in protein turnover in the body. This multistep process mainly involves the addition of a ubiquitin molecule to the translated proteins. After a number of ubiquitin molecules are covalently attached to a substrate protein, the ubiquitinated proteins are carried for degradation by the 26s proteasome complex. The entire process of ubiquitination combined with the degradation of the ubiquitinated proteins by the proteasome complex is known as the ubiquitin proteasome system (UPS). Ubiquitin-mediated degradation of various key regulatory proteins plays a vital role in numerous cellular processes such as transcription, cell cycle progression, signal transduction, receptor endocytosis, and antigen presentation.[1] For instance, ubiquitin-mediated proteolysis of cyclin, a regulatory protein involved in cell cycle, leads to rapid transition between different stages of cell cycle and also prevents regression to an earlier stage. In addition to this, ubiquitination helps to maintain the concentration of key signaling proteins.

Ubiquitin, a small 76-aminoacid long residue, conserves protein that gets attached to a substrate protein in monomeric or polymeric form through an energy-dependent enzymatic pathway. The process of

Protein Modificomics
https://doi.org/10.1016/B978-0-12-811913-6.00005-9

ubiquitination involves a sequence of enzymatic cascade reactions that are regulated by three main enzymes—ubiquitin-activating enzymes (E1), ubiquitin-conjugating enzymes (E2) and the ubiquitin ligases (E3).[2] Initially, prior to its attachment to the active site, activation of ubiquitin is accrued out by E1 enzymes, followed by transfer of high energy ubiquitin-thiol ester bond to E2 enzymes, resulting in the formation of an E2-thiol ester bond. Ubiquitin gets attached to the active site of the E2 enzyme and ATP is required for the transfer of activated ubiquitin to E2, the ubiquitin-conjugating enzyme. In the third and final step, E3 ligases catalyze the transfer of ubiquitin from the E2 enzymes to lysine residues of the substrate protein. So far, more than 600 E3 ligases have been identified, which indicates a high degree of substrate specificity.[3] For attachment, an isopeptide bond is formed between the conserved C-terminal glycine residue of ubiquitin and lysine residue of the substrate protein.[1] This multistep process is repeated a number of times, resulting in polyubiquitin chain formation on the substrate.[4] The so formed polyubiquitinated chains, consisting of more than 4 ubiquitin molecules, are recognized by the 26s proteasome, and the substrate protein gets degraded.[5] After their removal by ubiquitin carboxyl-terminal hydrolases, the released ubiquitin molecules are recycled for future use.[6]

The monoubiquitination, i.e., attachment of a single ubiquitin to the lysine amino acid residue of the substrate protein, serves as a signal for regulation of gene transcription, receptor endocytosis, DNA repair, and signal transduction.[7, 8] However, polyubiquitination, i.e., attachment of a ubiquitin chain to a specific lysine residue within the substrate protein leads to degradation of that protein by the UPS. Similar to protein phosphorylation, protein ubiquitination is also an important posttranslation modification that controls various signaling within the cells. However, in comparison to phosphorylation, ubiquitination is complex as it involves the formation of a ubiquitin chain with at least eight various linkages. Thereafter, the function of this ubiquitin modified protein totally depends on these linkages.[9] The presence of seven lysine residues, including lysine-6 (K6), lysine-11(K11), lysine-27(K27), lysine-29 (K29), lysine-33 (K33), lysine-48 (K48), and lysine-63 (K63), is responsible for the formation of polyubiquitin chains through ubiquitin-ubiquitin linkages.[10]

Linkage-specific polyubiquitinations have been shown to lead to distinct cellular events. Among these, K48 and K63-linked chains are the most abundant and well studied of all the Ub-linkages. While the K63-linked ubiquitin chains mediate recruitment of repair machineries to the DNA damage site, thereby assisting in DNA repair[11]; the K48-linked ubiquitin chains play a vital role in signal transduction as they are known to facilitate proteasomal degradation of major signaling proteins. K6-linked ubiquitination has been reported during DNA repair in response to UV radiation. However, intensive research is still required to fill the gap in understanding the specific role of K6 linked ubiquitination.[12] K11-linked ubiquitin chains are known to regulate the substrate during mitosis mainly on the anaphase-promoting complex.[13] K6, K27, and K29 ubiquitination have been reported in proteins, DJ-1 and alpha-synuclein, responsible for Parkinson's disease. In addition, K27 ubiquitination on IKKγ determines its function in NF-κB signaling.[14] K29 linkage acts as an inhibitor of Wnt signaling where polyubiquitinylation of Axin is responsible for repression of Wnt/β-catenin signaling.[15] Thus, the specific linkages affect the overall functions of the ubiquitin modified protein. Keeping in view the pathological importance/significance of ubiquitination, it is worthwhile discussing the role of ubiquitin in various disease processes and the component enzymes involved.

2 UBIQUITINATION IN INFLAMMATORY PATHWAYS

Ubiquitination plays a prominent role in regulation of various signaling pathways involved in a number of biological processes, including innate and adaptive immune responses. As Ubiquitination is responsible for turnover of various intracellular proteins, any aberrations in the multistep process leads to imbalance in cellular homeostasis, thereby resulting in development of multiple diseases.[16,17] A number of E3 ubiquitin ligases and deubiquitinases, involved in inflammatory pathways, resulting in diverse biological functions, have been identified. Inflammation is defined as a nonspecific host response via diverse functional and molecular mediators along with the recruitment and activation of immune cells.

Persistent low-grade chronic inflammation contributes to the underlying damage of several human diseases. However, the present understanding of the molecular mechanisms involved in the inflammatory process during the pathogenesis of various diseases needs to be expanded. Onset of inflammation has been reported to activate NF-κB with an increase in the nuclear translocation of p65 subunits and increased DNA-binding activity. The activation of NF-κB occurs through respective ubiquitin mediated activation and degradation of IκB kinase (IKK) and of inhibitor I kappa B (IκB). IKK activation in turn leads to K48 linked ubiquitination and degradation of IκBα. NF-κB is found to regulate multiple pathways, thereby disturbing expression of pro-inflammatory and antiinflammatory cytokines that are involved in the pathogenesis of many human diseases, e.g., inflammatory arthritis, psoriasis, Seronegative spondyloarthropathies (SpA), allergy, and asthma.[18] In addition to this, numerous ubiquitin enzymes have been found to be associated with development of autoimmune diseases. For instance, the A20 ubiquitin-editing enzyme, a key negative regulator of NF-κB, is associated with autoimmune diseases like rheumatoid arthritis (RA), psoriasis, Crohn's disease, inflammatory bowel disease, systemic lupus erythematosus (SLE), and type one diabetes.[19] A20 acts as a susceptibility gene for SLE by regulating B cell homeostasis and by preventing autoimmune responses caused by autoreactive B cells.[20] The E2 ubiquitin-conjugating enzyme, UBE2L3 (UbcH7) was recognized as a potential target in many autoimmune diseases such as SLE, RA, Crohn's disease and celiac disease.[21] The overexpression of UBE2L3 resulted in NF-κB activation, which is the major regulatory factor of various inflammatory and autoimmune diseases. Likewise, the E3 ubiquitin ligase TRIM21 (Ro52) was proposed as a potent target in SLE treatment due to its regulatory role in immune responses.[22] Inhibition of MDM2, an E3 ubiquitin ligase, has also been identified as a potent therapeutic strategy for SLE treatment, thus highlighting its role in promoting SLE and other autoimmune diseases.[23]

3 IMPLICATIONS OF UPS IN CANCER

UPS, being a key cell cycle regulator, promotes degradation and turnover of crucial proteins responsible for growth and functioning of cells. As oncogenesis results from abnormal cell cycle control, it is likely that aberrations within the UPS may lead to malignant transformation of cells.[24] Inappropriate ubiquitin-dependent proteolysis of oncogene products or tumor suppressors is known to contribute significantly to the etiology of several carcinomas. The possible roles of enzymes involved in ubiquitin-degradation pathways have been implicated in certain forms of cancers such as colorectal, cervical, pancreatic, breast, etc.[25, 26]

3.1 E3 Ubiquitin Ligases in Development of Cancer

The E3 ubiquitin ligases are capable of selectively recognizing a protein substrate via a specific protein recognition domain or through other cofactors. More than 600 E3 ligases have been identified and are grouped into four major classes depending on their specific mechanism in the ubiquitination process. Since E3 ligases play a crucial role in various cellular processes like cell proliferation, cell cycle arrest, and apoptosis, any aberrant expression of E3 ligases or mutations in the enzyme are linked to cancer development or suppression.

3.1.1 p53-Associated E3 Ligases

Transcription factor, p53, is an important tumor suppressor gene that prevents potential transformation of irreversibly damaged cells by prompting apoptosis. In unstressed conditions, Ubiquitin-mediated proteolysis plays a pivotal role in restraining the activity of p53. E3 ubiquitin ligases such as the Murine double minute-2 (Mdm2), E6-Associated Protein (E6-AP), Arf-BP1, constitutive photomorphogenesis protein 1 (COP1), and Pirh2 have been found to facilitate ubiquitination and subsequent degradation of p53.

3.1.1.1 Mdm2

Mdm2 is an E3 ligase that has been recognized as the principal cellular negative regulator of p53.[27] Mdm2 acts as a negative regulator of p53 by either binding to and inhibiting the transcriptional activity of p53 or by stimulating ubiquitin-mediated degradation of p53.[28, 29]

3.1.1.2 E6-AP

This was the first E3 ligase that was found to promote proteasome degradation of p53. The HECT (Homolog of E6-AP C-Terminus) domain of E6-AP catalyzes the degradation of p53 by assembling the Lys-48 linked polyubiquitin degradation signals for further degradation by the 26s proteasomes.

3.1.1.3 ARF-BP1

This is a HECT domain-containing, p53-associated E3 ligase that is known to be highly expressed in 80% of breast cancer cells and has been found to directly interact with and induce ubiquitin-mediated proteolysis of p53.[30]

3.1.1.4 COP1

This RING-finger-containing E3 ligase has been found to ubiquitinate p53 directly, thereby targeting it for proteasome degradation.[31] In another study, over expression of COP1 in breast and ovarian carcinomas was found to accelerate the degradation of p53, suggesting the role of COP1 as a tumor promoter.[32]

3.1.1.5 Pirh2

Pirh2 is a p53-induced ubiquitin protein ligase that interacts with p53 and promotes proteasome-mediated ubiquitination of p53 under both in vitro and in vivo conditions.[33]

3.1.2 Other E3 Ubiquitin Ligases Involved in Cancer Development

3.1.2.1 SKP1-CULLIN 1-F-BOX PROTEIN (SCF) E3 LIGASES

The SCF E3 ligases are composed of multiple subunits, namely the Skp I (S-phase kinase-associated protein (1), cullin, and F-box protein. The SCF ligases play a significant role in cell proliferation and survival. While Skp I and cullin are the invariable subunits of SCF E3 ligases, the F-box protein subunit is the variable and serves as recognition site for protein substrates.

Skp2 (S-phase kinase-associated protein (2) and Fbw7 (F-box and WD repeat domain-containing 7) are the two most extensively studied F-box proteins, and aberrations in their SCF complexes have been linked to tumorigenesis.[34]

Skp2 has been identified as a proto-oncogene that exerts its oncogenic functions primarily through ubiquitin-mediated proteasome degradation of its substrates, most of which are prominent tumor suppressor genes such as p21, p27, p57, p130, E-cadherin, FOXO1, etc.[35,36] Apparently, overexpression of SCF^{Skp2} has been linked with underexpression of the tumor suppressor gene, p27, in breast and colon cancer.[37,38] p27 is a cyclin-dependent kinase inhibitor and Skp2 was found to promote G1-S transition of cell cycle by specifically promoting the ubiquitin-mediated proteolysis of p27.[39] The role of Skp2 as a prognostic marker has been well established in some cancers. For instance, elevated levels of Skp2 has been associated with increased tumor size and histological grade in hepatocellular carcinoma.[40] Further, Skp2 was observed to be overexpressed in various cancers, such as breast cancer,[41] pancreatic cancer,[42] prostate cancer,[43] nasopharyngeal cancer,[44] etc., thereby, further validating its role as a tumor promoter.

SCF^{Fbw7}, an evolutionarily conserved F-box ubiquitin ligase, promotes ubiquitination of proto-oncogenes such as c-jun, notch, cyclin E, c-Myc, and mTOR.[45] As the substrates of SCF^{Fbw7} are dominant oncogenes, which are implicated in development of a wide range of human cancers, mutations in Fbw7 are linked with a number of human neoplasms. Fbw7 mutations have been found in early stages of colorectal cancers,[46] gastric carcinoma,[47] and in tumors of the blood, breast, bone, bile duct, colon, endometrium, lung, ovary, prostate, and pancreas,[48] thereby, emphasizing its key role as a general tumor suppressor in cancers.

3.1.2.2 BRCA1 /BARD1 AS AN E3 UBIQUITIN LIGASE

Breast cancer susceptibility gene 1 (BRCA1) is a tumor suppressor gene that has been found to be mutated in about 50% of inherited cases of breast cancer.[49] Moreover, it has also been implicated as a major regulator of cell cycle checkpoint control. BRCA1 codes for an N-terminal RING finger E3 ubiquitin ligase which upon heterodimerization with another RING finger protein known as BARD1 (BRCA1 Associated Ring Domain 1), acquired enhanced ubiquitin ligase activity. The heterodimeric RING finger complex of BRCA1/BARD1 is capable of catalyzing diverse ubiquitination reactions depending on the interacting E2 enzyme.[50] Mutations in the BRCA1 RING domain result in loss of its ligase activity, leading to tumorigenesis. Apart from breast and ovarian cancers, mutations in BRCA1 have also been linked to tumors of the stomach and esophagus.[51]

3.1.2.3 VON HIPPEL-LINDAU (VHL) TUMOR SUPPRESSOR GENE

The gene that codes for the VHL protein is located on chromosome 3p25.5 and is named after the hereditary cancer syndrome known as VHL syndrome. Patients with VHL disease were at increased risks of developing various cancers including tumors of the pancreas, central nervous system, and hemangioblastomas of the retina.[52] In addition to tumors associated with the VHL syndrome, the VHL gene has also been observed to be mutated in most of the sporadic renal carcinomas.[53] The VHL protein has been shown to act as a substrate recognition subunit in the VCB-Cul2-VHL ubiquitin ligase complex. HIF-1α (Hypoxia-inducible factor-1α) was identified as one of the key substrates of VHL. The oxygen-dependent polyubiquitinylation of hypoxia-inducible factor (HIF) that is mediated by VHL is known to play a critical role in the mammalian oxygen-sensing pathway.[54]

3.1.2.4 NEDD4-LIKE E3 LIGASES

The neural precursor cells-expressed developmentally downregulated 4 (Nedd4) family includes HECT-type ligases such as Nedd4-1, Smad ubiquitination regulatory factor 2 (Smurf2), and WW domain-containing protein 1 (WWP1), which have been genetically linked to tumor progression. The nine members of E3 ligases share a similar structure and consist of three functional domains including a C-terminal HECT domain for ubiquitin protein ligation, a WW2 domain, and an N-terminal C2 domain for membrane binding, which facilitates protein-protein interaction.[55]

Nedd4-1 has been proposed to act as a proto-oncogene, as it promotes polyubiquitinylation and degradation of Phosphatase and Tensin Homolog (PTEN) which is a well-known tumor suppressor. Moreover, Nedd4-1 also allows translocation of PTEN into the nucleus. In many cancers, increased expression of Nedd4-1 was found to correlate with lower levels of PTEN.[56]

The Nedd4-like E3 ligase, Smurf2 has been demonstrated to interact with Smad proteins, which are key signaling effectors of the TGF-β signaling cascade. The preferentially targeting of Smad1 to ubiquitin-mediated proteasome degradation may lead to attenuation of TGF-β signaling. For instance, in esophageal squamous cell cancer, increased expression of Smurf2 correlated with decreased Smad2 protein level. Smurf2 upregulation was also associated with higher invasiveness and poor prognosis.[57]

WWP1 is a Nedd4-like E3 ubiquitin ligase that acts as a potential oncogene in specific cancers.[58] It has been demonstrated to inactivate molecular components such as Smad2, Smad4, and TbetaR1, thereby negatively regulating the TGF-beta tumor suppressor pathway. In breast cancer cells, genomic aberrations of WWP1 were found to promote ubiquitin-mediated degradation of the protein RNF11, thereby contributing to breast cancer pathogenesis.[59] Also, WWP1 has been shown to target the transcription factor KLF5 (Kruppel-like factor 5) for ubiquitin-mediated proteasome degradation.[60] KLF5 is a tumor suppressor that suppresses the growth of cancer cells.[61] It has been shown to regulate cell proliferation, differentiation, and angiogenesis. Hence, the inactivation of WWP1 will prevent the over degradation of KLF5 in cancer cells, which will inhibit tumor growth. More significantly, WWP1 is reported to regulate the ubiquitination of p53 and promote its translocation to the cytoplasm, thereby, reducing its transcriptional activity.[62] Thus, the significance of WWP1 as a cancer drug target has been implicated, as inhibition of WWP1 will facilitate p53 transcriptional activity.

3.2 Role of Deubiquitinating Enzymes (DUBs) in Cancer

DUBs are enzymes that specifically cleave the polyubiquitin and monoubiquitin from targeted proteins. DUBs also play a significant role in regulation of proteasome-dependent degradation, turnover rate, localization, transcription, and endocytosis of multiple proteins. As DUBs control DNA repair, cell-cycle, chromatin remodeling, and several signaling pathways that are frequently altered in cancer, aberrations in DUBs have been found to be implicated in development of neoplastic diseases. Both oncogenic as well as tumor suppressor DUBs have been widely reported as discussed below.

3.2.1 Herpes Associated Ubiquitin Specific Protease (HAUSP) or USP7

This DUB was shown to specifically target p53 and its ubiquitin ligase, MDM2 through a complex mechanism.[63] HAUSP offsets ubiquitin-dependent proteasome degradation of p53 and MDM2 by independently interacting with and catalyzing deubiquitination of both the proteins.[64]

Further, DUBs have been demonstrated to deubiquitinate the tumor suppressor proteins Forkhead box O4 (FOXO4) and PTEN.[65] HAUSP-mediated deubiquitination of PTEN resulted in cytoplasmic accumulation of the protein, as monoubiqutination of PTEN was a requisite for its nuclear translocation. For instance, In human prostate cancer, increased expression of HAUSP correlated with PTEN nuclear exclusion and was found to be responsible for tumor invasiveness.[66]

3.2.2 Usp28

It has been proposed that the DUB Usp28 possess oncogenic potential as it deubiquitinates and hence stabilizes the oncogene c-Myc, thereby preventing the proteasome degradation of c-Myc. Usp28 was found to indirectly interact with the oncogene c-Myc through its binding to

SCFFBW7, which is the specific E3 ligase of c-Myc. Overexpression of Usp28 in breast and colon carcinomas further supports the notion that Usp28 may be involved in tumorigenesis.[67] Hence, it is quite evident that the inhibition of USP28 will result in accelerated degradation of c-Myc, which may prove to be a means of targeting malignancies.

3.2.3 *USP9x*

This DUB, belonging to the USP family of proteins, promotes deubiquitination of Smad4, thereby resulting in positive regulation of the TGF-β signaling pathway.[68] Since TGF- β signaling is like a double edged sword that can promote or inhibit tumor growth depending on the distinctive cell type, the relationship of USP9x with cancer development is still unclear.

3.2.4 *USP2a*

Aberrant expression of USP2a has been associated with progression of various cancers. This DUB acts as a significant regulator of p53 activity as ectopically expressed USP2a was shown to lead to accumulation of Mdm2, thereby promoting Mdm2-mediated degradation of p53.[69] By deubiquitinating Mdm2, USP2a indirectly promotes proteasome-degradation of p53 and hence, is considered as an effective oncogene. Increased expression of USP2a was demonstrated to prevent apoptosis of various cancer cells, which further highlights the role of USP2a as a proto-oncogene.

Further, USP2a was found to deubiquitinate the protein, fatty acid synthase (FAS), overexpressed in many cancers, like those of the breast and prostate. FAS has been recognized as a metabolic oncogene which promotes de novo lipid biosynthesis, which is a requisite for cancer cell growth and proliferation. USP2a was found to be overexpressed in 44% of prostate cancer samples and, in most cases, the overexpression correlated with increased expression of FAS. Significantly, USP2a was found to play an important role in prostate cancer cell survival by deubiquitinating and stabilizing the FAS protein against proteasomal degradation. Thus, USP2a has been regarded as a potential target in prostate cancer treatment.[70]

In another study, USP2a was demonstrated to act like a mediator of bladder cancer progression. By interacting with and deubiquitinating the protein cyclin D, USP2a aided in tumor progression.[71] Cyclin D is an important cell cycle regulator, the accumulation of which results in increased cell proliferation. Thus, USP2a can be considered as a therapeutic target in USP2a overexpressing cancers.

3.2.5 *A20*

The unique feature of this enzyme, which has been identified as a crucial tumor suppressor, is that it can act both an E3 ligase and DUB. The important substrates of this enzyme are the protein mediators of the NFκB signaling pathway. The OUT domain at the N-terminus catalyzes the removal of K63-linked polyubiquitin chains from Receptor Interacting Protein 1 (RIP1), RIP2, TRAF6, Nuclear factor kappa B Essential Modifier (NEMO), and Mucosa-associated lymphoid tissue lymphoma translocation protein 1 (MALT1), while the seven zinc fingers at its C-terminus catalyze conjugation of K48-linked polyubiquitin chains to RIP1 and TRAF2, thereby targeting them for proteasomal degradation.[72] Eventually, A20 inhibits activation of NFκB by deubiquitinating the K63 ubiquitin chains, as well as by K48 polyubiquitination and subsequent degradation of the intermediates of the NFκB pathway. The observation that A20 was deleted or mutated in several cell lymphomas and Hodgkin's lymphoma further highlights its role in tumor suppression. Inactivated A20 has been observed in B cell malignancies such as the MALT lymphoma, mantle cell lymphoma, and the marginal zone lymphoma.

3.2.6 CYLD

This protein was found to be mutated in the autosomally dominant inherited disease Familial Cylindromatosis, which is characterized by development of multiple skin tumors. CYLD, belonging to the UCH-family of DUBs, was found to regulate NFκB negatively by removing K63-linked polyubiquitin chains from critical mediators of I kappa kinase (IKK).[73] It further inhibits tumor cell proliferation by interacting with and removing K63-linked polyubiquitin chains Bcl-3, which acts as a transcription coactivator of NFκB.[74]

4 E3 LIGASES IN DIABETIC RETINOPATHY (DR)

DR has been reported to share similarities with chronic low grade inflammation and is characterized by leukocyte recruitment, tissue edema, increased inflammatory markers in the retina, vitreous, and plasma.[75] The cross talk between inflammation and progression of diabetic retinopathy has been widely accepted and provides opportunities for researchers to develop novel therapeutic approaches. Persistent low-grade chronic inflammation creates a hypoxic microenvironment in retinal vasculature, which leads to neovascularization in DR.[76] Onset of diabetes activates NF-κB resulting in an increase in the nuclear translocation of p65 subunits with enhanced DNA-binding activity in retinal endothelial cells of experimental animals. The activation of NF-κB results in synthesis of chemokines, cytokines, pro-inflammatory molecules and acute phase proteins.[77] Pro-inflammatory cytokines and chemokines including Tumor Necrosis Factor (TNF-α), Interleukin-6 (IL-6), Interleukin 1 beta (IL-1β), IL-8 and monocyte chemoattractant protein-1 (MCP-1) are significantly high in the vitreous fluids of human DR patients.[78] In fact, the expression of these pro-inflammatory mediators is under the direct control of several transcription factors, including Nuclear Factor kappa B (NF-kB). In addition, several intermediate molecules are involved that transmit the signal from the cell surface receptor to activation of these transcription factors.

Ubiquitin is present throughout the retina, where most of the endogenous ubiquitin is covalently attached to various target proteins. A number of E3 ubiquitin ligases, such as HERC6 and NEDD4, belong to the HECT family of E3 Ligase, Topoisomerase I-binding RS protein (TOPORS), Tripartite motif-containing protein 2 (TRIM2). Seven in absentia homolog (SIAH), Parkin, and UBR1 have been identified in the retina. Moreover, in view of the finding that there is altered glial-endothelial cell interaction at the blood-retinal-barrier (BRB), which contributes to DR, and is accompanied by a change in the protein amount and localization of occluding in the endothelial cells,[79] identifying the other potential substrates of ITCH and NEDD4 will be very useful in understanding the molecular mechanism involved in the pathogenesis of DR. It has been reported that NEDD4 and ITCH mainly interact through its WW domain and potentially recognize the PPXY motif in the target protein. Among the molecules identified, Peroxisome Proliferator-Activated Receptor-γ (PPAR-γ) and Protein Phosphatase 2A (PP2A) have been found to play a significant role in DR pathogenesis. PPAR-γ has been reported to inhibit diabetes-induced leukostasis and leakage, regulating inflammation, angiogenesis, and apoptosis in retinal and endothelial cells. Synthetic PPAR-γ agonists such as pioglitazone and rosiglitazone reduce inflammation biomarkers through various mechanisms involving inhibition of NF-κB activation. Similarly, PP2A has been shown to regulate important molecules, thereby contributing to the pathogenesis of DR. Interestingly both PPAR and PP2A have a PPXY motif. Moreover, few Retinal deubiquitinating enzymes that include ubiquitin-specific protease11 (USP-11) and ubiquitin specific protease12 (USP-12) have also been identified.[79]

5 IMPLICATIONS OF UPS IN NEURODEGENERATIVE DISEASES

Inappropriate ubiquitin-dependent proteolysis has been found to be involved in a number of pathological conditions. There are various reports that relate, directly or indirectly, the altered functioning of UPS to disease pathogenesis. While abnormalities in the ubiquitin-ligase enzymes and deubiquitinating enzymes may perturb substrate recognition and supply of ubiquitin, mutations in substrate proteins may directly lead to inefficient substrate recognition, thereby contributing to disease pathogenesis.[80] This includes neurodegenerative disorders like Alzheimer's disease, Parkinson's disease, Huntington's disease, etc., where protein misfolding may play a crucial role. These neurodegenerative disorders are mainly characterized by accumulation of toxic aggresomes or inclusion bodies that result due to alterations in the ubiquitin proteasome system.[81]

5.1 Alzheimer's Disease

Alzheimer's disease (AD), characterized by loss of neurons and synapses, is the most common neurodegenerative disorder that mostly occurs in the elderly. People with AD do suffer from memory and cognitive function loss, which leads to dementia. The typical aggregates that are associated with AD include the β-amyloid plaques (Aβ) and the neurofibrillary tangles. The neurofibrillary tangles that are present within the cells consist of hyperphosphorylated, ubiquitinated tau protein.[82] The presence of ubiquitin in plaques and tangles in the brain of AD patients provides evidence that the dysfunctioning of the UPS is probably involved in AD pathogenesis.[83, 84] Notably, ubiquitin carboxyl-terminal hydrolase L1 (UCH-L1), an enzyme that liberates ubiquitin from poly ubiquitinated proteins, was found to be oxidized in AD and downregulated in the some specific regions of brain in the early onset of AD.[85] Further, the accumulation of oxidized protein in the brain of AD patients may cause decrease in activity of proteasomes.[86]

There are also reports of direct evidence of aberrant proteasome activity in specific regions of the brain of AD patients. Keller et al., observed the activity of proteasomes to be decreased in specific regions of the brain that are more susceptible to AD, such as the hippocampus.[87] More importantly, when the proteasome activity was compared between the AD and controls, the activity was unchanged in other less susceptible brain regions such as the cerebellum.

The role of ubiquitination in turnover of tau protein further highlights the contribution of UPS in AD pathogenesis. The E3 ligase, CHIP (C terminus of Hsc70-interacting protein) has been shown to ubiquitinate tau protein, eventually leading to its degradation by the proteasomes.[88, 89] Further, accumulation of soluble phosphorylated tau in the brain of CHIP knockout mice supports the notion that the turnover of tau protein depends on the proteasome activity and that the accumulated tau protein inhibits the proteasome.[90]

5.2 Parkinson's Disease (PD)

Autosomal recessive juvenile parkinsonism (AR-JP) is a familial form of PD, characterized by complex clinical features like loss of dopaminergic neurons and motor deficits.[91] The protein aggregates of alpha-synuclein that are referred to as Lewy bodies are the major diagnostic feature of

PD. Furthermore, mitochondrial dysfunction and oxidative stress that eventually lead to cell death also contribute to the pathology of PD. Though mutations in one or more genes have been shown to be responsible for causing AR-JP, in approximately 50% of all cases, mutation in the gene PARK2, which encodes for Parkin protein, is found to cause AR-JP.[92] Parkin, functioning as an E3 ligase, has been found to link ubiquitin covalently to lysine residues of substrate proteins such as synphilin-1, CDCrel-1, far upstream element binding protein 1 (FBP-1), and alpha-synuclein.[93–96] Accumulation of substrates of parkin in the brain of PD patients interferes with the normal cellular functioning and induces neurotoxicity.[97]

5.3 Huntington's Disease

Huntington's disease (HD) is a hereditary, autosomal neurodegenerative disease that involves an expansion of CAG repeats in the huntingtin (htt) gene-coding region. The mutant htt gene causes selective neurodegeneration that primarily occurs in the striatum and further extends to other regions of the brain as the disease progresses. Global changes in the ubiquitin system has been shown to be associated with the pathology of HD.[98] While the ubiquitin-conjugating enzyme E2-25K and the E3 ligase Hrd1 have been shown to interact with and ubiquitinate both the wild type and mutant HTT protein, the E3 ligase parkin and the co-chaperone CHIP (C terminus of Hsc70-interacting protein) were found to specifically ubiquitinate the mutant protein.[99–101] In a later study, tumor necrosis factor receptor-associated factor 6 (TRAF6), an E3 ubiquitin ligase was observed to interact with and ubiquitinate the wild type and mutant HTT protein, thereby suggesting a novel role of TRAF6 in the HD pathogenesis.[102] The abnormal enrichment of inclusion bodies of HD with ubiquitin provides further indication that aberrations in ubiquitin metabolism contribute to the pathogenesis of HD.[98] Thus, finding ways to enhance the proteolytic processing that will decrease the accumulation of proteins might be beneficial. Increasing the degradation of the mutant huntingtin gene through drugs that alleviate autophagy has been reported to be an attractive strategy for the treatment of Huntington's disease.[103]

5.4 Amyotrophic Lateral Sclerosis (ALS)

ALS, a progressive neurodegenerative disorder, characterized by the presence of ubiquitinated inclusion bodies within the motor neurons, thereby indicating that aberrations in the UPS may contribute to the pathogenesis of ALS.[104] The cytoplasmic inclusions or aggregates are found in the spinal cord and other brain regions including the hippocampus, frontal and temporal cortices, and the cerebellum.[105] Abnormalities in the UPS caused by mutations in UPS-related proteins have been observed in ALS patients and in ALS animal models.[106]

5.5 Angelman Syndrome

This rare neurological disorder, characterized by mental retardation, abnormal gait, seizures, and absence of speech, occurs due to abnormalities in ubiquitin-mediated protein degradation. This syndrome occurs as a result of mutation in the E6-AP gene, which is a E3 ubiquitin ligase.[107] Although the specific targets of this enzyme are not known, it has been anticipated that the aberrant accumulation of substrates of E6-AP is responsible for Angelman syndrome.[108] Mutation in E6-AP enzyme is further known to reduce the frequency of nuclear inclusions, thereby accelerating polyglutamine-induced neurodegeneration.[109]

6 IMPLICATIONS OF UPS IN RENAL DISORDERS

Dysfunctioning of the UPS has been observed in pathogenesis of several renal diseases For instance, Liddle's syndrome and von Hippel-Lindau (VHL) are genetic disorders, with renal involvement that occurs due to dysfunction in the UPS pathway.[110] Also, renal pathologies such as ischemic acute renal failure, glomerulopathy in kidney transplants, tubulointerstitial fibrosis, class A immunoglobulin (IgA) nephropathy, lupus nephritis, and Diabetic nephropathy have also been shown to be directly or indirectly regulated by the UPS.[111]

6.1 Liddle Syndrome

This is a rare hereditary disorder, characterized by hypertension, salt sensitivity, metabolic alkalosis, and hypokalemia. Normally, Nedd4-2, which is an E3 ubiquitin ligase binds to the ENaC PY motif, thereby resulting in its ubiquitination and endocytosis. This in turn reduces the number of active channels at the plasma membrane. Meanwhile, in Liddle syndrome, the PY motif in ENac is mutated and hence does not properly bind to Nedd4-2, leading to accumulation of active channels at the cell surface.[112]

6.2 Ischemic Acute Renal Failure

In patients with acute renal failure (ARF), there is reduction in renal blood flow that leads to decline in renal function. The involvement of the proteasome-dependent proteolytic pathway in ARF pathogenesis was

TABLE 1 Summary of the List of UPS-Related Enzymes That Are Implicated in Various Diseases

Disease	Enzyme implicated
Rheumatoid arthritis (RA), systemic lupus erythematosus (SLE), Crohn's disease, psoriasis, type one diabetes, and inflammatory bowel disease	A20, an ubiquitin-editing enzyme
SLE, RA, Crohn's disease, and celiac disease	UBE2L3 (UbcH7), an E2 ubiquitin-conjugating enzyme
SLE	E3 ubiquitin ligases, TRIM21 (Ro52), and MDM2
Various cancers	P53-associated E3 ligases such as MDM2, E6-Associated Protein (E6-AP), Arf-BP1, constitutive photomorphogenesis protein 1 (COP1), and Pirh2 Other E3 ligases such as SKP1-Cullin 1-F-box protein (SCF), BRCA1 /BARD1, Von Hippel–Lindau (VHL) tumor suppressor gene, and Nedd4-like E3 ligases
Diabetic Retinopathy	E3 ubiquitin ligases such as HERC6 and NEDD4 belonging to HECT family of E3 Ligase, Topoisomerase I-binding RS protein (TOPORS), Tripartite motif-containing protein 2 (TRIM2), Seven inabsentia homolog (SIAH), Parkin, and UBR1
Alzheimer's disease	Ubiquitin carboxy-terminal hydrolase L1 (UCH-L1) and CHIP (C terminus of Hsc70-interacting protein)
Parkinson's disease	Parkin, an E3 ligase
Huntington's disease	E3 ligase parkin, CHIP
Angelman syndrome	E6-AP, an E3 ligase

demonstrated by using lactacystin, a selective proteasome inhibitor. The development of ischemic-induced ARF was found to be prevented in lactacystin administered in ischemic ARF rats.[113] A list of UPS-Related enzymes that are implicated in various diseases have been summarized (Table 1).

7 CONCLUSION AND FUTURE PERSPECTIVES

The ubiquitin-dependent proteolysis is implicated in various signaling pathways, thereby having a significant impact on various cellular processes like transcription, cell proliferation, cell cycle progression, cell differentiation, and apoptosis.[114] Thus, it is inevitable that any deregulation within the system will lead to various diseases/disorders. Aberrations in the expression of ubiquitin ligase and DUBs, as well as impairment of substrates of ubiquitin ligases, have been found to be involved in development of a number of immune disorders, pathogenesis of several genetic diseases (Angelman's syndrome, cystic fibrosis, and Liddle syndrome), in neurodegenerative diseases, and in many cancers. Hence, it is of significance to understand the dysregulation of UPS in various diseases and further manipulate the enzymes involved, in order to manage the consequent diseases. Additionally, the ability to regulate the ubiquitination at different levels provides new opportunities for targeting more components at multiple levels.

The involvement of the UPS in the degradation of crucial regulatory proteins has been exploited in the development of therapeutic strategies to treat various diseases. For instance, Bortezomib, a small molecule inhibitor of proteasomes, is being successfully used in treatment of both hematologic and solid tumors.[115] The clinical success of bortezomib revolutionized the clinical validation of proteasomes as a therapeutic target and led to the development of more drugs that target the UPS. Carfilzomib was the second proteasome inhibitor that was approved by the FDA for treatment of multiple myeloma,[116] following which, a third proteasome inhibitor, ixazomib, was approved in 2015 by the FDA. Ixazomib, which is administered as a prodrug, was the first orally available proteasome inhibitor.

Apart from inhibition of proteasome, therapeutic strategies that target the other components of the multilevel UPS are also being examined. This includes the E2, E3 ligase, and the DUBs. For instance, inhibiting the Ub-conjugating enzyme Cdc34 by using a small molecule, CC0651, was shown to suppress p27 ubiquitination effectively. However, it is yet to be pursued for therapeutic development.[117] Likewise, the efficacy of many compounds such as nutlin-3, serdemetan, and NSC-207895 to inhibit the E3 ligases MDM2 and MDMX have been studied. Both the ligases mediate the degradation of the p53 tumor suppressor.

The UPS is undoubtedly a well-characterized proteasome regulatory system. However, there are limitations in targeting specific components of the UPS, as the enzymes involved do not have a well-defined catalytic site. Identification of more substrates and their specific E2 and E3 enzymes will help to solve the enigmas associated with the mechanism behind the recognition of these substrates. This will further shed light on the mechanism underlying the pathogenesis of specific human diseases, thereby aiding in the development of potent and highly specific drugs that target single or few proteins.

References

1. Hershko A, Ciechanover A. The ubiquitin system. *Annu Rev Biochem* 1998;**67**:425–79.
2. Hershko A, Heller H, Elias S, Ciechanover A. Components of ubiquitin-protein ligase system. Resolution, affinity purification, and role in protein breakdown. *J Biol Chem* 1983;**258**:8206–14.
3. Iconomou M, Saunders DN. Systematic approaches to identify E3 ligase substrates. *Biochem J* 2016;**473**:4083–101.
4. Chau V, Tobias JW, Bachmair A, Marriott D, Ecker DJ, Gonda DK, Varshavsky A. A multiubiquitin chain is confined to specific lysine in a targeted short-lived protein. *Science* 1989;**243**:1576–83.
5. Gregori L, Poosch MS, Cousins G, Chau V. A uniform isopeptide-linked multiubiquitin chain is sufficient to target substrate for degradation in ubiquitin-mediated proteolysis. *J Biol Chem* 1990;**265**:8354–7.
6. Mayer AN, Wilkinson KD. Detection, resolution and nomenclature of multiple ubiquitin carboxyl-terminal esterases from bovine calf thymus. *Biochemistry* 1989;**28**:166–72.
7. Sigismund S, Polo S, Di Fiore PP. Signaling through monoubiquitination. *Curr Top Microbiol Immunol* 2004;**286**:149–85.
8. Ramanathan HN, Ye Y. Cellular strategies for making monoubiquitin signals. *Crit Rev Biochem Mol Biol* 2012;**47**:17–28.
9. Strieter ER, Korasick DA. Unraveling the complexity of ubiquitin signaling. *ACS Chem Biol* 2012;**7**:52–63.
10. Komander D. The emerging complexity of protein ubiquitination. *Biochem Soc Trans* 2009;**37**:937–53.
11. Panier S, Durocher D. Regulatory ubiquitylation in response to DNA double-strand breaks. *DNA Repair* 2009;**8**:436–43.
12. Akutsu M, Dikic I, Bremm A. Ubiquitin chain diversity at a glance. *J Cell Sci* 2016;**129**:875–80.
13. Wickliffe KE, Williamson A, Meyer HJ, Kelly A, Rape M. K11-linked ubiquitin chains as novel regulators of cell division. *Trends Cell Biol* 2011;**21**:656–63.
14. Malynn BA, Ma A. Ubiquitin makes its mark on immune regulation. *Immunity* 2010;**33**:843–52.
15. Fei C, Li Z, Li C, Chen Y, Chen Z, He X, Mao L, Wang X, Zeng R, Li L. Smurf1-mediated Lys29-linked nonproteolytic polyubiquitination of axin negatively regulates Wnt/beta-catenin signaling. *Mol Cell Biol* 2013;**33**:4095–105.
16. Elliott PJ, Zollner TM, Boehncke WH. Proteasome inhibition: a new anti-inflammatory strategy. *J Mol Med* 2003;**81**:235–45.
17. Basler M, Kirk CJ, Groettrup M. The immunoproteasome in antigen processing and other immunological functions. *Curr Opin Immunol* 2013;**25**:74–80.
18. Wang J, Maldonado MA. The ubiquitin-proteasome system and its role in inflammatory and autoimmune diseases. *Cell Mol Immunol* 2006;**3**:255–61.
19. Hymowitz SG, Wertz IE. A20: from ubiquitin editing to tumour suppression. *Nat Rev Cancer* 2010;**10**:332–41.
20. Tavares RM, Turer EE, Liu CL, Advincula R, Scapini P, Rhee L, Barrera J, Lowell CA, Utz PJ, Malynn BA, Ma A. The ubiquitin modifying enzyme A20 restricts B cell survival and prevents autoimmunity. *Immunity* 2010;**33**:181–91.
21. Lewis MJ, Vyse S, Shields AM, Boeltz S, Gordon PA, Spector TD, Lehner PJ, Walczak H, Vyse TJ. UBE2L3 polymorphism amplifies NF-kappaB activation and promotes plasma cell development, linking linear ubiquitination to multiple autoimmune diseases. *Am J Hum Genet* 2015;**96**:221–34.

22. Yoshimi R, Ishigatsubo Y, Ozato K. Autoantigen TRIM21/Ro52 as a possible target for treatment of systemic lupus erythematosus. *Int J Rheum* 2012;**2012**:11.
23. Allam R, Sayyed SG, Kulkarni OP, Lichtnekert J, Anders HJ. Mdm2 promotes systemic lupus erythematosus and lupus nephritis. *J Am Soc Nephrol* 2011;**22**:2016–27.
24. Bashir T, Pagano M. Aberrant ubiquitin-mediated proteolysis of cell cycle regulatory proteins and oncogenesis. *Adv Cancer Res* 2003;**88**:101–44.
25. Wang Y, Ren F, Feng Y, Wang D, Jia B, Qiu Y, Wang S, Yu J, Sung JJ, Xu J, Zeps N, Chang Z. CHIP/Stub1 functions as a tumor suppressor and represses NF-kappaB-mediated signaling in colorectal cancer. *Carcinogenesis* 2014;**35**:983–91.
26. Wang T, Yang J, Xu J, Li J, Cao Z, Zhou L, You L, Shu H, Lu Z, Li H, Li M, Zhang T, Zhao Y. CHIP is a novel tumor suppressor in pancreatic cancer through targeting EGFR. *Oncotarget* 2014;**5**:1969–86.
27. Honda R, Tanaka H, Yasuda H. Oncoprotein MDM2 is a ubiquitin ligase E3 for tumor suppressor p53. *FEBS Lett* 1997;**420**:25–7.
28. Momand J, Zambetti GP, Olson DC, George D, Levine AJ. The mdm-2 oncogene product forms a complex with the p53 protein and inhibits p53-mediated transactivation. *Cell* 1992;**69**:1237–45.
29. Haupt Y, Maya R, Kazaz A, Oren M. Mdm2 promotes the rapid degradation of p53. *Nature* 1997;**387**:296–9.
30. Chen D, Brooks CL, Gu W. ARF-BP1 as a potential therapeutic target. *Br J Cancer* 2006;**94**:1555–8.
31. Dornan D, Wertz I, Shimizu H, Arnott D, Frantz GD, Dowd P, O'Rourke K, Koeppen H, Dixit VM. The ubiquitin ligase COP1 is a critical negative regulator of p53. *Nature* 2004;**429**:86.
32. Dornan D, Bheddah S, Newton K, Ince W, Frantz GD, Dowd P, Koeppen H, Dixit VM, French DM. COP1, the negative regulator of p53, is overexpressed in breast and ovarian adenocarcinomas. *Cancer Res* 2004;**64**:7226–30.
33. Leng RP, Lin Y, Ma W, Wu H, Lemmers B, Chung S, Parant JM, Lozano G, Hakem R, Benchimol S. Pirh2, a p53-induced ubiquitin-protein ligase, promotes p53 degradation. *Cell* 2003;**112**:779–91.
34. Frescas D, Pagano M. Deregulated proteolysis by the F-box proteins SKP2 and β-TrCP: tipping the scales of cancer. *Nat Rev Cancer* 2008;**8**:438–49.
35. Bornstein G, Bloom J, Sitry-Shevah D, Nakayama K, Pagano M, Hershko A. Role of the SCFSkp2 ubiquitin ligase in the degradation of p21Cip1 in S phase. *J Biol Chem* 2003;**278**:25752–7.
36. Kamura T, Hara T, Kotoshiba S, Yada M, Ishida N, Imaki H, Hatakeyama S, Nakayama K, Nakayama KI. Degradation of p57Kip2 mediated by SCFSkp2-dependent ubiquitylation. *Proc Natl Acad Sci U S A* 2003;**100**:10231–6.
37. Traub F, Mengel M, Luck HJ, Kreipe HH, Von Wasielewski R. Prognostic impact of Skp2 and p27 in human breast cancer. *Breast Cancer Res Treat* 2006;**99**:185–91.
38. Shapira M, Ben-Izhak O, Linn S, Futerman B, Minkov I, Hershko DD. The prognostic impact of the ubiquitin ligase subunits Skp2 and Cks1 in colorectal carcinoma. *Cancer* 2005;**103**:1336–46.
39. Kossatz U, Dietrich N, Zender L, Buer J, Manns MP, Malek NP. Skp2-dependent degradation of p27kip1 is essential for cell cycle progression. *Genes Dev* 2004;**18**:2602–7.
40. Lu M, Ma J, Xue W, Cheng C, Wang Y, Zhao Y, Ke Q, Liu H, Liu Y, Li P, Cui X, He S, Shen A. The expression and prognosis of FOXO3a and Skp2 in human hepatocellular carcinoma. *Pathol Oncol Res* 2009;**15**:679–87.
41. Radke S, Pirkmaier A, Germain D. Differential expression of the F-box proteins Skp2 and Skp2B in breast cancer. *Oncogene* 2005;**24**:3448–58.
42. Schuler S, Diersch S, Hamacher R, Schmid RM, Saur D, Schneider G. SKP2 confers resistance of pancreatic cancer cells towards TRAIL-induced apoptosis. *Int J Oncol* 2011;**38**:219–25.

43. Wang Z, Gao D, Fukushima H, Inuzuka H, Liu P, Wan L, Sarkar FH, Wei W. Skp2: a novel potential therapeutic target for prostate cancer. *Biochim Biophys Acta* 2012;**1**:11–7.
44. Fang FM, Chien CY, Li CF, Shiu WY, Chen CH, Huang HY. Effect of S-phase kinase-associated protein 2 expression on distant metastasis and survival in nasopharyngeal carcinoma patients. *Int J Radiat Oncol Biol Phys* 2009;**73**:202–7.
45. Welcker M, Clurman BE. FBW7 ubiquitin ligase: a tumour suppressor at the crossroads of cell division, growth and differentiation. *Nat Rev Cancer* 2008;**8**:83–93.
46. Rajagopalan H, Jallepalli PV, Rago C, Velculescu VE, Kinzler KW, Vogelstein B, Lengauer C. Inactivation of hCDC4 can cause chromosomal instability. *Nature* 2004;**428**:77–81.
47. Lee JW, Soung YH, Kim HJ, Park WS, Nam SW, Kim SH, Lee JY, Yoo NJ, Lee SH. Mutational analysis of the hCDC4 gene in gastric carcinomas. *Eur J Cancer* 2006;**42**:2369–73.
48. Akhoondi S, Sun D, Von Der Lehr N, Apostolidou S, Klotz K, Maljukova A, Cepeda D, Fiegl H, Dofou D, Marth C, Mueller-Holzner E, Corcoran M, Dagnell M, Nejad SZ, Nayer BN, Zali MR, Hansson J, Egyhazi S, Petersson F, Sangfelt P, Nordgren H, Grander D, Reed SI, Widschwendter M, Sangfelt O, Spruck C. FBXW7/hCDC4 is a general tumor suppressor in human cancer. *Cancer Res* 2007;**67**:9006–12.
49. Ford D, Easton DF, Stratton M, Narod S, Goldgar D, Devilee P, Bishop DT, Weber B, Lenoir G, Chang-Claude J, Sobol H, Teare MD, Struewing J, Arason A, Scherneck S, Peto J, Rebbeck TR, Tonin P, Neuhausen S, Barkardottir R, Eyfjord J, Lynch H, Ponder BA, Gayther SA, Zelada-Hedman M, Et Al. Genetic heterogeneity and penetrance analysis of the BRCA1 and BRCA2 genes in breast cancer families. The breast cancer linkage consortium. *Am J Hum Genet* 1998;**62**:676–89.
50. Wu W, Koike A, Takeshita T, Ohta T. The ubiquitin E3 ligase activity of BRCA1 and its biological functions. *Cell Div* 2008;**3**:1.
51. Moran A, O'hara C, Khan S, Shack L, Woodward E, Maher ER, Lalloo F, Evans DG. Risk of cancer other than breast or ovarian in individuals with BRCA1 and BRCA2 mutations. *Familial Cancer* 2012;**11**:235–42.
52. Lonser RR, Glenn GM, Walther M, Chew EY, Libutti SK, Linehan WM, Oldfield EH. von Hippel-Lindau disease. *Lancet* 2003;**361**:2059–67.
53. Gnarra JR, Tory K, Weng Y, Schmidt L, Wei MH, Li H, Latif F, Liu S, Chen F, Duh FM, et al. Mutations of the VHL tumour suppressor gene in renal carcinoma. *Nat Genet* 1994;**7**:85–90.
54. Maxwell PH, Wiesener MS, Chang GW, Clifford SC, Vaux EC, Cockman ME, Wykoff CC, Pugh CW, Maher ER, Ratcliffe PJ. The tumour suppressor protein VHL targets hypoxia-inducible factors for oxygen-dependent proteolysis. *Nature* 1999;**399**:271–5.
55. Chen C, Matesic LE. The Nedd4-like family of E3 ubiquitin ligases and cancer. *Cancer Metastasis Rev* 2007;**26**:587–604.
56. Wang X, Trotman LC, Koppie T, Alimonti A, Chen Z, Gao Z, Wang J, Erdjument-Bromage H, Tempst P, Cordon-Cardo C, Pandolfi PP, Jiang X. NEDD4-1 is a proto-oncogenic ubiquitin ligase for PTEN. *Cell* 2007;**128**:129–39.
57. Fukuchi M, Fukai Y, Masuda N, Miyazaki T, Nakajima M, Sohda M, Manda R, Tsukada K, Kato H, Kuwano H. High-level expression of the Smad ubiquitin ligase Smurf2 correlates with poor prognosis in patients with esophageal squamous cell carcinoma. *Cancer Res* 2002;**62**:7162–5.
58. Chen C, Zhou Z, Sheehan CE, Slodkowska E, Sheehan CB, Boguniewicz A, Ross JS. Overexpression of WWP1 is associated with the estrogen receptor and insulin-like growth factor receptor 1 in breast carcinoma. *Int J Cancer* 2009;**124**:2829–36.
59. Chen C, Zhou Z, Zhou W, Dong J-T, Seth A. The WWP1 E3 ubiquitin ligase: a potential molecular target for breast cancer. *Cancer Res* 2007;**67**:353.
60. Chen C, Sun X, Guo P, Dong XY, Sethi P, Cheng X, Zhou J, Ling J, Simons JW, Lingrel JB, Dong JT. Human kruppel-like factor 5 is a target of the E3 ubiquitin ligase WWP1 for proteolysis in epithelial cells. *J Biol Chem* 2005;**280**:41553–61.

61. Bateman NW, Tan D, Pestell RG, Black JD, Black AR. Intestinal tumor progression is associated with altered function of KLF5. *J Biol Chem* 2004;**279**:12093–101.
62. Laine A, Ronai ZE. Regulation of p53 localization and transcription by the HECT domain E3 ligase WWP1. *Oncogene* 2007;**26**:1477–83.
63. Cummins JM, Rago C, Kohli M, Kinzler KW, Lengauer C, Vogelstein B. Tumour suppression: disruption of HAUSP gene stabilizes p53. *Nature* 2004;**428**:486–7.
64. Li M, Brooks CL, Kon N, Gu W. A dynamic role of HAUSP in the p53-Mdm2 pathway. *Mol Cell* 2004;**13**:879–86.
65. Van Der Horst A, De Vries-Smits AM, Brenkman AB, Van Triest MH, Van Den Broek N, Colland F, Maurice MM, Burgering BM. FOXO4 transcriptional activity is regulated by monoubiquitination and USP7/HAUSP. *Nat Cell Biol* 2006;**8**:1064–73.
66. Song MS, Salmena L, Carracedo A, Egia A, Lo-Coco F, Teruya-Feldstein J, Pandolfi PP. The deubiquitinylation and localization of PTEN are regulated by a HAUSP–PML network. *Nature* 2008;**455**:813–7.
67. Popov N, Wanzel M, Madiredjo M, Zhang D, Beijersbergen R, Bernards R, Moll R, Elledge SJ, Eilers M. The ubiquitin-specific protease USP28 is required for MYC stability. *Nat Cell Biol* 2007;**9**:765.
68. Dupont S, Mamidi A, Cordenonsi M, Montagner M, Zacchigna L, Adorno M, Martello G, Stinchfield MJ, Soligo S, Morsut L, Inui M, Moro S, Modena N, Argenton F, Newfeld SJ, Piccolo S. FAM/USP9x, a deubiquitinating enzyme essential for TGFbeta signaling, controls Smad4 monoubiquitination. *Cell* 2009;**136**:123–35.
69. Stevenson LF, Sparks A, Allende-Vega N, Xirodimas DP, Lane DP, Saville MK. The deubiquitinating enzyme USP2a regulates the p53 pathway by targeting Mdm2. *EMBO J* 2007;**26**:976–86.
70. Graner E, Tang D, Rossi S, Baron A, Migita T, Weinstein LJ, Lechpammer M, Huesken D, Zimmermann J, Signoretti S, Loda M. The isopeptidase USP2a regulates the stability of fatty acid synthase in prostate cancer. *Cancer Cell* 2004;**5**:253–61.
71. Kim J, Kim W-J, Liu Z, Loda MF, Freeman MR. The ubiquitin-specific protease USP2a enhances tumor progression by targeting cyclin A1 in bladder cancer. *Cell Cycle* 2012;**11**:1123–30.
72. Song HY, Rothe M, Goeddel DV. The tumor necrosis factor-inducible zinc finger protein A20 interacts with TRAF1/TRAF2 and inhibits NF-kappaB activation. *Proc Natl Acad Sci U S A* 1996;**93**:6721–5.
73. Trompouki E, Hatzivassiliou E, Tsichritzis T, Farmer H, Ashworth A, Mosialos G. CYLD is a deubiquitinating enzyme that negatively regulates NF-kappaB activation by TNFR family members. *Nature* 2003;**424**:793–6.
74. Massoumi R, Chmielarska K, Hennecke K, Pfeifer A, Fassler R. Cyld inhibits tumor cell proliferation by blocking Bcl-3-dependent NF-kappaB signaling. *Cell* 2006;**125**:665–77.
75. Noda K, Nakao S, Ishida S, Ishibashi T. Leukocyte adhesion molecules in diabetic retinopathy. *J Ophthalmol* 2012;**279037**:2.
76. Feenstra DJ, Yego EC, Mohr S. Modes of retinal cell death in diabetic retinopathy. *J Clin Exp Ophthalmol* 2013;**4**:2155–9570.
77. Tang J, Kern TS. Inflammation in diabetic retinopathy. *Prog Retin Eye Res* 2011;**30**:343–58.
78. Mohammad G, Mairaj SM, Imtiaz NM, Abu El-Asrar AM. The ERK1/2 inhibitor U0126 attenuates diabetes-induced upregulation of MMP-9 and biomarkers of inflammation in the retina. *J Diabetes Res* 2013;**658548**:10.
79. Campello L, Esteve-Rudd J, Cuenca N, Martin-Nieto J. The ubiquitin-proteasome system in retinal health and disease. *Mol Neurobiol* 2013;**47**:790–810.
80. Layfield R, Alban A, Mayer RJ, Lowe J. The ubiquitin protein catabolic disorders. *Neuropathol Appl Neurobiol* 2001;**27**:171–9.
81. Layfield R, Lowe J, Bedford L. The ubiquitin-proteasome system and neurodegenerative disorders. *Essays Biochem* 2005;**41**:157–71.

82. Grundke-Iqbal I, Iqbal K, Tung YC, Quinlan M, Wisniewski HM, Binder LI. Abnormal phosphorylation of the microtubule-associated protein tau (tau) in alzheimer cytoskeletal pathology. *Proc Natl Acad Sci U S A* 1986;**83**:4913–7.
83. Perry G, Friedman R, Shaw G, Chau V. Ubiquitin is detected in neurofibrillary tangles and senile plaque neurites of alzheimer disease brains. *Proc Natl Acad Sci U S A* 1987;**84**:3033–6.
84. Tabaton M, Cammarata S, Mancardi G, Manetto V, Autilio-Gambetti L, Perry G, Gambetti P. Ultrastructural localization of beta-amyloid, tau, and ubiquitin epitopes in extracellular neurofibrillary tangles. *Proc Natl Acad Sci U S A* 1991;**88**:2098–102.
85. Castegna A, Aksenov M, Aksenova M, Thongboonkerd V, Klein JB, Pierce WM, Booze R, Markesbery WR, Butterfield DA. Proteomic identification of oxidatively modified proteins in alzheimer's disease brain. Part I: creatine kinase BB, glutamine synthase, and ubiquitin carboxy-terminal hydrolase L-1. *Free Radic Biol Med* 2002;**33**:562–71.
86. Forero DA, Casadesus G, Perry G, Arboleda H. Synaptic dysfunction and oxidative stress in alzheimer's disease: emerging mechanisms. *J Cell Mol Med* 2006;**10**:796–805.
87. Keller JN, Hanni KB, Markesbery WR. Impaired proteasome function in alzheimer's disease. *J Neurochem* 2000;**75**:436–9.
88. Petrucelli L, Dickson D, Kehoe K, Taylor J, Snyder H, Grover A, De Lucia M, Mcgowan E, Lewis J, Prihar G, Kim J, Dillmann WH, Browne SE, Hall A, Voellmy R, Tsuboi Y, Dawson TM, Wolozin B, Hardy J, Hutton M. CHIP and Hsp70 regulate tau ubiquitination, degradation and aggregation. *Hum Mol Genet* 2004;**13**:703–14.
89. Shimura H, Schwartz D, Gygi SP, Kosik KS. CHIP-Hsc70 complex Ubiquitinates phosphorylated tau and enhances cell survival. *J Biol Chem* 2004;**279**:4869–76.
90. Dickey CA, Yue M, Lin WL, Dickson DW, Dunmore JH, Lee WC, Zehr C, West G, Cao S, Clark AM, Caldwell GA, Caldwell KA, Eckman C, Patterson C, Hutton M, Petrucelli L. Deletion of the ubiquitin ligase CHIP leads to the accumulation, but not the aggregation, of both endogenous phospho- and caspase-3-cleaved tau species. *J Neurosci* 2006;**26**:6985–96.
91. Parker Jr WD, Parks JK, Swerdlow RH. Complex I deficiency in parkinson's disease frontal cortex. *Brain Res* 2008;**16**:215–8.
92. Kitada T, Asakawa S, Hattori N, Matsumine H, Yamamura Y, Minoshima S, Yokochi M, Mizuno Y, Shimizu N. Mutations in the parkin gene cause autosomal recessive juvenile parkinsonism. *Nature* 1998;**392**:605–8.
93. Chung KK, Zhang Y, Lim KL, Tanaka Y, Huang H, Gao J, Ross CA, Dawson VL, Dawson TM. Parkin ubiquitinates the alpha-synuclein-interacting protein, synphilin-1: implications for lewy-body formation in parkinson disease. *Nat Med* 2001;**7**:1144–50.
94. Zhang Y, Gao J, Chung KK, Huang H, Dawson VL, Dawson TM. Parkin functions as an E2-dependent ubiquitin- protein ligase and promotes the degradation of the synaptic vesicle-associated protein, CDCrel-1. *Proc Natl Acad Sci U S A* 2000;**97**:13354–9.
95. Imam SZ, Zhou Q, Yamamoto A, Valente AJ, Ali SF, Bains M, Roberts JL, Kahle PJ, Clark RA, Li S. Novel regulation of parkin function through c-Abl-mediated tyrosine phosphorylation: implications for parkinson's disease. *J Neurosci* 2011;**31**:157–63.
96. Shimura H, Schlossmacher MG, Hattori N, Frosch MP, Trockenbacher A, Schneider R, Mizuno Y, Kosik KS, Selkoe DJ. Ubiquitination of a new form of alpha-synuclein by parkin from human brain: implications for parkinson's disease. *Science* 2001;**293**:263–9.
97. Lim KL, Lim TM. Molecular mechanisms of neurodegeneration in parkinson's disease: clues from mendelian syndromes. *IUBMB Life* 2003;**55**:315–22.
98. Bennett EJ, Shaler TA, Woodman B, Ryu KY, Zaitseva TS, Becker CH, Bates GP, Schulman H, Kopito RR. Global changes to the ubiquitin system in Huntington's disease. *Nature* 2007;**448**:704–8.

99. De Pril R, Fischer DF, Roos RA, Van Leeuwen FW. Ubiquitin-conjugating enzyme E2-25K increases aggregate formation and cell death in polyglutamine diseases. *Mol Cell Neurosci* 2007;**34**:10–9.
100. Yang H, Zhong X, Ballar P, Luo S, Shen Y, Rubinsztein DC, Monteiro MJ, Fang S. Ubiquitin ligase Hrd1 enhances the degradation and suppresses the toxicity of polyglutamine-expanded huntingtin. *Exp Cell Res* 2007;**313**:538–50.
101. Jana NR, Dikshit P, Goswami A, Kotliarova S, Murata S, Tanaka K, Nukina N. Co-chaperone CHIP associates with expanded polyglutamine protein and promotes their degradation by proteasomes. *J Biol Chem* 2005;**280**:11635–40.
102. Zucchelli S, Marcuzzi F, Codrich M, Agostoni E, Vilotti S, Biagioli M, Pinto M, Carnemolla A, Santoro C, Gustincich S, Persichetti F. Tumor necrosis factor receptor-associated factor 6 (TRAF6) associates with huntingtin protein and promotes its atypical ubiquitination to enhance aggregate formation. *J Biol Chem* 2011;**286**:25108–17.
103. Sarkar S, Rubinsztein DC. Huntington's disease: degradation of mutant huntingtin by autophagy. *FEBS J* 2008;**275**:4263–70.
104. Schmidt M, Finley D. Regulation of proteasome activity in health and disease. *Biochim Biophys Acta* 2014;**1843**. https://doi.org/10.1016/j.bbamcr.2013.1008.1012.
105. Al-Chalabi A, Jones A, Troakes C, King A, Al-Sarraj S, Van Den Berg LH. The genetics and neuropathology of amyotrophic lateral sclerosis. *Acta Neuropathol* 2012;**124**:339–52.
106. Cheroni C, Peviani M, Cascio P, Debiasi S, Monti C, Bendotti C. Accumulation of human SOD1 and ubiquitinated deposits in the spinal cord of SOD1G93A mice during motor neuron disease progression correlates with a decrease of proteasome. *Neurobiol Dis* 2005;**18**:509–22.
107. Matentzoglu K, Scheffner M. Ubiquitin ligase E6-AP and its role in human disease. *Biochem Soc Trans* 2008;**36**:797–801.
108. Kishino T, Lalande M, Wagstaff J. UBE3A/E6-AP mutations cause angelman syndrome. *Nat Genet* 1997;**15**:70–3.
109. Cummings CJ, Reinstein E, Sun Y, Antalffy B, Jiang Y, Ciechanover A, Orr HT, Beaudet AL, Zoghbi HY. Mutation of the E6-AP ubiquitin ligase reduces nuclear inclusion frequency while accelerating polyglutamine-induced pathology in SCA1 mice. *Neuron* 1999;**24**:879–92.
110. Reinstein E, Ciechanover A. Narrative review: protein degradation and human diseases: the ubiquitin connection. *Ann Intern Med* 2006;**145**:676–84.
111. Coppo R, Camilla R, Alfarano A, Balegno S, Mancuso D, Peruzzi L, Amore A, Dal CA, Sepe V, Tovo P. Upregulation of the immunoproteasome in peripheral blood mononuclear cells of patients with IgA nephropathy. *Kidney Int* 2009;**75**:536–41.
112. Rotin D. Role of the UPS in liddle syndrome. *BMC Biochem* 2008;**9**:S5.
113. Itoh M, Takaoka M, Shibata A, Ohkita M, Matsumura Y. Preventive effect of lactacystin, a selective proteasome inhibitor, on ischemic acute renal failure in rats. *J Pharmacol Exp Ther* 2001;**298**:501–7.
114. Voges D, Zwickl P, Baumeister W. The 26S proteasome: a molecular machine designed for controlled proteolysis. *Annu Rev Biochem* 1999;**68**:1015–68.
115. Roccaro AM, Vacca A, Ribatti D. Bortezomib in the treatment of cancer. *Recent Pat Anticancer Drug Discov* 2006;**1**:397–403.
116. Moreau P. The emerging role of carfilzomib combination therapy in the management of multiple myeloma. *Expert Rev Hematol* 2014;**7**:265–90.
117. Ceccarelli DF, Tang X, Pelletier B, Orlicky S, Xie W, Plantevin V, Neculai D, Chou YC, Ogunjimi A, Al-Hakim A, Varelas X, Koszela J, Wasney GA, Vedadi M, Dhe-Paganon S, Cox S, Xu S, Lopez-Girona A, Mercurio F, Wrana J, Durocher D, Meloche S, Webb DR, Tyers M, Sicheri F. An allosteric inhibitor of the human Cdc34 ubiquitin-conjugating enzyme. *Cell* 2011;**145**:1075–87.

CHAPTER

6

Role of Glycosylation in Modulating Therapeutic Efficiency of Protein Pharmaceuticals

Parvaiz Ahmad Dar, Usma Manzoor, Snowber Shabir Wani, Fasil Ali, Tanveer Ali Dar

Clinical Biochemistry, University of Kashmir, Srinagar, India

1 INTRODUCTION

The last three decades have been very important in the field of medicine, as proteins have been presented as a major new class of pharmaceuticals.[1] To date, more than 100 proteins have been acknowledged as therapeutics and a large number are under clinical trials.[2] These therapeutic proteins exhibit numerous favorable characteristics, particularly pharmacological potencies, low toxicity, and enhanced target specificities. All these do play a vital role in the treatment of diseases like cancers, autoimmune diseases, and various metabolic disorders. However the suboptimum therapeutic potential, due to intrinsic shortcomings in their pharmacological and physicochemical properties, delay their development and employment as effective therapeutics.[3–11] Low molecular stabilities, reduced pharmacodynamic response, and inadequate pharmacokinetic profiles are intrinsic limitations that not only influence the therapeutic potential of protein pharmaceutics, but also decrease their clinical effectiveness.[11–15]

So far, various strategies have been employed to stabilize the protein drugs and minimize their intrinsic limitations to optimize their therapeutic efficacy. Such strategies mainly involve some well-established technologies that significantly enhance the serum half-life and molecular stabilities of the

Protein Modificomics
https://doi.org/10.1016/B978-0-12-811913-6.00006-0

therapeutic protein through the generation of fusion proteins, targeted mutations, and glycosylation engineering.[4, 16–22] Among all these strategies, manipulation of surface glycosylation patterns of proteins through glycoengineering serves as a most promising approach.[10, 22–25] The presence of additional glycosylation sites on engineered proteins has been found to enhance their serum half-life and the target exposure.[10, 26] Glycoengineering has, in fact, been found to provide a diverse range of possibilities in improving therapeutic properties of protein drugs and ameliorating a majority of the constraints essential for optimization of therapeutic protein drug behavior. In addition, this approach considerably modulates solubility, immunogenicity, bioactivity, and clearance rate of therapeutic proteins from circulation. Keeping this in view, the present chapter was designed to provide an updated scenario on protein glycosylation as an important strategy for optimizing the therapeutic efficiency of protein pharmaceuticals. Future directions in this regard have also been discussed.

2 CHALLENGES ASSOCIATED WITH PROTEIN PHARMACEUTICALS

In comparison to conventional, nonprotein drugs, certain important properties of protein pharmaceuticals can be customized, like target specificity, stability, etc. These highly specific protein therapeutics do not interfere with normal biological processes and are usually well accepted within the human body.[27] They also surpass the need for gene therapy by offering effective replacement therapy in case of genetic diseases. They may also take less time to get approval by authorities.[1, 27] Even though the protein pharmaceuticals are highly advantageous, they put forward many challenges in their manufacturing, development, and clinical performance. Some of the challenges associated with these protein pharmaceuticals that have largely affected protein drug discovery are discussed below.

2.1 Physical Instability

The majority of proteins fold into a tertiary structure in the aqueous solution, so as to minimize the hydrophobic residues exposure toward polar aqueous environment.[28] Different types of atomic interactions (viz., hydrogen bonding, van der walls, and electrostatic interactions, etc.) play a key role in burying these hydrophobic residues within the protein core during folding and form an energetically stable compact hydrophobic core. However, because of the noncovalent nature of these interactions, the kinetic and thermodynamic stability of this state inclines to be intrinsically low and prone to structural changes. Disruption of these forces by any physio-chemical process will eventually lead to reduced pharmaceutical function of these protein drugs. In addition, conformationally altered protein species are at risk of interacting with themselves or other hydrophobic surfaces/molecules. This leads to added physical instabilities, such as, precipitation, aggregation, and adsorption during manufacturing, storage, and transportation of protein pharmaceuticals. Thus, rational design of engineering strategies to increase the stability and retaining therapeutic activity of protein during production, purification, storage, and administration is primarily important and challenging.

2.2 Aggregation

In some therapeutic proteins, i.e., antibodies and insulin, etc., used to treat many grave diseases like cancer, infectious diseases, metabolic disorders, and inflammation, aggregation has been observed to be a major concern.[29] Nearly all proteins are vulnerable to some sort of aggregation[30–33] and these aggregates lessen the pharmacological properties of protein in the context of therapeutics. The aggregates also stimulate an immune response and consequently lead to production of antibodies against native proteins, resulting in severe harmful side effects in the patients. Additionally, protein aggregates are cytotoxic on their own[34–36] and hence, prevention of aggregation is a challenging task in protein pharmaceutical development.

2.3 Solubility Issue

The development of a protein formulation with a particular concentration for dosing purposes is a very difficult job because of their limited solubility.[37–39] Development of highly concentrated protein drug formulations leads to their aggregation. In addition to this, reversible self-association of proteins and particulate formation may arise, adding up to a viscosity of protein drug solution that worsens their administration by injection.[38, 40, 41] Enhanced viscosity of protein drugs leads to impediment in the filtration processes during manufacturing. Hence, an integrated approach is required to achieve a suitable and desired protein drug formulation with a stable protein concentration that can be economically manufactured and successfully administered.

2.4 Chemical Degradation

Amino acids like Tyr, His, Cys, Met, and Trp in the pharmaceutical proteins are prone to oxidation, which in turn alters their bioactivity, especially during production and transportation. These oxidation events produce active oxygen-based radicals in the protein formulations.[12, 42] This form of chemical instability (oxidation) is a major concern in pharmaceutical proteins, responsible for the decrease in their therapeutic potential.[42] Therefore, there is a need to address this issue so that the pharmacological significance of therapeutic proteins can be improved.

2.5 Degradation by Proteases

Protein drugs administered via oral route are chemically damaged by proteases of the digestive system.[15] This approach leads to their reduced bioavailability and, therefore, other means of protein drug administration such as intravenous, intramuscular, etc. have been exploited. In fact, due to constitutive expression of proteases, therapeutic proteins are susceptible to proteolytic degradation irrespective of any means of drug administration.[15, 22]

3 PROTEIN GLYCOSYLATION AND ITS PHARMACOLOGICAL SIGNIFICANCE

Glycosylation, being the most common protein posttranslational modification, mainly involves covalent linkage of carbohydrate moieties with asparagine (N-linked glycosylation) and threonine or serine residues (O-linked glycosylation). This process has been known to influence several vital biochemical processes at both protein (molecular stability, protein-protein binding) and cellular levels (intracellular targeting). The favorable impact of glycosylation on proteins has been exploited as an important approach to improve the pharmacokinetic profiles of most of the protein drugs. So far, various technological advancements have been developed but glycoengineering is considered the most promising approach for enhancement of pharmacological properties of therapeutically important proteins. This approach involves the modification of glycosylational sites and patterns and hence ameliorates the majority of parameters important for optimization of protein drug behavior. Although the therapeutic potential of glycoengineered proteins is completely determined by the protein part,[22] the carbohydrate moieties do have a major role in solubility, immunogenicity, and serum half-life of the glycoproteins.[22] Similarly, other pharmacological advantages include increased bio availability in tissues, enhanced biological membranes penetration, increased stability, and reduced clearance rate, and also protection of side chains of amino acids and hydrogen backbone from oxidation.[16, 43–45] Table 1 summarizes effect of glycosylation on pharmacological properties of therapeutically important proteins.

TABLE 1 Effect of Glycosylation on the Structural and Pharmacological Properties of Therapeutic Proteins.

Protein	Effect	Application	Reference
IFN-2α	Increased serum half-life	Cancer, viral infections	46
Butyrylcholinesterase	PEGylation	Antidote against highly toxic, nerve agents	47
EGF	Sustained release of protein drugs	Wound healing	48
GLP-1	Increases half-life and stability	Diabetes	49
Chymotrypsin	Protects against protein aggregation	Adjunct therapy for multiple myeloma	50
Insulin	Protects against nondisulfide crosslinking and aggregation	Treatment of diabetes	51
RNAse	Prevents thermal denaturation and proteolytic degradation	Treatment of malignant mesothelioma	52,53
Bucelipase alfa (cholesterol esterase)	Protects against proteolytic degradation	Treatment of lipid malabsorption related to exocrine pancreatic	54
Drotrecogin alfa	Protects against proteolytic degradation	Treatment of severe sepsis	55

4 MODULATION OF THERAPEUTICALLY IMPORTANT PROPERTIES OF PROTEINS BY GLYCOENGINEERING

4.1 Physical and Chemical Properties

Glycoengineering enhances the resistance of therapeutic proteins against proteolytic degradation, in which the carbohydrate moieties impede the access of proteases to target amino acid sites, e.g., the carbohydrate in the fibronectin protects it against proteolytic degradation.[11, 21, 56] The hydrophilic carbohydrate molecules increase protein solubility by masking the hydrophobic residues and increasing the accessible surface area. This hydrophilic nature of carbohydrate moieties also helps in attaining a desirable protein concentration of drug without the formation of aggregates and precipitates.[56, 57] It also influences the adhesive properties of proteins that help in their sustainable release and proper bio distribution.[58] For example the heavily glycosylated sialic acids on mucins increase their gel-like properties. Similarly glycoengineered leptin showed more than a 15-fold increase in solubility as compared to wild type leptin.[58] Additionally, the precipitates and aggregates formed in unglycosylated proteins displayed a prominent immune response as compared to glycomodified protein. Due to this, manufacture and storage of protein pharmaceuticals resulted in reduced bioactivity due to oxidation of amino acids.[9, 10, 40, 56]

Glycosylation has been quite successful in ameliorating this kind of chemical instability in erythropoietin, where the loss of bioactivity correlated with tryptophan oxidation directly.[59] At extremes of pH, charge-charge interaction as well as internal electrostatic interactions are disrupted in the protein.[56] Furthermore, glycomodification has been shown to maintain the conformational stability of many proteins such as G-CSF, erythropoietin, glucose oxidase, fibronectin, acid phosphatase, and tripeptidyl peptidase against pH denaturation. Mechanistically, it enhances internal electrostatic interactions within the protein, which in turn stabilizes the protein conformation. A comparative in silico structural and energetic analysis conducted on a series of chemically glycosylated α-chymotrypsin conjugated with increased levels of glycosylation also supports this mechanism.[10] These glycans in protein act as molecular spacers, increasing the effective distance between solvent electrostatics and the protein electrostatics. Furthermore, the smaller dielectric screening by neighboring water molecules help in enhancing the strength of internal electrostatic force. Likewise, the interactions stabilizing the native state of protein are also temperature sensitive, and glycosylation influences this property in a positive way. In fact, as expected, the thermal denaturation susceptibility of protein is used as a principal stability indicator for a protein formulation strategy.

Protein solubility has been shown to be inversely proportional to its concentration, and while designing protein-based formulation, the achievement of desired concentrations of protein drugs in solution is a fundamental challenge. During formulation of protein drugs, precipitation becomes evident as the target concentration is increased. Many proteins have been known where glycosylation increased the solubility, i.e., interferon beta (REBIF1), alpha-galactosidase A (REPLAGAL1), and invertase. Tams et al.[60] reported that the solubility of the protein is directly proportional to the degree of glycosylation.[60] The greater hydration potential of glycans (glycosylation) is responsible for increased solubility of proteins, as these sugar residues possess higher aqueous solvent affinity than the polypeptide chain. Alternatively, simulation studies have also shown that glycosylation increases the overall molecular solvent accessible surface area of the glycosylated protein. This suggests that the possible interactions between the surrounding solvent molecules and glycan residues of the glycoprotein increase and thus lead to increased solubility of protein.

4.2 Aggregation

In addition to factors like protein concentration, pH, and temperature, the colloidal nature of therapeutic proteins makes them susceptible to aggregation.[61, 62] Such aggregation of therapeutic proteins, in turn, shows potential undesirable effects, not only on the patient, but also makes it noncost effective, due to refolding protocols and additional downstream protein recovery processes.[57] Glycosylation has been found to reduce or prevent the aggregate formation. For example, the physical stability of insulin has been shown to improve when a small sized glycan was chemically attached to it.[51] This glycan reduced the aggregation kinetics of insulin by preventing the cross linking transamidation reactions.[51] Another study has shown that glycosylation of α-galactosidase A (Replagal 1) at aspargine$_{215}$ prevents its aggregation by thwarting the exposure of hydrophobic patches present on the protein surface.[63] Similarly, deglycosylation of thyroid-stimulating hormone (Thyrogen1; Genzyme) increases its susceptibility to aggregation.[64] Hoiberg-Nielsen et al., also found that the colloidal stability of phytase increased to a considerable extent upon glycosylation.[65] In addition to this, Sola et al. found that small sized glycan molecules were less competent to inhibit protein aggregation than large sized glycans, which inhibited aggregation completely in the case of extreme conditions (protein concentration 20 mg/mL and temperature 60°C).[10] This study helped to understand the impact of systemic changes of glycosylation parameters on protein aggregation.[10, 11] From all these studies, we conclude that the shielding/steric hindrance by glycan molecules prevents aggregation-prone protein species from aggregation.

4.3 Half-Life Extension

Rapid clearance rates of therapeutic proteins because of several mechanisms, such as renal and hepatic elimination, proteolysis mediated elimination, and receptor-mediated endocytosis clearance, result in their low serum half-life.[26] Consequently, the enhancement of serum half-life would facilitate the reduction of protein doses and frequent injections, which, in turn, would lessen patient suffering and also be productive for therapeutic as well as economic reasons.[15, 26] Glycosylation results in increased in vivo potency of proteins due to increased circulating residence time or increased serum half-life. One such example is darbepoetin, which displayed threefold longer serum half-life when two additional molecules of carbohydrates were added, as compared to rHuEPO (recombinant human erythropoietin). Therefore hyperglycosylated darbepoetin, with threefold increase in mean terminal half-life compared to recombinant erythropoietin, has shown 2.5-fold lower clearance rate in EPO-naive human patients receiving peritoneal dialysis.[66, 67] In addition to this, out of different glycosylated isoforms of follicle stimulating hormone (FSH), some have been reported to have increased bioavailability and elevated in vivo potency. Such isoforms have been found to contain high sialic acid content.[68–70] Similarly, the glycosylated version of Interleukin 3 (IL-3) showed slow and sustained release into the circulation. The hydrophilic carbohydrates become trapped within the tissue by binding with ECM and enhance the volume of distribution of interleukins. The serum half-life of interleukin 3 improved twofold by glycosylation and also increased its ability to stimulate histidine carboxylase activity in bone marrow by about 30%–40% as compared to its nonglycosylated form.[26, 71]

4.4 Immunogenicity

Antibodies raised against therapeutic proteins may lead to harmful outcomes including loss of therapeutic potential through neutralization of activity, e.g., recombinant IFN-β and IFN-α.[72, 73] In some patients, antibodies are generated against rHuEPO that also neutralize endogenously produced erythropoietin. However, the highly glycosylated darbepoetin showed reduced antibody generation in these patients, as compared to rHuEPO.[74, 75] In fact, aggregation of protein drugs enhances the probability of antibody generation, which in turn reduces their serum half-life.[74, 75] Therefore, additional sugar molecules on therapeutic proteins have a positive effect on solubility, reducing aggregation and chances of antibody generation. Casadevall et al.[74] reported that these sugar molecules shield the underlying protein sequences (epitope sequence) and inhibit antibody generation.[74] In fact, the immune reactivity of these antibodies increases on removal of sugar molecules associated with the therapeutic proteins.

4.5 Protein Drug Delivery

Proper delivery of permeable blood-brain barrier protein or peptide drugs is required for successful treatment of various diseases like neurological disorders.[76] Therapeutic proteins/peptides designed by glycoengineering have shown enhanced potential to cross the blood-brain barrier.[1, 76] Glycosylation of opioid peptides such as endorphins, dynorphins, and enkephalins improves their diffusion into the brain through adsorptive endocytosis mechanism and hence increases their bioavailability.[1, 76, 77] This, in turn, increases the pharmacological activity of these peptides. Glycosylated opioid peptides (enkephalin and endomorphin-1), in comparison to the non-glycosylated form, have proved to be of improved analgesic potential.[1, 45]

4.6 Diagnostic Therapeutics

Radiolabel therapy is another approach, in which radio labeled peptides are used for therapeutic purposes and disease diagnosis. Radio probed peptide derivatives like bombesin have noteworthy potential in peptide radiotherapy and imaging of cancer cells.[1, 78, 79] However, some undesirable effects such as hepatobiliary excretion and hepatic accumulation are displayed by these radio labeled proteins. Glycosylation has been shown to reduce the hepatic accumulation of radio labeled bombesin as well as increase its uptake by cancer cells without compromising its pharmacological properties.[78] Similarly, glycosylation of radio labeled Tyr (3)-octreotide peptide improves its potential application in cancer radiotherapy and disease diagnostic imaging in somatostatin receptor expressing tumors due to its hydrophilicity.[80] Eventually, the improved pharmacokinetic properties of glycosylated peptides help it to make a potent compound for cancer cell imaging and tumor targeting.

5 CONCLUSION AND FUTURE DIRECTION

Currently, proteins are recognized as an important class of therapeutic agents both, in clinical as well as commercial arenas. Even though a number of favorable properties are displayed by therapeutic proteins, such as pharmacological efficiency, lower toxicity, and higher target specificities, their poor physical and chemical stability during purification, storage, and transport limit their utility. These instabilities, in turn, reduce their pharmacological potencies. Keeping such liabilities in mind, new technological approaches have been tested to improve the pharmacological efficiency of these pharmaceutical proteins. Glycoengineering is one such approach that helps to design therapeutic proteins with optimized therapeutic behavior. It takes advantage of the glycosylation phenomenon on proteins that confers important physical and biological properties to them, thereby providing diversified avenues to mankind in enhancing the therapeutic behavior of proteins. However, there is a need for proper glycosylation on therapeutic proteins, both in terms of position as well as pattern, for optimum therapeutic efficacy. In the near future, it is expected that therapeutic and commercial benefits of biotherapeutics will depend on glycosylation, as glycosylation technology holds a significant potential toward improving pharmacological properties of therapeutic proteins. Research in future is required for understanding the fundamentals of glycan effects on pharmacological and physicochemical properties of protein drugs.

References

1. Moradi SV, Hussein WM, Varamini P, Simerska P, Toth I. Glycosylation, an effective synthetic strategy to improve the bioavailability of therapeutic peptides. *Chem Sci* 2016; **7**(4):2492–500.
2. Craik DJ, Fairlie DP, Liras S, Price D. The future of peptide-based drugs. *Chem Biol Drug Des* 2013;**81**(1):136–47.
3. Andersen DC, Krummen L. Recombinant protein expression for therapeutic applications. *Curr Opin Biotechnol* 2002;**13**(2):117–23.
4. Beals JM, Shanafelt AB. Enhancing exposure of protein therapeutics. *Drug Discov Today Technol* 2006;**3**(1):87–94.
5. Carpenter JF, Manning MC, Randolph TW. Long-term storage of proteins. In: *Current protocols in protein science*; 2002. Suppl. 27: 4.6.1–4.6.6.
6. Davis GC. Protein stability: impact upon protein pharmaceuticals. *Biologicals* 1993; **21**(2):105.
7. Frokjaer S, Otzen DE. Protein drug stability: a formulation challenge. *Nat Rev Drug Discov* 2005;**4**(4):298–306.
8. Hawe A, Friess W. Formulation development for hydrophobic therapeutic proteins. *Pharm Dev Technol* 2007;**12**(3):223–37.
9. Manning MC, Chou DK, Murphy BM, Payne RW, Katayama DS. Stability of protein pharmaceuticals: an update. *Pharm Res* 2010;**27**(4):544–75.
10. Sola RJ, Griebenow K. Effects of glycosylation on the stability of protein pharmaceuticals. *J Pharm Sci* 2009;**98**(4):1223–45.
11. Sola RJ, Griebenow K. Glycosylation of therapeutic proteins: an effective strategy to optimize efficacy. *BioDrugs* 2010;**24**(1):9–21.
12. Arakawa T, Prestrelski SJ, Kenney WC, Carpenter JF. Factors affecting short-term and long-term stabilities of proteins. *Adv Drug Deliv Rev* 2001;**46**(1–3):307–26.
13. Brown LR. Commercial challenges of protein drug delivery. *Expert Opin Drug Deliv* 2005;**2**(1):29–42.
14. Mahmood I, Green MD. Pharmacokinetic and pharmacodynamic considerations in the development of therapeutic proteins. *Clin Pharmacokinet* 2005;**44**(4):331–47.
15. Tang L, Persky AM, Hochhaus G, Meibohm B. Pharmacokinetic aspects of biotechnology products. *J Pharm Sci* 2004;**93**(9):2184–204.
16. Byrne B, Donohoe GG, O'Kennedy R. Sialic acids: carbohydrate moieties that influence the biological and physical properties of biopharmaceutical proteins and living cells. *Drug Discov Today* 2007;**12**(7–8):319–26.
17. Grabenhorst E, Schlenke P, Pohl S, Nimtz M, Conradt HS. Genetic engineering of recombinant glycoproteins and the glycosylation pathway in mammalian host cells. *Glycoconj J* 1999;**16**(2):81–97.
18. Jefferis R. Glycosylation as a strategy to improve antibody-based therapeutics. *Nat Rev Drug Discov* 2009;**8**(3):226–34.
19. Jefferis R. Glycosylation of antibody therapeutics: optimisation for purpose. *Methods Mol Biol* 2009;**483**:223–38.
20. Lazar GA, Marshall SA, Plecs JJ, Mayo SL, Desjarlais JR. Designing proteins for therapeutic applications. *Current Opinion in Structural Biology* 2003;**13**(4):513–8.
21. Marshall SA, Lazar GA, Chirino AJ, Desjarlais JR. Rational design and engineering of therapeutic proteins. *Drug Discov Today* 2003;**8**(5):212–21.

22. Sinclair AM, Elliott S. Glycoengineering: the effect of glycosylation on the properties of therapeutic proteins. *J Pharm Sci* 2005;**94**(8):1626–35.
23. Elliott S, Lorenzini T, Asher S, Aoki K, Brankow D, Buck L, et al. Enhancement of therapeutic protein in vivo activities through glycoengineering. *Nat Biotechnol* 2003;**21**(4):414–21.
24. Koury MJ. Sugar coating extends half-lives and improves effectiveness of cytokine hormones. *Trends in Biotechnology* 2003;**21**(11):462–4.
25. Raju TS, Briggs JB, Chamow SM, Winkler ME, Jones AJ. Glycoengineering of therapeutic glycoproteins: in vitro galactosylation and sialylation of glycoproteins with terminal N-acetylglucosamine and galactose residues. *Biochemistry* 2001;**40**(30):8868–76.
26. Kontermann RE. Strategies for extended serum half-life of protein therapeutics. *Curr Opin Biotechnol* 2011;**22**(6):868–76.
27. Leader B, Baca QJ, Golan DE. Protein therapeutics: a summary and pharmacological classification. *Nat Rev Drug Discov* 2008;**7**(1):21–39.
28. Schiffter HA. Pharmaceutical proteins—structure, stability, and formulation. In: *Comprehensive biotechnology*. Pergamon; 2011. p. 521–41.
29. Redington JM, Breydo L, Uversky VN. When good goes awry: the aggregation of protein therapeutics. *Protein Pept Lett* 2017;**24**(4):340–7.
30. Dobson CM. Protein aggregation and its consequences for human disease. *Protein Pept Lett* 2006;**13**(3):219–27.
31. Fink AL. Protein aggregation: folding aggregates, inclusion bodies and amyloid. *Fold Des* 1998;**3**(1):R9–R23.
32. Pallares I, Ventura S. Understanding and predicting protein misfolding and aggregation: insights from proteomics. *Proteomics* 2016;**16**(19):2570–81.
33. Siddiqi MK, Alam P, Chaturvedi SK, Shahein YE, Khan RH. Mechanisms of protein aggregation and inhibition. *Front Biosci (Elite Ed)* 2017;**9**:1–20.
34. Breydo L, Uversky VN. Structural, morphological, and functional diversity of amyloid oligomers. *FEBS Lett* 2015;**589**(19 Pt A):2640–8.
35. Jin S, Kedia N, Illes-Toth E, Haralampiev I, Prisner S, Herrmann A, et al. Amyloid-beta(1-42) aggregation initiates its cellular uptake and cytotoxicity. *J Biol Chem* 2016;**291**(37):19590–606.
36. Mannini B, Mulvihill E, Sgromo C, Cascella R, Khodarahmi R, Ramazzotti M, et al. Toxicity of protein oligomers is rationalized by a function combining size and surface hydrophobicity. *ACS Chem Biol* 2014;**9**(10):2309–17.
37. Arakawa T, Timasheff SN. Theory of protein solubility. *Methods Enzymol* 1985;**114**:49–77.
38. Shire SJ, Shahrokh Z, Liu J. Challenges in the development of high protein concentration formulations. *J Pharm Sci* 2004;**93**(6):1390–402.
39. Volkin DB, Middaugh CR, Ahern TJ, Manning MC. *Stability of protein pharmaceuticals: Chemical and physical pathways of proteins degradation*, Plenum Press; 1992. pp. 1–462.
40. Cleland JL, Powell MF, Shire SJ. The development of stable protein formulations: a close look at protein aggregation, deamidation, and oxidation. *Crit Rev Ther Drug Carrier Syst* 1993;**10**(4):307–77.
41. Schein CH. Solubility as a function of protein structure and solvent components. *Biotechnology (N Y)* 1990;**8**(4):308–17.
42. Li S, Nguyen TH, Schoneich C, Borchardt RT. Aggregation and precipitation of human relaxin induced by metal-catalyzed oxidation. *Biochemistry* 1995;**34**(17):5762–72.
43. Costa AR, Rodrigues ME, Henriques M, Oliveira R, Azeredo J. Glycosylation: impact, control and improvement during therapeutic protein production. *Crit Rev Biotechnol* 2014;**34**(4):281–99.
44. Ho HH, Gilbert MT, Nussenzveig DR, Gershengorn MC. Glycosylation is important for binding to human calcitonin receptors. *Biochemistry* 1999;**38**(6):1866–72.

45. Varamini P, Mansfeld FM, Blanchfield JT, Wyse BD, Smith MT, Toth I. Synthesis and biological evaluation of an orally active glycosylated endomorphin-1. *J Med Chem* 2012;**55**(12):5859–67.
46. Ceaglio N, Etcheverrigaray M, Kratje R, Oggero M. Novel long-lasting interferon alpha derivatives designed by glycoengineering. *Biochimie* 2008;**90**(3):437–49.
47. Chilukuri N, Sun W, Naik RS, Parikh K, Tang L, Doctor BP, et al. Effect of polyethylene glycol modification on the circulatory stability and immunogenicity of recombinant human butyrylcholinesterase. *Chem Biol Interact* 2008;**175**(1–3):255–60.
48. Hardwicke J, Ferguson EL, Moseley R, Stephens P, Thomas DW, Duncan R. Dextrin-rhEGF conjugates as bioresponsive nanomedicines for wound repair. *J Control Release* 2008;**130**(3):275–83.
49. Ueda T, Tomita K, Notsu Y, Ito T, Fumoto M, Takakura T, et al. Chemoenzymatic synthesis of glycosylated glucagon-like peptide 1: effect of glycosylation on proteolytic resistance and in vivo blood glucose-lowering activity. *J Am Chem Soc* 2009;**131**(17):6237–45.
50. Sundaram PV, Venkatesh R. Retardation of thermal and urea induced inactivation of alpha-chymotrypsin by modification with carbohydrate polymers. *Protein Eng* 1998;**11**(8):699–705.
51. Baudys M, Uchio T, Mix D, Wilson D, Kim SW. Physical stabilization of insulin by glycosylation. *J Pharm Sci* 1995;**84**(1):28–33.
52. Kim BM, Kim H, Raines RT, Lee Y. Glycosylation of onconase increases its conformational stability and toxicity for cancer cells. *Biochem Biophys Res Commun* 2004;**315**(4):976–83.
53. Rudd PM, Joao HC, Coghill E, Fiten P, Saunders MR, Opdenakker G, et al. Glycoforms modify the dynamic stability and functional activity of an enzyme. *Biochemistry* 1994;**33**(1):17–22.
54. Wicker-Planquart C, Canaan S, Riviere M, Dupuis L. Site-directed removal of N-glycosylation sites in human gastric lipase. *Eur J Biochem* 1999;**262**(3):644–51.
55. Grinnell BW, Walls JD, Gerlitz B. Glycosylation of human protein-C affects its secretion, processing, functional activities, and activation by thrombin. *J Biol Chem* 1991;**266**(15):9778–85.
56. Wang W. Instability, stabilization, and formulation of liquid protein pharmaceuticals. *Int J Pharm* 1999;**185**(2):129–88.
57. Wang W. Protein aggregation and its inhibition in biopharmaceutics. *Int J Pharm* 2005;**289**(1–2):1–30.
58. Goochee CF, Gramer MJ, Andersen DC, Bahr JB, Rasmussen JR. The oligosaccharides of glycoproteins: bioprocess factors affecting oligosaccharide structure and their effect on glycoprotein properties. *Biotechnology (N Y)* 1991;**9**(12):1347–55.
59. Uchida E, Morimoto K, Kawasaki N, Izaki Y, Abdu Said A, Hayakawa T. Effect of active oxygen radicals on protein and carbohydrate moieties of recombinant human erythropoietin. *Free Radic Res* 1997;**27**(3):311–23.
60. Tams JW, Vind J, Welinder KG. Adapting protein solubility by glycosylation. N-glycosylation mutants of *Coprinus cinereus* peroxidase in salt and organic solutions. *Biochim Biophys Acta* 1999;**1432**(2):214–21.
61. Katsonis P, Brandon S, Vekilov PG. Corresponding-states laws for protein solutions. *J Phys Chem B* 2006;**110**(35):17638–44.
62. Valente JJ, Payne RW, Manning MC, Wilson WW, Henry CS. Colloidal behavior of proteins: effects of the second virial coefficient on solubility, crystallization and aggregation of proteins in aqueous solution. *Curr Pharm Biotechnol* 2005;**6**(6):427–36.
63. Ioannou YA, Zeidner KM, Grace ME, Desnick RJ. Human alpha-galactosidase A: glycosylation site 3 is essential for enzyme solubility. *Biochem J* 1998;**332**(Pt 3):789–97.

64. Weintraub BD, Stannard BS, Meyers L. Glycosylation of thyroid-stimulating hormone in pituitary tumor cells: influence of high mannose oligosaccharide units on subunit aggregation, combination, and intracellular degradation. *Endocrinology* 1983;**112**(4):1331–45.
65. Hoiberg-Nielsen R, Fuglsang CC, Arleth L, Westh P. Interrelationships of glycosylation and aggregation kinetics for *Peniophora lycii* phytase. *Biochemistry* 2006;**45**(15):5057–66.
66. Egrie JC, Dwyer E, Browne JK, Hitz A, Lykos MA. Darbepoetin alfa has a longer circulating half-life and greater in vivo potency than recombinant human erythropoietin. *Exp Hematol* 2003;**31**(4):290–9.
67. Macdougall IC, Gray SJ, Elston O, Breen C, Jenkins B, Browne J, et al. Pharmacokinetics of novel erythropoiesis stimulating protein compared with epoetin alfa in dialysis patients. *J Am Soc Nephrol* 1999;**10**(11):2392–5.
68. Creus S, Chaia Z, Pellizzari EH, Cigorraga SB, Ulloa-Aguirre A, Campo S. Human FSH isoforms: carbohydrate complexity as determinant of in-vitro bioactivity. *Molecular and Cellular Endocrinology* 2001;**174**(1–2):41–9.
69. D'Antonio M, Borrelli F, Datola A, Bucci R, Mascia M, Polletta P, et al. Biological characterization of recombinant human follicle stimulating hormone isoforms. *Hum Reprod* 1999;**14**(5):1160–7.
70. Perlman S, van den Hazel B, Christiansen J, Gram-Nielsen S, Jeppesen CB, Andersen KV, et al. Glycosylation of an N-terminal extension prolongs the half-life and increases the in vivo activity of follicle stimulating hormone. *J Clin Endocrinol Metab* 2003;**88**(7):3227–35.
71. Ziltener HJ, Clarklewis I, Jones AT, Dy M. Carbohydrate does not modulate the in-vivo effects of injected Interleukin-3. *Exp Hematol* 1994;**22**(11):1070–5.
72. Oberg K, Norheim I, Alm G. Treatment of malignant carcinoid-tumors—a randomized controlled-study of streptozocin plus 5-Fu and human-leukocyte interferon. *Eur J Cancer Clin Oncol* 1989;**25**(10):1475–9.
73. Zang YCQ, Yang D, Hong J, Tejada-Simon MV, Rivera VM, Zhang JZ. Immunoregulation and blocking antibodies induced by interferon beta treatment in MS. *Neurology* 2000; **55**(3):397–404.
74. Casadevall N, Nataf J, Viron B, Kolta A, Kiladjian JJ, Martin-Dupont P, et al. Pure red-cell aplasia and antierythropoietin antibodies in patients treated with recombinant erythropoietin. *N Engl J Med* 2002;**346**(7):469–75.
75. Mayeux P, Casadevall N. Antibodies to endogenous and recombinant erythropoietin. In: *Erythropoietins and erythropoiesis*. Springer; 2003. p. 229–40.
76. Egleton RD, Bilsky EJ, Tollin G, Dhanasekaran M, Lowery J, Alves I, et al. Biousian glycopeptides penetrate the blood-brain barrier. *Tetrahedron-Asymmetry* 2005;**16**(1):65–75.
77. Rodriguez MC, Cudic M. Peptide modifications to increase metabolic stability and activity. Springer; 2013. pp. 107–136.
78. Smith CJ, Volkert WA, Hoffman TJ. Radiolabeled peptide conjugates for targeting of the bombesin receptor superfamily subtypes. *Nucl Med Biol* 2005;**32**(7):73–740.
79. Watanabe A, Nishijima K, Zhao S, Tanaka Y, Itoh T, Takemoto H, et al. Effect of glycosylation on biodistribution of radiolabeled glucagon-like peptide 1. *Ann Nucl Med* 2012; **26**(2):184–91.
80. Schottelius M, Wester HJ, Reubi JC, Senekowitsch-Schmidtke R, Schwaiger M. Improvement of pharmacokinetics of radioiodinated Tyr(3)-octreotide by conjugation with carbohydrates. *Bioconjug Chem* 2002;**13**(5):1021–30.

CHAPTER

7

Posttranslational Modification of Heterologous Human Therapeutics in Plant Host Expression Systems

Ayyagari Archana, Lakshna Mahajan*, Safikur Rahman†, Rinki Minakshi**

Department of Microbiology, Swami Shraddhanand College, University of Delhi, New Delhi, India

†Department of Medical Biotechnology, Yeungnam University, Gyeongsan, South Korea

1 INTRODUCTION

Worthwhile production of human therapeutics using plant host expression systems has been attracting a lot of attention and proving to be a promising foothold in niche markets for quite some time now. A number of reasons could be attributed for this, namely, these hosts appear to be cheaper, safer, and more suited candidates than animal cell systems for producing biotherapeutics in huge quanta.[1–3] Plant systems, similar to their animal cell counterparts, are adequately equipped in bringing about posttranslational modifications, unlike their prokaryotic counterparts.[4–6] These modifications carried out by plant cell machinery have even been reported to be somewhat in the range of those carried out by mammalian cells.[1] However, there also exist cases wherein substantial variations in posttranslational modifications from those of mammalian host systems render the product useless for human usage.[7–9]

This challenge may be effectively handled by glycoengineering, a technology that imparts the appropriate type of glycosylation, yielding a

Protein Modificomics
https://doi.org/10.1016/B978-0-12-811913-6.00007-2

product that could be much more applicable for human usage.[10–12] It has been demonstrated that the possibility of inactivating plant specific glycosyltransferases and substituting them with human type glycosyltransferases could result in producing more "humanized" pharmaceuticals of plant origin. It is heartening that in some cases, products produced in such a way display effectiveness and functional value that is similar, or at times even greater than those obtained from human cell lines in vitro.[13] It is evident that there is immense scope of utilization of this particular field for humankind. In light of this, recently, a unique and powerful technology, named plant molecular farming (PMF), has fast emerged. The human biopharmaceutics using this technology are termed plant-made pharmaceuticals (PMPs).[3, 14–16] The countless advantages offered by this technology include presence of posttranslational modifications,[17, 18] speedy and bulk production unmatched by microbial and animal host systems,[19–21] and lower production costs.[5, 22–26]

Being relatively ecofriendly, plant-based host expression systems are promoted by the US government as "current good manufacturing practice."[27] All of these attributes eventually go on to equip the plant host expression systems to be an excellent medium for heterologous drug production in both developed and developing world alike.[28] Attempts to produce biopharmaceutics in plant factories are also in compliance with current good manufacturing practice norms.[29, 30] A tremendous safety point of utmost importance offered by plant hosts is that, unlike mammalian expression systems, they impart "natural protection" to human therapeutic recombinant products such as vaccines and antibodies, since they are incapable of harboring and propagating human pathogens within their system.[31, 32]

Currently, a number of recombinant human usage items are being industrially synthesized in microbial as well as mammalian host cell cultures. This requires bigger and customized fermenters, capital-intensive maintenance, aseptic conditions, expensive downstream processing, and unbroken cold chain in storage as well as transportation. These limitations led to the urgent need of development of alternative production systems.[4, 6]

Initially, when plant systems were explored as a workable option, a genuine matter of concern that was faced was that their protein yield was perceived to be rather low and inconsistent.[33] However, when their low maintenance cost, ease of handling, large numbers, and autotrophic mode of nutrition were considered, it was inferred that they could be ideal hosts for heterologous protein production. Furthermore, some reports of plant systems for fast productivity, excellent yields, ease of scale up, and quality products for clinical applications also exist.[34, 35] Certain governmental agencies have started recognizing the promise of this approach for safe, fast, and economical pharmaceutical manufacture, whose production had earlier been perceived to be rather challenging. This has paved the way toward huge financial and expertise inputs in in-depth research and large scale production of such quality end products.[35]

Plant host systems in which bulk production of human pharmacobiologics that has been expressed so far include tobacco, rice, canola, potato, tomato, carrot, *Arabidopsis*, algae, and mosses. Out of these, most of the work has been done on the tobacco plant.[36, 37]

2 N-GLYCOSYLATION AS A MAJOR POSTTRANSLATIONAL MODIFICATION

A number of posttranslational modifications (PTMs) of PMF proteins have been well studied.[38] These PTMs range from as minor as the addition of small phosphate or acetate ions, up to huge and complex alterations bringing about remarkable accentuation of protein size.[39, 40] Reversible protein phosphorylation is often witnessed in hydroxylated amino acids such as serine and threonine.[41] Some other common PTMs include sulfation and methylation of proteins. However, glycosylation of proteins soon after their synthesis is the most important event among these. Glycosylation refers to covalent addition of sugar residues to a nascent protein, and is indeed indispensable for functionality of native plant proteins during their growth and metabolism.[35, 42] About 80% of total proteins are reported to occur in their glycosylated form.[43]

The process of glycosylation may take place in a number of ways, of which the oligosaccharide linking to either the hydroxyl of threonine/serine or to the amide nitrogen of asparagine residues of the protein via N-glycosylation and O-glycosylation, respectively, are the most commonly seen among the therapeutically significant products.[44] Supplementation of a protein molecule with carbohydrates can't ever be overemphasized, as it protects the former from proteolytic attack, may enhance its life and stability, and serves to form functional recognition epitopes, etc.[43] Of the basic glycosylation types, the N-glycosylation of any protein is the most prominent in both plants and human beings, and contributes toward its correct folding and signals for quality control system.[45] The structure, conformation, and consequently, the biological activity of the therapeutic proteins depends majorly on their N-glycosylation profile. As an elegant proof, two distinct knockout mice for enzymes *N*-acetylglucosaminetransferase I and II (enzyme that is essential for processing complex type sugars) were constructed.[46, 47] Deletion of MgatI and MgatII genes in two independent knockout mice encoding these enzymes showed that the MgatI knockout mice could last in its embryo stage only for 9 days, and virtually all of the MgatII knockouts died within 7 days of birth.[46, 47] This data is suggestive of the paramount importance of proper N-glycosylation of proteins for its functionality, and therefore demonstrates its vital role among proteins to be produced from transgenic as well.[5, 41]

Synthesis of N-glycan commences in the interiors of endoplasmic reticulum (ER), where it is known to be rather conserved and alike in most of the eukaryotes. N-glycans emerging from ER, in both mammals and plants, possess a structure of mannose oligosaccharides (5–9 units of mannose sugar) linked with two residues of *N*-acetyl glucosamine ($(Man)_{5-9}$ $(GlcNAc)_2$).[48] However, further processing of N-glycosylated proteins, happening within interiors of Golgi complex, is often unique and kingdom specific.[12] It is this maturation process of N-glycosylation taking place in Golgi apparatus that results in acquiring variations and uniqueness in the resulting protein.[47] N-glycans in the plant type complex has β-(1,2)-xylose plus α-(1,3)-fucose modification on the glycan core, along with β-(1,3)-galactose plus α-(1,4)-fucose attached to the last *N*-acetylglucosamine residue.[49] Distinct from this, in mammals, N-glycans get modified with α-(1,6)-fucose plus β-(1,2) xylose residues attached to the glycan backbone, and β-(1,4)-galactose attached to α-(2,6)-sialic acid residues gets linked to terminal *N*-acetylglucosamine[3, 11] (Fig. 1).

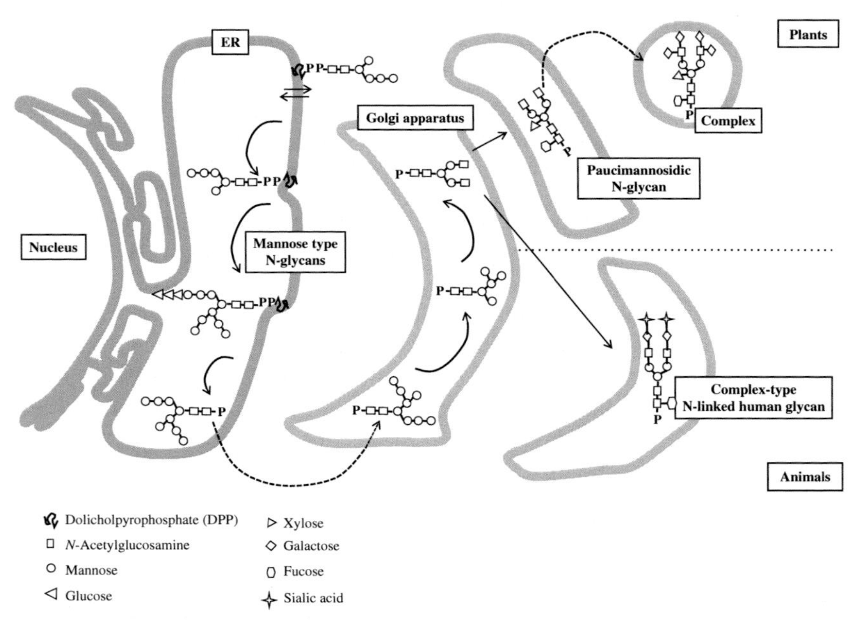

FIG. 1 Transport and processing of glycoproteins through endoplasmic reticulum (ER) and Golgi apparatus in plants and animals: The ER lumen provides environment for processing of attached oligosaccharides on the glycoprotein. The marked event is removal of glucose and mannose residues from the parent glycoprotein. This event is reported to be the same in both plants and animals. The addition of species-specific glycan residues is done in the Golgi apparatus, which is different in plants and animals. The trans Golgi of plants has Paucimannosidic N-glycan whereas in animals it is the Complex-type N-linked glycan.

In view of the above, exploitation of plant host expression systems for humanized protein production could be a workable option only after comprehending and amending the variations between the N-glycosylation patterns prevalent among the native proteins belonging to plant and mammalian hosts. There have been instances wherein human proteins produced in plants have triggered an immune response when injected in the human system, thereby not only diminishing their therapeutic efficacy, but also causing adverse side effects.[50] These responses have also been shown to be highly species specific. In a study, BALB/c mice showed no immunogenic response against plant-specific glycoepitopes {α-(1,3)-fucose and β-(1, 2)-xylose}, but rabbits, rats, and goats were observed to be producing IgE and IgG antibodies in response to the same glycoepitopes.[7, 51–54] About 20% of the total human population was estimated to be sensitive to the pharma-drug, which is definitely too high a percentage to be ignored.[55] Therefore, the immunogenicity of plant-derived molecules needs to be critically evaluated prior to planning for its utilization in the desired mammalian species. Still there is skepticism regarding such heterologously produced products, hence their industrial scale production hasn't yet taken off in a big way.[55, 56] Protein glycosylation and glycoengineering are rightly the major focal points in the production of therapeutics or vaccines in the biopharmaceutical industries and have been discussed in detail below. According to an estimate, about one-third of the total pharma-drugs that the current market demands are in their glycosylated versions.[57] Therefore, while designing the glycoengineering, utmost care must be taken to tailor the efficacy, stability, biomolecular structure, and evasion of immunogenicity of the drug.

3 FACTORS AFFECTING THE NATURAL GLYCOSYLATION OF PLANT-FARMED HUMAN PHARMACEUTICALS

3.1 Subcellular Compartmentalization

N-glycan structures have been shown to have a heterogenous profile through their subcellular compartmentalization within the mature plant cells[5, 43, 58] (Fig. 2A). This could be attributed to the presence of different and specific enzymes involved in N-glycan processing in various organelles of a eukaryotic cell. Glycoproteins localized in endoplasmic reticula harbor mainly the oligomannose type of N-glycans, while in the vacuoles the same proteins primarily remain as their paucimanosidic N-glycan version, and in the extracellular space, these majorly exist as rather complex N-glycans having β-(1,3)Gal[α-(1,4)Fuc], commonly known as Lewis a (Le^a)-containing N-glycans.[53, 59–61] When the target protein is aimed to be synthesized in recombinant form, it either may remain localized within

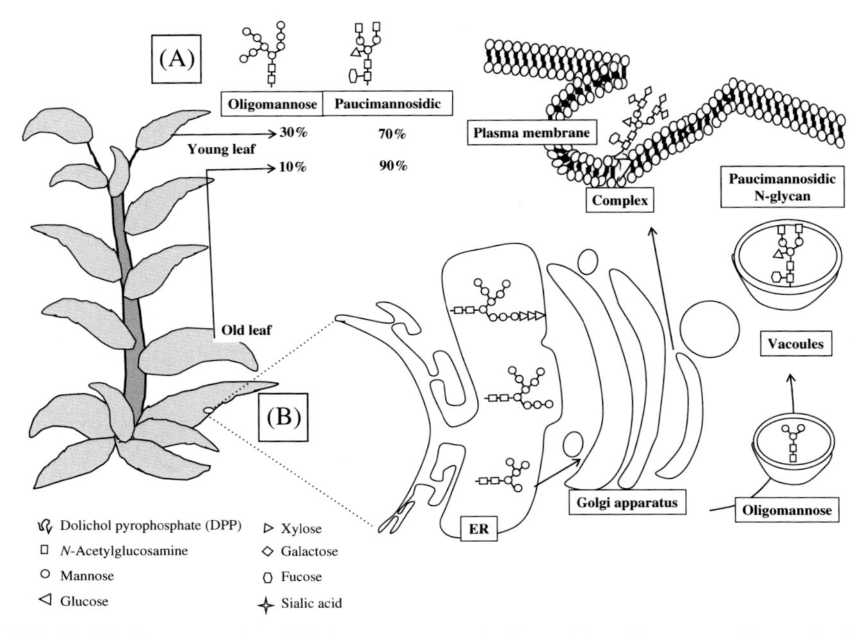

FIG. 2 The heterogeneity of N-glycan structures in the position of leaves on plant and subcellular localizations: (A) The oligomannose type is at 30% level in the young leaves as compared to the old leaves which show only 10% level. The levels of Paucimannosidic are 70% in the young leaves whereas 90% in the old leaves, (B) Processing and modification of glycoproteins in ER and Golgi apparatus.

a certain subcompartment in the plant cell or could be secreted. Any of these possibilities may be associated with certain beneficial or undesirable aspects.[62–64] The secretory pathway is well equipped for proper folding, followed by assembly and posttranslational alterations of nascent proteins. However, high yields of such proteins are mostly a remote possibility due to their confinement within cell compartments, which causes steric hindrance of the protein. The stability of secreted proteins in the apoplast has been shown to be lower as compared to those confined within endoplasmic reticulum, since the latter provides an oxidizing environment, protection from proteases, and the presence of chaperones to maintain the desired protein conformations.[63, 64] The routing of glycoproteins is expected to be different when proteins are compartmentalized in the endoplasmic reticulum, as they may not undergo complete N-glycan processing.[62]

A way of rectifying the problem of glycosylation variability of glycoproteins could be to attempt "aglycosylation" of plant derived products, accomplished by specific and pin-point alterations in the gene sequence of glycoproteins.[65] In some cases, this approach has worked, but it isn't successful in all of the usual functional roles played by the altered protein. For instance, an undesirable biological consequence of aglycosylation in human IgG1 was the loss of its capability to mount antibody dependent cell mediated cytotoxicity (ADCC) response, although its serum half-life and bio-distribution, direct interaction of its F_c portion with F_c receptors on cell surfaces, and complement dependent cytotoxicity were not adversely affected by the aglycosylation process.[11, 66] Therefore, this hit-and-miss approach may or may not be promising and fruitful in accomplishing homogenous quality of glycoproteins, depending upon whether or not their required function is unaltered or adversely affected by complete aglycosylation. The results definitely vary from case to case.[11]

3.2 The Protein Structures

It is already clear that N-glycans present on proteins are modified during their transport from ER to Golgi and further until their final destination, owing to the action of glycosyltransferases and glycosidases. As an example, N-glycans of paucimannosidic type are produced due to trimming of complex type glycans or terminal *N*-acetyl-glucosamine at post-Golgi transport stage.[67] However, glycan modification of oligosaccharides is subject to their accessibility to modifying enzymes. Normally, the protein composition begins to shape in ER, following the N-glycosylation of growing polypeptide chain. However, in some cases sugar chains may not get to form completely, since their scheduled location falls in the buried and inaccessible areas of proteins. Therefore, the final profile of N-glycan residues, as the protein passes through its secretory pathway, actually depends upon the protein structure too. The outcome is that a different glycoform profile is generated for different proteins.[68–70] It was demonstrated in the case of monoclonal antibodies generated in plants that the oligosaccharide chain was buried within the native structure of IgG, and was therefore inaccessible to modifying enzymes of transport system such as xylosyltransferases, resulting in lack of xylose and fucose residues in the resultant proteins.[9, 71, 72]

3.3 Environmental and Physiological Factors

The quality and the homogeneity of a biotherapeutic are the most important points of consideration for meeting its utilization, commercial success, and regulatory requirements. While its basic sequence and structure when expressed in transgenic hosts might be quite similar to its native counterpart, it is ultimately their N-glycosylation pattern that decides its final biological activity, which in turn is profoundly affected by environmental and physiological conditions of the culture system too. In a study of animal cell culture for monoclonal antibody production, any changes in pH by as little as one unit (6.8–7.8) resulted in as much as a 50% difference in galactosylation and sialylation patterns. Likewise, variations in culturing method and design of bioreactors (under similar steady-state conditions), dissolved oxygen tension, changes in nutrient concentration, and presence of accumulated product, may result in significant variations in the final glycosylated product.[73, 74] In contrast, it has been found that plants and the plant culture systems are not so sensitive to environmental changes, most probably because they are well adapted, given their constant natural exposure to a plethora of different environmental conditions.[75] In a study involving a transgenic tobacco plant, *Nicotiana tabacum cv. Samsun*, expressing a mouse IgG, it was found that the antibody remained rather consistent with respect to its glycosylation pattern, even when temperature and light intensity were varied a great deal.[9]

However, like their animal generated counterparts, intrinsic or extrinsic physiological parameters, e.g., age, developmental stage of the particular plant part being harvested, induction of senescence, or some other stress have been observed to be inducing heterogeneity in the protein products.[76–78] In tobacco host (*N. tabacum cv. Samsun NN*), the heavy chain of transgenic IgG degraded during leaf senescence.[79] Similarly, differences were also observed in the glycoform profile depending on the position of leaves on plant stem, which in turn depends on their age. While antibodies isolated from young/fresh leaves on newer branches showed a pronounced expression of mannose-type glycans, those from older leaves located at a lower position on the stem exhibited a higher terminal *N*-acetylglucosamine expression level[9] (Fig. 2B). Furthermore, in stressed conditions and during tissue senescence, the nutrients tend to get mobilized to the areas of higher demand, such as to sites of synthesis of stress gene products. These factors contribute immensely to the heterogeneity in glycosylation of target proteins.[76–78]

The effect of these factors on the plants grown in fields could be effectively addressed by culturing plant cells aseptically in strictly controlled conditions, similar to microbial *in vitro* cultivation. This is an elegant strategy to ensure uniform environmental conditions, and consequently, consistency in the product.[75]

4 PLANT CELL CULTURE VERSUS WHOLE PLANT CULTIVATION FOR BIOPHARMACEUTICAL PRODUCTION

There is no dearth of possibilities for plant-based production of recombinant pharmaceutical proteins via PMF.[3] Generating biopharmaceutics from transgenic plants is truly advantageous in many ways, as it is economical in investment as well as running cost of production, doesn't require skilled labor once the transgenic crop is established, survives on simple mineral solution, and possesses the product molecules "bioencapsulated" in seeds or other plant parts, without the need of preservation.[3] However, there are several disadvantages linked to this system too. The strategy for transgenic plant production involves random integration of the desired gene into the host genome, which may consume a period of 6–12 months or even more to obtain the required transgenic. Further, it may take a few generations of plants before consistent target products may be harvested from these. Random insertion of genes also makes it tough for the regulatory elements to yield efficacious, stable, and nonimmunogenic molecules with higher yields.[80] The time required for this standard procedure makes it difficult to design end products during time of urgent need, such as a fast spreading viral outbreak. In addition, various physiological, age related and subcellular localization mediated variations in PTMs are inevitable, reflecting as variations and heterogeneity in N-glycosylation profile of the product, and culminating in immunogenicity of the product, which is certainly counter-productive. These concerns are even more serious if the product is meant for oral/systemic administration rather than for topical application. The containment of genetically modified crops and the prevention of its free pollen dispersal are some of the other important factors of biosafety, which are to be stringently followed while opting for transgenic plants.[81–83]

Using plant-cell culture system may circumvent all of the aforementioned concerns. This is carried out in bioreactors in just the same way as the microbes or mammalian cells are cultured. Although such an arrangement no longer offers any of the advantages of using intact plants, it does hold more promise than developing whole transgenic plants in certain cases.[84, 85] Being less complex than the whole plant, the plant cell culture may be genetically manipulated more easily. Its downstream processing is simpler, as undesired plant metabolites and the complex plant fibers get automatically eliminated.[3, 86–88] This kind of cultivation of producer cells are already in the form of isolated, sterilized, and sealed bioreactors, which takes care of biosafety in a natural and inexpensive way. In such set ups, various undesirable physiological/senescence related PTMs can effortlessly be kept in check.[3, 84, 85] The first FDA approved PMF-based pharmaceutical, taliglucerase alfa, was synthesized in carrot cell suspension system, and was successfully utilized to manage Gaucher's disease.[89] There are, however, certain factors, which need to be kept in consideration while employing plant cell culture systems. The yield of the heterologous product may get compromised by host specific proteases present in the plant cell culture system, hence a way to prevent this ought to be devised to obtain great product yield.[85] Further, if the product is a secretory protein, its size, shape, and architecture, as well as the passage size through the cell wall of the host plant decides the success of its commercial production.[85] Plant cell culturing has been adopted by PMF-based industries for certain specific products, for which agriculture-based direct production is not found to be appropriate for various reasons.

5 DEVELOPING HUMANIZED AND IMPROVED GLYCOPROTEINS

A recent and innovative breakthrough that equips the host expression system via gene cloning to produce the target product with all the desired properties is termed "Host Engineering." Production of humanized biopharmaceuticals (proteins) showing an essential glycosylation pattern, much resembling that of the native human proteins, is achievable in plants through host engineering. It is specifically termed glycoengineering, as it exclusively involves manipulation of just the sugar residues.[90] The major factor responsible for differences in the glycosylation repertoire of plants from that of mammals is obviously the complete absence of mammalian glycosyltransferases responsible for galactosylation, sialation, and mammal-specific branching patterns, as well as the pathway required for synthesis of nucleotide sugar CMP-sialic acid.[91] By replacing plant-specific glycosylation imparting genes with their mammalian counterparts, glycoengineering yields recombinant plants that produce "humanized" monoclonal antibodies containing humanized N-glycans.[92] In addition, more of the so-called "glycovariants" may be created with a possibility that their bioactive levels may be much higher than their mammalian counter parts. Such products are called "Biobetters," discussed later in detail.[93] Thus, the produced degree of glycan homogeneity contributes to superior product quality.[94, 95] Antibodies comprising a single type of primary glycan, and without any plant-specific sugars, demonstrated an improved immune response.[4] Thus, a future vision of utilizing such production factories for bulk production of antibodies to be used for economical cancer immunotherapy appears to be indeed attractive.[4]

Plant specific glycosylation patterns could be attenuated by many approaches, e.g., by creating specific mutants or using antisense RNA technology toward the plant specific epitopes that could elicit an immune response in human beings.[96] One of the outstanding examples to explain this point is that of *Nicotiana benthamiana* (tobacco-related plant species), since it has proved to be among the most suitable candidates for heterologous production of recombinant proteins. RNA interference technology was exploited to prepare a glycosylation mutant of *N. benthamiana* (ΔXTFT), targeting the silencing of genes encoding α-(1, 3)-fucosyltransferase (ΔXTFT) and β-(1, 2)-xylosyltransferase.[97] Consequently, the composition of glycan in ΔXTFT showed a remarkable lessening of fucose and xylose content, which resulted in the biosynthesis of monoclonal immunoglobulins that were substantially humanized and possessed a high degree of homogeneity.[98]

Now many reports on production of human antibodies in recombinant plant origin exist, which display quite normal functionality.[1] However, it would be worthwhile to investigate the immunogenicity as well as allergenicity of N-glycans of a plant host expression system, which vary with different plant hosts.[99] Olive tree pollens are known to possess fucose and xylose rich glycoproteins, known to trigger allergic response in a few human beings, wherein basophils burst to release bioactive compounds such as histamines. Apparently, such plants would be most unsuitable candidates for bulk production of humanized biopharmaceuticals.[4] Although there exist frequent reports of PMF-derived injectable triggering an immune response in the patient's body,[50, 55, 100] topical application of this category of products on undamaged skin, as compared to oral consumption or injection, has been demonstrated to be much safer in this respect.[101] Many more side effects and challenges of drug production in plant systems are expected to keep cropping up and would have to be adequately addressed from time to time.

Nonimmunogenic glycoproteins derived from plants could be further humanized by cloning of genes related to mammalian glycosyltransferases, e.g., Golgi-specific α-(2,6)-sialyltransferase of mammalian systems targeted to the Golgi-glycosylation system of plants.[102–104] The combined glycoengineering, i.e., ΔXTFT and development of entire biosynthetic pathway of human like polysialylation in glycoproteins ($\Delta XTFT^{Sia}$) has been attempted though it was found to be inconsistent at this stage and showed transient expression for a few generations.[42, 105] Likewise, cocktails of MB-003 (recombinant chimeric monoclonal immunoglobulins) and murine antibodies (ZMAb) containing human constant portions were designed for production in *N. benthamiana* host, called glycoengineered monoclonal antibodies, ZMapp. This proved to be extremely useful in managing the dreaded Ebola virus infection among primates.[106] Clinical trials of ZMapp were also found to be safe and well-tolerated in a small-scale study in humans.[107]

Another approach for achieving homogeneity in the pattern of glycosylation of proteins is to channelize the glycoproteins into endoplasmic reticulum. The conserved glycosylation of eukaryotes in endoplasmic reticulum yields mannose rich glycans that evade plant-specific glycosylation patterns. Hence, plant-specific glycan structures are remarkably diminished (discussed as heterogenous N-glycan profile in plant subcellular compartments). This could be attained by fusion of glycoprotein to ER retrieval motif, which was identified to be KDEL (Lys-Asp-Glu-Leu), empowering the system to efficiently retain the protein within ER.[108] It is noteworthy that the fusion of this motif to both heavy and light chains resulted in high-mannose type N-glycans, and is essentially required if percent confinement of the product in the ER compartment is the goal. It was observed that if KDEL motif is fused with only heavy chains, still about 90% of the total antibodies remained within the ER compartment. However, since the remaining 10% of antibodies got secreted due to the fact that the light chain of the antibody wasn't modified with this motif, it resulted in glycosylation heterogeneity of the product, which is indeed undesirable.[49, 61, 109, 110] A bonus effect of confining protein products in the ER is an enhancement in the level of protein accumulation to as high as 2–10 time that of the secretory proteins, as ER provides protease free oxidizing environment and an abundance of molecular chaperones, all of these favor correct folding and assembly of candidate protein. It is implied that the stability of the glycoproteins in ER, thus, becomes much higher than that of secretory proteins in apoplast.[62, 63, 96, 97, 109, 111, 112] Such plant expression systems produce humanlike N-glycans with two terminal β-1,2-linked GlcNAc residues (GnGn, GlcNAc-2-Man-3-GlcNAc-2) with remarkable consistency.

6 GLYCOENGINEERING: CONCEPT OF "BIOBETTERS" AND "BIOSIMILARS"

Since glycosylation of biopharmaceutical drug products is the single most important factor that determines their utility, safety, or unsuitability for therapeutic purpose, it quite deservingly remains the focal point while strategizing to produce a target substance at a commercial level. In this light, a concept of "Biosimilars" has been developed, which are defined as the heterologously produced proteins possessing a glycosylation pattern very similar to that of natural product. This is to achieve excellent bioactivity, higher stability, enough serum-half lifetime, and, last but not least, nominal immunogenicity. These are the prerequisites for therapeutic applications of any biopharmaceutical. Therefore, it becomes a critical need to mold the plant glycosylation machinery for the required humanized glycoforms of the expressed glycoproteins.[53] However, it should be kept in mind that exact matching of glycoengineered products with their native human counterparts is not at all feasible, owing to the great molecular complexity of biomolecules. However, the fact that these products are adequately functional, stable, and nonimmunogenic is what actually counts. Therefore, the term "Biosimilar" is more apt instead of the term "Biogeneric" in its strict sense (EU Guideline CHMP/437/04).

Extensive research focusing upon "humanization" and homogeneous N-glycosylations has continuously been going on in a plethora of expression systems of plant origin.[4, 113–117] The biomolecules so obtained often display modified glycans that are compatible with therapeutic applications, and exhibit improved functionality. These studies have opened up new avenues for studying glycosylations, their modifications, and their biological significance and consequent effects on the efficacy and stability of these products. It has paved the way to another novel and up-trending concept of "Biobetters." As the name suggests, they are proving to be superior and even more promising than "Biosimilars." A Biobetter is defined as a biologically produced molecule that is formulated on the lines of an already existing and approved biological product, but is superior to any of the earlier in the series by one or more features.[118] Supported by the novel tools and technologies available online in this area, a new era of effective and affordable drugs for all is fast becoming a possibility. While formulating a drug, commercially, specialized glycoengineering companies may be consulted, which specifically focus on generating Biobetters and Biosimilars.

In a study, while the modification in human IgG glycosylation did not affect antigen binding in the case of a recombinant monoclonal IgG, there were significant differences in cellular interactions and immunity due to this change. IgG with altered glycosylations harboring bisecting NAG (*N*-acetylglucosamine) actually exhibited a better ADCC activity than natural IgG.[116] In another study, modification of primary *N*-acetylglucosamine by removal of fucose resulted in a remarkably enhanced affinity of F_c portion of IgG to the $F_c\gamma$RIIIa receptor found on the surface of a number of cells, leading to higher efficiency of this antibody in clearing the antigen.[119, 120] However, it has also been found that glycosylation of immunoglobulins may also modify them for their anti- and pro-inflammatory properties.[121, 122] It is hoped that such fine tuning of glycosylation modification of antibodies could also be exploited toward improving the effectiveness of antibody-based immunotherapy.

7 SUCCESS STORIES

Already many worthwhile products of human importance have been obtained from plant expression systems glycoengineered at posttranslational levels.[1, 123] These include a number of pathogen surface antigens used in various vaccine preparations, a series of monoclonal antibodies and cytokines indispensable for immunotherapy, erythropoietin for accelerating hematopoiesis, anticoagulants, antimicrobials, antihypertensives, immunodiagnostics, various enzymes and growth hormones, analgesics, collagen, human serum albumin, and food proteins such as lactoferrin and β-casein. The foremost plant derived substance of human significance was human growth hormone (hGH), which was produced in sunflower and tobacco.[124] This was followed by production of a hepatitis B surface antigen (HBsAg) in tobacco, which was similar to its yeast produced counterpart both physically and antigenically.[125] Such products are most useful for designing vaccines effective against many dreaded diseases. Some more examples that really made a difference are discussed below.

8 EBOLA VIRUS INFECTION

A great success story revolves around an experimental drug called ZMapp.[126] Ebola virus-infected American health aid workers were reported to be saved during an Ebola outbreak in Africa (2014) soon after being administered with ZMapp, which is an amalgamation of three monoclonal antibodies, synthesized in tobacco plants cultivated in green-house in accordance with current good manufacturing practice.[127] It would have been successful in saving Ebola infected individuals, provided the drug administration was done in a timely manner.[3] This story demonstrates the feasibility of producing a life-saving drug via such an approach.[2, 128] Optimization of the plant host system facilitated successful biosynthesis of these monoclonal antibodies, equaling their mammalian versions. These were reported to be even more effective as well as more stable than those prepared in mammalian host systems.[95, 129] The drugs were shown to be fully adequate and capable of curing the nonhuman primate experimental animals even when they were administered after 5 days of a lethal challenge.[106] Such developments indeed impart future promise and hope in conquering dreaded and lethal diseases like Ebola.

9 GAUCHER'S DISEASE

Another biopharmaceutical, enzyme glucocerebrosidase (GCD) that is useful for treating Gaucher's disease in humans was successfully produced by plant expression system.[13] Amazingly, this enzyme produced from a plant host system was already more functional and active as compared to its counterpart produced in mammalian cell lines. The latter required an in vitro N-glycan processing for its complete bioactivity unlike its plant counterpart.[13] Hence, the plant origin enzyme, commercially named ELELYSO gained the approval of the Food and Drug Administration (FDA) and other regulatory agencies for manufacture and treatment of Gaucher's disease.[89] A novel, hassle-free oral version of GCD produced in carrots has been proved to be successful in animal experimental models[130] and in preliminary human clinical trials.

10 MONOCLONAL ANTIBODIES

Therapeutic proteins such as monoclonal antibodies are being resorted to for immunotherapy in a plethora of immune disorders. Their production has been attempted in plant hosts for treating diseases such as HIV infection, certain cancerous conditions, malaria, rheumatoid arthritis, influenza, dental caries, herpes simplex, and cholera. The products have been demonstrated to be effective in prophylaxis or even therapeutic measures, and are in clinical trials. One of the most successful products produced in plant expression system are the antibodies against *Streptococcus mutans*, which causes dental caries.[131, 132]

11 VACCINES

It has proved very useful to clone and express protein antigens of dreaded pathogens in plant hosts for vaccine development. Vaccines derived from plants in such a way include those against Hepatitis B virus, Rabies virus, Norwalk virus, Cytomegalovirus, Rotavirus etc.[3]

12 BENEFITS OF PLANT EXPRESSION SYSTEM

Plant host systems may certainly be appealing vehicles for production of oral pharmaceuticals. Conventionally produced biologics by microbial and animal cell lines involve an expensive downstream processing component and after-expenses, as these need to be continuously maintained in a "cold chain" until administration. By contrast, a number of plant-cell-encapsulated drugs have been proposed with reasonable shelf stability. Oral delivery of cell-encapsulated drugs is very convenient as it doesn't require cumbersome downstream processing, as well as storage and transportation in cold chains.[133] Given the anatomy of plant cells, the expressed, bioencapsulated biologic proteins are protected naturally on oral administration, from hydrolytic enzymes and hydrochloric acid occurring in the human stomach, permitting them to be assimilated into the gut lining in an unbroken form.[134] Proposals to manage hypertension, diabetes, and many other metabolic disorders using such plant-based drugs are being rapidly explored.[134–136]

The plant expression systems exhibit a remarkably high flexibility in accommodating new genes and pathways, e.g., human glycosyltransferase gene and sialic acid pathway, respectively. At the same time, they can tolerate the diminished expression of their natural plant-specific glycosyltransferases.[137] The major plant-based protein production systems include *Arabidopsis thaliana*, *Lemna minor*, *N. benthamiana*, as the platforms that could tolerate the genetic manipulations without any obvious phenotypic change or any impact on development, when grown under standard conditions.[4, 24, 98, 138] Fewer studies showed a decreased seed production in XTFT^{Sia} lines of *N. benthamiana*, though this does not cause a major limitation. The rather narrow series of glycosylations occurring in plants also makes it possible to produce strikingly homogeneous plant-based biotherapeutics, which constitutes a mandatory regulatory requirement.[12]

13 LIMITATIONS OF PLANT EXPRESSION SYSTEM

Despite the numerous advantages offered by plant expression systems, there still remain a handful of technical concerns that ought to be addressed to tap the maximum potential of this promising platform. A major bottleneck working with such systems is that enormous amounts of debris are produced that have to be constantly eliminated, which renders the process quite messy as well as expensive. Hence, downstream processing costs associated with plant factories remain as high, or even higher, than animal or microbial cell culture systems[86–88]. A further problem is lesser product quantity produced per unit of mass.[1, 139–141] According to a current estimate, the downstream processing cost of PMF-based biopharmaceutics is about 80%–90% of the total cost of production,[142] and has to be seriously worked upon for PMF technology betterment.

Current plant biotechnology is not adequately equipped to ensure 100% transformation of all host plants, nor in obtaining uniform and consistent product levels in all plants parts, and even all progeny generations of the transformed plant. This heterogeneity and variation is a big deterring factor in PMF technology from the viewpoint of deciding the precise dosage of the product. Most of the purified protein of plant product is reported to be rendered useless due to proteolytic degradation by enzymes present in human gut.[143, 144]

Microbial and animal cell culturing methods have been standardized for their downstream processing protocols following bulk production, but purification of PMF-based production from different plant hosts under specific conditions demand for unique, customized, elaborate, and sometimes expensive protocols.[87, 88, 142, 145] Some residual quantities of fertilizers, pesticides, phenolics, alkaloids, plant pathogen extracts, and secondary metabolites in the product mixture further complicate the purification process.[3] Biosafety concerns may also arise as pollens of transgenic plants may reach out to the native ones.[81–83] Due to these challenges, currently, the FDA is really stringent in allowing the exploitation of food crops for the synthesis of recombinant biopharmaceutics (FDA Guidance for Industry). In the light of the aforementioned challenges, bulk production of these products has not really been wholeheartedly attempted.[3]

Following the advent of PMF products in the early 1990s, and initial teething troubles, these products have eventually started showing up in the actual market. Although it is indeed true that until now, ELELYSO is the sole plant-based pharmaceutical that has been given a green signal by the FDA, and ZMapp has been utilized to save the Ebola-infected workers, research and clinical trials of a number of them is in the pipeline.[3]

14 CONCLUSION AND FUTURE PROSPECTS

Due to two major limitations of PMF technology—immune response triggered by plant-based products within the human body and tedious/expensive downstream processing of the product from the heterologous host—large-scale plant-based production of human pharmaceuticals has not yet been ventured into significantly for a long time. In place of this, PMF technology using glycoengineered plant expression systems offer excellent possibilities of powerful and correct expression of human biopharmaceutics. Research is ongoing to characterize glycosylated and "humanized" proteins, possible favorable manipulations in the glycosylation process, and whether "Biosimilars" and "Biobetters" can deliver the results that are being expected of them. Since elaborate downstream processing is among prime prohibitive factors for producing PMF-based products, it could be great if transgenic fruits and vegetables, designed with therapeutic properties, could be grown and consumed directly so as to impart good health. Lumbrokinase is a unique enzyme found in earthworms that is capable of dissolving blood clots. When expressed in sunflower kernels, and directly fed to experimental mice containing blood clots, remarkable clot dissolution was witnessed. Hence, it was inferred that a health supplement based upon this PMF may be useful in improving the cardiovascular health of human beings.[146]

Since the rules and regulations for producing health supplements are much more relaxed as compared to those for medically prescribed drugs, and these supplements may be bought over the counter without any medical prescription, PMF-based crops whose extract could directly be ingested as health supplements may be a workable idea.[147] Gradually increasing the understanding and acceptance is certainly paving the way for a big time success of plant medium for pharmaceutical production. Support from the government and partnership of some leading pharmaceutical companies has also boosted up the possibility of plant expression systems being used for drug production.[27, 148] Even though very few success stories are available right now, it is quite possible that the new generation humanized healthcare products would certainly include more and more of these. However, total replacement of mammalian expression systems by plant factories is not so likely, owing to a few limitations stated in this article as well as a number of parameters, which remain to be standardized. Instead, plant-based host systems are most suited for special niche production of certain specific pharmaproducts. For really meaningful utilization of plant host expression technologies, it is crucial to come up with powerful strategies for effective and inexpensive downstream processing, and getting official approvals for manufacture of fine pharmaceuticals, such as monoclonal antibodies and vaccines in plant hosts. Indeed, it seems to be a ray of hope that the field of modern medicine may get a tremendous boost from next generation of humanized products in the forthcoming years.

References

1. Abiri R, Valdiani A, Maziah M, Shaharuddin NA, Sahebi M, Yusof ZN, et al. A critical review of the concept of transgenic plants: insights into pharmaceutical biotechnology and molecular farming. *Curr Issues Mol Biol* 2016;**18**:21–42.
2. Arntzen C. Plant-made pharmaceuticals: from "edible vaccines" to Ebola therapeutics. *Plant Biotechnol J* 2015;**13**(8):1013–6.
3. Yao J, Weng Y, Dickey A, Wang KY. Plants as factories for human pharmaceuticals: applications and challenges. *Int J Mol Sci* 2015;**16**(12):28549–65.
4. Cox KM, Sterling JD, Regan JT, Gasdaska JR, Frantz KK, Peele CG, et al. Glycan optimization of a human monoclonal antibody in the aquatic plant *Lemna minor*. *Nat Biotechnol* 2006;**24**(12):1591–7.
5. Gomord V, Faye L. Posttranslational modification of therapeutic proteins in plants. *Curr Opin Plant Biol* 2004;**7**(2):171–81.
6. Shaaltiel Y, Bartfeld D, Hashmueli S, Baum G, Brill-Almon E, Galili G, et al. Production of glucocerebrosidase with terminal mannose glycans for enzyme replacement therapy of Gaucher's disease using a plant cell system. *Plant Biotechnol J* 2007;**5**(5):579–90.
7. Bardor M, Loutelier-Bourhis C, Paccalet T, Cosette P, Fitchette AC, Vezina LP, et al. Monoclonal C5-1 antibody produced in transgenic alfalfa plants exhibits a N-glycosylation that is homogenous and suitable for glyco-engineering into human-compatible structures. *Plant Biotechnol J* 2003;**1**(6):451–62.
8. Cabanes-Macheteau M, Fitchette-Laine AC, Loutelier-Bourhis C, Lange C, Vine ND, Ma JK, et al. N-Glycosylation of a mouse IgG expressed in transgenic tobacco plants. *Glycobiology* 1999;**9**(4):365–72.
9. Elbers IJ, Stoopen GM, Bakker H, Stevens LH, Bardor M, Molthoff JW, et al. Influence of growth conditions and developmental stage on N-glycan heterogeneity of transgenic immunoglobulin G and endogenous proteins in tobacco leaves. *Plant Physiol* 2001;**126**(3):1314–22.
10. Chen Q. Glycoengineering of plants yields glycoproteins with polysialylation and other defined N-glycoforms. *Proc Natl Acad Sci U S A* 2016;**113**(34):9404–6.
11. Ko K, Ahn MH, Song M, Choo YK, Kim HS, Joung H. Glyco-engineering of biotherapeutic proteins in plants. *Mol Cells* 2008;**25**(4):494–503.
12. Steinkellner H, Castilho A. N-Glyco-engineering in plants: update on strategies and major achievements. *Methods Mol Biol* 2015;**1321**:195–212.
13. Zimran A, Brill-Almon E, Chertkoff R, Petakov M, Blanco-Favela F, Munoz ET, et al. Pivotal trial with plant cell-expressed recombinant glucocerebrosidase, taliglucerase alfa, a novel enzyme replacement therapy for Gaucher disease. *Blood* 2011;**118**(22):5767–73.
14. Moustafa K, Makhzoum A, Tremouillaux-Guiller J. Molecular farming on rescue of pharma industry for next generations. *Crit Rev Biotechnol* 2016;**36**(5):840–50.
15. Streatfield SJ. Approaches to achieve high-level heterologous protein production in plants. *Plant Biotechnol J* 2007;**5**(1):2–15.
16. Suslow TV, Thomas BR, Bradford KJ. *Biotechnology provides new tools for plant breeding.* UCANR Publications; 2002.
17. Daniell H, Streatfield SJ, Rybicki EP. Advances in molecular farming: key technologies, scaled up production and lead targets. *Plant Biotechnol J* 2015;**13**(8):1011–2.
18. Obembe OO, Popoola JO, Leelavathi S, Reddy SV. Advances in plant molecular farming. *Biotechnol Adv* 2011;**29**(2):210–22.

19. Chen Q, He J, Phoolcharoen W, Mason HS. Geminiviral vectors based on bean yellow dwarf virus for production of vaccine antigens and monoclonal antibodies in plants. *Hum Vaccin* 2011;**7**(3):331–8.
20. Klimyuk V, Pogue G, Herz S, Butler J, Haydon H. Production of recombinant antigens and antibodies in *Nicotiana benthamiana* using 'magnifection' technology: GMP-compliant facilities for small- and large-scale manufacturing. *Curr Top Microbiol Immunol* 2014;**375**:127–54.
21. Peyret H, Lomonossoff GP. When plant virology met agrobacterium: the rise of the deconstructed clones. *Plant Biotechnol J* 2015;**13**(8):1121–35.
22. Avesani L, Merlin M, Gecchele E, Capaldi S, Brozzetti A, Falorni A, et al. Comparative analysis of different biofactories for the production of a major diabetes autoantigen. *Transgenic Res* 2014;**23**(2):281–91.
23. Lai H, He J, Engle M, Diamond MS, Chen Q. Robust production of virus-like particles and monoclonal antibodies with geminiviral replicon vectors in lettuce. *Plant Biotechnol J* 2012;**10**(1):95–104.
24. Nagels B, Van Damme EJ, Pabst M, Callewaert N, Weterings K. Production of complex multiantennary N-glycans in *Nicotiana benthamiana* plants. *Plant Physiol* 2011;**155** (3):1103–12.
25. Tuse D, Ku N, Bendandi M, Becerra C, Collins Jr. R, Langford N, et al. Clinical safety and immunogenicity of tumor-targeted, plant-made id-KLH conjugate vaccines for follicular lymphoma. *Biomed Res Int* 2015;**2015**.
26. Tuse D, Tu T, McDonald KA. Manufacturing economics of plant-made biologics: case studies in therapeutic and industrial enzymes. *Biomed Res Int* 2014;**2014**.
27. Holtz BR, Berquist BR, Bennett LD, Kommineni VJ, Munigunti RK, White EL, et al. Commercial-scale biotherapeutics manufacturing facility for plant-made pharmaceuticals. *Plant Biotechnol J* 2015;**13**(8):1180–90.
28. Paul MJ, Teh AY, Twyman RM, Ma JK. Target product selection—where can Molecular Pharming make the difference? *Curr Pharm Des* 2013;**19**(31):5478–85.
29. Fischer R, Schillberg S, Hellwig S, Twyman RM, Drossard J. GMP issues for recombinant plant-derived pharmaceutical proteins. *Biotechnol Adv* 2012;**30**(2):434–9.
30. Merlin M, Gecchele E, Capaldi S, Pezzotti M, Avesani L. Comparative evaluation of recombinant protein production in different biofactories: the green perspective. *Biomed Res Int* 2014;**2014**.
31. Rybicki EP. Plant-made vaccines for humans and animals. *Plant Biotechnol J* 2010;**8** (5):620–37.
32. Tregoning J, Maliga P, Dougan G, Nixon PJ. New advances in the production of edible plant vaccines: chloroplast expression of a tetanus vaccine antigen, TetC. *Phytochemistry* 2004;**65**(8):989–94.
33. Bendandi M, Marillonnet S, Kandzia R, Thieme F, Nickstadt A, Herz S, et al. Rapid, high-yield production in plants of individualized idiotype vaccines for non-Hodgkin's lymphoma. *Ann Oncol* 2010;**21**(12):2420–7.
34. Gleba YY, Tuse D, Giritch A. Plant viral vectors for delivery by agrobacterium. *Curr Top Microbiol Immunol* 2014;**375**:155–92.
35. Stoger E, Fischer R, Moloney M, Ma JK. Plant molecular pharming for the treatment of chronic and infectious diseases. *Annu Rev Plant Biol* 2014;**65**:743–68.
36. Fahad S, Khan FA, Pandupuspitasari NS, Ahmed MM, Liao YC, Waheed MT, et al. Recent developments in therapeutic protein expression technologies in plants. *Biotechnol Lett* 2015;**37**(2):265–79.
37. Rigano MM, Walmsley AM. Expression systems and developments in plant-made vaccines. *Immunol Cell Biol* 2005;**83**(3):271–7.
38. Vukusic K, Sikic S, Balen B. Recombinant therapeutic proteins produced in plants: towards engineering of human-type O- and N-glycosylation. *Periodicum Biol* 2016;**118** (2):75–90.

39. Bosch D, Schots A. Plant glycans: friend or foe in vaccine development? *Expert Rev Vaccines* 2010;**9**(8):835–42.
40. Stulemeijer IJ, Joosten MH. Post-translational modification of host proteins in pathogen-triggered defence signalling in plants. *Mol Plant Pathol* 2008;**9**(4):545–60.
41. Hashiguchi A, Komatsu S. Impact of post-translational modifications of crop proteins under abiotic stress. *Proteomes* 2016;**4**(4):42.
42. Loos A, Gruber C, Altmann F, Mehofer U, Hensel F, Grandits M, et al. Expression and glycoengineering of functionally active heteromultimeric IgM in plants. *Proc Natl Acad Sci U S A* 2014;**111**(17):6263–8.
43. Rudd PM, Wormald MR, Dwek RA. Sugar-mediated ligand-receptor interactions in the immune system. *Trends Biotechnol* 2004;**22**(10):524–30.
44. Easton R, Leader T. Glycosylation of proteins—structure, function and analysis. *Life Sci* 2011;**48**:1–5.
45. Xu C, Ng DT. Glycosylation-directed quality control of protein folding. *Nat Rev Mol Cell Biol* 2015;**16**(12):742–52.
46. Ioffe E, Stanley P. Mice lacking N-acetylglucosaminyltransferase I activity die at mid-gestation, revealing an essential role for complex or hybrid N-linked carbohydrates. *Proc Natl Acad Sci U S A* 1994;**91**(2):728–32.
47. Wang Y, Tan J, Sutton-Smith M, Ditto D, Panico M, Campbell RM, et al. Modeling human congenital disorder of glycosylation type IIa in the mouse: conservation of asparagine-linked glycan-dependent functions in mammalian physiology and insights into disease pathogenesis. *Glycobiology* 2001;**11**(12):1051–70.
48. Kim SM, Lee JS, Lee YH, Kim WJ, Do SI, Choo YK, et al. Increased alpha2,3-sialylation and hyperglycosylation of N-glycans in embryonic rat cortical neurons during camptothecin-induced apoptosis. *Mol Cells* 2007;**24**(3):416–23.
49. Bardor M, Cabrera G, Rudd PM, Dwek RA, Cremata JA, Lerouge P. Analytical strategies to investigate plant N-glycan profiles in the context of plant-made pharmaceuticals. *Curr Opin Struct Biol* 2006;**16**(5):576–83.
50. Jin C, Altmann F, Strasser R, Mach L, Schahs M, Kunert R, et al. A plant-derived human monoclonal antibody induces an anti-carbohydrate immune response in rabbits. *Glycobiology* 2008;**18**(3):235–41.
51. Chargelegue D, Vine ND, van Dolleweerd CJ, Drake PM, Ma JK. A murine monoclonal antibody produced in transgenic plants with plant-specific glycans is not immunogenic in mice. *Transgenic Res* 2000;**9**(3):187–94.
52. Faye L, Gomord V, Fitchette-Laine AC, Chrispeels MJ. Affinity purification of antibodies specific for Asn-linked glycans containing alpha 1-->3 fucose or beta 1-->2 xylose. *Anal Biochem* 1993;**209**(1):104–8.
53. Gomord V, Chamberlain P, Jefferis R, Faye L. Biopharmaceutical production in plants: problems, solutions and opportunities. *Trends Biotechnol* 2005;**23**(11):559–65.
54. van Ree R, Cabanes-Macheteau M, Akkerdaas J, Milazzo JP, Loutelier-Bourhis C, Rayon C, et al. Beta(1,2)-xylose and alpha(1,3)-fucose residues have a strong contribution in IgE binding to plant glycoallergens. *J Biol Chem* 2000;**275**(15):11451–8.
55. Mari A. IgE to cross-reactive carbohydrate determinants: analysis of the distribution and appraisal of the in vivo and in vitro reactivity. *Int Arch Allergy Immunol* 2002;**129**(4):286–95.
56. Gomord V, Fitchette AC, Menu-Bouaouiche L, Saint-Jore-Dupas C, Plasson C, Michaud D, et al. Plant-specific glycosylation patterns in the context of therapeutic protein production. *Plant Biotechnol J* 2010;**8**(5):564–87.
57. Spok A, Karner S. Plant molecular farming: opportunities and challenges. In: Stein AJ, editor. *The institute for prospective technological studies*. Seville: European Commission; 2008.
58. Sturm A, Van Kuik JA, Vliegenthart JF, Chrispeels MJ. Structure, position, and biosynthesis of the high mannose and the complex oligosaccharide side chains of the bean storage protein phaseolin. *J Biol Chem* 1987;**262**(28):13392–403.

59. Fischer R, Stoger E, Schillberg S, Christou P, Twyman RM. Plant-based production of biopharmaceuticals. *Curr Opin Plant Biol* 2004;**7**(2):152–8.
60. Pagny S, Cabanes-Macheteau M, Gillikin JW, Leborgne-Castel N, Lerouge P, Boston RS, et al. Protein recycling from the Golgi apparatus to the endoplasmic reticulum in plants and its minor contribution to calreticulin retention. *Plant Cell* 2000;**12**(5):739–56.
61. Sriraman R, Bardor M, Sack M, Vaquero C, Faye L, Fischer R, et al. Recombinant anti-hCG antibodies retained in the endoplasmic reticulum of transformed plants lack core-xylose and core-alpha(1,3)-fucose residues. *Plant Biotechnol J* 2004;**2**(4):279–87.
62. Conrad U, Fiedler U. Compartment-specific accumulation of recombinant immunoglobulins in plant cells: an essential tool for antibody production and immunomodulation of physiological functions and pathogen activity. *Plant Mol Biol* 1998;**38**(1–2):101–9.
63. Nuttall J, Vine N, Hadlington JL, Drake P, Frigerio L, Ma JK. ER-resident chaperone interactions with recombinant antibodies in transgenic plants. *Eur J Biochem* 2002;**269**(24):6042–51.
64. Schillberg S, Zimmermann S, Voss A, Fischer R. Apoplastic and cytosolic expression of full-size antibodies and antibody fragments in *Nicotiana tabacum*. *Transgenic Res* 1999;**8**(4):255–63.
65. Walker MR, Lund J, Thompson KM, Jefferis R. Aglycosylation of human IgG1 and IgG3 monoclonal antibodies can eliminate recognition by human cells expressing Fc gamma RI and/or Fc gamma RII receptors. *Biochem J* 1989;**259**(2):347–53.
66. Dorai H, Mueller BM, Reisfeld RA, Gillies SD. Aglycosylated chimeric mouse/human IgG1 antibody retains some effector function. *Hybridoma* 1991;**10**(2):211–7.
67. Fitchette AC, Cabanes-Macheteau M, Marvin L, Martin B, Satiat-Jeunemaitre B, Gomord V, et al. Biosynthesis and immunolocalization of Lewis a-containing N-glycans in the plant cell. *Plant Physiol* 1999;**121**(2):333–44.
68. Faye L, Johnson KD, Chrispeels MJ. Oligosaccharide side chains of glycoproteins that remain in the high-mannose form are not accessible to glycosidases. *Plant Physiol* 1986;**81**(1):206–11.
69. Faye L, Sturm A, Bollini R, Vitale A, Chrispeels MJ. The position of the oligosaccharide side-chains of phytohemagglutinin and their accessibility to glycosidases determines their subsequent processing in the Golgi. *Eur J Biochem* 1986;**158**(3):655–61.
70. Vitale A, Chrispeels MJ. Transient N-acetylglucosamine in the biosynthesis of phytohemagglutinin: attachment in the Golgi apparatus and removal in protein bodies. *J Cell Biol* 1984;**99**(1 Pt 1):133–40.
71. Deisenhofer J. Crystallographic refinement and atomic models of a human Fc fragment and its complex with fragment B of protein a from Staphylococcus aureus at 2.9- and 2.8-A resolution. *Biochemistry* 1981;**20**(9):2361–70.
72. Ko K, Steplewski Z, Glogowska M, Koprowski H. Inhibition of tumor growth by plant-derived mAb. *Proc Natl Acad Sci U S A* 2005;**102**(19):7026–30.
73. Ivarsson M, Villiger TK, Morbidelli M, Soos M. Evaluating the impact of cell culture process parameters on monoclonal antibody N-glycosylation. *J Biotechnol* 2014;**188**:88–96.
74. Jenkins N, Curling EM. Glycosylation of recombinant proteins: problems and prospects. *Enzyme Microb Technol* 1994;**16**(5):354–64.
75. Tekoah Y, Shulman A, Kizhner T, Ruderfer I, Fux L, Nataf Y, et al. Large-scale production of pharmaceutical proteins in plant cell culture-the Protalix experience. *Plant Biotechnol J* 2015;**13**(8):1199–208.
76. Buchanan-Wollaston V. The molecular biology of leaf senescence. *J Exp Bot* 1997;**48**(2):181–99.
77. Noodén LD, Guiamét JJ, John I. Senescence mechanisms. *Physiol Plant* 1997;**101**(4):746–53.
78. Smart CM. Tansley review no. 64. Gene expression during leaf senescence. *New Phytol* 1994;**126**(3):419–48.

79. Stevens LH, Stoopen GM, Elbers IJ, Molthoff JW, Bakker HA, Lommen A, et al. Effect of climate conditions and plant developmental stage on the stability of antibodies expressed in transgenic tobacco. *Plant Physiol* 2000;**124**(1):173–82.
80. Sainsbury F, Lomonossoff GP. Transient expressions of synthetic biology in plants. *Curr Opin Plant Biol* 2014;**19**:1–7.
81. Breyer D, Goossens M, Herman P, Sneyers M. Biosafety considerations associated with molecular farming in genetically modified plants. *J Med Plant Res* 2009;**3**(11):825–38.
82. MacDonald J, Doshi K, Dussault M, Hall JC, Holbrook L, Jones G, et al. Bringing plant-based veterinary vaccines to market: managing regulatory and commercial hurdles. *Biotechnol Adv* 2015;**33**(8):1572–81.
83. Sparrow P, Broer I, Hood EE, Eversole K, Hartung F, Schiemann J. Risk assessment and regulation of molecular farming—a comparison between Europe and US. *Curr Pharm Des* 2013;**19**(31):5513–30.
84. Raven N, Rasche S, Kuehn C, Anderlei T, Klockner W, Schuster F, et al. Scaled-up manufacturing of recombinant antibodies produced by plant cells in a 200-L orbitally-shaken disposable bioreactor. *Biotechnol Bioeng* 2015;**112**(2):308–21.
85. Schillberg S, Raven N, Fischer R, Twyman RM, Schiermeyer A. Molecular farming of pharmaceutical proteins using plant suspension cell and tissue cultures. *Curr Pharm Des* 2013;**19**(31):5531–42.
86. Buyel JF, Fischer R. Downstream processing of biopharmaceutical proteins produced in plants: the pros and cons of flocculants. *Bioengineered* 2014;**5**(2):138–42.
87. Buyel JF, Twyman RM, Fischer R. Extraction and downstream processing of plant-derived recombinant proteins. *Biotechnol Adv* 2015;**33**(6 Pt 1):902–13.
88. Fischer R, Vasilev N, Twyman RM, Schillberg S. High-value products from plants: the challenges of process optimization. *Curr Opin Biotechnol* 2015;**32**:156–62.
89. Fox JL. First plant-made biologic approved. [News]. *Nat Biotechnol* 2012;**30**(6):472.
90. Maxmen A. Drug-making plant blooms. *Nature* 2012;**485**(7397):160.
91. Castilho A, Pabst M, Leonard R, Veit C, Altmann F, Mach L, et al. Construction of a functional CMP-sialic acid biosynthesis pathway in Arabidopsis. *Plant Physiol* 2008;**147**(1):331–9.
92. Strasser R, Altmann F, Steinkellner H. Controlled glycosylation of plant-produced recombinant proteins. *Curr Opin Biotechnol* 2014;**30**:95–100.
93. Saint-Jore-Dupas C, Faye L, Gomord V. From planta to pharma with glycosylation in the toolbox. *Trends Biotechnol* 2007;**25**(7):317–23.
94. Lai H, He J, Hurtado J, Stahnke J, Fuchs A, Mehlhop E, et al. Structural and functional characterization of an anti-West Nile virus monoclonal antibody and its single-chain variant produced in glycoengineered plants. *Plant Biotechnol J* 2014;**12**(8):1098–107.
95. Olinger Jr. GG, Pettitt J, Kim D, Working C, Bohorov O, Bratcher B, et al. Delayed treatment of Ebola virus infection with plant-derived monoclonal antibodies provides protection in rhesus macaques. *Proc Natl Acad Sci U S A* 2012;**109**(44):18030–5.
96. Koprivova A, Stemmer C, Altmann F, Hoffmann A, Kopriva S, Gorr G, et al. Targeted knockouts of Physcomitrella lacking plant-specific immunogenic N-glycans. *Plant Biotechnol J* 2004;**2**(6):517–23.
97. Schahs M, Strasser R, Stadlmann J, Kunert R, Rademacher T, Steinkellner H. Production of a monoclonal antibody in plants with a humanized N-glycosylation pattern. *Plant Biotechnol J* 2007;**5**(5):657–63.
98. Strasser R, Stadlmann J, Schahs M, Stiegler G, Quendler H, Mach L, et al. Generation of glyco-engineered *Nicotiana benthamiana* for the production of monoclonal antibodies with a homogeneous human-like N-glycan structure. *Plant Biotechnol J* 2008;**6**(4):392–402.
99. Pujol M, Gavilondo J, Ayala M, Rodriguez M, Gonzalez EM, Perez L. Fighting cancer with plant-expressed pharmaceuticals. *Trends Biotechnol* 2007;**25**(10):455–9.

100. Costa AR, Rodrigues ME, Henriques M, Oliveira R, Azeredo J. Glycosylation: impact, control and improvement during therapeutic protein production. *Crit Rev Biotechnol* 2014;**34**(4):281–99.
101. Ma JK, Hikmat BY, Wycoff K, Vine ND, Chargelegue D, Yu L, et al. Characterization of a recombinant plant monoclonal secretory antibody and preventive immunotherapy in humans. *Nat Med* 1998;**4**(5):601–6.
102. Bakker H, Bardor M, Molthoff JW, Gomord V, Elbers I, Stevens LH, et al. Galactose-extended glycans of antibodies produced by transgenic plants. *Proc Natl Acad Sci U S A* 2001;**98**(5):2899–904.
103. Bakker H, Schijlen E, de Vries T, Schiphorst WE, Jordi W, Lommen A, et al. Plant members of the alpha1–>3/4-fucosyltransferase gene family encode an alpha1–>4-fucosyltransferase, potentially involved in Lewis(a) biosynthesis, and two core alpha1–>3-fucosyltransferases. *FEBS Lett* 2001;**507**(3):307–12.
104. Wee EG, Sherrier DJ, Prime TA, Dupree P. Targeting of active sialyltransferase to the plant Golgi apparatus. *Plant Cell* 1998;**10**(10):1759–68.
105. Castilho A, Strasser R, Stadlmann J, Grass J, Jez J, Gattinger P, et al. In planta protein sialylation through overexpression of the respective mammalian pathway. *J Biol Chem* 2010;**285**(21):15923–30.
106. Qiu X, Wong G, Audet J, Bello A, Fernando L, Alimonti JB, et al. Reversion of advanced Ebola virus disease in nonhuman primates with ZMapp. *Nature* 2014;**514**(7520):47–53.
107. NIH. *Study finds Ebola treatment ZMapp holds promise, although results not definitive;2016.*
108. Schouten A, Roosien J, van Engelen FA, de Jong GA, Borst-Vrenssen AW, Zilverentant JF, et al. The C-terminal KDEL sequence increases the expression level of a single-chain antibody designed to be targeted to both the cytosol and the secretory pathway in transgenic tobacco. *Plant Mol Biol* 1996;**30**(4):781–93.
109. Ko K, Tekoah Y, Rudd PM, Harvey DJ, Dwek RA, Spitsin S, et al. Function and glycosylation of plant-derived antiviral monoclonal antibody. *Proc Natl Acad Sci U S A* 2003;**100**(13):8013–8.
110. Petruccelli S, Otegui MS, Lareu F, Tran Dinh O, Fitchette AC, Circosta A, et al. A KDEL-tagged monoclonal antibody is efficiently retained in the endoplasmic reticulum in leaves, but is both partially secreted and sorted to protein storage vacuoles in seeds. *Plant Biotechnol J* 2006;**4**(5):511–27.
111. Sharp JM, Doran PM. Characterization of monoclonal antibody fragments produced by plant cells. *Biotechnol Bioeng* 2001;**73**(5):338–46.
112. Wright KE, Prior F, Sardana R, Altosaar I, Dudani AK, Ganz PR, et al. Sorting of glycoprotein B from human cytomegalovirus to protein storage vesicles in seeds of transgenic tobacco. *Transgenic Res* 2001;**10**(2):177–81.
113. Li H, Sethuraman N, Stadheim TA, Zha D, Prinz B, Ballew N, et al. Optimization of humanized IgGs in glycoengineered Pichia pastoris. *Nat Biotechnol* 2006;**24**(2):210–5.
114. Meuris L, Santens F, Elson G, Festjens N, Boone M, Dos Santos A, et al. GlycoDelete engineering of mammalian cells simplifies N-glycosylation of recombinant proteins. *Nat Biotechnol* 2014;**32**(5):485–9.
115. Schuster M, Umana P, Ferrara C, Brunker P, Gerdes C, Waxenecker G, et al. Improved effector functions of a therapeutic monoclonal Lewis Y-specific antibody by glycoform engineering. *Cancer Res* 2005;**65**(17):7934–41.
116. Umana P, Jean-Mairet J, Bailey JE. Tetracycline-regulated overexpression of glycosyltransferases in Chinese hamster ovary cells. *Biotechnol Bioeng* 1999;**65**(5):542–9.
117. Yamane-Ohnuki N, Kinoshita S, Inoue-Urakubo M, Kusunoki M, Iida S, Nakano R, et al. Establishment of FUT8 knockout Chinese hamster ovary cells: an ideal host cell line for producing completely defucosylated antibodies with enhanced antibody-dependent cellular cytotoxicity. *Biotechnol Bioeng* 2004;**87**(5):614–22.
118. Anour R. Biosimilars versus "biobetters"—a regulator's perspective. *GaBI J* 2014;**3**(4):166–7.

119. Okazaki A, Shoji-Hosaka E, Nakamura K, Wakitani M, Uchida K, Kakita S, et al. Fucose depletion from human IgG1 oligosaccharide enhances binding enthalpy and association rate between IgG1 and FcgammaRIIIa. *J Mol Biol* 2004;**336**(5):1239–49.
120. Shields RL, Lai J, Keck R, O'Connell LY, Hong K, Meng YG, et al. Lack of fucose on human IgG1 N-linked oligosaccharide improves binding to human Fcgamma RIII and antibody-dependent cellular toxicity. *J Biol Chem* 2002;**277**(30):26733–40.
121. Beck A, Wurch T, Bailly C, Corvaia N. Strategies and challenges for the next generation of therapeutic antibodies. *Nat Rev Immunol* 2010;**10**(5):345–52.
122. Jefferis R. Glycosylation as a strategy to improve antibody-based therapeutics. *Nat Rev Drug Discov* 2009;**8**(3):226–34.
123. Rosales-Mendoza S, Salazar-Gonzalez JA, Decker EL, Reski R. Implications of plant glycans in the development of innovative vaccines. *Expert Rev Vaccines* 2016;**15**(7):915–25.
124. Barta A, Sommergruber K, Thompson D, Hartmuth K, Matzke MA, Matzke AJ. The expression of a nopaline synthase—human growth hormone chimaeric gene in transformed tobacco and sunflower callus tissue. *Plant Mol Biol* 1986;**6**(5):347–57.
125. Mason HS, Lam DM, Arntzen CJ. Expression of hepatitis B surface antigen in transgenic plants. *Proc Natl Acad Sci U S A* 1992;**89**(24):11745–9.
126. Chen Q, Davis KR. The potential of plants as a system for the development and production of human biologics. *F1000Research* 2016;**5**. F1000 Faculty Rev-1912.
127. Lyon GM, Mehta AK, Varkey JB, Brantly K, Plyler L, McElroy AK, et al. Clinical care of two patients with Ebola virus disease in the United States. *N Engl J Med* 2014;**371**(25):2402–9.
128. Castilho A, Bohorova N, Grass J, Bohorov O, Zeitlin L, Whaley K, et al. Rapid high yield production of different glycoforms of Ebola virus monoclonal antibody. *PLoS One* 2011;**6**(10):e26040.
129. Zeitlin L, Pettitt J, Scully C, Bohorova N, Kim D, Pauly M, et al. Enhanced potency of a fucose-free monoclonal antibody being developed as an Ebola virus immunoprotectant. *Proc Natl Acad Sci U S A* 2011;**108**(51):20690–4.
130. Shaaltiel Y, Gingis-Velitski S, Tzaban S, Fiks N, Tekoah Y, Aviezer D. Plant-based oral delivery of beta-glucocerebrosidase as an enzyme replacement therapy for Gaucher's disease. *Plant Biotechnol J* 2015;**13**(8):1033–40.
131. Gavilondo JV, Larrick JW. Antibody engineering at the millennium. *Biotechniques* 2000;**29**(1):128–32.
132. Larrick JW, Thomas DW. Producing proteins in transgenic plants and animals. *Curr Opin Biotechnol* 2001;**12**(4):411–8.
133. Lakshmi PS, Verma D, Yang X, Lloyd B, Daniell H. Low cost tuberculosis vaccine antigens in capsules: expression in chloroplasts, bio-encapsulation, stability and functional evaluation in vitro. *PLoS One* 2013;**8**(1):e54708.
134. Kwon KC, Daniell H. Low-cost oral delivery of protein drugs bioencapsulated in plant cells. *Plant Biotechnol J* 2015;**13**(8):1017–22.
135. Su J, Sherman A, Doerfler PA, Byrne BJ, Herzog RW, Daniell H. Oral delivery of Acid Alpha Glucosidase epitopes expressed in plant chloroplasts suppresses antibody formation in treatment of Pompe mice. *Plant Biotechnol J* 2015;**13**(8):1023–32.
136. Su J, Zhu L, Sherman A, Wang X, Lin S, Kamesh A, et al. Low cost industrial production of coagulation factor IX bioencapsulated in lettuce cells for oral tolerance induction in hemophilia B. *Biomaterials* 2015;**70**:84–93.
137. von Schaewen A, Sturm A, O'Neill J, Chrispeels MJ. Isolation of a mutant Arabidopsis plant that lacks N-acetyl glucosaminyl transferase I and is unable to synthesize Golgi-modified complex N-linked glycans. *Plant Physiol* 1993;**102**(4):1109–18.
138. Andrews LB, Curtis WR. Comparison of transient protein expression in tobacco leaves and plant suspension culture. *Biotechnol Prog* 2005;**21**(3):946–52.
139. Boothe J, Nykiforuk C, Shen Y, Zaplachinski S, Szarka S, Kuhlman P, et al. Seed-based expression systems for plant molecular farming. *Plant Biotechnol J* 2010;**8**(5):588–606.

140. Paul M, Ma JK. Plant-made pharmaceuticals: leading products and production platforms. *Biotechnol Appl Biochem* 2011;**58**(1):58–67.
141. Sack M, Rademacher T, Spiegel H, Boes A, Hellwig S, Drossard J, et al. From gene to harvest: insights into upstream process development for the GMP production of a monoclonal antibody in transgenic tobacco plants. *Plant Biotechnol J* 2015;**13**(8):1094–105.
142. Sabalza M, Christou P, Capell T. Recombinant plant-derived pharmaceutical proteins: current technical and economic bottlenecks. *Biotechnol Lett* 2014;**36**(12):2367–79.
143. Barzegari A, Saeedi N, Zarredar H, Barar J, Omidi Y. The search for a promising cell factory system for production of edible vaccine. *Hum Vaccin Immunother* 2014;**10**(8):2497–502.
144. Walmsley AM, Arntzen CJ. Plants for delivery of edible vaccines. *Curr Opin Biotechnol* 2000;**11**(2):126–9.
145. Lallemand J, Bouche F, Desiron C, Stautemas J, de Lemos Esteves F, Perilleux C, et al. Extracellular peptidase hunting for improvement of protein production in plant cells and roots. *Front Plant Sci* 2015;**6**:37.
146. Guan C, Du X, Wang G, Ji J, Jin C, Li X. Expression of biologically active anti-thrombosis protein lumbrokinase in edible sunflower seed kernel. *J Plant Biochem Biotechnol* 2014;**23**(3):257–65.
147. Wang KY, Tull L, Cooper E, Wang N, Liu D. Recombinant protein production of earthworm lumbrokinase for potential antithrombotic application. *Evid Based Complement Alternat Med* 2013;**2013**.
148. Paul MJ, Thangaraj H, Ma JK. Commercialization of new biotechnology: a systematic review of 16 commercial case studies in a novel manufacturing sector. *Plant Biotechnol J* 2015;**13**(8):1209–20.

CHAPTER

8

Protein Modification in Plants in Response to Abiotic Stress

Hilal Ahmad Qazi, Nelofer Jan, Salika Ramazan, Riffat John

Plant Molecular Biology Lab, Department of Botany, University of Kashmir, Srinagar, India

1 INTRODUCTION

Living organisms often undergo distinct environmental stress conditions, like salinity, extreme temperatures, altered nutrient availability, drought, and ultraviolet irradiation.[1] Plants, being sessile, strike equilibrium between growth, development, and survival, and as a result, adapt to stressful surroundings by reprogramming metabolic pathways and gene expression.[2] Beginning from perception of stress and concluding with particular transcriptional modifications, upshots of stress involve a complex cascade of multiple pathways.[3] Both specific and common outputs are observed, as concluding strategy of general response incorporates both stress-specific requirements and their crosstalk.[4] Morphological and physiological adjustments to environmental confinements are essentially linked with transcriptional and translational modifications regulated by complex molecular mechanisms.

Representative posttranslational modifications (PTMs) during plant stress adaptation, including glycosylation, acetylation, phosphorylation, and succinylation, have usually been deciphered using plant proteomic technologies. Posttranslationally, a plethora of molecules may adhere to different proteins and thereby alter their activity, half-life, and subcellular localization. One of the most significant PTMs, occurring during all known types of stress, is protein phosphorylation.[5] In almost all eukaryotes, small ubiquitin-like modifier (SUMO) conjugations and ubiquitin are emerging as stellar posttranslational regulatory mechanisms.[6]

Protein Modificomics
https://doi.org/10.1016/B978-0-12-811913-6.00008-4

Consequently, all these modifications in concert determine the ultimate target, as well as the effect on the related process/phenotypic traits. PTMs are of the essence, as they may be involved in refinement of protein half-life, function, localization, and interactions to extenuate the possible harm of environmental stresses.[7–9]

PTMs contribute extraordinarily in diversity of proteomes, enhance functionality, permit for speedy responses, play a vital role in expression of different genes, accelerating the process of signaling transduction, protein stability, and interactions, along with the enzyme kinetics.[10] More than 200 kinds of modifications ascertain indispensable facets of functions of proteins, comprising of turnover, subcellular localization, activity, or rest of protein interactions.[11, 12] Approximately 10% of the *Arabidopsis* genome is committed to the most predominant PTMs, such as protein phosphorylation and ubiquitination, thereby suggesting the wideness of regulation in plants.[13, 14] Moreover, it has been established that the protein kinase superfamily is vast in plants as compared to other eukaryotes,[13] directing toward the applicability of PTMs of proteins in the case of plants. The high count of protein kinases in plants can be attributed to the need of plants to adapt to dynamic surrounding circumstances.[15] Various transcriptional and PTMs of their specific transcript/protein targets, which are often regarded as regulators of cellular processes, comprising of various components in signaling and transcription factors (TFs) have been recognized. Hence, novel potent targets for establishing new genotypes with elevated stress tolerance are represented by posttranscriptional and posttranslational mechanisms. This chapter was designed to sum up our present understanding of the molecular players and their ordinance that cause modification in the posttranslational system of plant proteins in response to different kinds of stresses experienced by the plants.

2 ABIOTIC STRESS IN PLANTS

Plants unrelentingly fall upon a blanket of abiotic stresses, due to which agricultural productivity is trammeled. Abiotic stresses comprise of drought, salinity, temperature extremes, flood, radiation, heavy metals, and so on. Globally, it is considered a leading cause for decline of major crop plants.[16] Increment in the world's total terrestrial area toward desertification, acceleration of salinity content of water and soil, water resource deficit, and environmental pollution are adding fuel to the fire of crop loss. Plants, being sessile, cannot escape from such environmental constraints. Consequently, to combat this menace, plants have formulated different mechanisms to adapt to such settings for endurance. They are stimulated by first sensing the external stressful surroundings and then generating befitting responses at cellular level. Multiple signal transduction pathways assist the responses at cellular level, which act by passing the stimuli produced by sensors, positioned on the surface of cell or cytoplasm, to the transcriptional processes lying inside the nucleus. This directs the multiple transcriptional modifications conferring the stress tolerance to the plant (Fig. 1). A vital role is being played by the signaling webs, by linking up between the sensing of stressful surroundings and generation of proper biochemical and physiological response.[17]

Acclimation of a plant is achieved through effective changes in gene expression, which ultimately results in changes in plant transcriptome, proteome, and metabolome. Research studies carried out so far show that changes at transcriptomic level do not often match with those of proteomic level. Therefore, exploration of altered protein expression in in plant is very important because, unlike transcripts, these proteins are directly involved in response of plants toward stress. Plants' response to different abiotic stress conditions, such as, high/low temperature, drought, flooding, salinity, and heavy metal stresses are highly complex and involve drastic changes in their protein profiles.

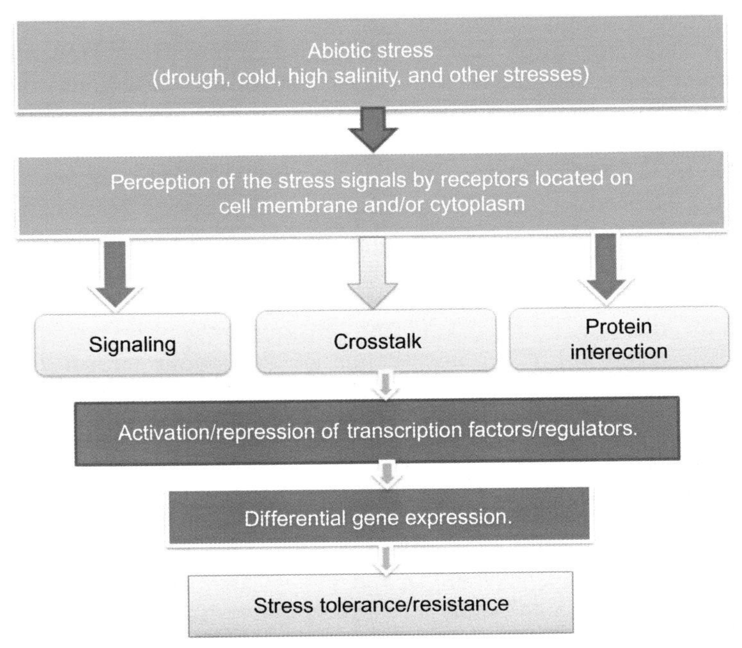

FIG. 1 Representation of cross-talk among stress signaling pathways.

3 TYPES OF ABIOTIC STRESS IN PLANTS AND THE PROTEINS INVOLVED THEREIN

3.1 Drought Stress and Associated Proteins

Whenever there is water shortage in the soil and atmosphere, the condition is said to be drought. During drought stress, plants pass through a series of biochemical and physiological events. These responses consequently cause closure of stomata and inhibition of cell growth and photosynthesis, diminish osmotic potential of plant tissues, lower the transpiration rate, and enhance respiration rate. Furthermore, adaptations take place also at cellular levels and molecular levels. Plants amass various types of osmoprotectants (viz. sorbitol, mannitol, fructans, trehalose), plant growth regulators, like abscisic acid, and also cause synthesis of novel stress combating proteins. Among the expression profiling of around 1300 genes in the case of *Arabidopsis*, an aggregate of 299 drought-inducible genes were recognized, and 73 stress inducible genes in the case of rice.[18, 19]

Based on the biological functionality, the drought-inducible genes may be classified into two categories. The first category comprises of the genes that code for proteins responsible for protein and membrane stabilization and homeostasis of cells, for example, late embryogenesis abundant (LEA) proteins, heat shock proteins (HSPs), and lipid transfer proteins (LTPs). The second category comprises of protein phosphatases class 2C (PP2C) and calcineurin-B-interacting protein kinases (CIPK16) that work as signaling molecules in different kinds of stress responses.[20, 21] Hence, based on the expression patterns of such genes that code for functional and regulatory proteins, the purported mechanism of drought tolerance can be apprehended. During drought stress, the two types of signaling pathways induced are ABA-dependent or ABA-independent.[16] During water scarcity, ABA is produced, causing plants to become very responsive and quite tolerant to drought stress and to high salinity as well. ABA-dependent signal transduction becomes visible as many genes that are induced during drought and cold stress are also elicited by exogenous supply of ABA. In contrast, different genes operate in response to cold and drought stress but are not elicited by exogenous supply of ABA, thereby pointing toward the occurrence of ABA-independent transduction of multiple stresses in case of *Arabidopsis*.[3, 17]

Upon experiencing water stress, a series of metabolic changes occur in a plant, including synthesis of new mRNA and protein. These transcripts encode polypeptides that closely resemble proteins like dehydrin and LEA, i.e., proteins involved during embryo maturation in seeds and in water stressed plants.[22] In one of the biochemical events, water stress causes accumulation of abscisic acid (ABA). Previously, it was considered that ABA inhibits RNA and protein synthesis. Now it is clear that ABA can promote formation of a set of its own proteins in certain plant tissues[23]; it regulates the response of plants to water stress by closing stomatal apertures, thereby decreasing transpiration rate, which ultimately leads to water conservation.[24] Water stress proteins are, therefore, divided into two groups based on participation of ABA: ABA-dependent and independent proteins.[25]

In plants, water stress responses cause changes in protein synthesis that share several similarities with those proteins occurring in developing seeds, e.g., RAB 21 (12 kDa) protein, a rice embryogenesis protein induced by water stress and ABA treatment in plants.[26] Late embryogenesis proteins or dehydrins are the family of polypeptides with glycine rich content. Drought-induced dehydrin proteins are hydrophilic and heat stable proteins found in many plant species.[27] These dehydrins were first identified in the late 1980s in rice seedlings induced by ABA.[28] *Arabidopsis thaliana* plasma membrane intrinsic protein[29] and cysteine protease[30] are also reported, and protect cells from dehydration. One of the dehydration stimulated NAC proteins, RD26, a 298 amino acid protein, is expressed in an ABA dependent stress-signaling pathway.[31] The overexpression of LOV KELCH PROTEIN 2 (LPK2) in *A. thaliana* induces dehydration tolerance by stimulating stomatal closure through expression of dehydration inducible genes.[32]

Many dehydration induced gene products are known to bestow protection to different cellular structures against water stress. These genes are commonly called "lea" and were actually identified as genes expressed during seed development.[33] The lea gene products popularly known as late LEA proteins were first identified in cotton and wheat.[34] Their expression is mainly associated with acquisition of desiccation tolerance by plants. These proteins are localized to cytoplasm with mainly hydrophilic nature lacking Cys and Trp. Based on their amino acid sequence and predicted protein structure, six groups of LEA proteins have been so far identified.[35] Among these different groups, group 3 is known to play a vital role in ion sequestration during cellular dehydration and has been already studied in *Deinococcus radiodurans*.[36] Similarly, other groups of LEA, groups 4 and 5, are reported to have a comparable role to group 3 during dehydration stress. LEA protein also functions as a chaperone and plays a vital role in protection of cellular and molecular structure from desiccation by hydration buffering, ion sequestration, and protection of other proteins.[34,37]

3.2 Cold Stress and Cold-Induced Proteins

Another common environmental stress that negatively influences the growth of plants and crop production is cold stress, i.e., low temperature. Even though most of the plants are very susceptible to below-zero temperatures, nevertheless, they can enhance their tolerance against freezing temperatures by exposing themselves to cold, nonfreezing temperatures. Such process is termed as cold acclimation, which reconstitutes the biochemical compositions of plasma membranes and their physical structures by altering the composition of lipids and brings out other nonenzymatic proteins. In the recent past, considerable studies related to transcriptomics have been conducted to explicate the sensing and regulatory responses and mechanisms of plants in response to chilling temperatures that renders them capable of developing cold acclimation. Multiple transcriptional cascades are triggered by cold stress as a majority of the early cold-responsive genes code for TFs, which in turn trigger late-induced cold responsive genes. Differential expression of genes that code for biosynthesis or signaling of plant growth regulators, for example auxin, ABA, and gibberellic acid occurs. In association with cold stress, DNA replication, mismatch repair, and spliceosome pathways have been reported.[38] Calcium plays an important role as messenger in cold temperature signal transduction. Speedy gain in the calcium levels in cytoplasm occurs in reaction to cold temperature. It has also been reported that calcium is vital for the perfect expression of a few of the cold responsive genes, including the COR6 genes and CRT/DRE controlled KIN1 in *Arabidopsis*.[16]

Protein quality is controlled by proteases and chaperones, which raise components of cell walls, playing a significant part in cold tolerance development. Some proteins show AU1 heightened degradation, particularly of proteins of photosynthesis, including larger subunits of rubisco, during cold stress.[39] Likewise, downregulation of Calvin cycle and Krebs cycle enzymes and upregulation of ascorbate recycling have been reported.[40] Rapid metabolic alterations elicited by cold stress provide instant protection to plants before temperature goes down below freezing point. Hence, an ample number of cellular proteins are expressed differentially in response to cold stress, thereby conferring cold acclimation.

Different genetic approaches have shown that cold acclimation is a quantitative feature that involves many genes with specific effective roles.[41, 42] In *Arabidopsis*, tobacco, and alfalfa, some of these genes have been identified, which encode proteins with known enzyme activities for freeze tolerance or as cryoprotectants or dehydrins.[42] The protein induction due to low temperature may act as functional enzymes, contributing to different biochemical pathways leading to accumulation of carbohydrate, amino acids, and other metabolites, thereby altering lipid composition of membranes. The transcription family factors, the C-repeat binding factor (CBF)/DREB1 (dehydration response element binding) proteins, has been identified in *Arabidopsis*, which controls cold induced (COR) gene expression for freezing tolerance.[42] MYC transcription factor ICE1 (Inducer of CBF Expression) are involved in regulation of around 40% of COR genes and 46% of cold regulated transcription factor genes, signifying the master role of ICE1 in regulation of CBF3/DREB1A and other COR genes.[43] The (OsICE) are reported to bestow cold tolerance to rice plants.[44] In *Brassica napus*, low temperature has been shown to induce HSPs like HSP 90 and HSP 70.[45] Polypeptides of 15 kDa and 39 kDa have been reported in *Arabidopsis* and wheat, respectively, to encode for cold tolerant genes.[42]

Glycerol-3-phosphate acyltransferase (GPAT), a chloroplast enzyme associated with fatty acid unsaturation, shows correlation with cold tolerance. Different experiments with *Arabidopsis* demonstrated that manipulation with GPAT caused change in plant cold tolerance.[46] In addition to this, cold shock domain proteins (CSDPs) have been extracted from many plant species.[47] Park et al.[48] reported that CSDP1 and CSDP2 give freezing tolerance to plants in addition to having a role in seed germination. In transgenic rice, overexpression of cold induced galactinol synthase confers chilling tolerance to these plants. Another protein, osmotin is well known to have a major role in cryo-protection during cold stress. During low temperature stress, expression of osmotin in endosperm and seed coat of olive and pollen grains of solanum is induced.[49, 50]

3.3 Salinity Stress and the Proteins Induced Thereby

Every year throughout the world, a large area of land is affected by salinity at an alarming rate, thereby making salinity a leading environmental stress limiting the crop yield globally. Plants cannot take up high salinity water as it readily causes diminution in growth rate, along with other groups of metabolic modifications similar to the changes caused by water stress. Premature senescence along with the reduction of the photosynthetic area of the plant on its leaf is the specific blow seen during excessive intake of salt inside the plant, due to which it cannot sustain growth. Among different spectacular features of saline stress responses studied in salt-tolerant barley, was the elucidation of jasmonic acid (JA) pathway genes along with ABA.[51] In addition, a huge number of abiotic stress-related (heat, cold, and drought) genes were found to be induced, thus supporting cross talk as being among the elements involved in abiotic stress responses. During the early salinity stress response genes in the case of tomato roots, genes encoding the inducible activators of transcription, stress sensors, HSPs, and upstream signal pathway components and other critical elements, e.g., MAPKKK and PP2C were upregulated.[52] When a plant senses salt stress, the JA pathway gets elicited, cytosolic Ca^{2+} levels become raised, and consequently, activate different signaling transduction pathways, which include SOS (salt overly sensitive). Other TFs, namely MYC/MYB, DREB/CBF, bZIP, and AREB/ABF, also get sparked off, with the outcome of downstream signaling cascade, which help in keeping the ion homeostasis and ameliorate the salt stress tolerance of plants.

Recently, a protein with salt stress-responsive role with a purported function in stress signaling has been reported in the apoplast of the roots of rice.[53] In accomplishing adaptation to salt stress, the metabolism-related proteins also act significantly in every organ (e.g., hypocotyls, roots, and leaves.[54] In fact, during seed germination, a range of proteins are generated that provide enough tolerance against salt stress.[55] Hence, plants amend their ability of salt stress tolerance by acquiring a more effective ionic and osmotic homeostasis, dominant capacity to withdraw toxic by-products, and ultimately, an improved potential for recovery of growth. All such processes encompass proteins involved in energy metabolism, ROS scavenging mechanisms, cytoskeleton integrity and stability, protein processing including structural and functional stability and folding, photosynthesis, photorespiration, and other processes of signal transduction. Conjointly with previously mentioned stresses, rest stresses experienced by plants include presence of heavy metals, high temperature, and flood.

Salt stress adversely effects protein synthesis, enzyme activities, photosynthetic pigments, membrane integrity, and yield.[56] To overcome salt stress, plants switch on multiple biochemical pathways, which result in synthesis of different metabolites, some proteins, and reactive oxygen species related enzymes controlling ion and water flow and hunting free radicals.[57] Proteins accumulated under salt stress are known as salt shock proteins (SSP).. Salt stress leads to increase in synthesis of SSP, which in turn enhances tissue dependent protein and enzyme activities.[56, 58] From 34 plant species, approximately 2171 different proteins have been identified and characterized as salt responsive proteins, which showed a change in their regulation under salt stress.[59] Salt stress proteins have been observed in various plant species like rice roots 14.5 kDa,[60] rice shoots 15 and 26 kDa,[61] cell cultures 26 and 27 kDa,[61], and germination seeds 23 kDa.[62] Similarly, such proteins in cultured tobacco cells 18, 19.5, 21, 26, 34, 35.5, 37, and 58 kDa[63] and in tomato roots 21, 21.5, 22, and 32 kDa[64] have been reported. Sobhanian et al.[54] reported 20 kDa abundant chaperonin in soybean leaves subjected to salt stress with the suggestive role of chaperonin in preventing protein misfolding during salt stress. Also, the overexpression of protein AtRab7 (AtRabG3e) in *A. thaliana* is reported to enhance tolerance against salt and osmotic stress.[65]

Genetic studies have proposed that salt stress triggers calcium signaling in cytosol, which activates calcium sensor proteins like salt overly sensitive (SOS3) protein regulating efflux of Na^+ and compartmentation of ions in vacuole.[66] In addition to this, osmotin is also known to play a vital role in salt tolerance by sequestration of Na^+ ions into vacuoles and intercellular spaces. The gene expression of osmotin causes enhanced antioxidant defense in *Capsicum annum* for its salt tolerance by accumulating different osmolytes and increased expression of some particular antioxidant enzymes.[67] Recently, some GmWRKY proteins identified in soybean genome are reported to be regulated positively in salt stress.[68]

3.4 Heat Stress and Heat Shock Induced Proteins

Heat stress, through altering cellular components and metabolism, has a drastic influence on physiology of a plant, development, and crop yield. In response to higher temperature, genomic arrangement, along with the profiling of transcripts of heat shock factors (HSFs), has been surveyed in rice.[69] Many HSF genes have been significantly stimulated, thereby suggesting their critical role in the transcriptional initiation of heat shock genes. These genes are related to metabolism, antioxidant, and heat shock, and are involved in several developmental pathways indicating considerable genomic differences in response to adaptation against elevated temperatures. Upregulation of different enzymes having a critical role in biosynthesis of starch, translation initiation factors, small HSPs, and carbohydrate metabolism have been reported in wheat grain by proteomic analysis.[70] Likewise, some small HSPs with low molecular weight and several enzymes involved in antioxidant pathways with intense induction were detected through proteomic approach in rice.[71] In rice, heat stress stimulates the phosphorylation of ATP-β and dephosphorylation of RuBisCo, causing the decreased activities of ATP synthase and RuBisCo[72], indicating the participation of phosphoproteins in the heat-stress signal transduction mechanisms.

Heat shock response is a universal phenomenon characterized by the increased expression of some particular genes, which accelerates synthesis of some defensive proteins upon exposure to some environmental stress conditions. These specific defensive proteins, known as HSPs, are primarily produced in response to heat shock in most of the organisms. Heat shock responses are mediated at transcriptional level,[73] which leads to synthesis of HSPs and postponing synthesis of other proteins. Depending upon the size, three groups of HSPs are present in a plant system:

(a) large HSPs, with a size ranging from 80 to 104 kDa, found in bacteria, animals, and plants;
(b) intermediate size HSPs, of size 60–80 kDa;
(c) small HSPs, about 8–20 kDa in size.

All these proteins work in coordination when an organism perceives heat stress. A drastic temperature upshift (39–41°C) is normally responsible for the induction of these proteins, however, gradual temperature increase (2.5°C per hour) can result in their induction. On average, the synthesis of HSPs can be detected from 3 to 20 minutes after heat shock.[74] The HSP gene structure, their consensus elements, TFs, their gene expression in transgenic plants, and structure of HSPs in plants has been described.[75] Heat stress causes complex disruption, which leads to generation of active heat shock transcription factor, later directed to nucleus, resulting in HSP transactivation by interaction of HSF trimers with heat shock elements in the promoters.[76] HSF1 and HSF3 play the main role in expression of HSP in *Arabidopsis*.[77] The presence of ROS may be detected by HSFs, which may act as molecular sensors, resulting in activation of downstream stress responsive genes. The activation of HSPs and HSFs after different stress treatments like drought, cold, heat, or salt exhibited some overlap in them, but showed a typical response to each stress condition.[78] Recent research showed that the growth of *Arabidopsis* seedlings was improved by a slight increase in environmental temperature by HSP90 protein helping in biosynthesis of auxin hormone.[76] It is reported in *Arabidopsis* that heat induced HSP90 controls temperature dependent seedling growth by stabilizing transport inhibitor response 1/auxin response F-box (TIR1/AFB) protein.

Heat shock response is mainly characterized by temporary expression of HSPs. The expression of HSPs has been found to be positively correlated with the acquirement of thermo-tolerance and the consequent over expression of HSPs often results in enhanced heat tolerance.[73] In higher plants, early reports of HSPs showed presence of HSP in stressed tobacco and soybean cell cultures.[79] HSPs have also been reported in tomato,[73, 80] green gram and mung bean,[81] barley,[82] *Lilium elogiform*,[83] soybean,[84] corn,[85] sunflower,[86] *Arabidopsis*,[87] and rice.[88] For heat shock response, the optimal induction temperature varies between different species but usually occurs from 10°C to 15°C above the temperature used for optimal plant growth. Some HSPs are known to function as molecular chaperons which reduce protein denaturation, cause proteasome degradation of denatured protein, enhance protein folding necessary for protein maturation, and control HSP gene expression by regulating HSF activity during thermos tolerance procurement.[89, 90] In Table 1, some important classes of HSP's and their functions are mentioned.[91]

3.5 Heavy Metal Stress and the Proteins Involved

Plants need heavy metals, including manganese (Mn), cobalt (Co), molybdenum (Mo), zinc (Zn), iron (Fe), and copper (Cu) in small quantities as essential micronutrients. They act as enzyme cofactors and are involved in indispensable biological processes like redox reactions; however, these metals have always proven toxic in excess amounts. In contrast, other heavy metals like lead (Pb), aluminum (Al), mercury (Hg),

TABLE 1 Heat Shock Proteins and Their Functions.

Class of HSPs	Functions of HSPs
HSP100	ATP controlled dissociation of assembled proteins.
HSP90	Regulates heat-associated signals through protein folding requiring ATP for functioning.
HSP70, HSP40	Newly synthesized proteins are stabilized by this protein, requiring ATP binding and release.
HSP60, HSP10	Helps in folding proteins in a specialized manner. It is also ATP dependent.
HSP20 or small HSP (sHSP)	Forming complexes that serve cellular matrix for stabilization of proteins. HSP (100, 70, 40) are required for this.

cadmium (Cd), and arsenic (As) result in ROS production and enzyme inactivation, thus have no nutrient functions and are very harmful to plants. For instance, through microanalysis studies, differential expression of many genes that code for HSPs, glutathione S-transferase, protein kinases, cytochrome P450 family proteins, transporters of ions, and different TFs, for instance NAC and DREB, have been seen to be produced in response to cadmium stress in rice, suggesting the potential part played by transporters in detoxification of Cd by Cd export from the cytosol.[92] Similarly, differential gene expression employed in primary metabolism of plant defense, transportation, transcription, and secondary metabolism was studied via microarray analysis after exposing *Brassica carinata* seedlings to lithium chloride.[93] Acute induction of several proteins that are involved in photosynthetic processes, transcriptional pathways, translation, and coded molecular chaperones was revealed in response to cadmium stress using various proteomic studies.[94]

Proteins that have high affinity toward different metal ions contain cysteine (Cys), methionine (Met), and histidine (His) residues play a role in plant cell homeostasis and tolerance. The cytosolic chelation of heavy metals by high affinity ligands like phytochelatins (PCs) and metallothioneins (MTs) is an important mechanism for detoxification of heavy metal and tolerance in plants during heavy metal stress.[95] Phytochelatins, a cysteine rich polypeptide family with the basic structure (α-Glu-Cys) n-X, where $n=2$–11 and X is Gly, α-Ala, Ser, Gln, or Glu, depending on the organism, with 2–4 peptides being most common.[96] Stress damaged proteins are repaired by different phytochelatins and, with the help of tonoplast located transporters, assist in compartmentation of metal ions in the vacuole.[97] Cell culture kinetic studies showed that, within minutes of Cd exposure, phytochelatins biosynthesis starts and is independent of de novo protein synthesis. Phytochelatins are produced from tripeptide glutathione (TSH) by phytochelatin synthase (PCS). In both conditions, in vivo and in vitro phytochelatin synthase are activated in presence of metal ions, particularly Pb, Cd, Cu, Ag, Hg, Sn, Zn, As, and Au.[98] This enzyme expresses itself independent of heavy metal exposure.[95] Phytochelatin are low molecular weight proteins (1.5–4 kDa), which plants employ to try to overcome heavy metal stress.[99] In addition to this, PCs are said to be important for normal tolerance to several unwanted metal ions.[100] Metallothioneins (MTs) are cysteine rich metal binding low molecular weight polypeptides with ability to bind wide array of heavy metals. Being very much diverse, plant MTs are classified into three groups, on the basis of cysteine residues.[101] The Cys-Cys, Cys-X-Cys, and Cys-X-X-Cys motifs are constituents and invariants for metallothioneins. The transcriptional regulation for the biosynthesis of MTs is enhanced by many factors like cytotoxic agents, different hormones, and metals ions, viz. Cd, Hg, Zn, Cu, Ag, Au, Co, Bi, and Ni.[96] Multiple studies have claimed the importance of MTs in maintaining cell homeostasis and heavy metal sequestration, thereby protecting cells from oxidative damage, but actual physiological functions of MTs are yet to be explained fully.[102] Effect of various abiotic stresses on plants and their response to such stresses are mentioned in Table 2.

3.6 Flooding Stress in Plants

Worldwide, flooding also limits the plant yield by causing hypoxia to plants, thus acting as another continual natural disaster. A huge number of genes acting in waterlogging response and associated with metabolism of amino acids and carbon, transport of ions, signal transduction, protein degradation, and transcriptional and translational mechanisms have been shown in case of maize roots through transcriptomic analysis.[103] Based on response processes, two groups of these genes have been found, i.e., early stage defense and late stage adaption. At late stage, amino acid metabolism plays two important roles, which are certified by crosstalk between amino acid and carbon metabolism. Many tolerance-related genes come into action through signal transduction for endurance of the plant, on facing extended waterlogging. More than 6000 flooding-responsive genes associated with ubiquitin-mediated protein degradation, transcriptional regulation, photosynthesis, glycolysis, amino acid metabolism, and cell death were reported by carrying out microanalysis of roots of soybean.[104] In contrast, secondary metabolite responsive genes, their transport, cell wall synthesis, cell organization, and synthesis of chromatin material and structure were found to get downregulated. In addition, proteins involved in storage functions and energy production were found to be

TABLE 2 Upshot of Different Abiotic Stresses on Plants and Their Response.

Stress type	Outcome	Response of plant
Heat	Greater temperature causes intense water deficit and rate of evaporation. The resultant enzyme turnover ultimately causes plants to die.	Competent repair processes of proteins and common stability of proteins support endurance, acclimation results from temperature.
Cold and chilling	Biochemical processes and reactions decelerate along with photosynthesis, slowdown of carbon dioxide fixation rate, causing oxygen free radical impairment; formation of ice crystals by freezing may disrupt cell membranes.	Alterations in metabolism occur in adaptable species to overcome cessation of growth. Formation of ice crystal can be precluded by accretion of osmolytes and hydrophilic synthesis of proteins.
Drought	Photosynthesis impaired by inability of water transport to leaves.	Different morphological adaptations including leaf rolling. Closure of stomata caused by ABA reduces evaporative transpiration. Water potential is lowered by accretion of metabolites and water attracting.
Flooding and submergence	Mitochondrial respiration gets disturbed by generation of anoxic or microaerobic ambience.	Cavities developed, usually within the roots, alleviate the oxygen and ethylene exchange between roots and shoot system.
Accretion of heavy metals	Reactions involved in detoxification processes can be depleted or retention capacity may surpass, in surplus.	Metal ions can generate oxygen radicals or can else be retorted by either export or deposition in vacuoles.
High light	Photo-oxidative damage may be caused by by-products of enhanced production of highly reactive intermediates due to excessive light leading to inhibition of photosynthesis.	Vulnerability of a plant to intense light exceeding the utilization in photochemistry results in deactivation of various functions involved in photosynthesis and ROS formation. Outcome of ROS produced can be protein oxidation, enzymes, and lipids important for the appropriate performance of chloroplast and the whole cell.

upregulated via proteomic analysis on soybean roots.[105] Hence, it can be concluded that during flooding, JA biosynthesis genes and reactive oxygen species often get downregulated and also lignification in the soybean roots and hypocotyl gets inhibited.[106]

4 PROTEIN MODIFICATION IN PLANTS IN RESPONSE TO STRESS

PTMs are the changes in a protein after translation. PTM occurs only in specific proteins of the entire proteome. PTMs fine-tune the protein localization, their interaction, and function to alleviate the damage of environmental stress.[107] PTMs increase protein functionality and diversity and play an important role in protein interactions, stability, and enzyme kinetics.[10] Various PTMs occurring in proteins in response to different environmental stresses include phosphorylation, acetylation, glycosylation, succinylation, ubiquitination, S-nitrosylation, and nitration.

4.1 Phosphorylation

Phosphorylation is a reversible PTM playing a key role in signal transduction and plant metabolism by affecting protein interactions, activities, and their subcellular location.[10] Phosphorylation occurs mostly in hydroxylated amino acids including tyrosine (Y), threonine (T), and serine (S) residues. Protein phosphorylation and dephosphorylation is regulated by a protein kinase gene family, which phosphorylates the target protein, and a protein phosphatase gene family, which removes the phosphate group from the protein.[108, 109] Y and S/T protein phosphorylation has been studied in several beans and wheat under drought stress,[110, 111] and in sugar beet, wheat, and maize under salt stress.[112, 113] These protein phosphorylations under these abiotic stresses improve crop productivity by retarding growth or imparting fertility. In plants, there is no specific tyrosine kinase gene but protein tyrosine phosphatases were found in tomato.[109, 114] Y phosphorylation has a key role in plant response to fluctuating environmental conditions. However, Y phosphorylated proteins are low abundant proteins—2.9% in rice, 1.3% in *Medicago*, and 4.2% in *Arabidopsis*.[115] Under drought stress, change in the response pattern of tobacco plants has been found upon the induction of a plant and fungi atypical dual-specificity phosphatase (PFA-DSP1) that is a drought stress-inducible phosphatase.[115] The upregulated expression of DSP4 in chestnut trees under cold stress suggests its important role in helping the plant to acclimate under cold stress.[116]

Nuclear protein phosphorylation by various protein kinases (PKs) helps the plant to respond and acclimate under various environmental cues (Table 3). MAPK acting in the nuclear compartment is involved in various abiotic stresses. AtMPK3 is activated in osmotic stress,[121] while AtMPK3 and AtMPK6 are involved in different abiotic stresses such as drought, salt, cold, ozone, and wounding stress.[118] SIPK (salicylic acid-induced protein kinase) and WIPK (wounding-induced protein kinase)

TABLE 3 Nuclear Kinases/Phosphatases Involved in Various Abiotic Stresses.

Stress	Kinase/phosphatase	References
Drought	SnRK2	117
	MPK6	118
Salt	MPK6	118
	MPK4	119, 120
	SnRK2	117
Cold	MPK4	119, 120
	MPK6	118
Osmotic	MPK3	117, 121
	SnRK2	119, 120
	MPK4	
Wounding	MPK6	118
Ozone	MPK6, MPK3	122

are also activated in response to various environmental stresses, such as osmotic stresses, salinity, ozone exposure, and wounding stress.[123, 124]

MAPKs help the plant to cope with the perceived abiotic stress by rearranging the transcription pattern. AtCPK3, acting at the plasma membrane level, has been found to be involved in salt stress.[125] AtHDA19, being expressed highly in imbibed and germinating seeds, is also accumulated highly in response to wounding stress,[126] light stress,[127, 128], or cold stress,[129] suggesting its role in various environmental stresses. The overexpression and interaction of AtHDA19 with bnKCP1 (a novel putative kinase-inducible domain containing protein) in *B. napus* demonstrates its role in cold stress.[130] In *A. thaliana*, AtHD2C over-expression increases tolerance to salt stress by affecting abscisic acid-responsive genes expression.[131]

4.2 Acetylation

Acetylation is a well-known PTM regulating gene expression. Acetylation occurring on histone/nonhistone proteins is regulated by various factors like ROS, infectious diseases, and physiological stresses.[107] Protein acetylation can be reversible, i.e., modification of amino group lysine (K) by K acetyl transferases (KATs) and K deacetylases (KDACs), or nonreversible, i.e., N-terminal modification by Nt-acetyltransferases.[132] Reversible acetylation occurs on histone proteins. The abiotic stresses involving the regulation/association of histone acetyltransferase (HAT) activities with different protein partners regulates various specific genes leading to the biological response. The histone acetylase activity of GCN5 HAT increases, while interacting with Ada2 (transcriptional coactivator), which enhances its ability to acetylate nucleosomal histones.[133, 134] It was also found that the mutants of ADA2b and ADA2 genes showed altered response to low temperature stress and hypersensitivity to salt stress.[135] GCN5 and ADA2b, while interacting with a coactivator, SGF29a, increase the acetylation activity in the promoter site of target genes including RD29b, RAB18, and COR6.6, which are cold responsive genes.[136] AtGCN5 HAT activity increases while interacting with AtEML (transcription factor) under cold stress.[137] AtEML controls the expression of various stress responsive genes through the recruitment of AtGCN5 to their respective promoters.[137]

4.3 Glycosylation

Glycosylation, being the most abundant PTM in plants, has wide-ranging biological significance. The glycosylation can be either N-glycosylation occurring at Asp (D) residue, or O-glycosylation, which occurs on S/T or Y residue.[138] N-glycosylation occurs in the proteins of the secretory pathway in the endoplasmic reticulum (ER), which are later efficiently folded with the help of calreticulin (CRT), calnexin (CNX), and HSPs.[138] Under salt stress, the overexpression of HSPs in tomato facilitates the protein folding in ER, thereby preventing cellular damage and increasing the stress tolerance of the plant.[139] Mutant rice with defective β1,2-xylosyltransferase (catalyzing the transfer of xylose to N-glycans) shows susceptibility to osmotic or low heat stress, suggesting its role in plant tolerance to abiotic stress.[140] In rice, the impaired α1, 3-fucosyltransferase (transferring fucose to N-glycan) activity reduces its gravitropic response. These results suggest that tolerance of rice to various abiotic stresses depends on the accurate transportation and localization of proteins using N-glycans as delivery tags.[107]

4.4 Succinylation

Succinylation involves the transfer of a succinyl group to a protein molecule at its lysine (K) residue.[107] Succinylation plays an important role in the growth and development of plants and is very dynamic under different cellular conditions.[141] In rice seed germination, succinylation leads to an altered proteomic pool, which alters the metabolic processes including carbon fixation, glycolysis/gluconeogenesis and citric acid cycle.[142] In tomato, the succinylation occurs on various sites of histone H3 (K122, K79, K56, and K14).[143] Succinylation has been found to be an important regulatory mechanism of various biological processes and is crucial for the adaptation of plants to various environmental stresses.[142]

4.5 Ubiquitination

Ubiquitination involves the addition of ubiquitin to a target protein, mediated by sequential actions of ubiquitin activating enzyme (E1), ubiquitin conjugating enzyme (E2), and ubiquitin ligase (E3).[144] The protein containing this multi-ubiquitin chain is degraded intracellularly through 26S proteasome, whose function is the proteolytic degradation of proteins. PTM by ubiquitination is involved in various developmental processes and various environmental stresses by regulating various signaling pathways.[145, 146] In various plant species under abiotic stress, ubiquitination-related proteins/transcripts become modified, showing the role of ubiquitination in plant stress tolerance (Table 4).[14, 158, 159] Ubiquitination regulates the transcriptional changes by modulation of activity of various regulatory proteins for abiotic stress tolerance (Fig. 2).[160] In plants, various ubiquitin proteins involved in abiotic stress are localized in the nucleus. AtHOS1 and AtDRIP1 (E3-ubiquitin ligases) are found to be localized in the nucleus of *A. thaliana*. AtHOS1 is involved in cold tolerance of *A. thaliana*, and AtDRIP1 increases drought tolerance in *A. thaliana* by interacting with AtDREB2A (Dehydration-responsive element binding protein 2A), and controlling the drought-inducible gene expression.[161] The overexpression of XERICO and SDIR1 genes encoding H2-type zinc-finger E3 ligase leads to an enhanced drought tolerance by enhancing ABA-induced closure of stomata.[162] Single mono-ubiquitin gene expression increases the tolerance of a plant to different stresses without affecting its growth and development under normal conditions.[163] However, the expression of polyubiquitin gene has been found to be induced by high temperature in maize tobacco and potato.[160]

TABLE 4 Ubiquitin Enzymes Involved in Plant Stress Tolerance.

Enzyme	Name	Biological function	References
E2	PUB1	Drought and salt stress tolerance	147
	PUB15	Oxidative stress tolerance	148
	PUB22/23	Salt and drought stress tolerance	149
	UBC2	Drought stress tolerance	150
E3	BIRF1	Drought and oxidative stress tolerance	151
	HOS1	Cold stress tolerance	152
	RFP1	Osmotic stress tolerance	153
	RING-1	Heat and drought stress tolerance	154
	RFP1	Drought, salinity, and cold tolerance	155
	SAP5	Salt and drought stress tolerance	156
	DDB1	UV radiation tolerance	157

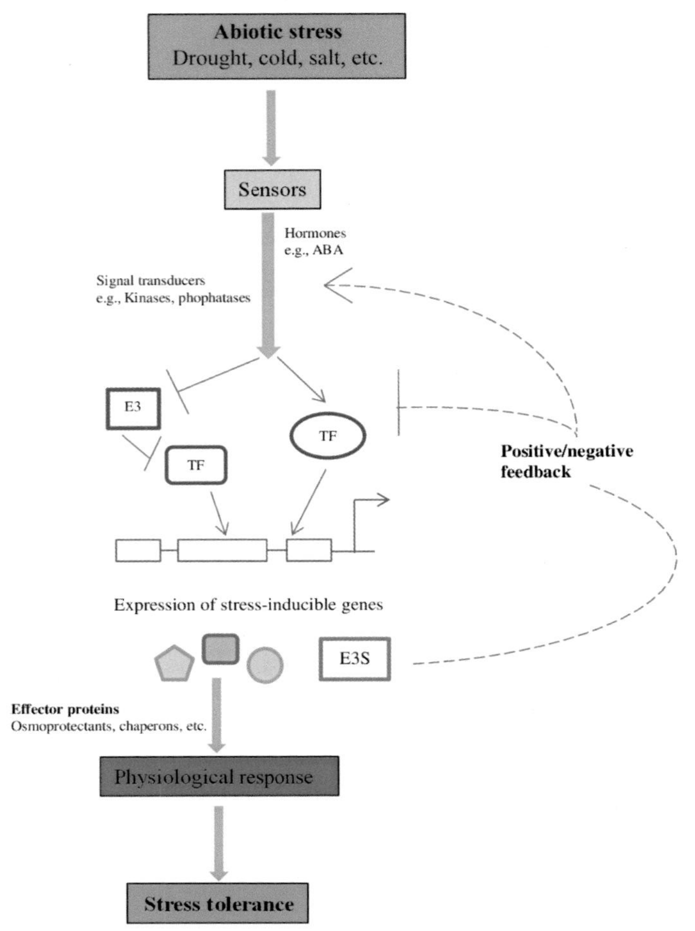

FIG. 2 E3 ligases regulating the abiotic stress signaling.[160]

4.6 S-Nitrosylation and Tyrosine Nitration

S-Nitrosylation is an important PTM that involves NO group binding to a cysteine residue, playing an important role in signaling mediated by NO.[164] Among the Cys residue, Cys-230 has been found to have maximum probability of being S-nitrosylated.[165] S-Nitrosylation regulates the function of photo-respiratory enzymes, metabolite flux of different pathways in peroxisomes, and prevents the oxidative damage caused by H_2O_2 accumulation.[166] In peas under abiotic stress, NO has been found to be produced in peroxisomes, which is involved in stress/signaling related to various nitrogen and oxygen species.[167] In plants under different stress conditions, various proteins are S-nitrosylated, which regulates the function of different proteins.[168] Through S-nitrosylation, NO regulates the antioxidant defense system and hence leads to the regulation of regulate H2O2 accumulation under various stress conditions.[166]

Tyrosine nitration involves the attachment of a nitro (—NO_2) group to aromatic ring of tyrosine residue at ortho-carbon position.[169] Tyrosine nitration is a PTM that helps the plant to acclimate under various abiotic stresses. In pepper plants, low temperature stress causing nitrosative stress leads to the tyrosine nitration of proteins and helps the plant to acclimate under low temperature stress.[170] It has been found in sunflower hypocotyls that mechanical wounding increases tyrosine nitration, which is due to S-nitrosoglutathione (GSNO). GSNO in the presence of O_2 is converted to glutathione radical ($GS^{\bullet}$) and ONOO resulting in an increased protein tyrosine nitration.[171] In *Arabidopsis* seedlings, heavy metal (arsenic) stress leads to protein nitration.[172] An increased tyrosine nitration was also observed in the roots of *Lotus japonicus* under water stress.[173] It has been found that NO-pretreatment helps citrus plants to tolerate drought stress.[174] These findings suggest that nitrosylation and nitration prepares the plant against various environmental stresses.

5 CONCLUSION AND FUTURE PERSPECTIVE

Protein PTMs affect the outcome of response of plants toward different kinds of stresses. This mechanism has been studied in some model plants by examining gene expression regulated by downstream-modified proteins.[175, 176] Momentous advancement has been reported in histone PTMs regarding their contribution to created "memory" of recurring stresses.[177, 178] PTM-governed systems as stress-responsive elements are acquirable for acetylation, phosphorylation, ubiquitination, and glycosylation in *Arabidopsis* and in some monocots, for instance, rice. Temporal and spatial feasible forms of downstream expression of genes are ensured by the crosstalk among such modifications. Important functional facets of various transcriptional factors, which include their turnover and promoter binding features,[179] cellular localization and partitioning within the cell,[180] transcriptional behavior,[181], and dimer forming ability with the rest of the TFs,[182] are determined by the PTMs and have not been much analyzed yet. Several PTMs are revocable, comprising of varied mechanisms that help plants to fine tune their adaptations to different environmental disputes.

The genetic engineering of some genes encoding stress proteins for production of stress tolerant plants is on its way. *A. thaliana, Thellungiella halophile, Craterostigma plantagineum*, etc. are the model plants used to study the effect of different abiotic stresses and their tolerance mechanism for genetic improvement of crops. At the same time, relatively scarce information regarding the posttranslational mechanisms supervising the adaptive response of plants hampers research progression. Nevertheless, selective information in the remaining PTMs among other plants, particularly dicots, is still restricted and descriptive, thereby requiring the further intensive study of PTMs for unraveling their ultimate and specific functions. Good understanding of all the posttranslational controls of all types of stress acclimation is indispensable to come up with novel crops having better stress tolerance. Moreover, radical development in laboratory proficiency and, as a whole, systems biology would boost future investigations and fast decoding of new protein PTM functions in plants that help them in combating environmental changes by adaptation. In addition, the exploration of genes and proteins in terms of their expression and regulation against different stress responses is vital to have a proper understanding of stress tolerance.

References

1. Mittler R. Abiotic stress, the field environment and stress combination. *Trends Plant Sci* 2006;**11**:15–9.
2. Golldack D, Li C, Mohan H, Probst N. Tolerance to drought and salt stress in plants: unraveling the signaling networks. *Front Plant Sci* 2014;**5**:151. https://doi.org/10.3389/fpls.2014.00151.
3. Yamaguchi-Shinozaki K, Shinozaki K. Transcriptional regulatory networks in cellular responses and tolerance to dehydration and cold stresses. *Annu Rev Plant Biol* 2006;**57**:781–803.
4. Chinnusamy V, Shumaker K, Zhu JK. Molecular genetic perspectives on cross-talk and specificity in abiotic stress signalling in plants. *J Exp Bot* 2004;**55**:225–36.
5. Romero-Puertas MC, Rodriguez-Serrano M, Sandalio LM. Protein nitrosylation in plants under abiotic stress: an overview. *Front Plant Sci* 2013;**4**:373.
6. Vierstra RD. The expanding universe of ubiquitin and ubiquitin-like modifiers. *Plant Physiol* 2012;**160**:2–14.
7. Xiong Y, Peng X, Cheng Z, Liu W, Wang GL. A comprehensive catalog of the lysine-acetylation targets in rice (*Oryza sativa*) based on proteomic analyses. *J Proteomics* 2016;**138**:20–9.
8. Yanagawa Y, Komatsu S. Ubiquitin/proteasome-mediated proteolysis is involved in the response to flooding stress in soybean roots, independent of oxygen limitation. *Plant Sci* 2012;**185–186**:250–8.
9. Yin X, Komatsu S. Quantitative proteomics of nuclear phosphoproteins in the root tip of soybean during the initial stages of flooding stress. *J Proteomics* 2015;**119**:183–95.
10. Friso G, van Wijk KJ. Post translational protein modifications in plant metabolism. *Plant Physiol* 2015;**169**:1469–87.
11. Mann O, Jensen N. Proteomic analysis of post-translational modifications. *Nat Biotechnol* 2003;**21**:255–61.
12. Mass AD. *A database of protein post translational modifications*. http://www.abrf.org/index.cfm/dm.home; 2010.
13. Lehti-Shiu MD, Shiu SH. Diversity, classification and function of the plant protein kinase superfamily. *Philos Trans R Soc Lond B* 2012;**367**:2619–39.

14. Mazzucotelli E, Mastrangelo AM, Crosatti C, Guerra D, Stanca AM, Luigi Cattivelli L. The E3 ubiquitin ligase gene family in plants: regulation by degradation. *Curr Genomics* 2006;**7**:509–22.
15. Hanada K, Zou C, Lehti-Shiu MD, Shinozaki K, Shiu SH. Importance of lineage-specific expansion of plant tandem duplicates in the adaptive response to environmental stimuli. *Plant Physiol* 2008;**148**:993–1003.
16. Verma S, Nizam S, Verma PK. Biotic and abiotic stress signaling in plants. In: *Stress signaling in plants: genomics and proteomics perspective*. New York: Springer Science+Business Media; 2013. https://doi.org/10.1007/978-1-4614-6372-6_2.
17. Zhu JK. Salt and drought stress signal transduction in plants. *Annu Rev Plant Biol* 2002;**53**:247–73.
18. Rabbani MA, Maruyama K, Abe H, Khan MA, Katsura K, Ito Y, Yoshiwara K, Seki M, Shinozaki K, Yamaguchi-Shinozaki K. Monitoring expression profiles of rice genes under cold, drought and high-salinity stresses, and abscisic acid application using cDNA microarray and RNA gel-blot analyses. *Plant Physiol* 2003;**133**:1755–67.
19. Seki M, Narusaka M, Ishida J, Nanjo T, Fujita M, Oono Y, Kamiya A, Nakajima M, Enju A, Sakurai T, Satou M, Akiyama K, Taji T, Yamaguchi-Shinozaki K, Carninci P, Kawai J, Hayashizaki Y, Shinozaki K. Monitoring the expression profiles of 7000 Arabidopsis genes under drought, cold and high salinity stresses using a full-length cDNA microarray. *Plant J* 2002;**31**(3):279–92.
20. Boominathan P, Shukla R, Kumar A, Manna D, Negi D, Verma PK, Chattopadhyay D. Long term transcript accumulation during the development of dehydration adaptation in *Cicer arietinum*. *Plant Physiol* 2004;**135**:1608–20.
21. Ok SH, Jeong HJ, Bae JM, Shin JS, Luan S, Kim KN. Novel CIPK1-associated proteins in Arabidopsis contain an evolutionarily conserved C-terminal region that mediates nuclear localization. *Plant Physiol* 2005;**139**:138–50.
22. Skriver K, Mundy J. Gene expression in response to abscisic acid and osmotic stress. *Plant Cell* 1990;**2**:503–12.
23. Zeevaart JAD, Creelman RA. Metabolism and physiology of abscisic acid. *Annu Rev Plant Physiol Plant Mol Biol* 1988;**39**:439–73.
24. Tumer NC, Kramer PJ. *Adaptations of plants to water and high temperature stress*. New York: Willey Interscience; 1980; 155–72.
25. Yamaguchi-Shinozaki K, Shinozaki K. A novel cis-acting element in an *Arabidopsis* gene is involved in responsiveness to drought, low temperature or high salt stress. *Plant Cell* 1994;**6**:251–64.
26. Mundy J, Chua NH. ABA and water stress induce the expression of novel rice gene. *EMBO J* 1988;**7**:2279–86.
27. Arora R, Pitchay DS, Bearce BC. Water stress induced heat tolerance in geranium leaf tissues: a possible linkage through stress proteins? *Physiol Plant* 1998;**103**:24–34.
28. Close TJ, Kortt AA, Cahndler PM. A cDNA based comparison of dehydration induced proteins (dehydrins) in barley and corn. *Plant Mol Biol* 1989;**13**:95–108.
29. Yamaguchi-Shinozaki K, Koizumi M, Urao S, Shinozaki K. Molecular cloning and characterization of 9 cDNAs for genes that are responsive to desiccation in *Arabidopsis thaliana*, sequence analysis of one cDNA clone that encodes a putative transmembrane channel protein. *Plant Cell Physiol* 1992;**33**:217–24.
30. Koizumi M, Yamaguchi-Shinozaki K, Tsuji H, Shinozaki K. Structure and expression of two genes that encode distinct drought inducible systeine proteinases in *Arabidopsis thalians*. *Gene* 1993;**129**:175–82.
31. Fujita M, Fujita Y, Maruyama K, Hiratsu K, Ohme-Takagi M, Lam-Son Tran P, Yamaguchi-Shinozaki K, Shinozaki K. A dehydration-induced NAC protein, RD26, is involved in a novel ABA-dependent stress signalling pathway. *Plant J* 2004;**39**:863–76.

32. Miyazaki Y, Abe H, Takase T, Kobayashi M, Kiyosue T. Overexpression of *LOV KELCH protein 2* confers dehydration tolerance and is associated with enhanced expression of dehydration-inducible genes in *Arabidopsis thaliana. Plant Cell Rep* 2015;**34**:843–52.
33. Backer J, Steel C, Dure L. Sequence and characterization of 6 LEA proteins and their genes from cotton. *Plant Mol Biol* 1988;**11**:277–91.
34. Cuming AC. LEA proteins. In: Shewry PR, Casey R, editors. *Seed proteins*. Dordrecht: Kluwer Academic Publishers; 1999. p. 753–80.
35. Dure III L. Structural motifs in lea proteins. In: Close TJ, Bray EA, editors. *Plants responses to cellular dehydration during environment stress. Current topics in plant physiology*, vol. 10. Rockville, MD: Amr Soc Plant Physiol; 1993. p. 91–103.
36. Battista JR, Park MJ, McLemore AE. Inactivation of two homologues of proteins presumed to be involved in the desiccation tolerance of plants sensitises *Deinococcus radiodurans* R1 to desiccation. *Cryobiology* 2001;**43**:133–9.
37. Ellis RJ. From chloroplasts to chaperones: how one thing led to another. *Photosynth Res* 2004;**80**:333–43.
38. Rong E, Zhao Z, Zhao Z, Ma J, Zang W, Wang L, Xie D, Yang W. Wheat cold and light stress analysis based on the *Arabidopsis* homology protein–protein interaction (PPI) network. *J Med Plant Res* 2011;**5**:5493–8.
39. Yan SP, Zhang QY, Tang ZC, Su WA, Sun WN. Comparative proteomic analysis provides new insights into chilling stress responses in rice. *Mol Cell Proteomics* 2006;**5** (3):484–96.
40. Rinalducci S, Egidi MG, Karimzadeh G, Jazii FR, Zolla L. Proteomic analysis of a spring wheat cultivar in response to prolonged cold stress. *Electrophoresis* 2011;**32**(14):1807–18.
41. Iba K. Acclimation response to temperature stress in higher plants: approaches of gene engineering for temperature tolerance. *Annu Rev Plant Biol* 2002;**53**:225–45.
42. Thomashow MF. Plant cold tolerance: freezing tolerance genes and regulatory mechanisms. *Annu Rev Plant Physiol Plant Mol Biol* 1999;**50**:571–99.
43. Lee BH, Henderson DA, Zhu JK. The *Arabidopsis* cold responsive transcriptome and its regulation by ICE1. *Plant Cell* 2005;**17**:3155–75.
44. Nakamura J, Yausa T, Huong TT, Harano K, Tanaka S, Iwata T, Phan T, Iwaya-Inoue M. Rice homologs of inducer of CBF expression (OsICE) are involved in cold acclimation. *Plant Biotechnol* 2011;**28**:303–9.
45. Krishna P, Sacco M, Cherutti JF, Hill S. Cold induced accumulation of HSP 90 transcripts in *Brassica napus*. *Plant Physiol* 1995;**107**:915–23.
46. Wolter FP, Schimidt R, Hein E. Chilling sensitivity of *Arabidopsis thaliana* with genetically engineered membrane liquids. *EMBO J* 1992;**11**:4685–92.
47. Karlson D, Imai R. Conservation of the cold shock domain protein family in plants. *Plant Physiol* 2003;**131**:112–5.
48. Park SJ, Kwak KJ, Oh TR, Kim YO, Kang H. Cold shock domain proteins affect seed germination and growth of *Arabidopsis thaliana* under abiotic stress conditions. *Plant Cell Physiol* 2009;**50**:869–78.
49. D'Angeli S, Altamura MM. Osmotin induces cold protection in olive trees by affecting programmed cell death and cytoskeleton organization. *Planta* 2007;**225**:1147–63.
50. Zhu B, Chen THH, Li PH. Expression of an ABA responsive osmotin like gene during the induction of freezing tolerance in *Solanum commersonii*. *Plant Mol Biol* 1993;**21**:729–35.
51. Walia H, Wilson C, Wahid A, Condamine P, Cui X, Close TJ. Expression analysis of barley (*Hordeum vulgare* L) during salinity stress. *Funct Integr Genomics* 2006;**6**(2):143–56.
52. Ouyang B, Yang T, Li H, Zhang L, Zhang Y, Zhang J, Fei Z, Ye Z. Identification of early salt stress response genes in tomato root by suppression subtractive hybridization and microarray analysis. *J Exp Bot* 2007;**58**:507–20.
53. Zhang L, Tian LH, Zhao JF, Song Y, Zhang CJ, Guo Y. Identification of an apoplastic protein involved in the initial phase of salt stress response in rice root by two-dimensional electrophoresis. *Plant Physiol* 2009;**149**(2):916–28.

54. Sobhanian H, Razavizadeh R, Nanjo Y, Ehsanpour AA, Jazii FR, Motamed N, Komatsu S. Proteome analysis of soybean leaves, hypocotyls and roots under salt stress. *Proteome Sci* 2010;**8**:19.
55. Xu XY, Fan R, Zheng R, Li CM, Yu DY. Proteomic analysis of seed germination under salt stress in soybeans. *J Zhejiang Univ Sci B* 2011;**12**(7):507–17.
56. Munns R, Tester M. Mechanisms of salinity tolerance. *Annu Rev Plant Biol* 2008;**59**:651–81.
57. Parida AK, Das AB. Salt tolerance and salinity effects on plants: a review. *Ecotoxicol Environ Saf* 2005;**60**:324–49.
58. Gupta B, Huang B. Mechanism of salinity tolerance in plants: physiological, biochemical and molecular characterization. *Int J Genomics* 2014.
59. Zhang H, Han B, Wang T, Chen S, Li H, Zhang Y, Dai S. Mechanism of plant salt response: insights from proteomics. *J Proteome Res* 2012;**11**:49–67.
60. Claes B, Dekeyser R, Villarroel R, Van den Bulcke M, Bauw G, Van Montagu M, Caplan A. Characterization of rice gene showing organ specific expression in response to salt stress and drought. *Plant Cell* 1990;**2**:19–27.
61. Shirata K, Takagishi H. Salt induced accumulation of 26 and 27 KD proteins in cultured cells of rice plant. *Soil Sci Plant Nutr* 1990;**36**:153–7.
62. Rani UR, Reddy AR. Salt stress responsive polypeptides in germinating seeds. *Plant Physiol* 1994;**143**:250–3.
63. Singh NK, Handa AK, Hasegawa PM, Bressen R. Proteins associated with glycoprotein mRNA in response to fungal elicitors and infection. *Proc Natl Acad Sci U S A* 1985;**82**:6551–5.
64. Chen CCS, Plant AL. Salt induced proteins synthesis in tomato roots: the role of ABA. *J Exp Bot* 1999;**50**:677–87.
65. Mazel A, Leshem Y, Tiwari BS, Levine A. Induction of salt and osmotic stress tolerance by overexpression of an intracellular vesicle trafficking protein AtRab7 (AtRabG3e). *Plant Physiol* 2004;**134**:118–28.
66. Chinnusamy V, Jagendorf A, Zhu J-K. Understanding and improving salt tolerance in plants. *Crop Sci* 2005;**45**:437–48.
67. Subramanyam K, Sailaja KV, Subramanyam K, Rao DM, Lakshmidevi K. Ectopic expression of an osmotin gene leads to enhanced salt tolerance in transgenic chilli pepper (*Capsicum annum* L.). *Plant Cell Tissue Organ Cult* 2011;**105**:181–92.
68. Song H, Wang P, Hou L, Zhao S, Zhao C, Xia H, Li P, Zhang Y, Bian X, Wang X. Global analysis of WRKY genes and their response to dehydration and salt stress in soybean. *Front Plant Sci* 2016;**7**:9.
69. Mittal D, Chakrabarti S, Sarkar A, Singh A, Grover A. Heat shock factor gene family in rice: genomic organization and transcript expression profiling in response to high temperature, low temperature and oxidative stresses. *Plant Physiol Biochem* 2009;**47**(9):785–95.
70. Majoul T, Bancel E, Triboı E, Hamida JB, Branlard G. Proteomic analysis of the effect of heat stress on hexaploid wheat grain: characterization of heat-responsive proteins from non-prolamins fraction. *Proteomics* 2004;**4**:505–13.
71. Lee DG, Ahsan N, Lee SH, Kang KY, Bahk JD, Lee IJ, Lee BH. A proteomic approach in analyzing heat-responsive proteins in rice leaves. *Proteomics* 2007;**7**:3369–83.
72. Chen X, Zhang W, Zhang B, Zhou J, Wang Y, Yang Q, Ke Y, He H. Phosphoproteins regulated by heat stress in rice leaves. *Proteome Sci* 2011;**9**:37.
73. Sun W, Van-Montagu M, Verbruggen N. Small heat shock protein and stress tolerance in plants. *Biochem Biophys Acta* 2002;**1577**:1–9.
74. Key JL, Kimpel JA, Vierheng E, Lin CV, Nagoic RT, Czarhecka E. Physiological and molecular analysis of the heat shock responses in plants. In: *Changes in eukaryotic gene expression in response to environmental stress*. New York: Academic Press; 1985. p. 327–48.

75. Nagoa RT, Key II. Heat shock protein genes of plants. In: Constable F, Vasil IK, editors. *Cell culture and osmotic cell genetics of plants*. vol. 6. Academic Press Inc.; 1989. p. 297–328.
76. Wang R, Zhang Y, Kleffer M, Yu H, Kepinski S, Estelle M. HSP90 regulates temperature dependent seedling growth genetic engineering for stress tolerance. *Planta* 2016;**218**:1–14.
77. Lohmann C, Eggers-Schumacher G, Wunderlich Mand Schoffl F. Two different heat shock transcription factors regulate immediate early expression of stress genes in *Arabidopsis*. *Mol Genet Genomics* 2004;**271**:11–21.
78. Hu W, Hu G, Han B. Genome-wide survey and expression profiling of heat shock proteins and heat shock factors revealed overlapped and stress specific response under abiotic stresses in rice. *Plant Sci* 2009;**176**:583–90.
79. Barnett T, Altschuler M, McDaniel CN, Mascarenhas JP. Heat shock induced proteins in plant cells. *Dev Genet* 1980;**1**:331–40.
80. Neta-Sharir I, Isaacsoon T, Lurie Sand Weiss D. Dual role of tomato heat shock protein 21: protecting photosystem II from oxidative stress and promoting color change during fruit maturation. *Plant Cell* 2005;**17**:1829–38.
81. Chakraborty U, Banset M, Tangden L. Biochemical response of *Cicer arietinum* L. and *Vigna radiata* (L. Wilczek) to elevated temperature stress. In: Chakraborty U, Chakraborty B, editors. *Stress biology*. Narosa Publishing House India; 2005. p. 106–11.
82. Belanger FC, Brodl MR, Ho THD. Heat shock causes destabilization of specific mRNA and destruction of endoplasmic reticulum in barley aleurone cells. *Proc Natl Acad Sci U S A* 1986;**83**:1554–8.
83. Hong-Qi Z, Croes AF, Linskens HF. Qualitative change in protein synthesis in germination pollen of *Lilium longiflorum* after a heat shock. *Plant Cell Environ* 1984;**7**:689–91.
84. Key JL, Tin CV, Chen YM. Heat shock proteins of higher plants. *Proc Natl Acad Sci U S A* 1981;**78**:3526–30.
85. Cooper P, Ho TLD, Hauptomann RH. Tissue specialty of the heat shock response in maize. *Plant Physiol* 1984;**101**:431–41.
86. Schoffl F, Sauman B. Thermo induced transcripts of a soybean heat shock sense after transfer into sunflower using a Ti-plasmid vector. *EMBO J* 1985;**4**:719–1124.
87. Hong SW, Vierling E. HSP101 is necessary for heat tolerance but dispensable for development and germination in the absence of stress. *Plant J* 2001;**27**:25–35.
88. Pareek A, Singla SL, Grover A. Immunological evidence for accumulation of two high molecular weight (90 and 104 KDa) HSPs in response to different stresses in rice and in response to high temperature stress in diverse plant genera. *Plant Mol Biol* 1995;**29**:293–301.
89. Kim SY, Sharma S, Hoskins JR, Wickner S. Interaction of the Dnak and Dnaj chaperone system with a native substrate, P1 Rep A. *J Biol Chem* 2002;**227**:44778–83.
90. Lee GJ, Vierling E. A small heat shock protein cooperates with heat shock protein 70 systems to reactivate a heat denatured protein. *Plant Physiol* 2000;**122**:189–98.
91. Hassanuzzaman M, Hanar K, Alam Md M, Roychowdhury R, Fujita M. Physiological, biochemical and molecular mechanisms of heat stress tolerance in plants. *Int J Mol Sci* 2013;**14**:9643–84.
92. Ogawa I, Nakanishi H, Mori S, Nishizawa NK. Time course analysis of gene regulation under cadmium stress in rice. *Plant Soil* 2009;**325**:97–108.
93. Li X, Gao P, Gjetvaj B, Westcott N, Gruber MY. Analysis of the metabolome and transcriptome of *Brassica carinata* seedlings after lithium chloride exposure. *Plant Sci* 2009;**177**:68–80.
94. Zhao L, Sun YL, Cui SX, Chen M, Yang HM, Liu HM, Chai TY, Huang F. Cd-induced changes in leaf proteome of the hyper-accumulator plant *Phytolacca americana*. *Chemosphere* 2011;**85**(1):56–66.
95. Cobbett CS, Goldsbrough P. Phytochelatins and metallothioneins: roles in heavy metal detoxification and homeostasis. *Annu Rev Plant Biol* 2002;**53**:159–82.

96. Yang XE, Jin XF, Feng Y, Islam E. Molecular mechanisms and genetic basis of heavy metal tolerance/hyper-accumulation in plants. *J Integr Plant Biol* 2005;**47**:1025–35.
97. Hall JL. Cellular mechanisms for heavy metal detoxification and tolerance. *J Exp Bot* 2002;**53**:1–11.
98. Cobbett CS. Phytochelatins and their roles in heavy metal detoxification. *Plant Physiol* 2000;**123**:825–32.
99. Prasad MNV. Cadmium toxicity and tolerance in vascular plants. *Environ Exp Bot* 1995;**35**:524–45.
100. Soborowiak R, Deckert J. Protein induced by cadmium in soybean cells. *J Plant Physiol* 2006;**163**:1203–6.
101. Zhou G, Xu Y, Li J, Yang L, Liu JY. Molecular analysis of the metallothionein gene family in rice (*Oryza sativa* L.). *J Biochem Mol Biol* 2006;**39**:595–606.
102. Hossain MA, Piyatida P, Jaime A, da Silva T, Fujita M. Molecular mechanism of heavy metal toxicity and tolerance in plants: central role of glutathione in detoxification of reactive oxygen species and methylglyoxal in heavy metal chelation. *J Bot* 2012;.
103. Zou X, Jiang Y, Liu L, Zhang Z, Zheng Y. Identification of transcriptome induced in roots of maize seedlings at the late stage of waterlogging. *BMC Plant Biol* 2010;**10**:189.
104. Nanjo Y, Maruyama K, Yasue H, Yamaguchi-Shinozaki K, Shinozaki K, Komatsu S. Transcriptional responses to flooding stress in roots including hypocotyl of soybean seedlings. *Plant Mol Biol* 2011;**77**:129–44.
105. Komatsu S, Kobayashi Y, Nishizawa K, Nanjo Y, Furukawa K. Comparative proteomics analysis of differentially expressed proteins in soybean cell wall during flooding stress. *Amino Acids* 2010;**39**(5):1435–49.
106. Komatsu S, Sugimoto T, Hoshino T, Nanjo Y, Furukawa K. Identification of flooding stress responsible cascades in root and hypocotyl of soybean using proteome analysis. *Amino Acids* 2010;**38**:729–38.
107. Hashiguchi A, Komatsu S. Impact of post-translational modifications of crop proteins under abiotic stress. *Proteome* 2016;**4**:42.
108. Singh A, Giri J, Kapoor S, Tyagi AK, Pandey GK. Protein phosphatase complement in rice: genome wide identification and transcriptional analysis under abiotic stress conditions and reproductive development. *BMC Genomics* 2014;**15**.
109. Wang X, Oh M, Sakata K, Komatsu S. Gel-free/label-free proteomic analysis of root tip of soybean over time under flooding and drought stresses. *J Proteomics* 2016;**130**:42–55.
110. Kumar R, Kumar A, Subba P, Gayali S, Barua P, Chakraborty S, Chakraborty N. Nuclear phosphoproteome of developing chickpea seedlings (*Cicer arietinum* L.) and protein kinase interaction network. *J Proteomics* 2014;**105**:58–73.
111. Zhang M, Lv D, Ge P, Bian Y, Chen G, Zhu G, Li X, Yan Y. Phosphoproteome analysis reveals new drought response and defense mechanisms of seedling leaves in bread wheat (*Triticum aestivum* L.). *J Proteomics* 2014;**109**:290–308.
112. Lv DW, Zhu GR, Zhu D, Bian YW, Liang XN, Cheng ZW, Deng X, Yan YM. Proteomic and phosphor proteomic analysis reveals the response and defense mechanism in leaves of diploid wheat *T. monococcum* under salt stress and recovery. *J Proteomics* 2016;**143**:93–105.
113. Yu B, Li J, Koh J, Dufresne C, Yang N, Qi S, Zhang Y, Ma C, Duong BV, Chen S, Li H. Quantitative proteomics and phosphor proteomics of sugar beet monosomic addition line M14 in response to salt stress. *J Proteomics* 2016;**143**:286–97.
114. Fordham-Skelton AP, Chilley P, Lumbreras V, Reignoux S, Fenton TR, Dahm CC, Pages M, Gatehouse JA. A novel higher plant protein tyrosine phosphatase interacts with SNF1-related protein kinases via a KIS (kinase interaction sequence) domain. *Plant J* 2002;**29**:705–15.
115. Liu P, Myo T, Ma W, Lan D, Qi T, Guo J, Song P, Guo J, Kang Z. TaTypA, a ribosome-binding GTPase protein, positively regulates wheat resistance to the stripe rust fungus. *Front Plant Sci* 2016;**7**:873.

116. Berrocal-Lobo M, Ibanez C, Acebo P, Ramos A, Perez-Solis E, Collada C, Casado R, Aragoncillo C, Allona I. Identification of a homolog of Arabidopsis DSP4 (SEX4) in chestnut: its induction and accumulation in stem amyloplasts during winter or in response to the cold. *Plant Cell Environ* 2011;**34**:1693–704.
117. Halford NG, Hey SJ. Snf1-related protein kinases (SnRKs) act within an intricate network that links metabolic and stress signalling in plants. *Biochem J* 2009;**419**:247–59.
118. Ichimura K, Mizoguchi T, Yoshida R, Yuasa T, Shinozaki K. Various abiotic stresses rapidly activate Arabidopsis MAP kinases ATMPK4 and ATMPK6. *Plant J* 2000;**24**:655–65.
119. Droillard M-J, Boudsocq M, Barbier-Brygoo H, Lauriere C. Involvement of MPK4 in osmotic stress response pathways in cell suspensions and plantlets of *Arabidopsis thaliana*: activation by hypoosmolarity and negative role in hyperosmolarity tolerance. *FEBS Lett* 2004;**574**:42–8.
120. Teige M, Scheikl E, Eugelm T, Doczi R, Ichimura K, Shinozaki K, Dangl J, Hirt H. The MKK2 pathway mediates cold and salt stress signaling in *Arabidopsis*. *Mol Cell* 2004;**15**:141–52.
121. Droillard MJ, Boudsocq M, Barbier-Brygoo H, Laurière C. Different protein kinase families are activated by osmotic stresses in *Arabidopsis thaliana* cell suspensions: involvement of the MAP kinases AtMPK3 and AtMPK6. *FEBS Lett* 2002;**527**:43–50.
122. Ahlfors R, Macioszek V, Rudd J, Brosché M, Schlichting R, Scheel D, Kangasjarvi J. Stress hormone-independent activation and nuclear translocation of mitogen-activated-protein kinases in *Arabidopsis thaliana* during ozone exposure. *Plant J* 2004;**40**:512–22.
123. Mikolajczyk M, Awotunde OS, Muszynska G, Klessig DF, Dobrowolska G. Osmotic stress induces rapid activation of a salicylic acid–induced protein kinase and a homolog of protein kinase ASK1 in tobacco cells. *Plant Cell* 2000;**12**:165–78.
124. Samuel MA, Ellis BE. Double jeopardy: both overexpression and suppression of a redox-activated plant mitogen-activated protein kinase render tobacco plants ozone sensitive. *Plant Cell* 2002;**14**:2059–69.
125. Mehlmer N, Wurzinger B, Stael S, Hofmann-Rodrigues D, Csaszar E, Pfister B, Bayer R, Teige M. The Ca^{2+}-dependent protein kinase CPK3 is required for MAPK-independent salt-stress acclimation in *Arabidopsis*. *Plant J* 2010;**63**:484–98.
126. Zhou C, Zhang L, Duan J, Miki B, Wu K. HISTONE DEACETYLASE 19 is involved in jasmonic acid and ethylene signaling of pathogen response in *Arabidopsis*. *Plant Cell* 2005;**17**:1196–204.
127. Tian L, Fong MP, Wang JJ, Wei NE, Jiang H, Doerge RW, Chen ZJ. Reversible histone acetylation and deacetylation mediate genome-wide, promoter-dependent and locus-specific changes in gene expression during plant development. *Genetics* 2005;**169**:337–45.
128. Tian L, Wang J, Fong MP, Chen M, Cao H, Gelvin SB, Chen ZJ. Genetic control of developmental changes induced by disruption of *Arabidopsis* histone deacetylase 1 (AtHD1) expression. *Genetics* 2003;**165**:399–409.
129. Alinsug M, Yu CW, Wu K. Phylogenetic analysis, subcellular localization, and expression patterns of RPD3/HDA1 family histone deacetylases in plants. *BMC Plant Biol* 2009;**9**:37.
130. Gao M-J, Schafer UA, Parkin IAP, Hegedus DD, Lydiate DJ, Hannoufa A. A novel protein from *Brassica napus* has a putative KID domain and responds to low temperature. *Plant J* 2003;**33**:1073–86.
131. Sridha S, Wu K. Identification of AtHD2C as a novel regulator of abscisic acid responses in *Arabidopsis*. *Plant J* 2006;**46**:124–33.
132. Nallamilli BR, Edelmann MJ, Zhong X, Tan F, Mujahid H, Zhang J, Nanduri B, Peng Z. Global analysis of lysine acetylation suggests the involvement of protein acetylation in diverse biological processes in rice (*Oryza sativa*). *PLoS One* 2014;**9**:e89283.
133. Mao Y, Pavangadkar KA, Thomashow MF, Triezenberg SJ. Physical and functional interactions of *Arabidopsis* ADA2 transcriptional coactivator proteins with the

acetyltransferase GCN5 and with the cold-induced transcription factor CBF1. *Biochem Biophys Acta (BBA) Gene Struc Expr* 2006;**1759**:69–79.
134. Stockinger EJ, Mao Y, Regier MK, Triezenberg SJ, Thomashow MF. Transcriptional adaptor and histone acetyltransferase proteins in *Arabidopsis* and their interactions with CBF1, a transcriptional activator involved in cold-regulated gene expression. *Nucleic Acids Res* 2001;**29**:1524–33.
135. Hark AT, Vlachonasios KE, Pavangadkar KA, Rao S, Gordon H, Adamakis I-D, Kaldis A, Thomashow MF, Triezenberg SJ. Two *Arabidopsis* orthologs of the transcriptional coactivator ADA2 have distinct biological functions. *Biochim Biophy Acta (BBA) Gene Reg Mech* 2009;**1789**:117–24.
136. Kaldis A, Tsementzi D, Tanriverdi O, Vlachonasios K. *Arabidopsis thaliana* transcriptional co-activators ADA2b and SGF29a are implicated in salt stress responses. *Planta* 2011;**233**:749–62.
137. Gao M-J, Hegedus D, Sharpe A, Robinson S, Lydiate D, Hannoufa A. Isolation and characterization of a GCN5-interacting protein from *Arabidopsis thaliana*. *Planta* 2007;**225**:1367–79.
138. Williams DB. Beyond lectins: the calnexin/calreticulin chaperone system of the endoplasmic reticulum. *J Cell Sci* 2006;**119**:615–23.
139. Fu C, Liu XX, Yang WW, Zhao CM, Liu J. Enhanced salt tolerance in tomato plants constitutively expressing heat-shock protein in the endoplasmic reticulum. *Genet Mol Res* 2016;**15**:.
140. Takano S, Matsuda S, Funabiki A, Furukawa J, Yamauchi T, Tokuji Y, Nakazono M, Shinohara Y, Takamure I, Kato K. The rice RCN11 gene encodes 1,2-xylosyl transferase and is required for plant responses to abiotic stresses and phytohormones. *Plant Sci* 2015;**236**:75–88.
141. He D, Wang Q, Li M, Damaris RN, Yi X, Cheng Z, Yang P. Global proteome analyses of lysine acetylation and succinylation reveal the widespread involvement of both modification in metabolism in the embryo of germinating rice seed. *J Proteome Res* 2016;**15**:879–90.
142. Zhen S, Deng X, Wang J, Zhu G, Cao H, Yuan L, Yan Y. First comprehensive proteome analyses of lysine acetylation and succinylation in seedling leaves of *Brachypodium distachyon* L. *Sci Rep* 2016;**6**:31576.
143. Jin W, Wu F. Proteome-wide identification of lysine succinylation in the proteins of tomato (Solanum lycopersicum). *PLoS ONE* 2016;**11**:e0147586.
144. Weissman AM. Themes and variations on ubiquitylation. *Nat Rev Mol Cell Biol* 2001;**2**:169–78.
145. Fu H, Lin YL, Fatimababy AS. Proteasomal recognition of ubiquitylated substrates. *Trends Plant Sci* 2010;**15**:375–86.
146. Trujillo M, Shirasu K. Ubiquitination in plant immunity. *Curr Opin Plant Biol* 2010;**13**:402–8.
147. Cho SK, Chung HS, Ryu MY, Park MJ, Lee MM, Bahk Y, Kim J, Pai HS, Kim WT. Heterologous expression and molecular and cellular characterization of CaPUB1 encoding a hot pepper U-BoxE3 ubiquitin ligase homolog. *Plant Physiol* 2006;**142**:1664–82.
148. Park JJ, Yi J, Yoon J, Cho LH, Ping J, Jeong HJ, Cho SK, Kim WT, An G. OsPUB15, an E3 ubiquitin ligase, functions to reduce cellular oxidative stress during seedling establishment. *Plant J* 2011;**65**:194–205.
149. Cho SK, Ryu MY, Song C, Kwak JM, Kim WT. *Arabidopsis* PUB22 and PUB23 are homologous U-BoxE3 ubiquitin ligases that play combinatory roles in response to drought stress. *Plant Cell* 2008;**20**:1899–914.

150. Wan X, Mo A, Liu S, Yang L, Li L. Constitutive expression of a peanut ubiquitin-conjugating enzyme gene in Arabidopsis confers improved water-stress tolerance through regulation of stress-responsive gene expression. *J Biosci Bioeng* 2010;**111**:478–84.
151. Liu H, Zhang H, Yang Y, Li G, Yang Y, Wang X, Basnayake BM, Li D, Song F. Functional analysis reveals pleiotropic effects of rice RING-H2 finger protein gene OsBIRF1 on regulation of growth and defense responses against abiotic and biotic stresses. *Plant Mol Biol* 2008;**68**:17–30.
152. Dong C, Agarwal M, Zhang Y, Xie Q, Zhu J. The negative regulator of plant cold responses, HOS1, is a RINGE3 ligase that mediates the ubiquitination and degradation of ICE1. *Proc Natl Acad Sci U S A* 2006;**103**:8281–6.
153. Hong J, Choi H, Hwang I, Hwang B. Role of a novel pathogen-induced pepper C3-H-C4 type RING-finger protein gene, CaRFP1, in disease susceptibility and osmotic stress tolerance. *Plant Mol Biol* 2007;**63**:571–88.
154. Meng XB, Zhao WS, Lin RM, Wang M, Peng YL. Molecular cloning and characterization of a rice blast-inducible RING-H2 type zinc finger gene. *DNA Seq* 2006;**17**:41–8.
155. Du Z, Zhou X, Li L, Su Z. Plants UPS: a database of plants' ubiquitin proteasome system. *BMC Genomics* 2009;**10**:227.
156. Kang M, Fokar M, Abdelmageed H, Allen RD. *Arabidopsis* SAP5 functions as a positive regulator of stress responses and exhibits E3 ubiquitin ligase activity. *Plant Mol Biol* 2011;**75**:451–66.
157. Castells E, Molinier J, Benvenuto G, Bourbousse C, Zabulon G, Zalc A, Cazzaniga S, Genschik P, Barneche F, Bowler C. The conserved factor DE-ETIOLATED1 cooperates with CUL4-DDB1DDB2 to maintain genome integrity upon UV stress. *EMBO J* 2011;**30**:1162–72.
158. Pandey A, Chakraborty S, Datta A, Chakraborty N. Proteomics approach to identify dehydration responsive nuclear proteins from chickpea (*Cicer arietinum* L.). *Mol Cell Proteomics* 2008;**7**:88–107.
159. Zang X, Komatsu S. A proteomics approach for identifying osmotic-stress related proteins in rice. *Phytochemistry* 2007;**68**:426–37.
160. Lyzenga WJ, Stone SL. Abiotic stress tolerance mediated by protein ubiquitination. *J Exp Bot* 2012;**63**:599–616.
161. Qin F, Sakuma Y, Tran LS, Maruyama K, Kidokoro S, Fujita Y, Fujita M, Umezawa T, Sawano Y, Miyazono K, Tanokura M, Shinozaki K, Yamaguchi-Shinozaki K. *Arabidopsis* DREB2A-interacting proteins function as RINGE3 ligases and negatively regulate plant drought stress-responsive gene expression. *Plant Cell* 2008;**20**:1693–707.
162. Mazzucotelli E, Mastrangelo AM, Crosatti C, Guerra D, Stanca AM, Cattivelli L. Abiotic stress response in plants: when post-transcriptional and post-translational regulations control transcription. *Plant Sci* 2008;**174**:420–31.
163. Guo Q, Zhang J, Gao Q, Xing S, Li F, Wang W. Drought tolerance through overexpression of monoubiquitin in transgenic tobacco. *J Plant Physiol* 2008;**165**:1745–55.
164. Stamler JS, Lamas S, Fang FC. Nitrosylation: the prototypic redox-based signaling mechanism. *Cell* 2001;**106**:675–83.
165. Xue Y, Liu Z, Gao X, Jin C, Wen L, Yao X, Ren J. GPS-SNO: computational prediction of protein S-nitrosylation sites with a modified GPS algorithm. *PLoS One* 2010;**5**:e11290.
166. Ortega-Galisteo AP, Rodrıguez-Serrano M, Pazmino DM, Gupta DK, Sandalio LM, Romero-Puertas MC. S-Nitrosylated proteins in pea (*Pisum sativum* L.) leaf peroxisomes: changes under abiotic stress. *J Exp Bot* 2012;**63**(5):2089–103.
167. del Río LA. Peroxisomes as a cellular source of reactive nitrogen species signal molecules. *Arch Biochem Biophys* 2011;**506**:1–11.
168. Lindermayr C, Durner J. S-Nitrosylation in plants: pattern and function. *J Proteomics* 2009;**73**:1–9.

169. Gow AJ, Farkouh CR, Munson DA, Posencheg MA, Ischiropoulos H. Biological significance of nitric oxide-mediated protein modifications. *Am J Physiol Lung Cell Mol Physiol* 2004;**287**:L262–8.
170. Airaki M, Leterrier M, Mateos RM, Valderrama R, Chaki M, Barroso JB, et al. Metabolism of reactive oxygen species and reactive nitrogen species in pepper (*Capsicum annuum* L.) plants under low temperature stress. *Plant Cell Environ* 2012;**35**:281–95.
171. Mata-Perez C, Begara-Morales JC, Chaki M, Sánchez-Calvo B, Valderrama R, Padilla MN, Corpas FJ, Barroso JB. Protein tyrosine nitration during development and abiotic stress response in plants. *Front Plant Sci* 2016;**7**:1699.
172. Leterrier M, Airaki M, Palma JM, Chaki M, Barroso JB, Corpas FJ. Arsenic triggers the nitric oxide (NO) and S-nitrosoglutathione (GSNO) metabolism in *Arabidopsis*. *Environ Pollut* 2012;**166**:136–43.
173. Signorelli S, Corpas FJ, Borsani O, Barroso JB, Monza J. Water stress induces a differential and spatially distributed nitro-oxidative stress response in roots and leaves of *Lotus japonicus*. *Plant Sci* 2013;**201**:137–46.
174. Ziogas V, Tanou G, Belghazi M, Filippou P, Fotopoulos V, Grigorios D, et al. Roles of sodium hydrosulfide and sodium nitroprusside as priming molecules during drought acclimation in citrus plants. *Plant Mol Biol* 2015;**89**:433–50.
175. Fujita S, Pytela J, Hotta T, Kato T, Hamada T, Akamatsu R, Ishida Y, Kutsuna N, Hasezawa S, Nomura Y, et al. An atypical tubulin kinase mediates stress-induced microtubule depolymerization in *Arabidopsis*. *Curr Biol* 2013;**23**:1969–78.
176. Kim JM, To TK, Ishida J, Morosawa T, Kawashima M, Matsui A, Toyoda T, Kimura H, Shinozaki K, Seki M. Alterations of lysine modifications on the histone H3 N-tail under drought stress conditions in *Arabidopsis thaliana*. *Plant Cell Physiol* 2008;**49**:1580–8.
177. Ding Y, Virlouvet L, Liu N, Riethoven JJ, Fromm M, Avramova Z. Dehydration stress memory genes of *Zea mays*; comparison with *Arabidopsis thaliana*. *BMC Plant Biol* 2014;**14**:.
178. Suter L, Widmer A. Environmental heat and salt stress induce transgenerational phenotypic changes in *Arabidopsis thaliana*. *PLoS ONE* 2013;**8**:e60364.
179. Hardtke CS, Gohda K, Osterlund MT, Oyama T, Okada K, Deng XW. HY5 stability and activity in *Arabidopsis* is regulated by phosphorylation in its COP1 binding domain. *EMBO J* 2000;**19**:4997–5006.
180. Djamei A, Pitzschke A, Nakagami H, Rajh I, Hirt H. Trojan horse strategy in *Agrobacterium* transformation: abusing MAPK defense signaling. *Science* 2007;**318**:453–6.
181. Furihata T, Maruyama K, Fujita Y, Umezawa T, Yoshida R, Shinozaki K, Yamaguchi-Shinozaki K. Abscisic acid-dependent multisite phosphorylation regulates the activity of a transcription activator AREB1. *Proc Natl Acad Sci U S A* 2003;**103**:1988–93.
182. Whitmarsh AJ, Davis RJ. Regulation of transcription factor function by phosphorylation. *Cell Mol Life Sci* 2000;**57**:1172–83.

Further Reading

183. Akula R, Ravishankar GA. Influence of abiotic stress signals on secondary metabolites in plants. *Plant Signal Behav* 2011;**6**(11):1720–31.
184. Boman HG. Peptide antibiotics and their role in innate immunity. *Annu Rev Immunol* 1995;**13**:16–92.
185. Bray EA, Bailey-Serres J, Weretilnyk E. Responses to abiotic stresses. In: Buchanan BB, Gruissem W, Jones RL, editors. *Biochemistry and molecular biology of plants*. Rockville, MD: Amer Soc Plant Physiol; 2000. p. 1158–203.
186. Broekaert WF, Terras FRC, Cammue BPA. Plant defensins: novel antimicrobial peptides as components of host defense system. *Plant Physiol* 1995;**108**:1353–8.

187. Cattivelli L, Bratels D. Biochemistry and molecular biology of cold inducible enzyme and proteins in higher plants. In: Wary JL, editor. *Inducible plant protein society for experimental biology seminar series 49*. Cambridge: Cambridge University Press; 1992. p. 267–88.
188. Cornelissen BJC, Melchers LS. Strategies for control of fungal diseases with transgenic plants. *Plant Physiol* 1993;**101**:709–12.
189. Creelman RA, Tierney ML, Mullet JE. Jasmonic acid/methyl jasmonate accumulate in wounded soybean hypocotyl and modulate wound gene expression. *Proc Natl Acad Sci U S A* 1992;**89**:4938–41.
190. Graham D, Petterson BD. Responses of plants to low, non-freezing temperatures; proteins, metabolism and acclimation. *Annu Rev Plant Physiol* 1982;**33**:347–72.
191. Iizumi T, Luo JJ, Challinor AJ, Sakurai G, Yokozawa M, Sakuma H, Brown ME, Yamagata T. Impacts of El Niño Southern Oscillation on the global yields of major crops. *Nat Commun* 2014;**5**:.
192. Karana S, Venkateshwaria JC, Kirtib PB, Chopra VL. Transgenic expression of havein, the rubber tree lectin in Indian mustard confers protection against *Alternaria brassicae*. *Plant Sci* 2002;**162**:441–8.
193. Kumari G, Reddy A, Naik S, Kumars S, Prasanthi J, Sriagganayakulu G, Reddy P, Sudhakar C. Jasmonic acid induced changes in protein pattern, antioxidative enzyme activities and peroxidase isoenzymes in peanut seedlings. *Biol Plant* 2006;**50**:219–26.
194. Larcher W. *Physiological plant ecology*. 4th ed. Berlin, Heidelberg: Springer Verlag; 2003.
195. Los DA, Murata N. Membrane fluidity and its role in the perception of environmental signals. *Bioch Biophys Acta Biomembr* 2004;**1666**:142–57.
196. Louda S, Mole S. Glucosinolates-chemistry and ecology. In: Rosenthal GA, Burndaum MR, editors. *Herbivores, their interaction with secondary plant metabolites*. 2nd ed. vol. 1. San Diego, CA: Acad. Press; 1991. p. 123–64.
197. Mendez E, Moreno A, Colilla F, Pelaez F, Limas GG, Mendez R, Soriano F, Salinas M, de Haro C. Primary structure and inhibition of protein synthesis in eukaryotic cell free system of a novel thionin, gamma hordothionin from barley endosperm. *Eur J Biochem* 1990;**194**(2):533–9.
198. Miura K, Furumoto T. Cold signalling and cold response in plants. *Int J Mol Sci* 2013;**14**:5312–37.
199. Molina A, Goy PA, Fraila A. Inhibition of bacterial and fungal plant pathogens by thionins of types I and II. *Plant Sci* 1993;**92**:169–77.
200. Negi NP, Shrivastava DC, Sharma V, Sarin NB. Overexpression of CuZnSOD from *Arachis hypogaea* alleviates salinity and drought stress in tobacco. *Plant Cell Rep* 2015;**34**:1109–26.
201. Ng TB. Antifungal proteins and peptides of leguminous and non-leguminous origins. *Peptides* 2004;**25**:1215–22.
202. Niess DH. Microbial heavy metal resistance. *Appl Microbiol Biotechnol* 1999;**51**:730–50.
203. Ohashi Y, Oshima M. Stress induced expression of genes for pathogenesis related proteins in plants. *Plant Cell Physiol* 1992;**7**:819–26.
204. Parthier B, Bruchnner C, Dathe W. Jasmonates, metabolism biological activities and mode of action in senescence and stress responses. In: Karseen CM, Van-Loon LC, Ureugdendi DD, editors. *Progress in plant growth regulation*. Dordrecht: Kluwer Academic Publishers; 1992. p. 276–85.
205. Pontoppidan B, Richard H, Lars R, Johan M. Infestation by cabbage aphid (*Brevicoryne brassicae*) on oilseed rape (*Brassica napus*) causes a long lasting induction of the myrosinase system. *Entomol Exp Appl* 2003;**109**:55–62.
206. Ramankutty N, Evan AT, Monfreda C, Foley JA. Farming the planet: geographic distribution of global agricultural lands in the year 2000. *Glob Biogeochem Cycles* 2008;**22**(1). GB1022.

207. Ryan CA, Pearce G. Systemin: a polypeptide signal for plant defensive genes. *Annu Rev Cell Dev Biol* 1998;**14**:1–17.
208. Schmidt S, Baldwin IT. Systemin in *Solanum nigrum*. The tomato homologous polypeptide does not mediate direct defense responses. *Plant Physiol* 2006;**142**:1751–8.
209. Singariya P. *Effect of sub-optimal environment and PGR's on metabolic pattern of certain species of* Cenchrus [Ph. D thesis]. India: J. N. Vyas, University of Jodhpur (Rajasthan); 2009.
210. Taipalensuu J, Falk A, Rask L. A wound and methyl jasmonate-inducible transcript coding for a myrosinase associated protein with similarities to an early nodulin. *Plant Physiol* 1996;**110**:483–91.
211. Vanloon LC, Van-Strien EA. The families of pathogenesis related protein, their activities and comparative analysis of PR-1 type proteins. *Physiol Mol Plant Pathol* 1999;**55**:85–97.
212. Wang W, Vincur B, Altman A. Plant responses to drought, salinity and extreme temperature: towards genetic engineering for stress tolerance. *Planta* 2003;**218**:1–14.
213. Wessels JGH, Sietsna JH. Fungal cell walls survey. In: Tanner W, Locures FA, editors. *Encyclopedia of plant physiology new series. Plant carbohydrates*, vol. 13B. Berlin: Springer; 1981. p. 352–94.
214. Woloshruk CP, Mewennoff JS, Selan Buutlaarge M. Pathogen induced proteins with inhibitory activity towards *Phytophthora infestans*. *Plant Cell* 1991;**3**:485–96.

CHAPTER

9

Posttranslational Modifications Associated With Cancer and Their Therapeutic Implications

Aniket Kumar Bansal, Laishram Rajendrakumar Singh*, Majid Rasool Kamli*[†,‡]

***Dr. B.R. Ambedkar Center for Biomedical Research, University of Delhi, Delhi, India**

[†]Center of Excellence in Bionanoscience Research, King Abdulaziz University, Jeddah, Saudi Arabia

[‡]Department of Biological Sciences, Faculty of Science, King Abdulaziz University, Jeddah, Saudi Arabia

1 INTRODUCTION

Inside the intracellular environment proteins are posttranslationally modified by various mechanisms (via enzymatic or nonenzymatic mechanisms) to achieve functional alterations. Phosphorylation, ubiquitylation, neddylation, SUMOylation, glycosylation, and prenylation are some of the most common posttranslational modifications (PTMs) mediated by enzymes. Among the nonenzymatic strategies, acetylation, homocysteinylation, glycation, prenylation, etc., are important modifications.[1] Significant development has been achieved in the identification of various PTMs in relation to their roles in different biological processes, in terms of their ability to form new protein complexes, enzyme activation/inactivation, regulation of protein stability, etc. It is understood that many of such PTMs are associated with the pathophysiology of several diseases. For instance, carbamoylation is associated with chronic kidney disease,[2] carbonylation and oxidation are linked with age-related diseases, hyperphosphorylation of tau leads to Alzheimer's disease,[3] and homocysteinylation and glycation is associated with hyperhomocystinuria and hyperglycemia

Protein Modificomics
https://doi.org/10.1016/B978-0-12-811913-6.00009-6

TABLE 1 Major Posttranslational Modifications Associated With Cancer

Posttranslational Modifications	References
Phosphorylation	7
Acetylation	8
Prenylation	9
Glycosylation	10
Methylation	11
Ubiquitylation	12

respectively.[4] Therefore understanding their mechanisms and the conformational allostery of the proteins involved due to their modification is important for the elucidation of their therapeutic functions.

Cancer is a term for diseases in which abnormal cells divide without control and can invade nearby tissues. In cancer, PTMs play a major role in regulating/dysregulating the various proteins involved in cell proliferation or apoptosis.[5] Cancer comprises six biological capabilities acquired during the multistep development of tumors. These capabilities are sustained proliferative signaling, evading growth suppressors, enabling replicative immortality, activation of invasion and metastasis, inducing angiogenesis, and resisting cell death.[6] All the given developmental steps involve one or another type of PTM. In general several PTMs are associated with cancer (see Table 1). These include phosphorylation, glycosylation, ubiquitylation, SUMOylation, prenylation, methylation, acetylation, etc.[13] Furthermore, some particular modifications are confined to the proteins present in the different compartments of the cells (e.g., cytoplasm, nucleus, secretory, or cell membrane, etc.).[13] At present there are few reviews that focus on particular modifications and their involvement in cancer[14] or that address all types of modifications related to cancer and their therapeutic implications. The present chapter is designed to address almost all of the major PTMs related to cancer. The target proteins, their involvement in the cancer progression, and the therapeutic implications have been addressed.

2 POSTTRANSLATIONAL MODIFICATIONS THAT REGULATE THE MDM2-P53 INTERACTION

p53 is a nuclear transcription factor with a proapoptotic function and it is also considered to be a classical type of tumor suppressor.[15] The p53 protein consists of 393 amino acids and is divided into three functional domains (N-terminal, central core, C-terminal) the subdomains and their properties are listed in Fig. 1.

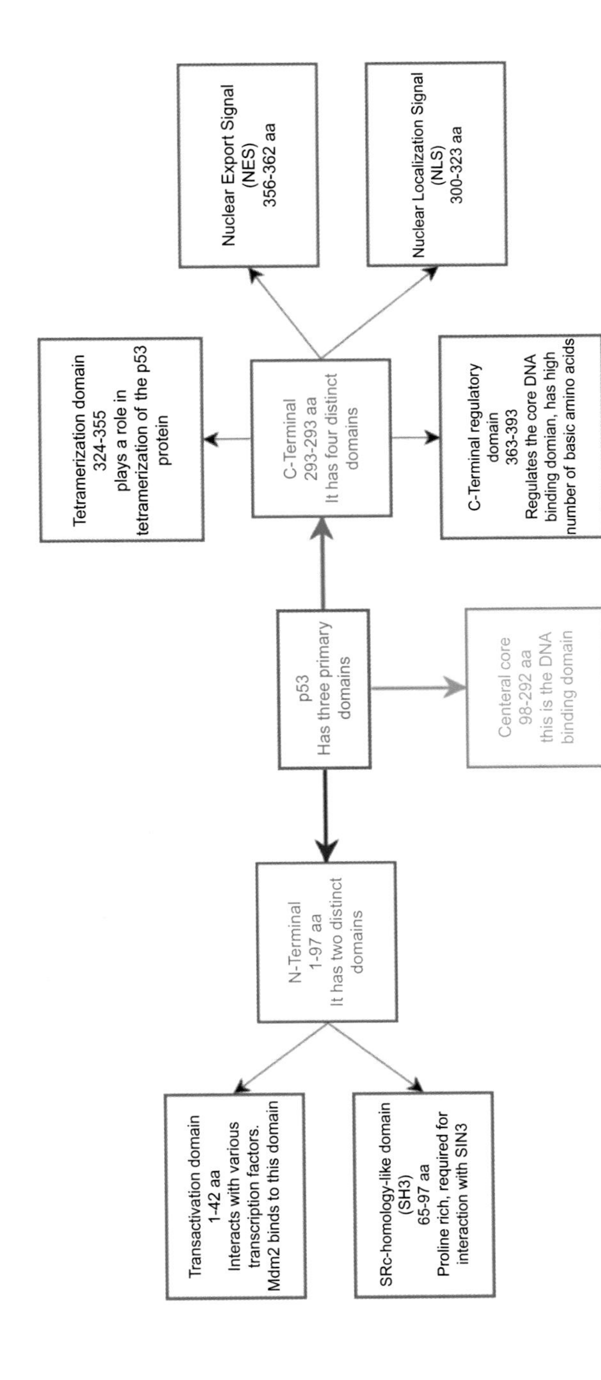

FIG. 1 Different domains and subdomains of p53 and their respective functions.

Under normal conditions p53 is functionally inactivated and has minimal expression, this is because of the proteosomal degradation caused by E3 ubiquitin ligase Mdm2. The targeted disruption of the Mdm2-p53 interaction occurs during stress conditions through various mechanisms, which may include PTMs.[12] The Mdm2-p53 interaction can be inhibited by stress-induced phosphorylation of Ser395 (by ATM kinase) and Tyr394 (by c-Abl kinase), which are present on Mdm2.[16] There are a number of phosphorylation sites present on p53 that serve to disrupt the above interaction and can be phosphorylated by stress-induced kinases, for example, phosphorylation of Thr18 present in the transactivating domain of p53 decreases binding to Mdm2[17] and the phosphorylation of Ser15 and Ser 20 in the transactivating domain leads to the stabilization of p53 and inhibition of the Mdm2 interaction.[18] Further studies have shown that phosphorylation, though playing a role in activation of p53, may be just a part of a series of posttranslational events that need to take place in order for p53 to be activated.[19] However the role of phosphorylation is not just limited to the stabilization and activation of p53, it also leads to an increase in its sequence-specific DNA binding ability.

Acetylation of Mdm2 by CBP/$_{p}$300 disrupts the Mdm2 and p53 interaction. The acetylation of eight C-terminal lysine residues of p53 inhibits the p53-Mdm2 interaction.[20] It has also been proven that purified acetylated p53 could not be ubiquitinated with Mdm2[21] and the levels of ubiquitinated p-53 decrease after induction of acetylation. However, in mice proline residues that were changed with arginine in p-53 no change in the level of p-53 was observed and it was still able to respond adequately to cellular stress and DNA damage.[22] p53 also shows other PTMs like glycosylation and ribosylation, but as yet there is no clear role apparent for these PTMs in cancer development and proliferation.[23]

3 MODIFICATION OF RETIONBLASTOMA TUMOR SUPPRESSOR PROTEIN (pRB) AND ITS ASSOCIATION TO CELL CYCLE PROGRESSION

Retionblastoma is a 110-kDa nuclear protein and has 928 amino acids. It has three domains: the N-terminal domain, pocket domain, and the C-terminal domain. The pocket domain is further subdivided into A and B domains that are separated by a spacer sequence and the integrity of the A and B subdomains is essential for the function of the tumor-suppressor protein (pRb) as the transcription factor E2F binds to these domains.[24] pRb is a part of the pocket proteins family, which includes p107 and p130. These proteins have a high degree of structural similarity to each other; however, they perform very distinct functions due to different binding partners.[25] pRb is the protein that regulates the G_1-S checkpoint and this protein is part of a pathway called the Rb pathway. pRb needs to undergo hyperphosphorylation in order for the cell to enter the S phase as hyperphosphorylation leads to the detachment of E2F, which is a transcription factor for various genes in the genome that are expressed in the S phase.[26] The pRB protein is capable of undergoing various phosphorylations by different serine/threonine kinases. These kinases include cyclin-dependent kinases (Cdks)[27] like Cdk9,[28] Cdk5,[29], and Cdk3[30], a total of 16 phosphorylation sites are found in three regions of pRB, out of which seven sites are clustered between S780 and T826 at the C-terminus domain, another six sites are in the N-terminus between T5 and T373, and the rest of the sites are in the small pocket but are not clustered, these are S567, S608, and S612.[31] This protein is also phosphorylated by p38 Map-kinase, which is activated by mitogens but is also associated with oxidative stress and DNA damage,[32] checkpoint kinase Chk1 and 2, which are activated by kinases upstream due to DNA breaks or replicative stress,[33] and aurora kinase, which helps in the maintenance of chromosome stability.[34]

Acetylation of pRB can also be regulated by acetylation of arginine residues and the acetylation is achieved by recruitment of pRB protein into p300/CPB transcriptional co-activator complex by binding to viral oncoproteins like E1A, and the pRB protein is acetylated at K873/K874 sites in the C-terminus domain.[35] Acetylation at these sites increases pRB's affinity to Mdm2 and this maintains pRb in an active state leading to the decrease in Cdk-dependent phosphorylation.[36]

Methylation of pRB can occur at both lysine and arginine residues. Methyltransferases have been shown to bind directly and indirectly to pRB and methylate at K873,[37] K810[38] and affect E2F transcriptional activity, which enhances cell-cycle arrest.[39] Ubiquitylation is the conjugation of lysine residues with ubiquitin and leads to the targeting and degradation of proteins. In the case of pRB it can be caused by interaction with viral proteins like the E7 proteins of the human papilloma virus.[40] This property of the viral proteins is associated with their ability to cause cancer. pRB has four ubiquitylation sites at K143, K256, K574, and K810.[41] The site of SUMOylation on pRB protein is K720, this lies in a cluster of lysine residues surrounding the LXCXE binding cleft in the B domain of the small pocket.[42] Although there has been a wide variety of research conducted on pRB protein and the RB pathway in general, there have been no breakthroughs in cancer therapy utilizing the PTM state of this protein.

4 REGULATION OF NF-κB BY VARIOUS POSTTRANSLATIONAL MODIFICATIONS

Nuclear-factor kappa-light-chain-enhancer of activated B cells (NF-κB) is a family of five different transcription factors. NF-κB can form various heterodimers or homodimers and bind to consensus DNA sequences at promoter regions of genes that may be responsive to it. NF-κB can be activated by various stimuli such as cytokines (TNFalpha, IL-1beta), growth factors like epidermal growth factor (EGF), viral and bacterial products like lipopolysaccharides and double-stranded RNA, UV and ionizing radiation, reactive oxygen species, DNA damage, and oncogenic stress.[43] All of these stimuli lead to a canonical pathway that involves the activation of the IKK (the inhibitor of κB (IκB) kinase),[44] which phosphorylates and marks the inhibitory factor IκB for proteasomal degradation. The stringent regulation of NF-κB is vital for maintaining the cellular function and requires the prompt activation and termination of NF-κB.[44] Altered activity of NF-κB is attributed to its deregulated phosphorylation and, recently, NF-κB activation has been found to play an important role in tumor development.[45] The role of NF-κB in cancer development was first studied when several members of the NF-κB family were found to be mutated in certain types of cancer, mostly in those that were hematopoietic in origin like the mutation of c-Rel gene, which can cause lymphoid malignancies in mammals, and its amplification is detected in various non-Hodgkin's lymphoma.[46] NF-κB2/p100 is frequently activated through chromosomal translocations in lymphomas as well as in leukemias and deletion of IKKβ and activation of NF-κB has signified its role in tumorigenesis, especially in patients with nonclassical glioblastoma.[47] NF-κB is the master regulator of crosstalk between inflammation and cancer at various levels, in tumors with elevated NF-κB activity the accumulation of proinflammatory cytokines at the tumor site directly contributes to the protumorogenic microenviornment.[48]

The ability to inhibit NF-κB activation is being studied as the primary strategy to improve the efficacy of currently available anticancer drugs. Blocking NF-κB has been seen to sensitize cancer cells to chemotherapy and radiotherapy.[49] IKK plays a central role in NF-κB metabolism and hence is a major drug target for its inhibition, the various inhibitors tested include BAY-11-7082,[50] MLN102B,[51] BMS-345541,[52] SC-514,[53], and CHS828.[54] These compounds bind to IKK directly or to an upstream molecule and inhibit its activation. Proteasome inhibition leads to the prevention of NF-κB activation as IκB is not degraded. Bortezomib is a reversible 26S inhibitor that has been approved by the FDA for the treatment of multiple myeloma[55] and other proteasomal inhibitors, which include RP-171, NPI-0052, and CEP-188770, are being examined and are in early clinical trials.[56] Blocking the nuclear transport machinery of NF-κB can be achieved by SN-50, a peptide of 41 amino acids that inactivates NF-κB making it susceptible to cisplatin.[57] Antiinflammatory drugs are also being studied as potential NF-κB inhibitors in addition to their effects when they are used as adjuvants to be used along with current cancer therapy.[58] However, no drug has been discovered that specifically targets the NF-κB pathway and long-term suppression will affect innate immunity.[59] Gene therapy to inhibit NF-κB is being explored as one option to develop as a credible therapy.[60]

5 REGULATION OF GENE EXPRESSION VIA HISTONE AND HISTONE DEACETYLASE MODIFICATIONS

Histone proteins play essential structural and functional roles in the transition between active and inactive chromatin states. The basic unit of chromatin is the nucleosome core particle, which comprises 147 base pairs of DNA wrapped around an octamer of histones. This octamer is composed of heterotetramer that has H3 and H4 dimers in the center, which are flanked by two heterodimers of histones, H2A and H2B, and each nucleosome is separated by 10–60 bp of linker DNA,[21] which is associated with the linker histone H1.[61] Histone PTMs influence DNA replication, transcription, repair, and recombination.[62] Nucleosome particles can be modified in their composition, structure, and location by a chromatin remodeling complex that introduces PTMs to the core histones. Histones are not considered to be simple DNA-packaging proteins, but are recognized as regulators of chromatin dynamics.[63] Histones can undergo a large number of PTMs like acetylation (lysine and arginine), methylation (lysine and arginine), phosphorylation (serine and threonine), ubiquitylation (lysine), SUMOylation (lysine), and ribosylation. In addition to the above PTMs, lysine can accept up to three methyl groups and arginine can accept one or two methyl groups.[21] New modification sites are being identified on histones every year. A comprehensive list of histones and their various modifications can be found at http://www.actrec.gov.in/histome/index.php. The abovementioned PTMs occur mostly in the amino tails of the histone protein that protrudes from the surface of the nucleosome as well as on the globular core region.[64]

There are two proposed mechanisms by which histone modification can affect chromosome function: (1) modification may alter the electrostatic charge of the histone, which result in structural changes in the histone or a change in the way they are bound to the DNA and (2) the modifications are sites for binding of the protein recognition modules, such as chromodomains and bromodomains, which recognize the methylated lysines and the acetylated lysines respectively.[63] The primary factors in histone modifications are enzymes, histone acetyltransferases, which acetylate histone tails and induce chromatin decondensation. Histone deacetylases (HDACs) on the other hand deacetylate histone, which leads to its tighter binding to DNA. Histone methyltranferases (HMTs) also promote or inhibit transcription, depending on the site on which it methylates a particular amino acid and histone demethylases (HDMs), which counteract HMTs.[65] Some of the important histone modifications that have been observed in cancers are described in the following section.

H4K16Ac and H4K20me3

H4K16Ac is the acetylation of H4 at lysine 6 and H4K20me3 indicates the loss of trimethylation at lysine 20 of histone H4. It has been reported that these two modifications lead to genomic instability and also influence the sensitivity of cancers to chemotherapy.[66]

Trimethylation of H4K20

This is associated with depressed chromatin.[67] Cancer cells exhibit a global decrease in H4K20me3 and prevent the silencing of various genes.[68] This reduction occurs in repetitive DNA sequences and is associated with a global loss of methylation.[69]

Di and trimethylation of H3K4 (H3K4me2/me3)

Methylation of H3K4 is associated with transcriptional activation observed near the start sites of highly expressed genes.[70] A decrease in this methylation is observed in a range of neoplastic tissues[71] and can act as a predictive factor for treatment outcome. Tumor reoccurrence is observed in low-grade prostate carcinoma patients with low H3K4me2, this is independent of other pathological parameters.[72]

Acetylation/trimethylation of H3K9 histone H3 lysine 9 trimethylation (H3K9me3)

This is an important epigenetic marker that has been associated with transcriptional repression and the acetylation of H3K9 is a transcriptional activator. A decrease in acetylation leads to a poor cancer prognosis, except in hepatocellular carcinoma where an increase in acetylation is observed.[72]

Trimethylation of H3K27me3

Trimethylation occurs at lysine 27 of histone H3. This type of methylation spreads over large regions and targets many genes and negatively regulates their transcription by promoting a compact chromatin structure. It is also responsible for the repression of unwanted differentiation programs.[73] It has been implicated in the prognosis of breast, ovarian, pancreatic, and esophageal cancers.[74]

Acetylation of H3K18ac

This is a general marker for active transcription. Loss of this acetylation is correlated with a poor prognosis in patients with prostrate, pancreatic, lung, kidney, and breast cancer.[75]

Acetylation of H4K12ac, which is a hypoacetylation, can be used as biomarker for some cancers and can also be used to determine the stage in colorectal cancer.[71]

Histone modifications are not directly targeted for cancer therapy; however, the enzymes that control histone modifications can become good targets for cancer therapy. Like the enzyme H3K79 methyltransferase or DOT1L, H3K79 methylation is required to activate and maintain transcription state.[76] DOT1L has been found to be a drug target for mixed-lineage leukemia, in fact various inhibitors of DOT1L have been found.[77] These inhibitors have been proven to be effective and selective.[78]

H3K4 methylation has been described as being brought about by an enzyme LSD1, this enzyme can also be targeted by its various inhibitors like tranylcypromine, GSK2879553, GSK690, and SP2509.[79] H3K27 methylation is carried out by EZH2 and EZH1, and the inhibitors for the enzyme provide selective cancer therapy and a better prognosis for the cancer patients. The inhibitors of EZH2 are GSK129, EPZ-6438, and CPI-169. There are other enzyme inhibitors that are available, but these have not been used for therapeutic purposes.

6 DEREGULATION OF TYROSINE KINASE BY POSTTRANSLATIONAL MODIFICATIONS

These are a family of enzymes that are activated by growth factors (e.g., EGF, PDGF, VEGF) and hormones.[80] They get autophosphorylated then phosphorylate other downstream proteins and use ATP as energy and for the transfer of the Pi group. Deregulation of the tyrosine kinase may cause uncontrolled growth and division in the cell. Tyrosine kinases are primarily classified as receptor tyrosine kinase (RTK), they are not only cell-surface, transmembrane proteins, but also enzymes with kinase activity.[81]

An important mechanism leading to tyrosine kinase deregulation is a mutation. Mutations within the extracellular domains such as EGFRv III mutant lacks amino acid 6-273 give rise to the constitutive activity of RTK and lead to cell proliferation even in the absence of (protein) ligands in the matrix and can lead to glioblastomas, ovarian tumors, and nonsmall-cell lung carcinoma.[82] Point mutations in the extracellular domain of FGFR 3 result in myeloma.[83] Somatic mutations in the EGFR2 and 3 have been associated with human bladder and cervical carcinomas.[84] Breakpoints of abnormal chromosomal translocation are also an important source of mutation, an important example of which is Bcr-A human leukemia. The breakpoint cluster region (BCR) sequences of chromosome-22 on translocation are placed with c-ABL tyrosine kinase of chromosome 9. The Bcr-Abl chimeric gene product has a tyrosine kinase activity several fold higher than its normal counterpart[35] and correlates with the disease phenotype. Tel-Abl tyrosine kinase is also constitutively phosphorylated due to reciprocal translocation.[85] Other important translocations include t(5, 12) in CMML and NPM-ALK fusion in anaplastic large cell lymphoma.[38]

Paracrine stimulation is an important mechanism for constitutive activation of tyrosine kinase.[86] This happens when an RTK gets abnormally expressed or overexpressed in the presence of its associated ligand or when there is an overexpression of the ligand in the presence of its receptor, e.g., EGFR (bladder cancer, breast cancer, and glioblastoma multiforme), PDGFR (brain tumor and gliomas) and IGF (breast cancer and prostate cancer).[87]

Several attempts have been made over the years to target tyrosine kinases, which include the inhibition of the low-molecular-weight compounds capable of interfering with binding sites in the case of RTK or the protein substrate in the NRTK, a bi-substrate inhibitor approach that offered promise, and efforts to generate noncompetitive or allosteric inhibitors. However, all of the abovementioned approaches have mostly failed. The current targets of choice are the competitive inhibitors of ATP.[88]

7 POSTTRANSLATIONAL MODIFICATIONS IN RAS/RAF/MEK/ERK IN MAPK PATHWAY

Mitogen-activated protein kinases (MAPKs) are protein Ser/Thr kinases that convert extracellular signals into a wide range of cellular responses and Ras/Raf/MEK/ERKs are part of the MAPK cascade.[89] MAPK pathways are kinase models that are evolutionarily conserved[90] and link extracellular signals to the machinery that controls the fundamental processes like growth proliferation, differentiation, migration, and apoptotic processes. These pathways are comprised of three-tier kinase module in which a MAP kinase is activated by phosphorylation by mitogen-activated protein kinase kinase (MAPKK), which in turn is activated when phosphorylated by MAPKK. The ERK pathway is deregulated in approximately one-third of all human cancers.[91] The ERK pathway is linked to various facets of the cancer phenotype and cell proliferation is just one of the phenotypes. This pathway is also linked with RTKs and their constitutive expression can lead to activation of this pathway.[92]

ERK is activated by phosphorylation from MEK[93] and MEK is activated when it is phosphorylated by Raf, which in turn is activated when Ras GTPase (activated by receptor tyrosine kinase) recruits Raf kinases to the plasma membrane for activation. Therefore constitutive activation of RTK will automatically constitutively activate this pathway, methods of which are discussed in the previous section. Ras and Raf mutations can also lead to the constitutive activation of the ERK pathway.[94] The deregulation of the ERK also affects its downstream nuclear transcription factors, most notably AP-1 and myc.[95]

Ras GTPases

These act as molecular switches that control the activity of many signaling pathways. The mutations in Ras are found on codons 12, 13, or 61[96] and these prevent efficient GTP (guanosine-5′-triphosphate) hydrolysis rendering Ras inactive. The GTP-bound state thus can constitutively bind to its downstream effectors including Raf. For Ras to function as a signal transducer, it has to associate with the plasma membrane.[97] This step requires isoprenylation (farnesylation or geranylation) near the Ras C-terminus. Consequently, isoprenylation inhibitors targeted Ras. However, the clinical trials of such inhibitors were unsuccessful.[98]

Raf

These are just downstream of Ras and upstream of the ERK module. The structures of the three Raf molecules are similar but there are substantial differences in how they are activated.[99] Genetic silencing of these isoforms shows that they serve in redundant (unique) roles in the body and differ considerably in their modes of regulation, their tissue distribution, and their ability to activate MEK.[100] Once bound to Ras, Raf kinases are activated by a sequence of events, which involves phosphorylation, protein-protein interactions, and protein-lipid interactions. These events increase the catalytic ability of Raf by neutralizing autoinhibition and facilitating activation of the kinase domain. Raf-1 activation involves a complex series of changes in phosphorylation, which entail the dephosphorylation of an inhibitory site, S259, and the phosphorylation of the N-region including a critical activating site, S338,[101] as well as phosphorylation of the activation loop for maximal activation These sites are conserved in A-Raf, and activation seems to follow a similar pattern to Raf-1. However, B-Raf has already a negative charge in the N-region due to twin aspartic acids and the equivalent of Raf-1 S338 is constitutively phosphorylated. Additionally, Ras alone is sufficient to activate B-Raf, whereas Raf-1 requires other factors in addition. However, it is still unclear which of the Raf isoforms is required to activate ERK, and this may be different depending on the cellular context and the stoichiometries of Raf isoforms. B-Raf is found to be mutated in 66%[102] of malignant melanomas and at a lower frequency in many other human malignancies. The most common mutation that is found in 90%[103] of cases is a V600E change in the activation loop that induces the activation of the catalytic activity of B-Raf. B-Raf mutations are lethal unless a certain genetic and biochemical microenvironment permits such cells to survive.[104] As mentioned earlier, Raf activates MEK1 and MEK2 by phosphorylation of serine 218 and 222 in the activation loop. The three Raf isoforms differ in their abilities to activate MEK1 and MEK2. B-Raf is the strongest activator followed by RAF-1 and A-Raf is a weak MEK activator and preferentially activates MEK1. B-Raf has a constitutive negative charge on its binding site and hence is a better activator of MEK in comparison to the other Raf proteins.[105]

The ERK pathway has been a focus of drug discovery with Ras, Raf, and MEK as the main targets. Sorafenib is a drug that targets Raf kinases, but its monotherapy has little or no antitumor activity.[106] HSP90 is an obligatory chaperone for several signaling proteins including Raf kinase

AKT and EGF receptors and without its binding the proteins are degraded. Drugs like geldanamycin exploit this fact. MEK inhibitors are also being explored with CI-1040, which is the first MEK inhibitor to enter clinical trials with some showing antiproliferative activity.[107] HDAC proteins interact with the cellular proteins implicated in cancer development and behave as oncogenes in various cancer types including the retinoblastoma (Rb) tumor suppressor, metastasis-associated protein2 (MTA2), and nuclear hormone receptors like the retinoic acid receptor. The phosphorylation of HDAc2 is essential for the formation of multiprotein complexes. The N-terminal regions of class 2 HDACs have two or three conserved serine residues that are subjected to phosphorylation and this leads to the binding of 14-3-3 proteins, the nuclear export of HDACs, and the depression of their target genes. It should be noted that HDAC phosphorylation is dynamic and dependent on the balance of opposing forces involving kinases and phosphatases.

8 MODIFICATION OF SIGNAL TRANSDUCERS AND ACTIVATORS OF TRANSCRIPTION

Signal transducers and activators of transcription (STATs) are transcription factors that under basal conditions remain inactive in the cytoplasm.[108] The Janus kinase (JAK) family of kinases are activated by cytokines, which in turn phosphorylates STATs, which then translocate to the nucleus and bind to specific regulatory genes and activate the transcription of the genes that control cellular processes such as survival, proliferation, and invasion.[109] The activation of STAT is transient and rapid but in cancer, through the action of mutated kinases or autocrine or paracrine mechanisms,[110] STAT3 facilitates cell-cycle progression by promoting activation of cyclin-dependent kinases (CDKs). It increases the transcription of positive regulators such as cyclin D2 and downregulates the transcription of CDK inhibitors such as p21.[111] STAT5 confers protection from apoptosis, by activating transcription of Bcl-x, to produce the antiapoptotic protein Bcl-xL.

In many chronic and acute hematological malignancies various point mutations have been identified in the JAK family that cause aberrant JAK/STAT signaling and thus lead to aberrant behavior, which includes their constitutive expression.[112] There are also mutations that can be caused by fusion proteins, e.g., BCR-JAK2 in CML,[113] PCM1-JAK2[114] in AML, T-ALL.[115] In many solid tumors JAK/STAT activation is a feature. In head and neck squamous cell carcinoma phosphorylation of STAT3 is a consequence of the increased production of IL-6 by tumor cells.[116] In ovarian cancer an increased expression of G-CSF receptor is observed, which leads to the activation of the JAK/STAT pathway.[117]

Activation mutations in STATs are rare but are present in 40% of patients with granular lymphocytic leukemia[118] with the mutation affecting the SH2 domain of STAT3.[118] Reduced expression of negative regulators for the pathway can also result in increased activation, for example, in nonsmall-cell lung cancer the expression of SOCs3 is lost due to promoter hypermethylation.[119] Preclinical therapies that inhibit STAT activity decrease proliferation and increase apoptosis in cell culture studies and tumor xenograft models.[120]

9 PHOSPHORYLATION OF CYCLIN-DEPENDENT KINASES AND THEIR THERAPEUTIC IMPLICATIONS

CDKs are a family of heterodimeric serine/threonine protein kinases comprising of a CDK subunit and activating cyclin subunit. CDKs require the presence of cyclins to become active. These cyclins are a family of proteins that have no enzymatic activity of their own but activate CDKs by binding to them.[121] CDKs must also be in a specific phosphorylation state as there are two inhibitory sites and one activating phosphorylation site. In higher eukaryotes inhibitory phosphorylation regulates the timing of entry into the different phases of the cell cycle. Therefore the CDK can only be activated with the phosphorylation of some sites (activating sites) and the dephosphorylation of others (inactivation sites) in order for activation to occur. Correct phosphorylation depends on the action of other kinases and a second class of enzymes called phosphatases that are responsible for removing phosphate groups from proteins.[122] The activity of the cyclin subunit depends on its association with the CDK and there have been reports of about 25 different cyclin and 13 different cyclin-dependent kinase.

Cyclin-dependent kinases are critical regulatory enzymes that derive all the cell-cycle transitions, and their activity is under stringent control to ensure cell-cycle division.[122] The quantities of all CDKs remain stable throughout the cell cycle. In contrast, the synthesis and breakdown of cyclin varies by stage with cell-cycle progression dependent on the synthesis of new cyclin molecules. Thus cells synthesize G1- and G1/S-cyclins during the G1 phase, and they produce M-cyclin molecules during the G2 phase. Cyclin degradation is essential for progression through the cell cycle.[122] Specific enzymes break down cyclins at predefined times during the cell cycle. As the cyclin decreases the corresponding CDK becomes inactive. Failure of cyclins to degrade can result in cell-cycle arrest. As cyclin-CDK complexes recognize multiple substrates, they are able to coordinate the multiple events that occur during each phase of the cell cycle. For example, at the beginning of S phase, S-CDK catalyzes the phosphorylation of the proteins that initiate DNA replication by allowing DNA replication complexes to form.[123] Later, during mitosis, M-CDKs phosphorylate a wide range of proteins. These include condensin proteins, which are essential for the extensive condensation of mitotic chromosomes, and lamin proteins, which form a stabilizing network under the nuclear membrane that disassembles during mitosis. M-CDKs also influence the assembly of the mitotic spindle by phosphorylating proteins that[124] regulate microtubule behavior. The net effect of these coordinated phosphorylation reactions is the accurate separation of chromosomes during mitosis.[124]

The CDK4/6-RB axis is essential for cell-cycle entry; therefore, it has been observed that the vast majority of cancers subvert this axis to promote proliferation.[125] In order to induce cell-cycle arrest most of the oncogenes induce p16INK4A as an intrinsic break to deregulate proliferation. Overexpression of p16INK4A eventually engages RB to suppress growth and cell cycle progression and promote oncogene-induced senescence.[126] CDK2 is structurally and functionally similar to CDK1, but it has a considerably broader substrate profile than CDK4 and CDK6, as it phosphorylates a large number of proteins involved in cell-cycle progression (for example, p27KIP1 and RB), DNA replication (for example, replication factors A and C), histone synthesis (for example, NPAT (nuclear protein, coactivator of histone transcription)), centrosome duplication (for example, nucleophosmin (NPM)), among other processes.[127] In vitro, CDK2 and its preferred E-type and A-type cyclin partners assemble spontaneously to form active kinase complexes. Much of the control over CDK2 involves the synthesis and availability of the cyclins, with RB and E2F regulating the abundance of CDK2, cyclin E1, and cyclin E2 transcripts and proteins.[128] CDK1 is essential for cell-cycle progression in mammals, as it is the only CDK that can initiate the M phase in the cell cycle has also been shown through mouse knock out studies.[129] Premature initiation of mitosis before completion of the S-phase results in chromosomal shattering and cell death. Multiple factors put a hold on the activity of CDK1 until DNA replication is complete and ensure that there is minimal DNA damage.[130] This integration of DNA replication and CDK1 activity is mediated by checkpoint signaling kinases such as CHK1 and WEE1 (initially discovered in yeast), which suppress the activity of CDK1 through inhibitory phosphorylation (as discussed previously), as well as through the cell division cycle 25 (CDC25) family of phosphatases. At the onset of mitosis, activation of CDK1 occurs rapidly generating a positive feedback loop that enables CDK1 to phosphorylate and inactivate WEE1. CDK1 subsequently phosphorylates multiple substrates, leading to a nuclear envelope breakdown, chromosome condensation, and mitotic spindle assembly. The subsequent progression from metaphase to anaphase is controlled by the spindle assembly checkpoints, and progression through anaphase is dependent on the attenuation of CDK1 activity through the degradation of cyclin B1 by the anaphase-promoting complex.[131]

Several CDK inhibitors have been developed as potential anticancer drugs. Numerous trials targeting several tumor types have been conducted. The first generation of CDK inhibitors was nonspecific and was referred to as "pan-CDK" inhibitors, for example, flavopiridol. However, some drugs, such as olomucine and roscovitine, have a comparatively low affinity to CDK4 and CDK6. Our understanding of CDK biology has progressed in tandem with the development of these anticancer drugs and so we have over time gained a deeper understanding of how these drugs work.[132] It was believed initially that roscovitine was a specific inhibitor of CDK1, CDK2, and CDK5, but on further study and experimentation it was found that roscovitine also inhibited transcription through the inhibition of CDK7 and CDK9.[133]

Flavopiridol is a first-generation inhibitor that is a semisynthetic flavonoid derived from rohitukine, a chromone alkaloid. It has been most extensively studied and investigated as a CDK inhibitor and has been show to inhibit CDK1, CDK2, CDK4, CDK6, CDK7, and CDK9. More than 60 clinical trials were carried out between 1998 and 2014. Flavopiridol can induce cell-cycle arrest in G1 and G2 phases in certain cases, it can also induce a cytotoxic response. However, flavopiridol did not meet the initial expectations for a CDK inhibitor, and low levels of clinical activity were seen in phase-2 studies in several solid tumor types. Later several other CDK inhibitors were developed and synthesized with the primary aim of increasing selectivity for CDK1 and CDK2 and/or increasing overall potency. Only a few of these drugs, even after showing a lot of promise initially, have progressed on to phase-1 trials.[134]

Dinaciclib is a second-generation inhibitor of CDK and it is also known as MK-7965 and SCH727965 (developed by Merck), it was developed as an potent inhibitor of CDK1, CDK2, CDK5, and CDK9 with IC50 (half inhibitory concentration value) in the range of 1–4 nM with comparatively less activity against CDK4, CDK6, and CDK7, with IC50 values 25 times greater. Dinaciclib has superior activity compared to the first-generation inhibitor Flavopiridol. It was also more successfully able to suppress the phosphorylation of RB in cell-based assays. It was able to inhibit cell-cycle progression in more than 100 cell lines and established solid tumor models in mice. As can be noted, the initial results for this drug seemed promising, which included the phase-1 studies. However, the phase-2 randomized trials could not keep up this trend of promising results and have largely been disappointing.[135]

There are various other CDK inhibitors including AT7519, which is a 3-carboxyamide compound that acts as an inhibitor of CDK1, CDK2, CDK4, CDK6, and CDK9. It has been evaluated in a phase-2 trial in combination with bortezomib. Patients in the trial had been treated previously for multiple myeloma. In the trial greater than one-third of the patients had a partial response and the combination overall was well tolerated.[136] R547 is an inhibitor of CDK1, CDK2, and CDK4 and this drug was tested in a phase-1 study of weekly intravenous injections and was reported to have manageable side effects, although no further clinical trials have been conducted.[137] The compound SnS-032 was initially thought to have greater selectivity for CDK2 in comparison to CDK1 and CDK4, but it is now recognized to target CDK7 and CDK9.[138] Two phase-1 clinical studies with SNS-032, one in 2010 in advanced lymphoid malignancies[130] and one in 2008 in selected advanced tumors,[131] have been reported, but there are no further reports of any developments regarding these compounds.[139] AZD5438 is an orally administered compound and is a potent inhibitor of CDK1, CDK2, and CDK9 with less selectivity for CDK5 and CDK6, but this compound was not further studied or administered when it was reported to be intolerable for patients with advanced solid tumours.[140] AG-024322 was a potent inhibitor of CDK1, CDK2, and CDK4 but was also discontinued in 2007 after the phase-1 study was terminated, as it failed to achieve an acceptable clinical end point.[141] As CDK4 and CDK6 inhibitors continued to be studied for their mechanisms and points of action, three expectations arose from the inhibitors of CDK4 and CDK6 in the laboratory: (1) it was expected that a pure CDK4 or CDK6 inhibitor would give a cytostatic response, i.e., G0/G1 arrest in all tumors, (2) the abovementioned effect would be the result of these inhibitors acting on or engaging with the RB to suppress gene expression and further cellular proliferation, and (3) these effects would be more prevalent in tumors that exhibit deregulated CDK4 and CDK6 activity in comparison to tumors which have other deregulated CDKs.[142] There are various studies that are being conducted using dual CDK4 and CDK6 inhibitors. One of the phase-1 studies with palbociclib monotherapy indicated promising clinical efficacy and a well-tolerated toxicity profile in patients with RB-positive advanced solid tumors and non-Hodgkin lymphoma.[143]

10 POSTTRANSLATIONAL MODIFICATION IN CADHERINS

The presence of a functional E-cadherin/catenin cell-cell adhesion complex is a prerequisite for the normal development and maintenance of epithelial structures. One of the hallmarks of cancer cells is the change in cell to cell adhesion properties in EMT to MET transition. Cadherins are type-1 transmembrane glycoproteins, which play a critical role in cell adhesion. Phosphorylation is essential in maintaining the functional integrity of the cadherin-catanin complex.[144] The dysregulation of cadherin is strongly associated with cell-adhesion defects in carcinomas. E-cadherin is phosphorylated by the casein kinase ii and glycognin synthase kinase-3β within the cytoplasmic domain serine cluster. It has been reported that in transfected mouse. Lickert and co-workers found that phosphorylation of E-cadherin by protein kinase D1 play a pivotal role in regulating the cell shape and tumor cell invasion. The upregulation and overexpression of PKD1 and increased kinase activity shows a drastic increase in cell aggregation and a marked decrease in cell motility. In a study of green tea catechins antiangiogenic activity was shown by decreasing the phosphorylation of VE-cadherin tyrosine and it also inhibits AKT activation.[145]

CD44 is a membrane receptor involved in matrix-dependent cell motility and migration, the primary ligand of CD44 is hyaluronic acid. CD44 is a cell-surface molecule with various functions, which may include cell proliferation, differentiation, migration, and signaling. It has 20 splice variants that have been associated with tumor development and progression.[146] Changes in the glycosylation state of CD44 can markedly influence hyaluronic acid ligand recognition and binding, thus modifying cancer cell signaling. The modification of glycosylation in CD44 leads to enhanced cell motility and carcinogenicity in rat carcinoma cells. N-cadherins have been implicated in blood vessel formation in cancer and its antagonists like ADH1 are being used to prevent cancer progression in primary as well as secondary sites.[147]

11 PRENYLATION AND G-PROTEIN COUPLED RECEPTORS

Heterotrimeric G-protein signaling is versatile and is used for a variety of signal transductions in eukaryotes, such as those involved in the pathways controlling sight, neurotransmission, and cell migration. They consist of an alpha (α), beta (β), and gamma (γ) subunit, these are heterotrimeric membrane-bound G-proteins. During heterotrimeric G-protein signal transduction, ligand binds a G-protein-coupled receptor (GPCR) resulting in the activation of the Gα subunit to dissociate its bound GDP (guanosine diphosphate) and bind the more highly concentrated free cytosolic GTP.[148] The activated Gα subunit dissociates from the Gβγ dimer allowing both components to influence separate downstream effectors such as adenylyl cyclase phospholipase Cβphosphatidylinositol 3-kinase,[78] β-adrenergic receptor kinase, and ion channels.

GPCR proteins have a carboxy terminal pattern CaaX found in Gγ subunits. Prenylation results in the increase in the hydrophobicity of the protein and inserts it into cell membranes with the association of other hydrophobic proteins. The translocation of G-protein into the cell membrane is done by prenylation of Gγ subunits and palmitoylation of Gα subunits. Prenylation of Gγ subunit is also important for Gβγ dimer because it interacts with a Gα subunit receptor (rhodopsin, A1 adenosine receptor) and effectors (phosducin (PDC), phospholipaseCβ (PLCβ and PLCβ2), adenylyl cyclase) efficiently.[149] GPCR is not only one of the major pharmaceutical targets, but also has a surprising clinical target in cancer treatment, which includes only a few drugs that can act upon GPCR-mediated signaling. Testosterone production via the signaling cascade begins with the secretion of gonadotropin-releasing hormone (GnRH) from the hypothalamus, which results in the progression and growth of prostate cancer cells. Luteinizing hormone (LH) and follicle-stimulating hormone (FSH) are two GPCR agonists from the pituitary gland that are synthesized and secreted by GnRH. This results in steroidogenesis, which is induced in the adrenal glands and testes. Testosterone is next released and reaches the prostate where it stimulates cancer cell growth. Testosterone levels can be reduced via molecules acting on GnRH. Octreotide and pegvisomant are two other GPCRs ligands prescribed for cancer treatment. The pituitary gland secretes growth hormone (GH) and insulin-like growth factor1 (IGF-1) by inhibition of synthetic somatostatin (SST) agonist. GH antagonists and SST agonists are highly effective antiproliferative drugs. Somatotroph adenomas associated with acromegaly are a result of the overproduction of GH by the pituitary gland, which can be prevented by octreotide. In addition, it can be utilized to prevent the effects of SST secreting malignant gastroenteropatic neuroendocrine tumors. The actions of a natural ligand can resemble cyclohexapeptide, an analog of octreotide (pasireotide) that binds to a greater number of SST receptor isotypes. If IGF-1 levels failed to normalize other treatment, then pegvisomant (pegylated peptide) acting on GH antagonist may be considered as another supplementary treatment. Both approaches described above act indirectly to inhibit cell growth or to prevent the secondary effects caused by peptides released from the tumor. However, there is a wealth of opportunities for directly targeting GPCRs expressed on tumor cells.[150]

12 SUMMARY AND FUTURE PERSPECTIVES

It is clearly understood that a large number of PTMs play a major role in regulating gene expression, cell cycle, cell proliferation, and apoptosis. Some proteins are limited to one or fewer modifications while other proteins are employed in multiple different modifications. Importantly, there exists numerous common modifications that target a large set of different proteins (e.g., phosphorylation or acetylation). It would therefore be justified to look for a common strategy to target such phosphorylation or acetylation so that multiple biological pathways could be influenced. Interestingly, prenylation turns out to be a unique strategy to modify different receptors of the GPCR superfamily indicating that one may target prenylation, which will ultimately affect most of the GPCR receptors connected to cancer. In the future, research should try to explore the more abundant and unique PTMs associated with specific types of cancer (different tissues) or in different stages of cancer. Nevertheless, identifying PTMs as drug targets or tumor biomarkers for specific cancers will remain a very exciting and expanding field of research.

References

1. Harmel R, Fiedler D. Features and regulation of non-enzymatic post-translational modifications. *Nat Chem Biol* 2018;**14**:244–52.
2. Gillery P, Jaisson S, Gorisse L, Pietrement C. Rôle de la carbamylation des protéines dans les complications de l'insuffisance rénale chronique. *Néphrol Thér* 2015;**11**:129–34.
3. Šimić G, et al. Tau protein hyperphosphorylation and aggregation in Alzheimer's disease and other tauopathies, and possible neuroprotective strategies. *Biomolecules* 2016;**6**:6.
4. Negre-Salvayre A, Salvayre R, Augé N, Pamplona R, Portero-Otín M. Hyperglycemia and glycation in diabetic complications. *Antioxid Redox Signal* 2009;**11**:3071–109.
5. Cooper J, Maupin K, Merrill N. *Origins of cancer symposium 2015: posttranslational modifications and cancer*. *Genes Cancer* 2015;**6**:7–8. www.impactjournals.com/Genes&Cancer.
6. Hanahan D, Weinberg RA. Hallmarks of cancer: the next generation. *Cell* 2011;**144**:646–74.
7. Rossetto D, Avvakumov N, Côté J. Histone phosphorylation: a chromatin modification involved in diverse nuclear events. *Epigenetics* 2012;**7**(10):1098–108.
8. Zhen L, Gui-lan L, Ping Y, Jin H, Ya-li W. The expression of H3K9Ac, H3K14Ac, and H4K20TriMe in epithelial ovarian tumors and the clinical significance. *Int J Gynecol Cancer* 2010;**20**:82–6.
9. Wang M, Casey PJ. Protein prenylation: unique fats make their mark on biology. *Nat Rev Mol Cell Biol* 2016;**17**:110–22.
10. Reis CA, Osorio H, Silva L, Gomes C, David L. Alterations in glycosylation as biomarkers for cancer detection. *J Clin Pathol* 2010;**63**:322–9.
11. Song Y, Wu F, Wu J. Targeting histone methylation for cancer therapy: enzymes, inhibitors, biological activity and perspectives. *J Hematol Oncol* 2016;**9**(1):49.
12. Brooks CL, Gu W. p53 ubiquitination: Mdm2 and beyond. *Mol Cell* 2006;**21**:307–15.
13. Krueger KE, Srivastava S. Posttranslational protein modifications: current implications for cancer detection, prevention, and therapeutics. *Mol Cell Proteomics* 2006;**5**:1799–810.
14. Singh V, et al. Phosphorylation: implications in cancer. *Protein J* 2017;**36**:1–6.
15. Ozaki T, Nakagawara A. Role of p53 in cell death and human cancers. *Cancers (Basel)* 2011;**3**:994–1013.
16. Meek DW. The p53 response to DNA damage. *DNA Repair (Amst)* 2004;**3**:1049–56.
17. Teufel DP, Bycroft M, Fersht AR. Regulation by phosphorylation of the relative affinities of the N-terminal transactivation domains of p53 for p300 domains and Mdm2. *Oncogene* 2009;**28**:2112–8.
18. Kruse J-P, Gu W. Modes of p53 regulation. *Cell* 2009;**137**:609–22.
19. Brooks CL, Gu W. New insights into p53 activation. *Cell Res* 2010;**2053**:614–21.
20. Tang Y, Zhao W, Chen Y, Zhao Y, Gu W. Acetylation is indispensable for p53 activation. *Cell* 2008;**133**:612–26.
21. Krummel KA, Lee CJ, Toledo F, Wahl GM. The C-terminal lysines fine-tune P53 stress responses in a mouse model but are not required for stability control or transactivation. *Proc Natl Acad Sci U S A* 2005;**102**:10188–93.
22. Krummel KA, Lee CJ, Toledo F, Wahl GM. The C-terminal lysines fine-tune P53 stress responses in a mouse model but are not required for stability control or transactivation. *Proc Natl Acad Sci U S A* 2005;**102**:10188–93.
23. Wesierska-Gadek J, Bugajska-Schretter A, Cerni C. ADP-ribosylation of p53 tumor suppressor protein: mutant but not wild-type p53 is modified. *J Cell Biochem* 1996;**62**:90–101.
24. Nevins JR. The Rb/E2F pathway and cancer. *Hum Mol Genet* 2001;**10**:699–703.
25. Classon M, Harlow E. The retinoblastoma tumour suppressor in development and cancer. *Nat Rev Cancer* 2002;**2**:910–7.

26. Knudsen E, Wang JY. Differential regulation of retinoblastoma protein function by specific Cdk phosphorylation sites. *J Biol Chem* 1996;**271**:8313–20.
27. Ezhevsky SA, Ho A, Becker-Hapak M, Davis PK, Dowdy SF. Differential regulation of retinoblastoma tumor suppressor protein by G1 cyclin-dependent kinase complexes in vivo. *Mol Cell Biol* 2001;**21**(14):4773–84.
28. Romano G, Giordano A. Role of the cyclin-dependent kinase 9-related pathway in mammalian gene expression and human diseases. *Cell Cycle* 2008;**7**(23):3664–8.
29. Bonda DJ, Lee HP, Kudo W, Zhu X, Smith MA, Lee HG. Pathological implications of cell cycle re-entry in Alzheimer disease. *Expert Rev Mol Med* 2010;**12**.
30. Ren S, Rollins BJ. Cyclin C/cdk3 promotes Rb-dependent G0 exit. *Cell* 2004;**117**(2):239–51.
31. Lents NH, Gorges LL, Baldassare JJ. Reverse mutational analysis reveals threonine-373 as a potentially sufficient phosphorylation site for inactivation of the retinoblastoma tumor suppressor protein (pRB). *Cell Cycle* 2006;**5**:1699–707.
32. Faust D, Schmitt C, Oesch F, Oesch-Bartlomowicz B, Schreck I, Weiss C, Dietrich C. Differential p38-dependent signalling in response to cellular stress and mitogenic stimulation in fibroblasts. *Cell Commun Signal* 2012;**10**(1):6.
33. Sørensen CS, Syljuåsen RG. Safeguarding genome integrity: the checkpoint kinases ATR, CHK1 and WEE1 restrain CDK activity during normal DNA replication. *Nucleic Acids Res* 2011;**40**(2):477–86.
34. Bakhoum SF, Compton DA. Chromosomal instability and cancer: a complex relationship with therapeutic potential. *J Clin Invest* 2012;**122**(4):1138–43.
35. Deininger MW, et al. BCR-ABL tyrosine kinase activity regulates the expression of multiple genes implicated in the pathogenesis of chronic myeloid leukemia. *Cancer Res* 2000;**60**:2049–55.
36. Chan HM, La Thangue NB. p300/CBP proteins: HATs for transcriptional bridges and scaffolds. *J Cell Sci* 2001;**114**(13):2363–73.
37. Knight JS, Sharma N, Robertson ES. Epstein-Barr virus latent antigen 3C can mediate the degradation of the retinoblastoma protein through an SCF cellular ubiquitin ligase. *Proc Natl Acad Sci U S A* 2005;**102**:18562–6.
38. Gu T-L, et al. NPM-ALK fusion kinase of anaplastic large-cell lymphoma regulates survival and proliferative signaling through modulation of FOXO3a. *Blood* 2004;**103**:4622–9.
39. Carr SM, Munro S, Kessler B, Oppermann U, La Thangue NB. Interplay between lysine methylation and Cdk phosphorylation in growth control by the retinoblastoma protein. *EMBO J* 2011;**30**:317–27.
40. Knight JS, Sharma N, Robertson ES. Epstein-Barr virus latent antigen 3C can mediate the degradation of the retinoblastoma protein through an SCF cellular ubiquitin ligase. *Proc Natl Acad Sci U S A* 2005;**102**:18562–6.
41. Hornbeck PV, Kornhauser JM, Tkachev S, Zhang B, Skrzypek E, Murray B, Latham V, Sullivan M. PhosphoSitePlus: a comprehensive resource for investigating the structure and function of experimentally determined post-translational modifications in man and mouse. *Nucleic Acids Res* 2011;**40**:261–70.
42. Ledl A, Schmidt D, Müller S. Viral oncoproteins E1A and E7 and cellular LxCxE proteins repress SUMO modification of the retinoblastoma tumor suppressor. *Oncogene* 2005;**24**:3810–8.
43. Preeminent, I. S. From www.bloodjournal.org by guest on March 18, 2015. For personal use only. Nonthrombogenic 2015;**1971**:509–524.
44. Oeckinghaus A, Ghosh S. The NF-κB family of transcription factors and its regulation. *Cold Spring Harb Perspect Biol* 2009;a000034.
45. Karin M, Cao Y, Greten FR, Li Z-W. NF-κB in cancer: from innocent bystander to major culprit. *Nat Rev Cancer* 2002;**2**:301–10.
46. Gilmore TD, Kalaitzidis D, Liang M-C, Starczynowski DT. The c-Rel transcription factor and B-cell proliferation: a deal with the devil. *Oncogene* 2004;**23**:2275–86.

47. Neri A, et al. B cell lymphoma-associated chromosomal translocation involves candidate oncogene lyt-10, homologous to NF-kappa B p50. *Cell* 1991;**67**:1075–87.
48. Bredel M, et al. *NFKBIA* deletion in glioblastomas. *N Engl J Med* 2011;**364**:627–37.
49. Ahmed KM, Li JJ. ATM-NF-kappaB connection as a target for tumor radiosensitization. *Curr Cancer Drug Targets* 2007;**7**:335–42.
50. García MG, et al. Inhibition of NF-κB activity by BAY 11-7082 increases apoptosis in multidrug resistant leukemic T-cell lines. *Leuk Res* 2005;**29**:1425–34.
51. Hideshima T, et al. Biologic sequelae of I B kinase (IKK) inhibition in multiple myeloma: therapeutic implications. *Blood* 2009;**113**:5228–36.
52. Yang J, Amiri KI, Burke JR, Schmid JA, Richmond A. BMS-345541 targets inhibitor of B kinase and induces apoptosis in melanoma: involvement of nuclear factor B and mitochondria pathways. *Clin Cancer Res* 2006;**12**:950–60.
53. Choo M-K, Sakurai H, Kim D-H, Saiki I. A ginseng saponin metabolite suppresses tumor necrosis factor-alpha-promoted metastasis by suppressing nuclear factor-kappaB signaling in murine colon cancer cells. *Oncol Rep* 2008;**19**:595–600.
54. Ravaud A, et al. Phase I study and pharmacokinetic of CHS-828, a guanidino-containing compound, administered orally as a single dose every 3weeks in solid tumours: an ECSG/EORTC study. *Eur J Cancer* 2005;**41**:702–7.
55. Gasparian AV, et al. Targeting transcription factor NFkappaB: comparative analysis of proteasome and IKK inhibitors. *Cell Cycle* 2009;**8**:1559–66.
56. Shen H-M, Tergaonkar V. NFkappaB signaling in carcinogenesis and as a potential molecular target for cancer therapy. *Apoptosis* 2009;**14**:348–63.
57. Mabuchi S, et al. Inhibition of NFκB increases the efficacy of cisplatin in *in vitro* and *in vivo* ovarian cancer models. *J Biol Chem* 2004;**279**:23477–85.
58. Sethi G, Sung B, Aggarwal BB. Nuclear factor-kappaB activation: from bench to bedside. *Exp Biol Med (Maywood)* 2008;**233**:21–31.
59. Lin Y, Bai L, Chen W, Xu S. The NF-kappaB activation pathways, emerging molecular targets for cancer prevention and therapy. *Expert Opin Ther Targets* 2010;**14**:45–55.
60. Tas SW, Vervoordeldonk MJBM, Tak PP. Gene therapy targeting nuclear factor-kappaB: towards clinical application in inflammatory diseases and cancer. *Curr Gene Ther* 2009;**9**:160–70.
61. Luger K, Mäder AW, Richmond RK, Sargent DF, Richmond TJ. Crystal structure of the nucleosome core particle at 2.8 Å resolution. *Nature* 1997;**389**:251–60.
62. Jenuwein T, Allis CD. Translating the histone code. *Science (80-)* 2001;**293**:1074–80.
63. Füllgrabe J, Kavanagh E, Joseph B. Histone onco-modifications. *Oncogene* 2011;**30**:3391–403.
64. Cosgrove MS, Boeke JD, Wolberger C. Regulated nucleosome mobility and the histone code. *Nat Struct Mol Biol* 2004;**11**:1037–43.
65. Füllgrabe J, Hajji N, Joseph B. Cracking the death code: apoptosis-related histone modifications. *Cell Death Differ* 2010;**17**:1238–43.
66. Hajji N, et al. Opposing effects of hMOF and SIRT1 on H4K16 acetylation and the sensitivity to the topoisomerase II inhibitor etoposide. *Oncogene* 2010;**29**:2192–204.
67. Schotta G, et al. A silencing pathway to induce H3-K9 and H4-K20 trimethylation at constitutive heterochromatin. *Genes Dev* 2004;**18**:1251–62.
68. Henckel A, et al. Histone methylation is mechanistically linked to DNA methylation at imprinting control regions in mammals. *Hum Mol Genet* 2009;**18**:3375–83.
69. Tryndyak VP, Kovalchuk O, Pogribny IP. Loss of DNA methylation and histone H4 lysine 20 trimethylation in human breast cancer cells is associated with aberrant expression of DNA methyltransferase 1, Suv4-20h2 histone methyltransferase and methyl-binding proteins. *Cancer Biol Ther* 2006;**5**:65–70.
70. Shi Y, et al. Histone demethylation mediated by the nuclear amine oxidase homolog LSD1. *Cell* 2004;**119**:941–53.

71. Seligson DB, et al. Global histone modification patterns predict risk of prostate cancer recurrence. *Nature* 2005;**435**:1262–6.
72. Barlési F, et al. Global histone modifications predict prognosis of resected non-small-cell lung cancer. *J Clin Oncol* 2007;**25**:4358–64.
73. Mikkelsen TS, et al. Genome-wide maps of chromatin state in pluripotent and lineage-committed cells. *Nature* 2007;**448**:553–60.
74. Tzao C, et al. Prognostic significance of global histone modifications in resected squamous cell carcinoma of the esophagus. *Mod Pathol* 2009;**22**:252–60.
75. Seligson DB, et al. Global levels of histone modifications predict prognosis in different cancers. *Am J Pathol* 2009;**174**:1619–28.
76. van Leeuwen F, Gafken PR, Gottschling DE. Dot1p modulates silencing in yeast by methylation of the nucleosome core. *Cell* 2002;**109**:745–56.
77. Liedtke M, Cleary ML. Therapeutic targeting of MLL. *Blood* 2009;**113**:6061–8.
78. Stephens L, Smrcka A, Cooke FT, Jackson TR, Sternweis PC, Hawkins PT. A novel phosphoinositide 3 kinase activity in myeloid-derived cells is activated by G protein βγ subunits. *Cell* 1994;**77**(1):83–93.
79. Mohammad HP, et al. A DNA Hypomethylation signature predicts antitumor activity of LSD1 inhibitors in SCLC. *Cancer Cell* 2015;**28**:57–69.
80. Blume-Jensen P, Hunter T. Oncogenic kinase signalling. *Nature* 2001;**411**:355–65.
81. Wagner JP, Wolf-Yadlin A, Sevecka M, Grenier JK, Root DE, Lauffenburger DA, MacBeath G. Receptor tyrosine kinases fall into distinct classes based on their inferred signaling networks. *Sci Signal* 2013;**6**(284):ra58.
82. Nishikawa R, et al. A mutant epidermal growth factor receptor common in human glioma confers enhanced tumorigenicity. *Proc Natl Acad Sci U S A* 1994;**91**:7727–31.
83. Intini D, et al. Analysis of FGFR3 gene mutations in multiple myeloma patients with t (4;14). *Br J Haematol* 2001;**114**:362–4.
84. Qureshi R, et al. Mutation analysis of EGFR and its correlation with the HPV in Indian cervical cancer patients. *Tumor Biol* 2016;**37**:9089–98.
85. Million RP, Aster J, Gilliland DG, Van Etten RA. The Tel-Abl (ETV6-Abl) tyrosine kinase, product of complex (9;12) translocations in human leukemia, induces distinct myeloproliferative disease in mice. *Blood* 2002;**99**:4568–77.
86. Renne C, Willenbrock K, Küppers R, Hansmann M-L, Bräuninger A. Autocrine- and paracrine-activated receptor tyrosine kinases in classic Hodgkin lymphoma. *Blood* 2005;**105**:4051–9.
87. Paul MK, Mukhopadhyay AK. Tyrosine kinase—role and significance in Cancer. *Int J Med Sci* 2004;**1**:101–15.
88. Zwick E, Bange J, Ullrich A. Receptor tyrosine kinases as targets for anticancer drugs. *Trends Mol Med* 2002;**8**:17–23.
89. Li L, et al. The Ras/Raf/MEK/ERK signaling pathway and its role in the occurrence and development of HCC. *Oncol Lett* 2016;**12**:3045–50.
90. Xu C, et al. The diversification of evolutionarily conserved MAPK cascades correlates with the evolution of fungal species and development of lifestyles. *Genome Biol Evol* 2017;**9**:311–22.
91. Park J-I. Growth arrest signaling of the Raf/MEK/ERK pathway in cancer. *Front Biol (Beijing)* 2014;**9**:95–103.
92. Kohno M, Pouyssegur J. Targeting the ERK signaling pathway in cancer therapy. *Ann Med* 2006;**38**:200–11.
93. Mebratu Y, Tesfaigzi Y. How ERK1/2 activation controls cell proliferation and cell death: is subcellular localization the answer? *Cell Cycle* 2009;**8**:1168–75.
94. Houben R, et al. Constitutive activation of the Ras-Raf signaling pathway in metastatic melanoma is associated with poor prognosis. *J Carcinog* 2004;**3**:6.

95. Marampon F, Ciccarelli C, Zani BM. Down-regulation of c-Myc following MEK/ERK inhibition halts the expression of malignant phenotype in rhabdomyosarcoma and in non muscle-derived human tumors. *Mol Cancer* 2006;**5**:31.
96. Chang Y-S, et al. Detection of N-, H-, and KRAS codons 12, 13, and 61 mutations with universal RAS primer multiplex PCR and N-, H-, and KRAS-specific primer extension. *Clin Biochem* 2010;**43**:296–301.
97. Lu S, Jang H, Nussinov R, Zhang J. The structural basis of oncogenic mutations G12, G13 and Q61 in small GTPase K-Ras4B. *Sci Rep* 2016;**6**(21949).
98. Li S, De Souza P. Ras isoprenylation and pAkt inhibition by zoledronic acid and fluvastatin enhances paclitaxel activity in T24 bladder cancer cells. *Cancers (Basel)* 2011;**3**:662–74.
99. Leicht DT, et al. Raf kinases: function, regulation and role in human cancer. *Biochim Biophys Acta* 2007;**1773**:1196–212.
100. Matallanas D, et al. Raf family kinases: old dogs have learned new tricks. *Genes Cancer* 2011;**2**:232–60.
101. Zang M, Gong J, Luo L, Zhou J, Xiang X, Huang W, Huang Q, Luo X, Olbrot M, Peng Y, Chen C. Characterization of Ser338 phosphorylation for Raf-1 activation. *J Biol Chem* 2008;**283**(46):31429–37.
102. Sullivan RJ, Flaherty KT. BRAF in melanoma: pathogenesis, diagnosis, inhibition, and resistance. *J Skin Cancer* 2011;**2011**.
103. Ascierto PA, et al. The role of BRAF V600 mutation in melanoma. *J Transl Med* 2012;**10**:85.
104. Ziai J, Hui P. *BRAF* mutation testing in clinical practice. *Expert Rev Mol Diagn* 2012;**12**:127–38.
105. O'Neill E, Kolch W. Conferring specificity on the ubiquitous Raf/MEK signalling pathway. *Br J Cancer* 2004;**90**:283–8.
106. Ramakrishnan V, et al. Sorafenib, a dual Raf kinase/vascular endothelial growth factor receptor inhibitor has significant anti-myeloma activity and synergizes with common anti-myeloma drugs. *Oncogene* 2010;**29**:1190–202.
107. Allen LF, Sebolt-Leopold J, Meyer MB. CI-1040 (PD184352), a targeted signal transduction inhibitor of MEK (MAPKK). *Semin Oncol* 2003;**30**:105–16.
108. Frank DA. Transcription factor STAT3 as a prognostic marker and therapeutic target in cancer. *J Clin Oncol* 2013;**31**:4560–1.
109. Moustakas A, Rosler KM, Harrison DA. Smad signalling network. *J Cell Sci* 2002;**115**:3355–6.
110. Bromberg J. Stat proteins and oncogenesis. *J Clin Invest* 2002;**109**:1139–42.
111. Barré B, Vigneron A, Coqueret O. The STAT3 transcription factor is a target for the Myc and Rb proteins on the Cdc25A promoter. *J Biol Chem* 2005;.
112. Scott LM, Rebel VI. JAK2 and genomic instability in the myeloproliferative neoplasms: a case of the chicken or the egg? *Am J Hematol* 2012;**87**:1028–36.
113. Chakraborty S, et al. Activation of Jak2 in patients with blast crisis chronic myelogenous leukemia: inhibition of Jak2 inactivates Lyn kinase. *Blood Cancer J* 2013;**3**:e142.
114. Lee J-M, et al. PCM1—JAK2 fusion in a patient with acute myeloid leukemia. *Ann Lab Med* 2018;**38**:492.
115. Van Vlierberghe P, Ferrando A. The molecular basis of T cell acute lymphoblastic leukemia. *J Clin Invest* 2012;**122**:3398–406.
116. Choudhary MM, France TJ, Teknos TN, Kumar P. Interleukin-6 role in head and neck squamous cell carcinoma progression. *World J Otorhinolaryngol Head Neck Surg* 2016;**2**:90–7.
117. Kumar J, et al. Granulocyte colony-stimulating factor receptor signalling via Janus kinase 2/signal transducer and activator of transcription 3 in ovarian cancer. *Br J Cancer* 2014;**110**:133–45.
118. Koskela HLM, et al. Somatic STAT3 mutations in large granular lymphocytic leukemia. *N Engl J Med* 2012;**366**:1905–13.

119. Huang L, et al. Transcriptional repression of SOCS3 mediated by IL-6/STAT3 signaling via DNMT1 promotes pancreatic cancer growth and metastasis. *J Exp Clin Cancer Res* 2016;**35**(27).
120. Johnston PA, Grandis JR. STAT3 signaling: anticancer strategies and challenges. *Mol Interv* 2011;**11**:18–26.
121. De Vivo M, et al. Cyclin-dependent kinases: bridging their structure and function through computations. *Future Med Chem* 2011;**3**:1551–9.
122. Noble MEM, Endicott JA. Cyclin-dependent kinases. In: *Encyclopedia of Genetics*; 2001. p. 500–6.
123. Muramatsu S, Hirai K, Tak Y-S, Kamimura Y, Araki H. CDK-dependent complex formation between replication proteins Dpb11, Sld2, Pol ε, and GINS in budding yeast. *Genes Dev* 2010;**24**:602–12.
124. Morgan DO. SnapShot: cell-cycle regulators II. *Cell* 2008;**135**.
125. Harris AL, et al. Targeting the cyclin dependent kinase and retinoblastoma axis overcomes standard of care resistance in BRAF V600-mutant melanoma. *Oncotarget* 2018;**9**:10905–19.
126. Romagosa C, et al. p16Ink4a overexpression in cancer: a tumor suppressor gene associated with senescence and high-grade tumors. *Oncogene* 2011;**30**:2087–97.
127. Tanudji M, Xiao LR, Schebye M, L'Italien L. Unmasking the redundancy between Cdk1 and Cdk2 at G 2 phase in human cancer cell lines. *Cell Cycle* 2006;**5**.
128. Cardoso MC, Leonhardt H, Nadal-Ginard B. Reversal of terminal differentiation and control of DNA replication: cyclin A and cdk2 specifically localize at subnuclear sites of DNA replication. *Cell* 1993;**74**:979–92.
129. Sherr CJ, Roberts JM. CDK inhibitors: positive and negative regulators of G1-phase progression. *Genes Dev* 1999;**13**:1501–12.
130. Satyanarayana A, Hilton MB, Kaldis P. p21 inhibits Cdk1 in the absence of Cdk2 to maintain the G1/S phase DNA damage checkpoint. *Mol Biol Cell* 2008;**19**:65–77.
131. Enserink JM, Kolodner RD. An overview of Cdk1-controlled targets and processes. *Cell Div* 2010;**5**:11.
132. Sarosiek T. Inhibitors of cyclin-dependent kinases (CDK)—a new group of medicines in therapy of advanced breast cancer. *Pol Merkur Lekarski* 2018;**44**:5–9.
133. Cicenas J, Kalyan K, Sorokinas A, Stankunas E, Levy J, Meskinyte I, Stankevicius V, Kaupinis A, Valius M. Roscovitine in cancer and other diseases. *Ann Transl Med* 2015;**3**(10).
134. Tan AR, Swain SM. Review of flavopiridol, a cyclin-dependent kinase inhibitor, as breast cancer therapy. *Semin Oncol* 2002;**29**:77–85.
135. Nemunaitis JJ, Small KA, Kirschmeier P, Zhang D, Zhu Y, Jou YM, Statkevich P, Yao SL, Bannerji R. A first-in-human, phase 1, dose-escalation study of dinaciclib, a novel cyclin-dependent kinase inhibitor, administered weekly in subjects with advanced malignancies. *J Transl Med* 2013;**11**(1):259.
136. Santo L, et al. AT7519, a novel small molecule multi-cyclin-dependent kinase inhibitor, induces apoptosis in multiple myeloma via GSK-3β activation and RNA polymerase II inhibition. *Oncogene* 2010;**29**:2325–36.
137. DePinto W, et al. In vitro and in vivo activity of R547: a potent and selective cyclin-dependent kinase inhibitor currently in phase I clinical trials. *Mol Cancer Ther* 2006;**5**:2644–58.
138. Chen R, et al. Mechanism of action of SNS-032, a novel cyclin-dependent kinase inhibitor, in chronic lymphocytic leukemia. *Blood* 2009;**113**:4637–45.
139. Conroy A, et al. SNS-032 is a potent and selective CDK 2, 7 and 9 inhibitor that drives target modulation in patient samples. *Cancer Chemother Pharmacol* 2009;**64**:723–32.
140. Byth KF, et al. AZD5438, a potent oral inhibitor of cyclin-dependent kinases 1, 2, and 9, leads to pharmacodynamic changes and potent antitumor effects in human tumor xenografts. *Mol Cancer Ther* 2009;**8**:1856–66.

141. Zhong W-Z, Lalovic B, Zhan J. Characterization of in vitro and in vivo metabolism of AG-024322, a novel cyclin-dependent kinase (CDK). *Health (Irvine Calif)* 2009;**01**:249–62.
142. Sherr CJ, Beach D, Shapiro GI. Targeting CDK4 and CDK6: from discovery to therapy. *Cancer Discov* 2016;**6**:353–67.
143. Malorni L, Curigliano G, Minisini AM, Cinieri S, Tondini CA, D'Hollander K, Arpino G, Bernardo A, Martignetti A, Criscitiello C, Puglisi F. Palbociclib as single agent or in combination with the endocrine therapy received before disease progression for estrogen receptor-positive, HER2-negative metastatic breast cancer: TREnd trial. *Ann Oncol* 2018;.
144. Ivanov DB, Philippova MP, Tkachuk VA. Structure and functions of classical cadherins. *Biochemistry (Mosc)* 2001;**66**:1174–86.
145. Jeanes A, Gottardi CJ, Yap AS. Cadherins and cancer: how does cadherin dysfunction promote tumor progression? *Oncogene* 2008;**27**:6920–9.
146. da Cunha CB, et al. De novo expression of CD44 variants in sporadic and hereditary gastric cancer. *Lab Investig* 2010;**90**:1604–14.
147. Blaschuk OW. N-cadherin antagonists as oncology therapeutics. *Philos Trans R Soc Lond Ser B Biol Sci* 2015;**370**:20140039.
148. Federman AD, Conklin BR, Schrader KA, Reed RR, Bourne HR. Hormonal stimulation of adenylyl cyclase through Gi-protein beta gamma subunits. *Nature* 1992;**356**:159–61.
149. Nogués L, et al. G protein-coupled receptor kinases (GRKs) in tumorigenesis and cancer progression: GPCR regulators and signaling hubs. *Semin Cancer Biol* 2018;**48**:78–90.
150. Innamorati G, et al. Molecular approaches to target GPCRs in cancer therapy. *Pharmaceuticals* 2011;**4**:567–89.

CHAPTER

10

Nonenzymatic Posttranslational Protein Modifications: Mechanism and Associated Disease Pathologies

Sheeza Khan, Ajaz A. Bhat*[†]

*School of Life Science, B. S. Abdur Rahman Crescent Institute of Science and Technology, Chennai, India

[†]Division of Translational Medicine, Research Branch, Sidra Medicine, Doha, Qatar

1 INTRODUCTION

Protein posttranslational modification (PTM) is a very common phenomenon in eukaryotic organisms. These modifications of proteins at the posttranslational level may involve chemical alterations of proteins that may be either reversible or irreversible. In addition to the limited proteolysis, such modifications encompass specific covalent addition of lipids (palmitic acid, etc.), chemical groups (phosphoryl, etc.), carbohydrates (glucose, etc.), or sometimes full length proteins (e.g., ubiquitin) to the side chains of amino acids.[1] PTM, in exceptional cases (e.g., hydroxylation), occurs at protein side chains that may act as weak (N, Q) or strong (S, C, T, M, Y, H, E, R, K, D) nucleophiles. The P, F, G, L, A, I, V, W residue side chains undergo covalent modifications. These modifications at the posttranslational level routinely affect the function of proteins (via bringing changes in the dynamics and structure of proteins). It is frequently possible that the residue that has been modified acts as a binding region for other proteins. For instance, phosphotyrosines are binding regions for SH2 domains[2] and bromo domains are targets for acetyllysines.[3] The modification at the posttranslational level has an important biological

https://doi.org/10.1016/B978-0-12-811913-6.00010-2

role in regulating expression of genes, conferring stability to the proteins or marking them for degradation, activating or deactivating enzymatic activity, and in mediating interactions among proteins, etc.[1]

Strikingly, a huge fraction of these PTMs have been found to be catalyzed by modifying enzymes. Five percent of the genes in humans are considered to be modifying enzymes.[1] It is known that 518 kinases and more than 15 phosphatases exist in the human genome.[4] These enzymes are universally found in all kingdoms of life (eukaryotes). In *Arabidopsis thaliana*, 300 phosphatase and 1019 kinase coding genes are present. The genome of yeast is also known to code for 119 different kinases.[5] In similar fashion, the genome of humans is known to code for approximately 80 deubiquitinases, in addition to 600 E3 ubiquitinating ligases.[6] Though an increase in the recognition of PTMs' importance has occurred over the years, still a complete understanding of the functional aspects of PTMs as well as their common occurrences are not known. PTMs are usually associated with turnover of proteins. However, a few may have pathological consequences whose persistence may lead to disorder.[7]

In addition to PTMs, which are catalyzed by enzymes, there are many nonenzymatic PTMs, which lead to changes in the structure and function domain of proteins in living systems. These include carbamylation, carbonylation, nitrosylation, glycation, and homocysteinylation.[8] These types of modifications comprise of a small metabolite binding to free reactive groups present on proteins, which results in molecular rearrangements in the protein. Such modifications are cumulative and irreversible in nature.[9] The consequences of PTMs due to enzymatic mechanisms have been reviewed in detail.[10–12] This chapter is dedicated to precisely reviewing mechanisms of almost all of the PTMs due to nonenzymatic mechanisms. We have also extensively discussed the associated disease physiology due to the modifications.

2 CARBAMYLATION

These reactions are a type of PTM that are nonenzymatic and are capable of changing both the structural properties as well as function of particular proteins. Friedrich Wöhler, in 1828, found that by making cyanate react with ammonia, urea could be synthesized. Later, in 1895, it was discovered that, in a physiological environment, urea dissociated, though slowly, into cyanate and its tautomeric form, isocyanate.[13, 14] A tremendous electrophile, isocyanate, immediately undergoes reaction with the nucleophile groups; for instance, amines (primary) and sulfhydryl groups (free). It was in 1949 that F. Schutz proposed that cyanate derived from urea reacts with the free sulfhydryl and primary amine groups of proteins, as well as of the amino acids present freely in solution.[15] George Stark proved this proposition true in 1960 with his observation of ribonuclease incubated with concentrated urea. He found that an altered and less positively charged form of the protein accumulated after urea treatment. The experiments that followed this finding demonstrated that lysine residues present in the three dimensional structure of polypeptide were modified in an irreversible manner to form N(6)-carbamoyl-L-lysine (homocitrulline).[16] Stark and co-workers showed that cyanate has the capability to cause modifications of primary amines that are of irreversible nature and also could modify hydroxyls, thiols, imidazole, and phenols, groups which were reversible.[16–20] All these reactions taken together came to be known as carbamylation—the insertion of a "carbamoyl" moiety (—$CONH_2$) to a functional group.

2.1 Mechanism

Carbamylation, previously known as carbamoylation, occurs by the covalent adduct formation of the electrophile (isocyanic acid) with specific nucleophilic functional groups. The crucial position where carbamylation occurs consists of Nα-amino moiety at N-terminus of proteins and the N^{ε}e-amino moiety (of protein lysine residues). Additionally, carbamylation has been found to happen at the arginine's guanidine moiety, and the reduced thiol groups of amino acid residue, cysteine. The crucial pathways (biochemical) demonstrated to cause protein carbamoylation in vivo are:

(i) The protein catabolism waste product, urea, is abundantly present in the human body. It slowly and spontaneously decomposes to cyanic acid in aqueous solutions (to cyanate, its conjugate base) depending on the equilibrium, which favors urea >99%.[21] Cyanic acid remains in swift equilibrium with isocyanic acid (its form that is reactive).[21] The isocyanic acid plasma concentrations (in individuals who are healthy) are approximately 50 nmol/L. However, in chronic kidney disease (CKD) affected individuals it can increase up to 150 nmol.

(ii) Recent studies have shown the generation of cyanate from pseudohalide thiocyanate (SCN^-) undergoing oxidation by myeloperoxidase (MPO) induced catalysis.[22–28] MPO is an abundantly found protein in monocytes and neutrophils (leukocytes). It is also abundant and active (catalytically) in atherosclerotic lesions.[22–27] Additionally, MPO has been operationally related to the progression of plaques of vulnerable nature and atherosclerosis in humans.[22, 25, 27, 28]

2.2 Contribution to Diseases

Since it is already reported that any protein can be carbamylated, it comes as no surprise to find it being implicated in many disease pathogeneses and progressions.

2.2.1 *Chronic Kidney Disease*

With the progression of CKD, there is a decline in the normal functioning of kidneys; therefore, metabolic waste products such as urea accumulate to very high levels in the patient's tissues as well as blood. The thiocyanate and urea concentrations in the serum of healthy people are, more or less, 0.29 mg/L and 0.4 g/L, respectively, whereas in CKD subjects, the values (the maximum reported) were found to be 1.86 mg/L and 4.60 g/L, respectively.[29] As a consequence of increased concentrations of urea and thiocyanate in CKD patients' blood, the rate of carbamoylation increases to significant extents. The quantity of carbamylated plasma proteins, as well as carbamylated hemoglobin (valine hydantoin), are accelerated in CKD stage 5 patients put on dialysis, in comparison to normal people, and are proposed indicators of uremia.[30] The posttranslational carbamylation of collagen protein present in the extracellular matrix (type I) causes its destabilization at particular locations on the triple helix, thereby decreasing its polymerizing propensity to regular fibrils. This results in the alteration of the inflammatory cells (neutrophils) oxidative function and the way they interact with the extracellular matrix. This decrease in neutrophil's oxidation function causes deterioration of the antioxidant capability of the living system.[31, 32]

Albumin that has been carbamylated causes inhibition of the polymorphonuclear neutrophil's respiratory burst, and thus is responsible for infection-related complications of CKD patients.[33] In a study, high concentrations of cyanate were used (comparable with those of uremic patients) for in vitro investigations performed on CKD affected individuals. The limitations of organism-caused oxidation stress increments with specific antioxidative procedures (e.g., the caeruloplasmin's ferroxidase activity). Interestingly, caeruloplasmin becoming carbamylated decreases the activity of ferroxidase in individuals affected by CKD, leading to increments in oxidation stress (in these individuals). Moreover, cysteine and glutathione were implicated in oxidative stress associated with thiol, causing protection of low-density lipoproteins (LDLs) from apolipoprotein modifications and oxidation of lipids. In individuals with CKD, the carbamylated LDL (cLDL) does not interact with its receptor thus is not cleared from circulation. Various studies have proposed, cLDL as a marker for CKD monitoring. Carbamylation of cysteine's sulfide groups (in addition to glutathione) causes a decline in the protective ability of glutathione and cysteine.[34]

Hydrogen sulfide (H2S), which is a popular signaling messenger molecule, is also a powerful antioxidant and cytoprotectant. H_2S hinders the NADH oxidase activity, thereby causing a decrease in ROS formation.[35] These protective properties of H_2S are however, curbed by S-carbamylation.[36] Consequently, the safeguarding mechanism does not exist and oxidative stress increases in CKD patients. The additional augmentation of anemia (in CKD affected individuals) results from the loss of activity (biological) of erythropoietin, as a consequence of carbamoylation.[37] Consequently, epidemiological studies have confirmed posttranslational carbamylated proteins to be a self-contained risk factor for mortality and morbidity in CKD affected individuals.[38]

2.2.2 *Cardiovascular Disease*

Carbamylated proteins are produced particularly during process of inflammation (by MPO-mediated cyanate formation); these are very strongly implicated in the genesis of cardiovascular diseases (CVD). The LCAT (lecithin-cholesterol acyl transferase) activity for esterification of cholesterol and maturation of HDL was demonstrated to be reduced by HDL carbamylation. HDL forms lipid droplets in macrophages, which results in cholesterol accumulation and thus, the formation of lipid droplet. The major activator of LCAT is apoA-I, which is a component of HDL, and the when the positive charge at its active site is eliminated, this leads to suppression of the LCAT activity, causing dysfunction of HDL.[39] Since apoA-I carbamyl-lysine levels are increased with the progression of lesions stage (in atherosclerotic lesions), critical targets for carbamoylation are HDLs. Carbamylation of lysine residues causes alteration of (well defined) properties (biological) of proteins that has been clearly explained by LDL carbamoylation studies. In 2005, Ok et al.[40] showed cLDL to be the inducing factor for growth and spread of muscle cells (smooth type) found in the vasculature and endothelial apoptosis, both of which are implicated in the pathogenic progression of CVD (atherosclerotic). It was further demonstrated that such carbamoylation altered LDL associated with receptor A1 of macrophages, leading to a buildup of cholesterol foam production in cells.[41]

In addition to their effects observed in macrophages, the vasoprotective properties of endothelium were also found to be altered. The expression of scavenger receptors on endothelium cell surfaces, for instance, scavenger receptor A1, lectin-like oxidized LDL receptor-1 (LOX-1), scavenger receptor of endothelial cells-1(SREC-1), and CD36 have been found to mediate a swift internalization (endothelial) as well as transcytosis of cLDL.[42] Additionally, cLDL caused stimulation of endothelial surface expression of LOX-1.[41] Furthermore, cLDL caused induction of phenotype associated with endothelial pro-inflammation by causing an increase in the expression of the vascular adhesion molecules (VCAM-1) and intercellular adhesion molecules (ICAM-1), which mediated the monocytes adhering to the endothelial cell surface by LOX-1-dependent fashion.[43] The central task of LOX-1 that caused the mediation of the unfavorable vascular cLDL effects was supported by the finding that demonstrated that LOX-1 caused stimulation of cLDL mediated endothelial apoptosis by DNA fragmentation (mediated by endonuclease G) in a mitogen-activated protein kinase

(MAPK)-dependent fashion.[44] The endothelial dysfunction mediated by cLDL directly was demonstrated recently.[45] Therefore, it has been found that when cLDL interacts with LOX-1, it leads to formation of ROS (by oxidation of NADPH of endothelial origin), which is p38-MAPK-dependent.

The S-glutathionylation caused LOX-1 mediated endothelial nitric oxide synthase (eNOS) uncoupling reaction results in reduction of nitric oxide (NO) bioavailability and impairment in the vasodilatation in vivo. The research group of Jankowski[29] have shown that carbamylated lysine residues are substantially increased in number in LDL obtained from patients undergoing dialysis, but not in LDL obtained from individuals who were healthy. But these studies only show that LDL carbamylation causes an alteration of lipoprotein's vascular properties and conveys to a pro-atherogenic particle. Moreover, there is an ongoing confrontation on the concentration levels of cLDL found in CKD patients; the difference might result from measurement using distinct methods and may also result from inclusion of fewer patients in the studies.[29] Thornalley and Rabbani[46] have described the ways to identify sensitive spots that can undergo protein damage considering PTMs, citing a few examples of their impact on pathophysiology.

2.2.3 *Sickle Cell Disease and Cataracts*

In 1970, the first finding relating to the disadvantageous effects of carbamoylation in vivo surfaced. It was believed, during that time, that sickle cell associated distress could be circumvented if sickle cell disease patients could be treated with cyanate/urea. This novel strategy of treatment was justified based on the fact that the hemoglobin that had undergone carbamoylation was found to possess more affinity for oxygen compared to the uncarbamylated form. It was also found that the carbamylated hemoglobin S sickles were slow compared to the unmodified form.[47–52] However, the patients who underwent such procedures subsequently developed cataracts. The reason for cataract development was found to be carbamylation.[53] Additional studies on mammalian species further supported the earlier finding, which has clearly reported that crystallins, which are the protein component of lens, when are subjected to cyanate treatment, form interchain disulfide bonds and protein aggregates. Thus, proving that carbamoylation of lens protein led to the development of cataracts.[54–62]

2.2.4 *Rheumatoid Arthritis*

Proteins that have been modified posttranslationally have been found to possess the capability to evade tolerance mediated by the immune system and cause induction of responses mediated by autoantibody.[63–65] Steinbrecher et al.[66] suggested for the first time that proteins modified by carbamylation could also elicit such a response, and proved this by performing immunization experiments using proteins that were carbamylated. These findings were later confirmed in patients as well other animal models.[67–69] Furthermore, anticarbamylated protein antibodies (anti-CarP antibodies) have been found in patients with rheumatoid arthritis (RA).[69] RA, which is an autoimmune, chronic, and systemic disease, affects 0.5%–1% of the population, commonly in the synovial joints.[70] The symptoms of this disease include pain, stiffness, swelling, warmth, and redness, and may ultimately lead to loss of joint function.[71]

Though until now, the specific disease development of RA is unspecified, anticitrullinated protein antibodies (ACPA) have been used as crucial markers for serology of RA. These ACPA categorize individuals affected with RA into two subclasses, with distinct genetic risk profile and disease course.[72] ACPA, which are a group of auto antibodies targeting citrullinated proteins, are specifically identified in patients in "pre-RA" stages and RA.[64] ACPA are found to occur in approximately 60% of early RA subjects and 70% of people with confirmed RA.[73] The major critical risk element of genetic origin for RA involves the existence of shared epitopes (SE) on class II major histocompatibility complex (MHC). This enhances the vulnerability to RA, which is positive for ACPA.[74] RA subjects who are positive for ACPA have been found to have more damage in joints compared to RA individuals who are negative for ACPA.[64, 75, 76] ACPA, along with other autoantibodies (the rheumatoid factor (RF)), independently demonstrate prediction of greater risk for RA development in "pre-RA" symptoms (clinical); for instance, in indistinct arthralgia and arthritis. These are also found to be associated with lesser chances of achieving drug-free remission.[77–79] RF and ACPA are possible to identify in healthy subjects, years before they progress to RA.[80–82]

Since ACPA exists in approximately 60% of the individuals affected with RA, and a fraction of the RA individuals who are negative for ACPA also show joint damage of a severe nature, there is an urgent need for identification of more biomarkers for individuals who are negative for ACPA, but require more aggressive intervention.[83] As homocitrulline possesses high structural similarity to citrulline, it has been hypothesized that anti-homocitrulline containing (carbamylated) protein antibodies might exist in some RA individuals. The anti-CarP antibodies present in the models previously described proved this presumption. It was demonstrated that rabbits[84] and mice[68] developed antibody response to antigens containing homocitrulline when they underwent immunization with peptides (containing homocitrulline). Shi et al.[85] found that anticarbamylated fetal calf serum (Ca-FCS) IgA and IgG antibodies to be present in both ACPA positive (IgG $\sim$73%, IgA $\sim$51%) and ACPA negative (IgG $\sim$16%, IgA $\sim$30%) RA affected individuals. Additionally, they also found anticarbamylated human fibrinogen (Ca-Fib) antibodies in the individuals affected with RA, which indicated this to be a response mediated by autoantibody. Furthermore, they demonstrated that anti-CarP antibodies bind to carbamylated antigens, induced by IgG's F(ab)2 domains.[69] It was found that anti-CarP antibodies in RA individuals who are negative for ACPA demonstrated more serious joint damage.[69]

The models that have been declared previously demonstrate cross-reactivity (to a high degree) between anti-CarP antibodies and antigens that were citrullinated. Interestingly, the cross reactivity of anti-CarP antibodies to RA in patient's citrullinated antigens is comparatively found to be lower, since there exist anti-CarP single positive individuals in addition to ACPA single positive individuals. Furthermore, a fraction of the ACPA/anti-CarP double positive individuals are found to contain anti-CarP antibodies that are found not to cross-react to proteins that are citrullinated.[69, 86] It has been observed that sera of RA individuals who are double positive, contain ACPA and anti-CarP antibodies that are able to interact with either carbamylated or citrullinated proteins, and might contain antibodies that interact with both. In a compact cohort of double positive[85] RA individuals the median percentage of noncross-reacting anti-Ca-FCS antibody was measured to be 70% (interquartile range (IQR) ~47%–87%).[86] Anti-CarP antibodies do not only exist in RA subjects, but are also found to occur in 40% of individuals that suffer from arthralgia.[87] Joint pain is found to be the crucial arthralgia symptom (clinical) and it, over the course of time, may progress into RA in a few of these subjects. When compared to ACPA, the anti-CarP antibodies that exist in patients affected with arthralgia are found to be related independently to the risk of developing RA.[87] The anti-CarP antibodies are not only present in adult subjects, but are also found to exist in juvenile idiopathic arthritis patients (JIA). JIA is the main occurring disease of rheumatic origin in children (affecting 0.01%–0.4% of children), and consists of 8 heterogeneous subgroups.[88] In a JIA cohort, 39 out of 234 (~16.7%) individuals tested, where found anti-CarP antibodies positive (either anti-Ca-FCS or anti-Ca-Fib antibodies), which is higher compared to RF-IgM (~8.1%) and ACPA (~6.4%). Demonstrating similarity with ACPA positive JIA subjects, anti-CarP antibody positive JIA individuals are enhanced in polyarticular RF positive subgroup.[89] The existence of carbamylated antigens has not yet been demonstrated, though in the joints of RA subjects, citrullinated antigens have been detected.[90–92] The citrullinated proteins have been mostly identified using the "AMC-Senshu" method. However, this method does not differentiate between homocitrulline and citrulline.[84, 86] Hence, more work needs to be done for identification of the carbamylated and citrullinated proteins' presence and composition (nature) in various organs.[85]

2.2.5 *Other Pathologies*

Glomerular mesangial cells have been found to activate to profibrogenic phenotype and cause stimulation of collagen deposition, induced by carbamylated proteins present at concentrations under uremia.[93] Therefore, an increase in carbamylation as a result of kidney failure could conceivably increase the advancement of kidney disease, which could cause the induction of a feedback loop (positive) that could be detrimental to organ function. Many such instances are observed in other diseases. In the case of Alzheimer's, it has been observed that if the τ protein, which is a very significant protein in the brain, gets carbamylated, it causes the formation of tubulin polymers of abnormal structure. This is due to the decreased ability of tubulin to self-assemble to microtubules.[94] Again when the albumin that has been carbamylated was added to the culture of mesangial cells, increased expression of specific microRNAs, particularly those that were over expressed in kidney carcinoma of humans was observed.[95] In another instance it was observed that the carbamoylation of caeruloplasmin lead to the loss of antioxidant activity of ferroxidase, which in turn resulted in cardiovascular complications.[96] Dissociation of urea, in humans, also leads to production of cyanate. In humans, the source of cyanate has also been linked to cassava used as food. The cyanogens present in cassava are metabolized to give cyanate. There is a report that links use of cassava in diet with higher rates of carbamoylation of protein and incidence of brain disease (i.e., the paralytic disease Konzo).[97–100] This relationship between cassava in diet leading to carbamoylation, which causes the development of neurologic disease, is very challenging and interesting. This gives the reason to implicate chronic uremia in the progression of neurologic complications.

3 CARBONYLATION

Carbonylation of proteins that is mediated by reactive oxygen species (ROS) is a nonenzymatic, irreversible protein modification.[101–104] ROS are continuously formed in cells at lower concentrations and they play a crucial part in regulating the redox reactions in cells. ROS is produced by nonenzymatic and enzymatic reactions.[105, 106] The electron transport chain present in the mitochondria contributes majorly to ROS levels in cells, with plasma membrane as a crucial origin site of ROS. Enzymes such as the cytochrome P-450 and b_5 families, present in the smooth endoplasmic reticulum (ER), act as catalysts for reactions that are responsible for detoxifying drugs (lipid-soluble) and harmful products of metabolism. Peroxisomes are considered to be an important cellular H_2O_2 production source.[107] They are known to have within them various enzymes of the oxidase family that generate H_2O_2. Moreover, various soluble (cytoplasmic) enzymes, for example, aldehyde oxidase, xanthine oxidase, tryptophan dioxygenase, and flavoprotein dehydrogenase, also produce ROS in catalytic cycling. Small molecules (dopamine, flavins, hydroquinones, and epinephrine) auto-oxidation also often leads to ROS (intracellular) production.[107] Proteins are known to be the major targeted for ROS and oxidative stress by-products under in vivo conditions, since they are the crucial components of biological systems and are responsible for removing ~50%–75% of reactive radicals, for instance, OH.[108] Few of these protein modifications induced by ROS cause unfolding or changes in structure of protein, while others are not harmful.[109]

3.1 Mechanism

Reactive derivatives of carbonyl compounds (ketones and aldehydes) are formed when arginine, threonine, lysine, and proline amino acid residues of the polypeptide side chains react with the oxidative species.[101] Furthermore, when three-dimensional polypeptides, directly react with ROS, it leads to the generation of derivatives of protein or peptide fragments having highly reactive carbonyls. Additionally, secondary reactions that take place between lysine residue's primary amino groups and reducing sugars, or the primary amino groups of lysine residues and the oxidation products (glycation and/or glycoxidation reactions) of reducing sugars, both may cause formation of reactive carbonyls in proteins.[103, 110] The other mechanisms by which the reactive carbonyl groups are formed occur by the reactions of Michael addition, where cysteine, lysine, or histidine residues react with α or β-unsaturated aldehydes formed when the fatty acids (polyunsaturated) present in the membranes undergo peroxidation.[103, 110, 111] The other additional forms of protein oxidations consist of the oxidation reactions involving aromatic amino acid residues, cross-link formation among proteins, oxidation reactions that are cyclic in pathway, reduction reactions of methionine, reactions of chlorination, free amino acids undergoing oxidation, and proteins getting modified by reactive nitrogen species (RNS).

3.2 Contribution in Diseases

3.2.1 Alzheimer's Disease

Alzheimer's disease (AD), wherein aging serves as a risk factor, is the common dementia of adults. Stress due to oxidative species is considered a factor in AD beginning and in its advancement, manifested by oxidation of proteins, lipid peroxidation, and ROS formation.[112–114] In the aging brain, oxidative damage is a common event, but is pronounced in AD. In AD, protein carbonyl derivatives (PCO) levels were observed to be elevated by approximately 42% in the hippocampus and by approximately 37% in the inferior parietal lobule, compared to the cerebellum, a region of the brain that demonstrates minor degenerative changes.[115] Interestingly, hippocampus and parietal lobule are the two regions that are rich in β amyloid plaque deposits. However, the PCO concentrations in the inferior parietal lobule and control hippocampus were the same as in control cerebral levels. Therefore, it would not be wrong to conclude that AD brain proteins are more carbonylated compared to the controls (age-matched), and also are concentrated in regions of severe histopathological alterations[115] (although PCOs are found to be elevated in the cell bodies of neurons that do not have visible pathomorphological changes, as well as in the of neuron's cell body, which demonstrates abnormalities of neurofibrillary origin).[116] Additionally, carbonyl levels are found to show good correlation with neurofibrillary tangle (NFT).[117]

The carbonylation of NFH (neurofilament heavy) subunit, under in vivo conditions, demonstrates carbonyl caused modification to be linked with cytoskeletal abnormality of generalized nature, which can be critical to AD neurofibrillary pathology. Nevertheless, it has been proposed that this carbonyl mediated NFH subunit modification and the pathological lesions associated with the disease (for instance, formation of senile plaques and NFT) might be cell defense mechanisms, basically functioning to shield neurons from oxidation.[118] Fascinatingly, NFH has an extended half-life but the carbonyl modification extent is found not to change much during the normal procedure of aging and also along the axon length. Moreover, NFH also gets adapted (uniquely) as scavenger of carbonyl due to its excessive lysine content. Therefore, it is very appealing to contemplate carbonylation of NFH to be an additional neuronal defense mechanism to protect the vital site of neurofilaments (axon) from aldehydes that are highly reactive, since these are oxidation products that are deleteriously toxic. The rate of protein turnover, which is slow, taking approximately many years in axon, might be required for this protection. Breakdown of proteins that get oxidized is also elevated (normally), mainly via the proteasomal system.[119] In addition to this, the neuronal proteins can also undergo synthesis localized in the dendrites (closer to sites of synapses and the axon).[120, 121]

These findings, taken together, suggest that carbonylation may be crucially required to protect proteins involved in the neuronal-process, since there is a possibility that their synthesis occurs locally, which does not depend upon synthesis of protein occurring in cell-bodies. Nevertheless, the proteasomal system might get restricted in activation by oxidants, specifically, during the process of aging,[119] and the machinery of translation may be influenced by process of carbonylation.[122] Thus, the protecting role of protein carbonylation of neuronal origin cannot be ignored. Therefore, identifying particular targets that undergo oxidation of proteins are critical to establish a correlation between neuron death and modifications originating from oxidation. Creatine kinase BB and β-actin (both proteins in the brain) have been identified to particularly undergo carbonylation, using two-dimensional (2D) gel fingerprinting coupled with immunochemical PCOs identification, in the brains of AD patients.[116, 123] The other targets (specific) of carbonylation of proteins, for example, ubiquitin C-terminal hydrolase L-1, α-enolase, glutamine synthase, and dihydropyrimidinase-related-protein-2 (which functions in growth and guidance of axons), have been identified by study of proteomics.[124, 125] The recognition of the targets of carbonylation proposes feasible mechanisms that induce degeneration of neurons in AD affected brains. These include retardation of proteasome mediated degeneration of damaged or malfolded proteins, proteins that have formed aggregates (ubiquitin C-terminal hydrolase L-1), reduction or exhaustion of energy (α-enolase and creatine kinase), the reduced length of dendrites (which results in decrement in interneuronal communication, i.e., impaired memory), and excitation related toxicity of glutamine synthase.

3.2.2 *Chronic Lung Disease*

Various factors related to environment and agents of infection have been shown to be involved in causing the commencement of injury to the lungs that causes CLD (chronic lung disease), but the critical or key factors are oxygen-induced toxicity and mechanical ventilation. Infants born preterm are found to be frequently subjected to high levels of oxidative stress, a result of hypoxia (supra-physiological concentrations of oxygen), combined with decreased levels of surfactant, decreased antioxidant defense mechanisms, and reduced ability to produce induction antioxidants. The molecular mechanisms that underlie injury to the lung (also the vascular endothelium) and death of cells, which are induced by hypoxia, are very complex, and also are found to be under regulation of ROS generation (in high concentrations), inflammation mediated by cytokines, loss of antioxidants, and modification or alteration of pathways of signal transduction (causing controlled expression of genes responsible for apoptosis and stress response).[126] In lungs of babies born preterm who are put under ventilation support and also develop CLD, oxidative stress has been observed in bronchoalveolar lavage (BAL) fluid.[127] Premature babies demonstrate elevated levels of ascorbate, urate, and PCOs during the first 72 hours of life, which then falls off in a progressive manner over the upcoming 6 days. Increase in PCOs, in similar fashion, is observed in aspirates of trachea, which correlates firmly with myeloperoxidase activity, proposing contribution of neutrophil-derived ROS to lung injury.[128, 129]

3.2.3 Disuse Muscle Atrophy

During extended duration of disuse of muscle, which can be due to physical inactivity, immobilization, unloading of diaphragm via mechanical ventilation, spaceflight, or chronic bed rest, significant loss of muscle mass and function has been observed.[130] Therefore, identifying mechanisms (biological) that lead to disuse muscle atrophy is an issue of great significance for development of therapeutic strategies that prevent this form of wasting of muscle and promote recovery of function.[130] Moreover, muscle atrophy has also been found to be linked with chronic and acute conditions such as cancer, sepsis, various diseases, and aging.[130] Immobilization based animal models have demonstrated that loss in performance and muscle mass is caused by decrement in the anabolic pathway of muscle protein and increment in the proteolysis rate.[131, 132]

Recently, experiments have demonstrated that oxidation injury occurs in disuse durations in muscles of skeletal system of animals that have been subjected to hind limb suspension[133–136], during elongated mechanical ventilation in rodents with unloaded diaphragms,[137, 138] Similar oxidative stress induced injuries are also observed during muscle disuse periods in human skeletal system.[139] Though identifying sources that produce ROS within the myofibers (during extended periods of inactivity) still needs to be investigated, several landmark studies have demonstrated mitochondria and xanthine oxidase to be the critical contributors in oxidant production in atrophy of diaphragm resulting from mechanical ventilation.[140, 141] Moreover, elevated concentrations of various markers of oxidation stress (including carbonylation of proteins) were found to be elevated to significant concentrations, while antioxidant levels showed a decrement (in various organs and hearts of rodent's subjected to a protocol of immobilization, for 2 weeks, of all four limbs).[142, 143]

Oxidation related stress has been implicated in disuse muscle atrophy by causing enhancement in catabolism of proteins by causing increased activity of different systems that are proteolytic in nature, including caspase-3, calpains, and the ubiquitin-proteasome system of skeletal muscles.[144] Disruption of homeostasis related to calcium induced due to ROS may cause calcium overload in cells, using different mechanisms, which may activate calpains.[132, 144] Various cell types showing an increase in concentrations of ROS may also cause activation of caspase-3,[145] which causes degradation of actin-myosin complexes that remain intact. Moreover, oxidation stress also demonstrates promotion in proteolysis using the ubiquitin-proteasome system of myotubes in skeletal muscle.[146] Therefore, based on recent evidence, elevated production of ROS seems to participate in disuse muscle atrophy pathophysiology.[130] Nevertheless, further studies leading to identification of sources (molecular) of ROS synthesis along with their targets (cell structures) in myofibers in various models of disuse, need to be performed.[130]

3.2.4 *Chronic Renal Failure*

The vital role of oxidation stress has been proposed in renal disease progression as well as in patients who undergo dialysis, where massive levels of ROS are produced with each session of dialysis (with persistent insufficiency in the crucial antioxidation systems). In individuals with CRF (chronic renal failure) and also in people who are on chronic hemodialysis therapy, oxidative stress manifests as an elevation in oxidation of plasma proteins (including formation of PCO and oxidation of thiol-group).[147] Albumin is the critical target in these cases.[148] Oxidation of albumin (plasma antioxidant) leads to a decrease in plasma antioxidant defense mechanisms, and therefore, such patients are at increased risk of oxidant-stress caused injury to tissues and diseases of cardiovascular origin. Additionally, patients affected by CRF combined to proteinuria demonstrate elevated carbonylation of proteins in urine in comparison to proteins in plasma, and albumin present in urine is found to be a crucial target, demonstrating ~71% carbonylation increment compared to plasma albumin.[149] Weakness of muscle and reduction in exercising capability are usual complaints mentioned by individuals suffering from uremia (chronic). Elevation in levels of oxidation stress induced damage to lipid and protein constituent of skeletal muscle have been shown to occur in individuals affected by uremia undergoing hemodialysis, which suggests a role for oxidation induced destruction in the pathogenesis associated with skeletal myopathy (in hemodialysed individuals).[150] AGE and ALE (advanced lipoxidation end-product) levels in proteins of plasma in addition to proteins in matrix have been observed to be increased significantly compared to normal individuals. This assembly or buildup of glycoxidation, lipoxidation products, and dicarbonyl compounds (which are reactive) is termed "carbonyl stress."[151] Hence, uremia has been found to be correlated with overload of carbonyl species (carbonyl stress), which is a result of increased production of carbonyl groups (highly reactive) in carbohydrates, proteins, and lipids, and the resulting modification of proteins that are of irreversible nature and may then give rise to long-term issues associated with CRF and hemodialysis, for example, dialysis-related arteriosclerosis and amyloidosis.[152, 153]

3.2.5 *Diabetes*

The most commonly occurring chronic disease identified or distinguished by high blood glucose levels and glucose excretion in urine is diabetes mellitus. Diabetic patients are inclined toward increments in mortality due to cardiovascular diseases and are at increased risk of developing neuropathy, nephropathy, and retinopathy. Type-II (noninsulin dependent) diabetes prevails among adults in the range of 5%–40% among different populations. The less prevalent form is type-I (juvenile onset) diabetes.[154]

Elevated levels of PCO have been found in both types of diabetes,[155–158] in the thickened arterial wall's intimae, in arteriosclerotic tissues, and are also found to co-localize with products of lipoxidation and glycoxidation. Moreover, AGEs and ALEs are found to build up in the glomerular lesions that are characteristic of diabetes; for instance, the expanded regions of matrix of mesangial cells and lesions in nodular regions. The ALE and AGE build up rates are correlated to the complication's severity (neuropathy, lens disorders, nephropathy, and retinopathy), and their levels show direct correlation with age and also with disease severity of micro vascular origin.[152] Though AGEs are found to be present in healthy subjects (6 years) the diabetic subjects have been found to possess many times higher concentrations.[153]

Carbonyl associated stress can be caused due to oxidation stress, hyperglycemia (lipidemia), and impairment in detoxification mechanisms of RCOs. Strategies to treat diabetes carbonyl stress, as well as carbonyl stress in uremia and arteriosclerosis have been found to offer new approaches of a therapeutic nature, which comprise redox modulation, detoxification of RCO, and prevention of carbonyl stress. Hypotensive agents that have been widely employed, such as antagonist of angiotensin-II-receptor and inhibitor of angiotensin-converting-enzyme, have been found to be potentially useful treatments, since they have been observed not to cause side effects, for instance, deficiency of vitamin B6 and neurotoxicity, which are characteristic of the very first developed of inhibitors of carbonyl-stress, such as amino guanidine, whose function is trapping of RCO. The compounds that have been recently discovered are found to act on the steps leading to RCO precursor formation by searching for and eliminating different radicals involved and causing changes in oxidation induced stress.

An imbalance has been found to be present in the oxidant to antioxidant ratio (systemic oxidative stress) with type-I diabetes, which begins very early and then progresses to early adulthood. PCO concentrations in plasma are found to be significantly high in children suffering with diabetes and in subjects during adolescence who do not have complications compared to individuals set as control, which indicates that oxidation induced damage to protein starts at the beginning of disease and finally shows increment during later levels. Additionally, a decrease in antioxidant defense mechanism may further add to diabetic patients becoming susceptible to oxidation-induced injury.[156] Studies in low age group type-I diabetic patients who were free (clinically) of complications, confirmed protein carbonylation.[158] This work demonstrated that type-I diabetic patients harboring complications had increased levels of plasma PCO compared to subjects that did not harbor complications.[158] No other significant alterations were found between diabetic patients with and without complications based on other markers (levels) of oxidation-induced damage to protein (protein thiols and nitrotyrosine).[155] Type-II diabetes patients (without complications) demonstrated elevated levels of PCOs (plasma) in comparison to control, but differences in plasma thiol concentration observed were insignificant.[157] Increased PCO levels correlate well with the complications of ophthalmic nature, observed in diabetes.[159]

4 N-HOMOCYSTEINYLATION

A phenomenon termed "protein N-homocysteinylation" has emerged, which is related to an important cause of neurodegeneration diseases (NDs) in humans. Various reports have demonstrated that Hcy thiolactone (HTL), which is a reactive thioester of Hcy[160–163] and is a product of methionyl tRNA synthetase activity[160, 164, 165] in a reaction that edits error, causes covalent modification of lysine's of proteins. This process is known as "protein N-homocysteinylation."[162]

4.1 Mechanism

It is well known that HTL binds particularly to the lysine/arginine residue in the protein's polypeptide chain and, therefore, proteins enriched in lysine are the critical target for the HTL induced covalent modulation.[161] The intracellular rate of HTL production is determined specifically by the accessibility of Hcy (precursor metabolite), as well as the levels of Hcy (under conditions of hyperhomocystinuria).[164, 166] Through works of Jakubowski,[161] it has been concluded that the HTL reactivity to proteins is determined by the lysine content of the proteins in consideration. A good correlation is found to exist between the rate at which protein N-homocysteinylation occurs and the protein's lysine content (i.e., number of lysine residues).[161] It has previously been demonstrated that in those proteins that vary in amino acid residues, 104–698, the content of lysine shows a significant negative interrelationship in their ability to react with HTL.[161] Interestingly, for larger proteins, no such correlation is found to exist. For example, for proteins of size 3500–6000 amino acid residues (e.g., α2-macroglobulin, LDL, and fibrinogen) are not homocysteinylated effectively compared to their lysine residues content or size. This can be explained by the fact that many lysines in these proteins are not available to the solvent. These affirmations advocate that, in addition to the overall lysine content in the protein, other determining characteristics exist that are responsible for both the degree and rate of N-homocysteinylation of proteins.

Recently, the reliance of the degree of N-homocysteinylation on the content of lysine in proteins has been disputed.[167–169] In protein, cytochrome *c* (cyt c), it was demonstrated that if the four lysine amino acid residues (i.e., Lys-86/Lys-87, Lys-8/Lys-13, Lys-100, and Lys-99) are modified, this confers no significant changes on the tertiary structure of its native state.[168] But, if a single residue is modified in insulin, complete protein unfolding occurs as a consequence, which ultimately leads to production of aggregates.[167, 170] Another study demonstrated differential reactivity of lysines in hemoglobin to HTL.[169] In the case of human albumin found in serum, the Lys-525 has been found to be more susceptible to N-homocysteinylation. It has also been observed that the ability to react of albumin's lysines (including Lys-525) is influenced by Cys-34 status.[171] Furthermore, the reactivity of HTL has also been found to be dependent on p*K*a values of amino acid side chains. Moreover, the pKa (10.6) of lysine residues which constitute proteins differ significantly from lysine which are present free in buffer. Additionally, pKa also shows dependence on the microenvironment of the buried lysine. Amines, which have a p*K*a value of 7.7, have been observed to react 6.5 times faster with HTL compared to free lysines in solutions.[172] Very recently, it has been demonstrated that lysine's reactivity and the resulting changes in structure and function due to protein N-homocysteinylation may also be determined by the protein's isoelectric point (pI).[173] Therefore, a meticulous study was performed with proteins having dissimilar pI values. The study demonstrated that proteins that were acidic demonstrated pronounced structural changes, while proteins that were basic demonstrated insignificant structural and functional changes on being subjected to HTL-induced modification.[173] This clearly suggests that proteins of acidic nature will remain susceptible to HTL-induced changes in proteins at native state.

4.2 Contribution to Diseases

4.2.1 Tau Homocysteinylation

Tau, a microtubule-associated protein (MAP), is a constituent of neuronal cell's cytoskeleton. They are found to occur in different isoforms.[174, 175] Tau proteins are known to bind microtubule proteins (MTP) and, thereby, help in maintaining stability of neuronal cells. A category of pathologies termed as tauopathies, are known to be caused by the tau proteins undergoing hyperphosphorylation[176] and the most familiar tauopathy is AD. In tau's microtubule binding repeats (MTBRs), the positively charged lysine is found to interact, in a sequence-specific manner, with the negatively charged tubulin's C-terminal.[177] The tau's binding ability to MTP is distorted fairly by mutation or covalent modulations of lysine residues (particular lysine in MTBRs).[178] Recent studies have demonstrated that if tau's lysine residues undergo N-linked-HTL modification, then it results in the alteration in the tubulin assembly dynamics in vitro.[179] Tau is also found to play a crucial part in regulating the assembly and nucleation of tubulin, as well as in stabilizing MTP against becoming disassembled. If the interactions between N-homocysteinylated tau and MTP are eliminated, it prevents association of tau with MTP. After release, tau self-associates and aggregates; this results in formation of paired helical filaments,[176, 180] which further causes stabilization to form aggregates of higher order, thereby, indulging in an early beginning of AD associated pathology. This study presented crucial evidence that suggests prominent roles of tau-HTL modification in AD pathologies.

4.2.2 *Prion Protein*

Prion diseases or TSE's (transmissible spongiform encephalopathies) are a diverse group of neurodegenerative diseases (NDs). These have been reported to occur in animals and humans. The critical symptoms accompanying these diseases comprise of insomnia, paraplegia, ataxia, paresthesias, dementia, etc. These originate from protein that is infectious, called "prion." Prions are known to replicate by assembling PrPC (normal prion protein) and thus, trigger them to transform to PrPSc (the disease associated (scrapie) isoform).[181] Correlation between AD and prion diseases was found when it was demonstrated that amyloid-β peptide has the ability to make interactions with prion (found on cell surface of neurons).[182, 183] The disease expansion of AD is observed to be governed by the reactions of Aβ (or its precursors) with the normal forms or the prion forms (infectious). In a model of prion (ovine) studied, it was demonstrated that N-homocysteinylation induced changes in structure of the normal PrPC and also caused induction of formation of insoluble multimers, in addition to conversion to amyloidogenic forms.[184] Structural analyses have found that HTL-mediated modulations of prion proteins results in increment in content of β-sheet compared to the proteins which are unmodified.[184] When 100 μM HTL concentrations were used (which corresponded to the HTL concentrations that exist under conditions of mild homocystinuria), HTL was found capable of invoking PrPC to convert to amyloidogenic forms. Protein dimer, oligomer formation, and further production of aggregates of prion were observed to progress in minutes after incubating with HTL. On the other hand the conversions to amyloidogenic forms took hours.[184]

5 S-NITROSYLATION

A type of PTM, S-nitrosylation, can be found regulating a wide array of functions of protein, such as phosphorylation. This process involves the NO moiety becoming covalently attached to a thiol group of cysteine, thereby, forming an SNO complexed protein.

5.1 Mechanism

The reaction process of S-nitrosylation consists of the involvement of cation of nitrosonium (NO^+) intermediate, which reacts with anion (RS^-) of thiolate, which requires free radical NO (•NO) oxidation, mediated by transition metal as catalyst.[185–187] Recently, the phrase "S-nitrosylation" has been generally utilized to denote the biological outcome of the S-nitrosylation chemical process. Additionally or alternatively, the nitrosylation mechanism that causes the production of R-SNO by NO group might involve a radical recombination between •NO, which owns only one electron, the pi orbital, in addition to thiyl radical (RS•).[186–188] S-Nitrosothiol formed in this way causes modifications in the target protein's function by conferring changes in conformational, protein activity modulations, or protein-protein interactions modifications.[189, 190] Theoretically, it is accepted that a free thiol group possess the capability to get S-nitrosylated, in the presence of adequate NO (exogenous) levels. However, under both physiological and pathophysiological conditions (that is, in practice), it has been observed that the admissible levels of NO that are present are found to affect the thiol groups only.[189–192] This specificity is to some extent dictated by protein structures in the neighborhood; i.e., thiol groups present in those areas that involve in direct interaction with NOS readily get S-nitrosylated, since they are in very close proximity to NO formation sources.

It is worth mentioning that the "SNO-motif" which is characterized by the presence of acidic or basic amino acids present at 6–8 Å distance from the cysteines (target) is responsible for facilitating SNO modification.[193] These groups of acid or base amino acid lead to thiol group deprotonation, which in turn promotes target thiols getting S-nitrosylated. Under conditions of base physiologic levels, low concentrations of NO supports the neuronal functions (normal), which include energy homeostasis of energy, neuronal survival, and transmission of synapses via regulating particular target proteins, which is mediated by SNO. These SNO-proteins might remain restricted close to NOSs, and may harbor thiol groups (sensitive to SNO), which are enclosed by full SNO motifs.[192] However, under diseased conditions, high concentrations of NO may cause induction of cysteine thiols nitrosylation (which consists of limited SNO motif and are present at a distance from the source of NO) getting S-nitrosylated. These proteins that are S-nitrosylated aberrantly activate cell destruction processes, causing neurodegeneration promotion.[187] The extent to which a protein gets S-nitrosylated depends on the rate of S-nitrosylation as well as on the grade of denitrosylation. Since very often NO acts as better "leaving group," few SNO-proteins drop NO groups from their thiols groups of cysteine, spontaneously, in a nonenzymatic pathway.

Interestingly, investigations have shown that few specific SNO-proteins groups (i.e., that harbor complete SNO-motifs or those that are produced under conditions of disease) are stable, relatively.[194–196] Huge presence of SNO-proteins which are stable, and hence, the regulation of SNO signaling cascades is counter balanced by thioredoxin and S-nitrosoglutathione reductase, the denitrosylating enzymes. Interestingly, due to resulting protein-protein trans-nitrosylation, the NO-donating proteins get denitrosylated simultaneously to the S-nitrosylation of companion protein. S-Nitrosylation has been often observed to impact conformation of protein. It was found that when a protein gets denitrosylated, the thiol group that is exposed often displays an elevated ability to interact with ROS and forms sulfinic, sulfenic, or sulfonic acids derivatives, in an empirical manner.[195, 196] S-Nitrosylation of single thiol also promotes the superficial production of disulfide bond with a vicinal thiol and hence initial thiol (reactive) gets denitrosylated. Moreover, when concentrations of NO are found to be increased, NO was found to inhibit the formation of disulfide bond between the two nearby cysteine thiols via S-nitrosylation of both thiol groups.[196, 197]

5.2 Contribution to Diseases

5.2.1 Neurodegeneration

nNOS actively produces NO as signal molecules in the excitatory synapse of neurons. Numerous investigations have indicated that excessively produced quantities of NO are a huge source of oxidation stress, which causes neurodegeneration. This becomes specifically crucial since cell death of neurons is usually related to inflammatory response that causes the induced iNOS expression, and therefore, leads to more increases in the levels of NO in areas affected. Peroxynitrite types of RNS are formed when unrestricted amounts of NO reacts with other radicals (free), for example, superoxide anion. These free radicals cause induction of peroxidation of lipids, protein nitration, and damage to DNA, which can ultimately cause neuronal degeneration.[198]

Such a disease causing mechanism has been observed in many neurodegenerative disorders. Interestingly, most recent studies demonstrate proteins undergoing S-nitrosylation to be a contributing causative agent in the majority of neurodegenerative disorders, for example, AD, PD, and ALS. The typically occurring disorder of motor neurons, ALS, is identified by degeneration of motor neurons in the cortex, brain stem, and spinal cord.[199] In a few individuals with familial history of ALS, various mutations in superoxide dismutase 1 (SOD1) gene are observed, although most ALS is sporadic in occurrence.[199] How SOD1 mutations lead to development of ALS is still not completely understood, but investigations demonstrated that mutations in SOD1 showed an increment in activity of denitrosylase compared to WT SOD1 (wild-type).[200] The cells that express SOD1 mutants demonstrate a decrement in the levels of S-nitrosothiol (SNO) in the mitochondria.[200] Similarly, the levels of SNO are known to be decreased in the spinal cords of SOD1 mutant transgenic mice.[200] These results, therefore, give a mechanistic hypothesis that mutations in SOD1 lead to imbalance in levels of SNO and this causes degeneration of motor neurons in ALS.

Though the task of S-nitrosylation is not well charted in AD, recently it has been reported that NO affects the fission process of mitochondria and induces injury to neurons by causing dynamin-related protein 1 (Drp1) to become S-nitrosylated and causes the pathogenic progression of AD.[201] AD, which is a frequently occurring neurodegenerative disorder, affects the increasing population of elderly people. Various mechanistic pathways have already been put forward for the beginning and progression of AD, and the β-amyloid (Aβ) mediated degeneration of neurons is the widely accepted contributor.[201] The mechanism of Aβ induced injury to neurons has been intensively investigated and many theories have been proposed, which include toxicity associated with excitation, oxidation stress, dysfunction of mitochondria, and apoptosis. Precisely, stress associated with oxidative processes in addition to dysfunction of mitochondria has been suggested to be a crucial mechanistic factor leading to the neurodegenerative process of AD.

Recently it was demonstrated that a disease mechanism of Aβ induces nitrosative stress that finally causes abnormal mitochondrial fission or fusion, which ultimately affects survival of neurons.[201] It is known that Aβ induces increased nitrosative stress, which has been linked to dysfunction of mitochondria. The correlation between nitrosative stress and dysfunction of mitochondria has been found to be a consequence of Drp1 undergoing S-nitrosylation.[201] Drp1 is considered to be a critical molecule that causes regulation of mitochondrial fission, and dysfunction of Drp1 is found to be involved in neurodegenerative disorders.[201] Aβ causes the induced formation of NO and this causes Drp1 getting S-nitrosylated.[201] The S-nitrosylation at Cys644 of Drp-1 enhances dimerization and leads to increased GTPase activity of Drp-1.[201] When the Drp-1 is S-nitrosylated, this leads to induced fragmentation of mitochondria with respect to neurodegenerative processes of AD.[201] Therefore, these outcomes give a new mechanistic insight into how nitrosative process associated stress (by Drp1 getting S-nitrosylated) contributes to the disease progression of AD.

In PD, S-nitrosylation has recently emerged as a critical cause in the disease beginning and progressing. The other most frequent disorder associated with neurodegeneration, PD is also marked by impairment of movement. The disease is a result of specific destruction of dopaminergic neurons (in the substantia nigra (SNc)) with observed aggregation of protein in the intra neuronal regions, the Lewy bodies (LB).[202] Why cell death of neurons occurs in the SNc is still not completely understood, but investigations propose that oxidation stress, dysfunction of mitochondria, aggregation of protein, and UPS dysfunction contribute majorly to neurodegeneration in PD.[202, 203] The recently identified gene mutations that lead to a rare form of familial PD (FPD) have strengthened our knowledge of the mechanism of PD. For example, α-synuclein (α-syn) and mutations in LRRK2 lead to development of autosomal dominant FPD, while mutations in PINK1, DJ1 and parkin cause induction of FPD recessive forms.[204] The functional studies performed with these genes are in agreement with the accepted disease mechanism of PD. For example, α-syn, which immediately undergoes aggregation. Directly after it was established as the product of gene which is related to FPD, α-syn was observed as major component of LB.[202] Additionally, parkin was found to function as an E3 ligase of UPS, DJ1 was found to be a chaperone associated with oxidation process, and PINK1 was recognized as a kinase in mitochondria, crucial for proper functioning of mitochondria.[202, 203]

These observations clearly suggest the significance of aggregation of proteins, oxidation stress, mitochondrial or dysfunction of UPS in the development of PD. Since, aggregates of nitrated protein are frequently cited to be a distinguished feature in the brain tissues of PD patients, stress associated with nitrosylation is considered a key source of PD.[205] This hypothesis gets further support from the studies that demonstrate that α-syn (nitrated) is aggregation susceptible, and α-syn (which is nitrated) is frequently identified in the LB.[203, 205] These studies demonstrated that stress associated with nitrosylation can cause α-syn aggregation, which can consequently mediate LB production. Interestingly, recent investigations have shown that NO contributes also to PD by inducing S-nitrosylation of distinct constituents in pathways of neuroprotection. It was previously reported that parkin protein gets S-nitrosylated and this modulation compromises on neuroprotective functions of protein parkin.[206] Parkin is an E3 ligase of the UPS and mutations of parkin were first observed in patients of Japanese origin who developed an early onset form of autosomal recessive FPD.[202]

Many different studies have indicated the protective ability of parkin targeted to various types of cellular assaults.[207] For instance, parkin causes ubiquitination (as an E3 ligase) of its substrates and causes them to undergo degradation, which is dependent on UPS, thus preventing the buildup of malfolded and aggregated toxic compounds of protein in the neurons.[208] Chung et al.[206] and Yao et al.[209] reported for the first time the S-nitrosylation of parkin under both in vivo and in vitro conditions. The S-nitrosylated parkin has inhibited E3 ligase activity and compromised protective function.[206, 209] The more important finding was when Chung et al.[206] observed the increased concentrations of S-nitrosylated parkin in the tissues of brains of PD individuals and animal models. The other FPD linked gene found to undergo S-nitrosylation was DJ-1. A study demonstrated DJ-1 S-nitrosylation at residues Cys46 and Cys53.[210] Later studies showed Cys46 to be an important residue for the DJ-1 dimerization, but how DJ-1 S-nitrosylation affected its dimer formation could not be clarified.[210]

Studies have also demonstrated DJ-1 dimerization to be crucial for it to function normally as a chaperone; hence, S-nitrosylation of DJ-1 might be affecting its protective role in neurons (dopaminergic). Taken together, these studies conclude that S-nitrosylation-caused modification of gene products of FAD due to nitrosative stress contribute majorly in the pathogenesis of PD. Many different proteins with neuroprotective functions, in addition to parkin and DJ-1, have shown to be S-nitrosylation specific. For instance, under high nitrosative stress conditions, a protein of endoplasmic reticulum (ER) catalyzes disulfide bond formation to assist appropriate folding of protein and its maturation, known as protein-disulfide isomerase (PDI), is also found to be S-nitrosylated.[196] S-Nitrosylation causes the suppression of the chaperone as well as the isomerase activities of the enzyme PDI.[196] Under normal circumstances, PDI has been found to inhibit the formation of aggregates (which are Lewy body-like) in PD cell model.[196, 211] PDI S-nitrosylation causes a reduction in the antiaggregate production property of PDI.[196] In neuroblastoma of SH-SY5Y, PDI S-nitrosylation, overrides its neuron protecting role against stress of ER mediated by proteins (unfolded) or proteasomal inhibition.[196] In the case of neurons that were exposed to excitation mediated toxicity (which is NMDA-induced), S-nitrosylated PDI levels were found to increase, with accompanying buildup of unfolded and polyubiquitinated proteins leading to death of neurons.[196]

Elevated concentrations of PDI that was S-nitrosylated were seen in the postmortem tissues (brain) taken from AD and PD patients. This scientific study, hence, establishes a correlation between stress mediated by S-nitrosylation and protein malfolding in disorders of neurodegeneration through PDI dysfunction. Since parkin, DJ-1, and PDI proteins protect cells against protein malfolding or aggregation, their S-nitrosylation enhances cellular insults to neurons. Additional studies have shown that NO affects various other neuroprotective pathways, such as antioxidation stress protein or proteins that have antiapoptotic function. For example, an intracellular peroxidase, peroxiredoxin 2 (Prx2), was also identified to be S-nitrosylated thereby, affecting its role as an enzyme of antioxidation stress enzyme.[212] Prx2, which is the most abundantly found peroxidase present in neuron cells, causes metabolism of peroxides and, thereby, safeguards neuronal cells from oxidation injuries.[212] Prx2 has been identified to be S-nitrosylated at residues Cys51 and Cys172, which are crucial residues present at the catalytic Prx2 domain.[212] The Prx2 peroxidase activity is diminished after its S-nitrosylation, which modulates its neuron protecting functions of preventing oxidation induced stress in neuronal cells.[212] The S-nitrosylated Prx2 levels are found to be increased in tissues (brain) of PD patients suggesting the vulnerability of neurons to oxidative stress under such circumstances.[212]

Tsang and Chung[203] have recently established X-linked inhibitor of apoptosis (XIAP) as another protein who's S-nitrosylation has been shown to be correlated with the neurodegeneration development in PD. XIAP belongs to a protein family that is highly conserved and involved in promoting survival of cell via their IAP repeat (BIR) domains of baculovirus.[213, 214] Resembling parkin, XIAP has a ring-like domain present at C-terminal which entitles XIAP to act as an E3 ligase (of UPS) to identify various substrates (inclusive of XIAP).[213, 214] Investigations have demonstrated that the antiapoptotic function induced by XIAP is the ability of its strictly conserved BIR domains, which antagonize the pro-apoptotic function of caspases.[213, 214] Caspases execute apoptosis, which mediates the procedure of programmed cell death. It is not surprising that XIAP is over expressed frequently in tumors, the vital contributors of tumorigenesis development.

Tsang and Chung[203] found that S-nitrosylation of XIAP occurs both in vitro and in vivo. Contrary to parkin, XIAP is S-nitrosylated at the BIR domains by NO. XIAP undergoing S-nitrosylation hinders its ability to cause inhibition of its caspase-3 activity, thereby downregulating its antiapoptotic functions. Contrary to parkin, the E3 ligase ability of XIAP remains unaffected by S-nitrosylation.[203] In PD (striatum) models, increased S-nitrosylated levels of XIAP were identified in comparison to the control. Additionally, higher XIAPS-nitrosylation was also identified in PD patients' brains.[203] These results clearly suggest that the XIAP-mediated pro-survival feature in neuronal cells is halted by the S-nitrosylation during PD pathogenesis. Summarizing all the above demonstrates that nitrosative stress contributes to PD via a number of mechanistic pathways. For example, nitration of proteins, or damages in lipids and DNA, which are mediated by peroxy nitrite, can cause death of neurons. Additionally, proteins (with neuroprotective function) that are S-nitrosylated (e.g., parkin, XIAP, Prx2, and DJ-1) could cause deleterious effects on neuronal cell survival under conditions of cytotoxic assaults thus, making them susceptible to degeneration.

5.2.2 Cancer

Many recent studies have made indications toward the role of abnormal S-nitrosylation development of cancer development and its progression, and also as an outcome of certain therapeutics and treatments.[215] Interestingly, literary evidence exists that suggests divergent S-nitrosylation to be a critical event in the onset of cancer and may play an important and dramatic role in increasing cancer risk.[216–220] Bentz et al.[221] had showed that the levels of nitrotyrosine are high to significant figures, in the dysplastic and reactive forms of head and neck squamous cell carcinoma (HNSCC) lesions when in comparison to the normal mucosa (squamous). Abnormalities of S-nitrosylation have also been implicated in manipulating the process of carcinogenesis by causing an alteration in checkpoints of cell-cycle,[222] apoptosis,[223] and DNA repair.[224]

Sustained inflammation has been implicated in ~25% cancers throughout world.[225, 226] The inflammatory reaction involves increased concentrations of activated macrophage cells.[227] These macrophage cells have been found to contain the active form of the iNOS gene, which induces the generation of increased levels of nitric oxide (NO). Higher intracellular levels of NO, thus, results in increased levels of proteins that are S-nitrosylated. This activity is found to drive carcinogenesis by causing an alteration in targets and pathways that are critical for progression of cancer at fast rates in healthy tissue.[228, 229] The three NOS isoforms are known to own similar modes of catalysis and structures. These are distinct in their mechanisms of action for control of activities related to temporal and spatial function.[229] The intracellular localization of NOS depends upon their activity. Various researchers have suggested the existence of distinct compartments where full NOS activation can occur, thus, providing unrestrained availability to substrates.[230] A large number of evidences suggest that the specificity of S-nitrosylation is decided by the proximity (spatial) of specific targets and NOSs.[231, 232] It is thought that the iNOS expression that is induced by stimuli generated by inflammation, in addition to nNOS and eNOS (constitutive) expression, contributes to elevated risk of cancer.[215]

Though it is known that specific cellular and molecular components present in the cancer microenvironment play a part, it is NOS over expression (especially over expression of iNOS)[233] or abnormally high levels of NOS-derived NO that have been identified to play a crucial role in cancer beginning in humans. These effects are known to comprise of malignant transformations, metastasis, angiogenesis, and a number of other cellular effects.[227] Additionally, activity of NOS has been determined in several tumors, highlighting correlation between grade of cancer and rate of proliferation that occurs at the molecular level.[219]

6 GLYCATION

This reaction occurs when reducing sugars act by covalently binding to amino-containing compounds, thereby, producing products of early glycation.[234] These are then subjected to multistep reactions (in series) leading to the formation of irreversibly modified AGEs.[235, 236]

6.1 Mechanism

Fructoselysine (Amadori products) and adducts of Schiff base, are formed when the amino groups (free) undergo reaction with sugars (reducing). These formed intermediates are found to be freely reversible in biological systems and, hence, remain in an equilibrium, which correlates with free glucose concentrations. The concentrations of Amadori products remain only two to threefold greater in diabetic patients compared to patients who are nondiabetic. This demonstrates that these products are freely reversible in nature, and equilibrium always exists between the protein that has undergone modulation and nonmodulated proteins. Consequently, these (Amadori) modulations are not found to assemble on macromolecules and, hence, no relation has been found to be present between adduct formation on tissues and complications associated with diabetes.[237]

The condensation reactions (which are nonenzymatic) that occur between sugars (reducing) and N terminal groups or ε-amino groups cause production of AGE. The glycation modulations involve amino acids, arginine, and lysine preferentially (although lipids and DNA containing free amines are also involved) and advance spontaneously through a series of complex rearrangements (chemical) to produce reactive products with differing in cross linking, fluorescence properties, and pigmentation.[238] A major portion of such products produces (from a variety of precursor molecules) various chemical structures, which constitute the heterogeneity of AGE. For instance, the Amadori intermediates tend to undergo oxidative reactions (metal-catalyzed) producing products of "glycoxidation," for instance, *N*-ε-(carboxyethyl) lysine (CEL) or *N*-ε-carboxymethylated lysine (CML), which are irreversible[239, 240] and, thus, tend to build up on the substrates to which they get attached and/or also can cause the production of dicarbonyl species that are very reactive. Dicarbonyls (for example 1-, 3-, or 4-deoxyglucosones, glyoxal, and methylglyoxal) are extremely reactive intermediates, which are found to be upregulated during hyperglycemia, and thus are found to undergo reaction with proteins and cause propagation of intermolecular or intramolecular formation of AGE.[241, 242] These pathways act as important sources of AGEs (extra and intracellular) and also lead to rapid formation of AGEs, since they arise from highly reactive "AGE intermediates."[242]

6.2 Contribution to Diseases

6.2.1 Diabetic Complications

Diabetes and the complications associated with it are fast becoming our planet's major cause of mortality and morbidity.[243, 244] The deleterious effects of constantly escalating concentrations of glucose in plasma on various parts of the body differ depending on the cell type involved.

6.2.1.1 RETINOPATHY

Retinopathy is a major complication of microvasculature associated with diabetes and is known to be a leading reason for blindness in patients in the 30–70 years age group.[245, 246] The pathogenic condition involves vascular occlusion, angiogenesis, increased proliferation of blood vessels, thickening of the capillary basement membrane, pericyte loss from retinal capillaries, micro aneurysms, hemorrhages, elevated permeability of retinal capillaries, and infarction affecting the eye (retina).[246] Since upregulation of RAGE causes induction of responses that increase inflammation (involving Müller glia cells of retina), blocking RAGE has been used in clinical therapeutics and has been found to shown a decrement in the retinopathy progression.[247] AGE has been observed to play an important part in retinopathy advancement, which causes retinal cell dysfunction and death.[248] The various AGE-RAGE axis components, which comprise the end-point effectors, ligand production, and signal transduction of signals have been proposed as feasible targets to treat retinopathy associated with diabetes.[249]

A few investigations have shown that Müller glial cells in rats becoming nonfunctional (during retinopathy associated with diabetes) are correlated to AGEs build up.[250] Many clinical and preclinical reports exhibit that AGEs (inclusive of MGO) are responsible in influencing diabetic retinopathy by multiple aspects.[251] Recent experiments have demonstrated that detoxifying MGO causes a reduction in the accumulation of AGEs, thereby, preventing vascular lesions and retinal neuroglial and formation.[252] The interaction between AGE and RAGE seems to play a significant part in the persistent inflammation and dysfunction of microvasculature of the retina and neurodegeneration occurring during retinopathy associated with diabetes. By inducing retinal Müller cells to produce bFGF (basic fibroblast growth factor), AGEs have been reported multiple times to be a causative agent for diabetic retinopathy development.[253] Many studies have provided strong literary evidence to show proteins that are components of the mammalian eye to be highly affected by AGEs formation (interaction of sugars and carbonyl compounds). AGEs then progressively are accumulated in the retina and lens, which ultimately has adverse effects on the vision.[254]

Recent findings have implicated N-(ε) CML and AGEs to be important modifiers in the occurrence of nonproliferative retinopathy among type 2 diabetic patients.[255] Cross linking formation among proteins in the wall of vessels caused by AGEs results in an increase in vascular stiffness and ECM proteins modification, which leads to decreased adherence of pericyte.[256] The extracellular proteins modified by AGEs are known to be implicated in injury to the retina via RAGE interaction.[257] Interaction with RAGE causes the activation of various pathways associated with signaling that cause an increase in oxidation stress and causing cytokines, local growth factors, and adhesion molecules to be synthesized.[258] AGEs are known to be deleterious for the receptors of AGE that possess pericytes and damage them (as reported in diabetic retinopathy).[259] Upregulation of RAGE mRNA levels by AGEs has been demonstrated by various studies, in micro vascular endothelial cells and pericytes.[260] Upregulated RAGE may lead to an increment in transduction signals followed by AGEs stimulation, which might worsen the pericyte's loss in retinopathy associated with diabetes. AGEs have also been found to cause upregulation of expression of ICAM-1 (in retinal (bovine) endothelial cell culture) and causes micro vascular leukostasis of retina associated with diabetes.[261, 262]

When the retinal cells are exposed to AGEs, upregulation of the mitogen (potent), vascular endothelial cell growth factor (VEGF) by causing an increment in gene expression of VEGF occurs. VEGF induces stimulation of angiogenesis and neovascularization, known to be causing pathogenesis associated with proliferative retinopathy. The VEGF concentrations in ocular fluid are found to tally with neovascularization activity (in retinopathy) and are also found to be linked with faults in the blood-retinal barrier, known to be involved in elevated micro vascular permeability (seen in retinopathy). The reduction in the perfusion of the retina, which is an outcome of capillary loss, induces ischemia, which acts as stimulant for faults in blood-retinal barrier and the retinal neovascularization.[263] Elevated levels of IL-6 and AGEs have been demonstrated in the eyes (vitreous) of individuals with retinopathy associated to diabetes, recently. AGEs cause stimulation of IL-6 secretion from the cells of the retina, which then causes angiogenesis induction by increment in VEGF expression.[264] The elevation in level of adhesion of leukocytes to the endothelium of the retina in experimentally induced diabetes leads to apoptosis of cells of endothelium utilizing Fas-Fas ligand pathway, has been demonstrated.[265] Increase in vascular permeability has been linked with local elevation in VEGF levels.[266] VEGF is formed (within the retina) by cells of retinal pigment epithelium, astrocytes, pericytes, endothelial, and Muller glial cells. VEGF gene expression can be induced by local hypoxic conditions and an increment in concentrations of inflammatory reactive oxygen species, cytokines, and AGEs characteristic of diabetes.[267]

6.2.1.2 CARDIOMYOPATHY

Cardiomyopathy manifests itself as hypertrophy of myocardial cells and fibrosis of myocardia leading to abruption in normal diastolic function and with high prevalence of heart failure in patients with diabetes.[268] Diastolic dysfunction is found in 50%–60% of patients with diabetes and is also found in (most cases) diabetic patients having microalbuminuria, which later advances into systolic dysfunction.[268] Diastolic dysfunction, which is correlated to concentrations of HbA1c, results from buildup of

AGEs in the myocardium.[269] It has already been shown that in experimental diabetic models, amino guanidine effectively prevents cardiac hypertrophy and arterial stiffening, thus, emphasizing the pathogenic part played by AGEs in cardiomyopathy associated with diabetes.[270]

AGEs derived from methylglyoxal (MG-RAGE) cause upregulation of mRNA of cardiac RAGE, which triggers the cardiomyocyte contractile dysfunction. AGEs are known to contribute to heart failure development via two pathways. AGEs may affect the properties (physiological) of ECM proteins causing the generation of cross-links between them. AGEs may induce intracellular changes in both myocardial and vascular tissue by interacting with AGE receptors.[271] Diabetes has been known to cause noteworthy increment in pre-AGE molecule (e.g., MGO, AGEs, and RAGE) concentrations in cardiomyocytes. The drop in contractility of the left ventricle induced by diabetes has been found to be checked by RAGE knockdown. Dysfunction of cardiac cells (induced by AGE) has been correlated to depolarization of mitochondrial membrane, which in turn has been found, can be prevented by RAGE expression knockdown mediated by RNA interference.[272] Under hyperglycemic conditions, appreciable increases in intracellular sugars levels have been found to occur (such as fructose 3-phosphate, fructose, and glucose-6-phosphate), which are known to be more reactive compared to glucose. Moreover, an increase in intracellular buildup of dicarbonyls (such as glyoxal and MGO, the potent cross-linking agents) is also observed.[268] Continued increment in blood glucose concentrations lead to increment in glycation of proteins of nonenzymatic type, including the ryanodine receptor and SERCA2a.[273] Studies have shown that when the cardiac myocytes (that were isolated) are exposed to high glucose levels, this results in damaged calcium handling and contractility.[274]

AGEs are known to add to complications linked to diabetes through two pathways: (i) by forming cross-links among important molecules in the ECM basement membrane, thus, causing permanent change architecture and structure of cell; and (ii) reaction between AGEs and RAGE on the surface of cells, which changes the signaling cascades and function of cells. AGEs cause increment in cross-linking (much above the basal physiological concentrations) with proteins of matrix proteins, for instance, laminin, collagen, elastin, and vitronectin.[275] This unrestricted cross-link formation causes disturbances in the characteristic flexible nature of matrix proteins, thus, increasing their rigidity. This causes reduction in the contractile nature of cardiac cells and leads to development of diastolic dysfunction. Collagen and elastin, cross linked by AGE, cause an increment in the ECM surface area, thereby, increasing the vasculature stiffness.[276, 277] Laminin, in addition to type I and type IV collagens are the critical molecules of basement membrane; when these become glycated, this reduces their adherence capability to cells of endothelium for both the glycoproteins of the matrix.

It has been proposed that production of AGE causes decrease in collagen and heparin's binding to vitronectin (adhesive matrix molecule).[276] LDL that has undergone glycation has been shown to reduce the production of NO and cause suppression of LDL uptake/clearance by endothelial cell receptor.[278] Another pathway by which AGEs contribute to developing diastolic dysfunction is by activating RAGE.[279] RAGE is known to effect fibrosis induction by causing transforming growth factor-β (TGF-β) upregulation.[280] Glycation also causes increase in α 3(IV) collagen, type III collagen, type VI collagen, type V collagen, fibronectin, and laminin synthesis in the ECM, most likely by upregulating TGF-β.[281] Petrova et al.[282] cultured myocytes (cardiac) from the RAGE-transgenic and analyzed them for Ca^{2+} transients, demonstrating that RAGE over expression caused reduction in the systolic and diastolic intracellular calcium concentration. This clarified that AGE and RAGE play a significantly dynamic role in beginning and progression of diabetes-induced dysfunction of cardiac cells.[282]

6.2.2 Alzheimer's Disease

Type 2 diabetes mellitus (T2DM) has recently been identified as a critical risk element for developing AD, which increases AD risk up to 60%.[283] Intriguingly, in cases of AD and T2DM both, desensitization of receptors for insulin has been observed in the brain.[284–286] This leads to the impairment of glucose homeostasis impairment as well as the impairment of insulin's neuroprotective functions, thus inducing the development of AD.[234] Surprisingly, when insulin was administered to AD patients, an improvement in the ratio of Aβ peptides 1-40/1-42 (which are markers of reduced toxicity) was observed, together with enhancement in cortical activation; improvement in performing cognitive tasks, was also seen.[287] Additionally, T2DM drugs were found to cause a reduction in the formation of plaques of amyloid-β, and inflammation of the brain. Interestingly, a diet rich in sucrose was also found to modulate AD. For instance, an AD model of mouse (triple transgenic) diet rich in sucrose (that induced several metabolic changes found in T2DM) was found to cause a significant increase in Aβ levels.[288] This study further lends support to the important interplay that exist between diabetes and AD.[234]

Various studies have demonstrated the glycation of proteins that are crucially involved in AD, such as Aβ and tau. Plaques that were extracted from brains affected with AD were found to contain three times more AGEs compared to preparations from healthy controls (age matched).[289] Glycation has also been found to occur in both NFTs and senile plaques (in AD brain tissues).[290] Various studies have supported this observation[291–293] and proposed that glycation leads to stabilization and promotes the production of Aβ and tau aggregates.[289, 294–297] In AD, glycation not only causes direct modulation of protein aggregation, but also upregulation of APP expression, thus, elevating the concentrations of Aβ peptides 1-42[298] and also causes increased tau phosphorylation,[299–301] an crucial disease hallmark. Strikingly enough, an age as well as stage-dependent AGEs accumulation has been observed in AD brains. AGEs tend to co-localize, with phosphorylated form of tau, along with other neurodegeneration markers, for example, nNOS and caspase-3.[302] Furthermore, the levels of proteins that underwent glycation are found to be increased in an AD individual's cerebrospinal fluid (CSF)[303–305] and might be considered promising biomarkers for AD. In order to gain better understanding of glycation role in Aβ, a study assessed normal and glycated Aβ toxicity in primary hippocampal neurons. The investigators then established the hypothesis elucidating the role of glycation in exacerbating the Aβ toxicity, and noticed increments in concentrations of RAGE. These findings, taken together, highlight glycation's importance in AD and propose that protein glycation inhibition may be used as a criterion for the treatment of this pathophysiology.[234]

6.2.3 *Parkinson's Disease*

Recent literary evidence suggests the importance of diabetes as a risk factor for PD. Many reports show that large numbers of PD patients do not have normal glucose tolerance[306–308] and, thus, exhibit hyperglycemia.[309] To ameliorate the understanding of these observations, studies where performed to confirm positive correlation between PD and T2DM.[310, 311] Furthermore, diabetes was often found to be linked with severe forms of Parkinson's disease,[312] and diabetes onset before the appearance of PD acted as a critical feature for more severe PD symptoms.[313] However, other studies have failed to give any indisputable evidence to find correlation between diabetes and PD.[314–317] Therefore, new experiments need to be carried out to clarify this issue.[234] One of the direct consequences of hyperglycemia is the increment in the production of glycation agents, particularly, MGO. Interestingly enough, it has been found that a diet with a high glycemic index causes a 34 fold increment in the formation of AGEs in the brain (primarily in the substantia nigra).[318] Additionally, in a study on a PD rat model, when 6-hydroxydopamine was injected, a diet rich in fat caused increased numbers of dopaminergic neurons to be degenerated in the striatum and substantia nigra. This caused the authors to hypothesize that such a diet may lead to lowering of the threshold to develop PD.[319] Altogether, these findings suggest that a comprehensive study should be performed based on the effect of diet in PD to get a better understanding of the effect of dietary pattern in influencing the progression of disease.

In previous findings, the contribution (feasible) of MGO and glycation to the observed phenotypes has been ignored, which was crucially significant.[234] For instance, it was shown that MGO levels were found to increase to two to five times in diabetic patients.[28, 320, 321] Such increase was mimicked by providing the model systems with a diet rich in fructose, which induced an increment in the concentrations of MGO.[322–324] It has also been observed that in diabetic rats (streptozotocin-induced), elevated concentrations of MGO and other glycation agents (~15-fold) were found.[325] Correspondingly, a diet rich (high) in fat causes significant increment in the concentration levels of glucose and, thus, may also cause the MGO concentration levels to increase. Therefore, increments in glycation of proteins likely take place under different conditions. The first published reports that glycation is involved in PD recognized AGEs present at Lewy body's periphery in PD individuals.[326] Subsequently, elevated glycation levels were demonstrated in the amygdala, cerebral cortex, and substantia nigra of individuals with PD.[327] The discovery of presence of AGEs in the Lewy bodies formed (in incidental Lewy body disease) indicates that AGEs might be inducing the production of Lewy body in individuals considered to be pre-PD individuals.[328]

Intriguingly, glycation has also been reported in various Parkinson model systems. For instance, CML and CEL AGEs were found in the dopaminergic neurons of Parkinson MPTP model of mouse. Furthermore, the oligomeric species of α-Syn were found to be glycated in control brains as well as MPTP-treated animal brains.[329] This agrees well with the AGEs capacity of inducing protein cross-kinking. Additionally, glycation was found to cause acceleration of aggregation of α-Syn by induction of cross-links[330] and thus causing the oligomers of α-Syn to form, in vitro. Lee et al.,[331] and others, demonstrated α-Syn oligomeric species to be more toxic compared to large assembly of aggregates, thus, playing a key part in the disease progression.[332–335] However, no direct evidence demonstrating the role of glycation in PD exists as yet. Therefore, investigating what effect glycation demonstrates in PD animal model systems will help answer if elevated glycation levels elicit immediate changes in α-Syn.[234] If such a correlation can be found, this may aggravate toxicity and offer an additional clarification for diabetes and PD connection. A good correlation between α-Syn and MGO already exists. Studies on α-Syn knockout mice demonstrate an increase in the concentrations of MGO, thus, proposing a role of α-Syn in modulating brain glucose levels as well as Glo1 concentrations.[336] Although it is known that Glo1 possess the capability to cause prevention of the production of agents glycation but still, increased levels of Glo1 could not halt the increment in the production of AGEs.[336] Therefore, it is reasonable enough to propose α-Syn to be a target for glycation, thereby removing MGO.

It is of utmost significance to keep it in mind that Glo1 activity depends on reduced glutathione (GSH) concentrations.[337] Interestingly, there is a drastic drop in the concentration of GSH in PD.[338, 339] Furthermore, Glo1 levels are found to show a decrease with increasing age. Kuhla et al.[340] gave the hypothesis that Glo1 might play a significant role in PD, thereby, giving justification for further studies in this field.[234] MGO might play various parts in the context of PD. Crucially, MGO causes interference with the ubiquitin-proteasome system (UPS), which is known to play a significant part in the degradation of malfolded proteins such as α-Syn.[234] For instance, it is known that glycated proteins resist proteasomal degradation. Additionally, ubiquitin is known to be a target for MGO glycation, and this may hinder the complex formation of ubiquitin to its (target) proteins.[318] The UPS failure accounts for an increment in α-Syn concentrations, thereby, promoting a vicious cycle since, buildup of α-Syn leads to impairment of proteasome function[294, 341] and, consequently, contributes additionally to the oligomer formation, aggregation, and toxicity of α-Syn. Thus, there is a possibility that if UPS gets impaired due to glycation reactions, this may play a crucial part in PD.[234]

Surprisingly, the PARK7 gene has been linked with PD familial forms, and has been associated with the MGO metabolism. Though, DJ-1 (which is a PARK7 gene product) function is still not completely understood but it was demonstrated to cause upregulation of the concentrations of Glo1 by regulating the concentrations of the nuclear transcription factor E2-related factor.[342, 343] DJ-1 has also been described recently as an anti-MGO enzyme that has glyoxalase activity,[344] and also as a protein deglycase that is known to repair proteins that are MGO-glycated.[345] Nonfunctional DJ-1, thus, may be linked with elevated concentrations of MGO, thereby adding potential to its damaging effects, and, thus, may serve as an interesting target for treatment strategies.[234] The other enzyme that has been suggested to add to elevated concentrations of MGO is GAP dehydrogenase (GAPDH). This is known to be important for the catabolism of GAP in the glycolytic pathway, which is the major precursor of MGO. A decrement in the activity of GAPDH has been found to be accompanied by a sixfold increase in the concentrations of MGO.[346] Very recently, studies have demonstrated that genetic variations in GAPDH could cause modulation of susceptibility of sporadic PD. The gene encoding for GAPDH, undergoes single-nucleotide polymorphisms and was demonstrated to either cause reduction or increment in the risk for PD.[347] Therefore, it is very critical to be able to assess the polymorphisms role in GAPDH activity in order to ascertain if the risk for PD is linked to its activity.[234]

A recent discovery, which demonstrates that MGO interacts with dopamine, and leads to production of 1-acetyl-6,7-dihydroxy-1,2,3,4-tetrahydroisoquinoline (ADTIQ), again establishes a connection between glycation and development of PD. This metabolite is found to be available in human brain tissue (including substantia nigra). Additionally, the ADTIQ concentrations show an increment in PD individuals.[348] Interestingly enough, the structure of ADTIQ has been found to show resemblance to MPTP (which causes PD associated symptoms).[349]

Therefore, elevated production of ADTIQ in PD has been hypothesized to be associated with greater rate of dopaminergic degeneration. Models of PD (in mice) that were established on the over expression of A30P or A53T mutants of α-Syn, were consistently found to show an increment in the concentrations of ADTIQ. In the same study, it was demonstrated that ADTIQ acted as neurotoxin that is endogenous in cells of neuroblastoma (SH-SY5Y) and that the concentrations of both ADTIQ and MGO are elevated in the striatum of a diabetic rat model induced by streptozotocin.[350]

From the above studies, it can be concluded that AGEs are identified by specific receptors (multiligand receptor for AGEs is RAGE). In PD individuals, RAGE concentrations are found to be elevated as compared to those in age-matched individuals.[327] Consequently, the elevated concentrations of AGEs cause persistent inflammation, since glycated proteins cause an increment in the nuclear factor kappa-light-chain-enhancer, a transcription factor present in activated B concentrations on reacting with RAGEs, thus, further causing RAGE expression induction. In turn, this vicious cycle leads to death of neurons.[332] Conclusively, it can be said that for understanding the part played by glycation in the progression and pathogenesis of PD and also to conclude that modification and regulation of glycation might act as instrumental approach for intervening therapeutically, further studies need to be performed.[234]

7 SUMMARY

It is evident from this review that PTMs due to nonenzymatic mechanisms are associated with various human disease pathologies. Therefore, strategies that can suppress or prevent such modifications will be of great use for the therapeutic intervention in large number of human diseases. Such modifications are diverse and include modification of lysine, tyrosine, arginine, cysteine, threonine, or proline. Although modifications lead to functional alterations in the proteins, the basic cause of other cellular toxicities due to different type of modifications (or targeting different amino acid residues) have not been understood in detail. Future research should try to focus on unfolding the cellular level toxicities due to the different types of modifications. Nevertheless, such PTMs will be good biomarkers for the early onset of various diseases.

Acknowledgment

SK is grateful to Department of Science and Technology for financial support through DST WOS-A scheme.

References

1. Walsh C. *Posttranslational modification of proteins: Expanding nature's inventory.* Englewood, CO: Roberts and Company Publishers; 2006.
2. Felder S, Zhou M, Hu P, Urena J, Ullrich A, Chaudhuri M, White M, Shoelson SE, Schlessinger J. SH2 domains exhibit high-affinity binding to tyrosine-phosphorylated peptides yet also exhibit rapid dissociation and exchange. *Mol Cell Biol* 1993;**13**:1449–55.
3. Yang XJ. Lysine acetylation and the bromodomain: a new partnership for signaling. *Bioessays* 2004;**26**:1076–87.
4. Manning G, Whyte DB, Martinez R, Hunter T, Sudarsanam S. The protein kinase complement of the human genome. *Science* 2002;**298**:1912–34.
5. Wang D, Harper JF, Gribskov M. Systematic trans-genomic comparison of protein kinases between Arabidopsis and *Saccharomyces cerevisiae. Plant Physiol* 2003; **132**:2152–65.
6. Komander D, Clague MJ, Urbe S. Breaking the chains: structure and function of the deubiquitinases. *Nat Rev Mol Cell Biol* 2009;**10**:550–63.
7. Karve TM, Cheema AK. Small changes huge impact: the role of protein posttranslational modifications in cellular homeostasis and disease. *J Amino Acids* 2011;**2011**.
8. Soskic V, Groebe K, Schrattenholz A. Nonenzymatic posttranslational protein modifications in ageing. *Exp Gerontol* 2008;**43**:247–57.
9. Jaisson S, Gillery P. Evaluation of nonenzymatic posttranslational modification-derived products as biomarkers of molecular aging of proteins. *Clin Chem* 2010;**56**:1401–12.
10. Xu M, Xie L, Yu Z, Xie J. Roles of protein N-myristoylation and translational medicine applications. *Crit Rev Eukaryot Gene Expr* 2015;**25**:259–68.
11. Rowland EA, Snowden CK, M., C. I.. Protein lipoylation: an evolutionarily conserved metabolic regulator of health and disease. *Curr Opin Chem Biol* 2018;**42**:76–85.
12. Wang M, Casey PJ. Protein prenylation: unique fats make their mark on biology. *Nat Rev Mol Cell Biol* 2016;**17**:110–22.
13. Walker J, Hambly FJ. Transformation of ammonium cyanate into urea. *J Chem Soc* 1895;**67**:746–67.
14. Wohler F. Ueber künstliche Bildung des Harnstoffs. *Ann Phys* 1828;**87**:253–6.
15. Schutz F. Cyanate. *Experientia* 1949;**5**:133–41.
16. Stark GR, Stein WH, Moore S. Reaction of the cyanate present in aqueous urea with amino acids and proteins. *J Biol Chem* 1960;**235**:3177–81.
17. Nyc JF, Mitchell HK. Synthesis of orotic acid from aspartic acid. *J Am Chem Soc* 1947;**69**:1382–4.
18. Smyth DG. Carbamylation of amino and tyrosine hydroxyl groups. Preparation of an inhibitor of oxytocin with no intrinsic activity on the isolated uterus. *J Biol Chem* 1967;**242**:1579–91.
19. Stark GR. On the reversible reaction of cyanate with sulfhydryl groups and the determination of Nh2-terminal cysteine and cystine in proteins. *J Biol Chem* 1964; **239**:1411–4.
20. Stark GR. Reactions of cyanate with functional groups of proteins. 3. Reactions with amino and carboxyl groups. *Biochemistry* 1965;**4**:1030–6.
21. Kraus LM, Kraus Jr AP. Carbamoylation of amino acids and proteins in uremia. *Kidney Int Suppl* 2001;**78**:S102–7.
22. Brennan ML, Penn MS, Van Lente F, Nambi V, Shishehbor MH, Aviles RJ, Goormastic M, Pepoy ML, McErlean ES, Topol EJ, Nissen SE, Hazen SL. Prognostic value of myeloperoxidase in patients with chest pain. *N Engl J Med* 2003;**349**:1595–604.
23. Daugherty A, Dunn JL, Rateri DL, Heinecke JW. Myeloperoxidase, a catalyst for lipoprotein oxidation, is expressed in human atherosclerotic lesions. *J Clin Invest* 1994;**94**:437–44.

24. Hazen SL, Heinecke JW. 3-Chlorotyrosine, a specific marker of myeloperoxidase-catalyzed oxidation, is markedly elevated in low density lipoprotein isolated from human atherosclerotic intima. *J Clin Invest* 1997;**99**:2075–81.
25. Huang Y, DiDonato JA, Levison BS, Schmitt D, Li L, Wu Y, Buffa J, Kim T, Gerstenecker GS, Gu X, Kadiyala CS, Wang Z, Culley MK, Hazen JE, Didonato AJ, Fu X, Berisha SZ, Peng D, Nguyen TT, Liang S, Chuang CC, Cho L, Plow EF, Fox PL, Gogonea V, Tang WH, Parks JS, Fisher EA, Smith JD, Hazen SL. An abundant dysfunctional apolipoprotein A1 in human atheroma. *Nat Med* 2014;**20**:193–203.
26. Podrez EA, Schmitt D, Hoff HF, Hazen SL. Myeloperoxidase-generated reactive nitrogen species convert LDL into an atherogenic form in vitro. *J Clin Invest* 1999;**103**:1547–60.
27. Sugiyama S, Kugiyama K, Aikawa M, Nakamura S, Ogawa H, Libby P. Hypochlorous acid, a macrophage product, induces endothelial apoptosis and tissue factor expression: involvement of myeloperoxidase-mediated oxidant in plaque erosion and thrombogenesis. *Arterioscler Thromb Vasc Biol* 2004;**24**:1309–14.
28. Wang Z, Nicholls SJ, Rodriguez ER, Kummu O, Horkko S, Barnard J, Reynolds WF, Topol EJ, DiDonato JA, Hazen SL. Protein carbamylation links inflammation, smoking, uremia and atherogenesis. *Nat Med* 2007;**13**:1176–84.
29. Gajjala PR, Fliser D, Speer T, Jankowski V, Jankowski J. Emerging role of post-translational modifications in chronic kidney disease and cardiovascular disease. *Nephrol Dial Transplant* 2015;**30**:1814–24.
30. Balion CM, Draisey TF, Thibert RJ. Carbamylated hemoglobin and carbamylated plasma protein in hemodialyzed patients. *Kidney Int* 1998;**53**:488–95.
31. Jaisson S, Larreta-Garde V, Bellon G, Hornebeck W, Garnotel R, Gillery P. Carbamylation differentially alters type I collagen sensitivity to various collagenases. *Matrix Biol* 2007;**26**:190–6.
32. Jaisson S, Lorimier S, Ricard-Blum S, Sockalingum GD, Delevallee-Forte C, Kegelaer G, Manfait M, Garnotel R, Gillery P. Impact of carbamylation on type I collagen conformational structure and its ability to activate human polymorphonuclear neutrophils. *Chem Biol* 2006;**13**:149–59.
33. Jaisson S, Delevallee-Forte C, Toure F, Rieu P, Garnotel R, Gillery P. Carbamylated albumin is a potent inhibitor of polymorphonuclear neutrophil respiratory burst. *FEBS Lett* 2007;**581**:1509–13.
34. Schreier SM, Steinkellner H, Jirovetz L, Hermann M, Exner M, Gmeiner BM, Kapiotis S, Laggner H. S-carbamoylation impairs the oxidant scavenging activity of cysteine: its possible impact on increased LDL modification in uraemia. *Biochimie* 2011;**93**:772–7.
35. Samhan-Arias AK, Garcia-Bereguiain MA, Gutierrez-Merino C. Hydrogen sulfide is a reversible inhibitor of the NADH oxidase activity of synaptic plasma membranes. *Biochem Biophys Res Commun* 2009;**388**:718–22.
36. Praschberger M, Hermann M, Laggner C, Jirovetz L, Exner M, Kapiotis S, Gmeiner BM, Laggner H. Carbamoylation abrogates the antioxidant potential of hydrogen sulfide. *Biochimie* 2013;**95**:2069–75.
37. Jin K. Effects of amino acids and albumin on erythropoietin carbamoylation. *Clin Exp Nephrol* 2013;**17**:575–81.
38. Koeth RA, Kalantar-Zadeh K, Wang Z, Fu X, Tang WH, Hazen SL. Protein carbamylation predicts mortality in ESRD. *J Am Soc Nephrol* 2013;**24**:853–61.
39. Holzer M, Zangger K, El-Gamal D, Binder V, Curcic S, Konya V, Schuligoi R, Heinemann A, Marsche G. Myeloperoxidase-derived chlorinating species induce protein carbamylation through decomposition of thiocyanate and urea: novel pathways generating dysfunctional high-density lipoprotein. *Antioxid Redox Signal* 2012;**17**:1043–52.
40. Ok E, Basnakian AG, Apostolov EO, Barri YM, Shah SV. Carbamylated low-density lipoprotein induces death of endothelial cells: a link to atherosclerosis in patients with kidney disease. *Kidney Int* 2005;**68**:173–8.

41. Apostolov EO, Shah SV, Ray D, Basnakian AG. Scavenger receptors of endothelial cells mediate the uptake and cellular proatherogenic effects of carbamylated LDL. *Arterioscler Thromb Vasc Biol* 2009;**29**:1622–30.
42. Tarng DC, Huang TP, Wei YH, Liu TY, Chen HW, Wen Chen T, Yang WC. 8-hydroxy-2′-deoxyguanosine of leukocyte DNA as a marker of oxidative stress in chronic hemodialysis patients. *Am J Kidney Dis* 2000;**36**:934–44.
43. Apostolov EO, Shah SV, Ok E, Basnakian AG. Carbamylated low-density lipoprotein induces monocyte adhesion to endothelial cells through intercellular adhesion molecule-1 and vascular cell adhesion molecule-1. *Arterioscler Thromb Vasc Biol* 2007;**27**:826–32.
44. Apostolov EO, Ray D, Alobuia WM, Mikhailova MV, Wang X, Basnakian AG, Shah SV. Endonuclease G mediates endothelial cell death induced by carbamylated LDL. *Am J Physiol Heart Circ Physiol* 2011;**300**:H1997–2004.
45. Speer T, Owala FO, Holy EW, Zewinger S, Frenzel FL, Stahli BE, Razavi M, Triem S, Cvija H, Rohrer L, Seiler S, Heine GH, Jankowski V, Jankowski J, Camici GG, Akhmedov A, Fliser D, Luscher TF, Tanner FC. Carbamylated low-density lipoprotein induces endothelial dysfunction. *Eur Heart J* 2014;**35**:3021–32.
46. Thornalley PJ, Rabbani N. Protein damage in diabetes and uremia—identifying hot-spots of proteome damage where minimal modification is amplified to marked pathophysiological effect. *Free Radic Res* 2011;**45**:89–100.
47. Dean J, Schechter AN. Sickle-cell anemia: molecular and cellular bases of therapeutic approaches (first of three parts). *N Engl J Med* 1978;**299**:752–63.
48. Gillette PN, Peterson CM, Lu YS, Cerami A. Sodium cyanate as a potential treatment for sickle-cell disease. *N Engl J Med* 1974;**290**:654–60.
49. Harkness DR, Roth S. Clinical evaluation of cyanate in sickle cell anemia. *Prog Hematol* 1975;**9**:157–84.
50. Nalbandian RM, Henry RL, Barnhart MI, Camp Jr. FR. Sickle cell disease: clinical advances by the Murayama molecular hypothesis. *Mil Med* 1972;**137**:215–20.
51. Nalbandian RM, Nichols BM, Stehouwer EJ, Camp Jr. FR. Urea, urease, cyanate, and the sickling of hemoglobin S. *Clin Chem* 1972;**18**:961–4.
52. Nigen AM, Njikam N, Lee CK, Manning JM. Studies on the mechanism of action of cyanate in sickle cell disease. Oxygen affinity and gelling properties of hemoglobin S carbamylated on specific chains. *J Biol Chem* 1974;**249**:6611–6.
53. Nicholson DH, Harkness DR, Benson WE, Peterson CM. Cyanate-induced cataracts in patients with sickle-cell hemoglobinopathies. *Arch Ophthalmol* 1976;**94**:927–30.
54. Beswick HT, Harding JJ. Conformational changes induced in bovine lens alpha-crystallin by carbamylation. Relevance to cataract. *Biochem J* 1984;**223**:221–7.
55. Beswick HT, Harding JJ. High-molecular-weight crystallin aggregate formation resulting from non-enzymic carbamylation of lens crystallins: relevance to cataract formation. *Exp Eye Res* 1987;**45**:569–78.
56. Derham BK, Harding JJ. Alpha-crystallin as a molecular chaperone. *Prog Retin Eye Res* 1999;**18**:463–509.
57. Harding JJ, Rixon KC. Carbamylation of lens proteins: a possible factor in cataractogenesis in some tropical countries. *Exp Eye Res* 1980;**31**:567–71.
58. Kern HL, Bellhorn RW, Peterson CM. Sodium cyanate-induced ocular lesions in the beagle. *J Pharmacol Exp Ther* 1977;**200**:10–6.
59. Lapko VN, Smith DL, Smith JB. In vivo carbamylation and acetylation of water-soluble human lens alphaB-crystallin lysine 92. *Protein Sci* 2001;**10**:1130–6.
60. Liu X, Li S. Carbamylation of human lens gamma-crystallins: relevance to cataract formation. *Yan Ke Xue Bao* 1993;**9**:136–142, 157.
61. Yan H, Zhang J, Harding JJ. Identification of the preferentially targeted proteins by carbamylation during whole lens incubation by using radio-labelled potassium cyanate and mass spectrometry. *Int J Ophthalmol* 2010;**3**:104–11.

62. Zhang J, Yan H, Harding JJ, Liu ZX, Wang X, Ruan YS. Identification of the primary targets of carbamylation in bovine lens proteins by mass spectrometry. *Curr Eye Res* 2008;**33**:963–76.
63. Ahmad J, Gupta A, Alam K, Farooqui KJ. Detection of autoantibodies against glycosylated-DNA in diabetic subjects: its possible correlation with HbA(1C). *Dis Markers* 2011;**30**:235–43.
64. Willemze A, Trouw LA, Toes RE, Huizinga TW. The influence of ACPA status and characteristics on the course of RA. *Nat Rev Rheumatol* 2012;**8**:144–52.
65. Wu JT, Wu LL. Autoantibodies against oxidized LDL. A potential marker for atherosclerosis. *Clin Lab Med* 1997;**17**:595–604.
66. Steinbrecher UP, Fisher M, Witztum JL, Curtiss LK. Immunogenicity of homologous low density lipoprotein after methylation, ethylation, acetylation, or carbamylation: generation of antibodies specific for derivatized lysine. *J Lipid Res* 1984;**25**:1109–16.
67. Kummu O, Turunen SP, Wang C, Lehtimaki J, Veneskoski M, Kastarinen H, Koivula MK, Risteli J, Kesaniemi YA, Horkko S. Carbamyl adducts on low-density lipoprotein induce IgG response in LDLR-/- mice and bind plasma autoantibodies in humans under enhanced carbamylation. *Antioxid Redox Signal* 2013;**19**:1047–62.
68. Mydel P, Wang Z, Brisslert M, Hellvard A, Dahlberg LE, Hazen SL, Bokarewa M. Carbamylation-dependent activation of T cells: a novel mechanism in the pathogenesis of autoimmune arthritis. *J Immunol* 2010;**184**:6882–90.
69. Shi J, Knevel R, Suwannalai P, van der Linden MP, Janssen GM, van Veelen PA, Levarht NE, van der Helm-van Mil AH, Cerami A, Huizinga TW, Toes RE, Trouw LA. Autoantibodies recognizing carbamylated proteins are present in sera of patients with rheumatoid arthritis and predict joint damage. *Proc Natl Acad Sci U S A* 2011;**108**:17372–7.
70. Alamanos Y, Drosos AA. Epidemiology of adult rheumatoid arthritis. *Autoimmun Rev* 2005;**4**:130–6.
71. Aletaha D, Neogi T, Silman AJ, Funovits J, Felson DT, Bingham 3rd CO, Birnbaum NS, Burmester GR, Bykerk VP, Cohen MD, Combe B, Costenbader KH, Dougados M, Emery P, Ferraccioli G, Hazes JM, Hobbs K, Huizinga TW, Kavanaugh A, Kay J, Kvien TK, Laing T, Mease P, Menard HA, Moreland LW, Naden RL, Pincus T, Smolen JS, Stanislawska-Biernat E, Symmons D, Tak PP, Upchurch KS, Vencovsky J, Wolfe F, Hawker G. 2010 rheumatoid arthritis classification criteria: an American College of Rheumatology/European League Against Rheumatism collaborative initiative. *Ann Rheum Dis* 2010;**69**:1580–8.
72. Scott IC, Steer S, Lewis CM, Cope AP. Precipitating and perpetuating factors of rheumatoid arthritis immunopathology: linking the triad of genetic predisposition, environmental risk factors and autoimmunity to disease pathogenesis. *Best Pract Res Clin Rheumatol* 2011;**25**:447–68.
73. van Venrooij WJ, Zendman AJ, Pruijn GJ. Autoantibodies to citrullinated antigens in (early) rheumatoid arthritis. *Autoimmun Rev* 2006;**6**:37–41.
74. van der Woude D, Houwing-Duistermaat JJ, Toes RE, Huizinga TW, Thomson W, Worthington J, van der Helm-van Mil AH, de Vries RR. Quantitative heritability of anti-citrullinated protein antibody-positive and anti-citrullinated protein antibody-negative rheumatoid arthritis. *Arthritis Rheum* 2009;**60**:916–23.
75. Nielen MM, van der Horst AR, van Schaardenburg D, van der Horst-Bruinsma IE, van de Stadt RJ, Aarden L, Dijkmans BA, Hamann D. Antibodies to citrullinated human fibrinogen (ACF) have diagnostic and prognostic value in early arthritis. *Ann Rheum Dis* 2005;**64**:1199–204.
76. van der Helm-van Mil AH, Verpoort KN, Breedveld FC, Toes RE, Huizinga TW. Antibodies to citrullinated proteins and differences in clinical progression of rheumatoid arthritis. *Arthritis Res Ther* 2005;**7**:R949–58.

77. Bos WH, Wolbink GJ, Boers M, Tijhuis GJ, de Vries N, van der Horst-Bruinsma IE, Tak PP, van de Stadt RJ, van der Laken CJ, Dijkmans BA, van Schaardenburg D. Arthritis development in patients with arthralgia is strongly associated with anti-citrullinated protein antibody status: a prospective cohort study. *Ann Rheum Dis* 2010;**69**:490–4.
78. Mjaavatten MD, van der Heijde D, Uhlig T, Haugen AJ, Nygaard H, Sidenvall G, Helgetveit K, Kvien TK. The likelihood of persistent arthritis increases with the level of anti-citrullinated peptide antibody and immunoglobulin M rheumatoid factor: a longitudinal study of 376 patients with very early undifferentiated arthritis. *Arthritis Res Ther* 2010;**12**:R76.
79. van der Woude D, Young A, Jayakumar K, Mertens BJ, Toes RE, van der Heijde D, Huizinga TW, van der Helm-van Mil AH. Prevalence of and predictive factors for sustained disease-modifying antirheumatic drug-free remission in rheumatoid arthritis: results from two large early arthritis cohorts. *Arthritis Rheum* 2009;**60**:2262–71.
80. Majka DS, Deane KD, Parrish LA, Lazar AA, Baron AE, Walker CW, Rubertone MV, Gilliland WR, Norris JM, Holers VM. Duration of preclinical rheumatoid arthritis-related autoantibody positivity increases in subjects with older age at time of disease diagnosis. *Ann Rheum Dis* 2008;**67**:801–7.
81. Nielen MM, van Schaardenburg D, Reesink HW, van de Stadt RJ, van der Horst-Bruinsma IE, de Koning MH, Habibuw MR, Vandenbroucke JP, Dijkmans BA. Specific autoantibodies precede the symptoms of rheumatoid arthritis: a study of serial measurements in blood donors. *Arthritis Rheum* 2004;**50**:380–6.
82. Rantapaa-Dahlqvist S, de Jong BA, Berglin E, Hallmans G, Wadell G, Stenlund H, Sundin U, van Venrooij WJ. Antibodies against cyclic citrullinated peptide and IgA rheumatoid factor predict the development of rheumatoid arthritis. *Arthritis Rheum* 2003;**48**:2741–9.
83. Trouw LA, Mahler M. Closing the serological gap: promising novel biomarkers for the early diagnosis of rheumatoid arthritis. *Autoimmun Rev* 2012;**12**:318–22.
84. Turunen S, Koivula MK, Risteli L, Risteli J. Anticitrulline antibodies can be caused by homocitrulline-containing proteins in rabbits. *Arthritis Rheum* 2010;**62**:3345–52.
85. Shi J, van Veelen PA, Mahler M, Janssen GM, Drijfhout JW, Huizinga TW, Toes RE, Trouw LA. Carbamylation and antibodies against carbamylated proteins in autoimmunity and other pathologies. *Autoimmun Rev* 2014;**13**:225–30.
86. Shi J, Willemze A, Janssen GM, van Veelen PA, Drijfhout JW, Cerami A, Huizinga TW, Trouw LA, Toes RE. Recognition of citrullinated and carbamylated proteins by human antibodies: specificity, cross-reactivity and the 'AMC-Senshu' method. *Ann Rheum Dis* 2013;**72**:148–50.
87. Shi J, van de Stadt LA, Levarht EW, Huizinga TW, Toes RE, Trouw LA, van Schaardenburg D. Anti-carbamylated protein antibodies are present in arthralgia patients and predict the development of rheumatoid arthritis. *Arthritis Rheum* 2013;**65**:911–5.
88. Manners PJ, Bower C. Worldwide prevalence of juvenile arthritis why does it vary so much? *J Rheumatol* 2002;**29**:1520–30.
89. Muller PC, Anink J, Shi J, Levarht EW, Reinards TH, Otten MH, van Tol MJ, Jol-van der Zijde CM, Brinkman DM, Allaart CF, Hoppenreijs EP, Koopman-Keemink Y, Kamphuis SS, Dolman K, van den Berg JM, van Rossum MA, van Suijlekom-Smit LW, Schilham MW, Huizinga TW, Toes RE, Ten Cate R, Trouw LA. Anticarbamylated protein (anti-CarP) antibodies are present in sera of juvenile idiopathic arthritis (JIA) patients. *Ann Rheum Dis* 2013;**72**:2053–5.
90. Asaga H, Senshu T. Combined biochemical and immunocytochemical analyses of postmortem protein deimination in the rat spinal cord. *Cell Biol Int* 1993;**17**:525–32.
91. Demoruelle MK, Deane K. Antibodies to citrullinated protein antigens (ACPAs): clinical and pathophysiologic significance. *Curr Rheumatol Rep* 2011;**13**:421–30.

92. Senshu T, Sato T, Inoue T, Akiyama K, Asaga H. Detection of citrulline residues in deiminated proteins on polyvinylidene difluoride membrane. *Anal Biochem* 1992;**203**:94–100.
93. Shaykh M, Pegoraro AA, Mo W, Arruda JA, Dunea G, Singh AK. Carbamylated proteins activate glomerular mesangial cells and stimulate collagen deposition. *J Lab Clin Med* 1999;**133**:302–8.
94. Farias G, Gonzalez-Billault C, Maccioni RB. Immunological characterization of epitopes on tau of Alzheimer's type and chemically modified tau. *Mol Cell Biochem* 1997;**168**:59–66.
95. Ha E, Bang JH, Son JN, Cho HC, Mun KC. Carbamylated albumin stimulates microRNA-146, which is increased in human renal cell carcinoma. *Mol Med Rep* 2010;**3**:275–9.
96. Roxborough HE, Millar CA, McEneny J, Young IS. Carbamylation inhibits the ferroxidase activity of caeruloplasmin. *Biochem Biophys Res Commun* 1995;**214**:1073–8.
97. Kassa RM, Kasensa NL, Monterroso VH, Kayton RJ, Klimek JE, David LL, Lunganza KR, Kayembe KT, Bentivoglio M, Juliano SL, Tshala-Katumbay DD. On the biomarkers and mechanisms of konzo, a distinct upper motor neuron disease associated with food (cassava) cyanogenic exposure. *Food Chem Toxicol* 2011;**49**:571–8.
98. Kimani S, Moterroso V, Lasarev M, Kipruto S, Bukachi F, Maitai C, David L, Tshala-Katumbay D. Carbamoylation correlates of cyanate neuropathy and cyanide poisoning: relevance to the biomarkers of cassava cyanogenesis and motor system toxicity. *Springerplus* 2013;**2**:647.
99. Kimani S, Sinei K, Bukachi F, Tshala-Katumbay D, Maitai C. Memory deficits associated with sublethal cyanide poisoning relative to cyanate toxicity in rodents. *Metab Brain Dis* 2014;**29**:105–12.
100. Tor-Agbidye J, Palmer VS, Lasarev MR, Craig AM, Blythe LL, Sabri MI, Spencer PS. Bioactivation of cyanide to cyanate in sulfur amino acid deficiency: relevance to neurological disease in humans subsisting on cassava. *Toxicol Sci* 1999;**50**:228–35.
101. Stadtman ER. Metal ion-catalyzed oxidation of proteins: biochemical mechanism and biological consequences. *Free Radic Biol Med* 1990;**9**:315–25.
102. Stadtman ER, Berlett BS. Fenton chemistry. Amino acid oxidation. *J Biol Chem* 1991;**266**:17201–11.
103. Stadtman ER, Levine RL. Free radical-mediated oxidation of free amino acids and amino acid residues in proteins. *Amino Acids* 2003;**25**:207–18.
104. Stadtman ER, Levine RL. Chemical modification of proteins by reactive oxygen species. In: Dalle-Donne I, Scaloni A, Butterfield DA, editors. *Redox proteomics: From protein modifications to cellular dysfunction and disease*. Hoboken, NJ: John Wiley & Sons, Inc.; 2006. p. 3–23.
105. Moldovan L, Moldovan NI. Oxygen free radicals and redox biology of organelles. *Histochem Cell Biol* 2004;**122**:395–412.
106. Thannickal VJ, Fanburg BL. Reactive oxygen species in cell signaling. *Am J Physiol Lung Cell Mol Physiol* 2000;**279**:L1005–28.
107. Dalle-Donne I, Rossi R, Colombo R, Giustarini D, Milzani A. Biomarkers of oxidative damage in human disease. *Clin Chem* 2006;**52**:601–23.
108. Davies MJ, Fu S, Wang H, Dean RT. Stable markers of oxidant damage to proteins and their application in the study of human disease. *Free Radic Biol Med* 1999;**27**:1151–63.
109. Cabiscol E, Ros J. Oxidative damage to proteins: structural modifications and consequences in cell function. In: Dalle-Donne I, Scaloni A, Butterfield DA, editors. *Redox proteomics: From protein modifications to cellular dysfunction and disease*. Hoboken, NJ: John Wiley & Sons, Inc.; 2006. p. 399–471.
110. Friguet B, Stadtman ER, Szweda LI. Modification of glucose-6-phosphate dehydrogenase by 4-hydroxy-2-nonenal. Formation of cross-linked protein that inhibits the multicatalytic protease. *J Biol Chem* 1994;**269**:21639–43.

111. Requena JR, Fu MX, Ahmed MU, Jenkins AJ, Lyons TJ, Thorpe SR. Lipoxidation products as biomarkers of oxidative damage to proteins during lipid peroxidation reactions. *Nephrol Dial Transplant* 1996;**11**(Suppl. 5):48–53.
112. Butterfield DA, Castegna A, Lauderback CM, Drake J. Evidence that amyloid beta-peptide-induced lipid peroxidation and its sequelae in Alzheimer's disease brain contribute to neuronal death. *Neurobiol Aging* 2002;**23**:655–64.
113. Butterfield DA, Drake J, Pocernich C, Castegna A. Evidence of oxidative damage in Alzheimer's disease brain: central role for amyloid beta-peptide. *Trends Mol Med* 2001;**7**:548–54.
114. Butterfield DA, Lauderback CM. Lipid peroxidation and protein oxidation in Alzheimer's disease brain: potential causes and consequences involving amyloid beta-peptide-associated free radical oxidative stress. *Free Radic Biol Med* 2002;**32**:1050–60.
115. Hensley K, Hall N, Subramaniam R, Cole P, Harris M, Aksenov M, Aksenova M, Gabbita SP, Wu JF, Carney JM, et al. Brain regional correspondence between Alzheimer's disease histopathology and biomarkers of protein oxidation. *J Neurochem* 1995;**65**:2146–56.
116. Aksenov MY, Aksenova MV, Butterfield DA, Geddes JW, Markesbery WR. Protein oxidation in the brain in Alzheimer's disease. *Neuroscience* 2001;**103**:373–83.
117. Smith MA, Perry G, Richey PL, Sayre LM, Anderson VE, Beal MF, Kowall N. Oxidative damage in Alzheimer's. *Nature* 1996;**382**:120–1.
118. Perry G, Raina AK, Nunomura A, Wataya T, Sayre LM, Smith MA. How important is oxidative damage? Lessons from Alzheimer's disease. *Free Radic Biol Med* 2000;**28**:831–4.
119. Stolzing A, Grune T. The proteasome and its function in the ageing process. *Clin Exp Dermatol* 2001;**26**:566–72.
120. Jiang C, Schuman EM. Regulation and function of local protein synthesis in neuronal dendrites. *Trends Biochem Sci* 2002;**27**:506–13.
121. Job C, Eberwine J. Localization and translation of mRNA in dendrites and axons. *Nat Rev Neurosci* 2001;**2**:889–98.
122. Piwien-Pilipuk G, Ayala A, Machado A, Galigniana MD. Impairment of mineralocorticoid receptor (MR)-dependent biological response by oxidative stress and aging: correlation with post-translational modification of MR and decreased ADP-ribosylatable level of elongating factor 2 in kidney cells. *J Biol Chem* 2002;**277**:11896–903.
123. Aksenov M, Aksenova M, Butterfield DA, Markesbery WR. Oxidative modification of creatine kinase BB in Alzheimer's disease brain. *J Neurochem* 2000;**74**:2520–7.
124. Castegna A, Aksenov M, Aksenova M, Thongboonkerd V, Klein JB, Pierce WM, Booze R, Markesbery WR, Butterfield DA. Proteomic identification of oxidatively modified proteins in Alzheimer's disease brain. Part I: Creatine kinase BB, glutamine synthase, and ubiquitin carboxy-terminal hydrolase L-1. *Free Radic Biol Med* 2002;**33**:562–71.
125. Castegna A, Aksenov M, Thongboonkerd V, Klein JB, Pierce WM, Booze R, Markesbery WR, Butterfield DA. Proteomic identification of oxidatively modified proteins in Alzheimer's disease brain. Part II: Dihydropyrimidinase-related protein 2, alpha-enolase and heat shock cognate 71. *J Neurochem* 2002;**82**:1524–32.
126. Parinandi NL, Kleinberg MA, Usatyuk PV, Cummings RJ, Pennathur A, Cardounel AJ, Zweier JL, Garcia JG, Natarajan V. Hyperoxia-induced NAD(P)H oxidase activation and regulation by MAP kinases in human lung endothelial cells. *Am J Physiol Lung Cell Mol Physiol* 2003;**284**:L26–38.
127. Schock BC, Sweet DG, Halliday HL, Young IS, Ennis M. Oxidative stress in lavage fluid of preterm infants at risk of chronic lung disease. *Am J Physiol Lung Cell Mol Physiol* 2001;**281**:L1386–91.
128. Buss IH, Darlow BA, Winterbourn CC. Elevated protein carbonyls and lipid peroxidation products correlating with myeloperoxidase in tracheal aspirates from premature infants. *Pediatr Res* 2000;**47**:640–5.

129. Dalle-Donne I, Giustarini D, Colombo R, Rossi R, Milzani A. Protein carbonylation in human diseases. *Trends Mol Med* 2003;**9**:169–76.
130. Barreiro E. Role of protein carbonylation in skeletal muscle mass loss associated with chronic conditions. *Proteomes* 2016;**4**.
131. Booth FW. Effect of limb immobilization on skeletal muscle. *J Appl Physiol Respir Environ Exerc Physiol* 1982;**52**:1113–8.
132. Powers SK, Kavazis AN, DeRuisseau KC. Mechanisms of disuse muscle atrophy: role of oxidative stress. *Am J Physiol Regul Integr Comp Physiol* 2005;**288**:R337–44.
133. Kondo H, Miura M, Itokawa Y. Oxidative stress in skeletal muscle atrophied by immobilization. *Acta Physiol Scand* 1991;**142**:527–8.
134. Kondo H, Miura M, Nakagaki I, Sasaki S, Itokawa Y. Trace element movement and oxidative stress in skeletal muscle atrophied by immobilization. *Am J Physiol* 1992;**262**: E583–90.
135. Kondo H, Nakagaki I, Sasaki S, Hori S, Itokawa Y. Mechanism of oxidative stress in skeletal muscle atrophied by immobilization. *Am J Physiol* 1993;**265**:E839–44.
136. Lawler JM, Song W, Demaree SR. Hindlimb unloading increases oxidative stress and disrupts antioxidant capacity in skeletal muscle. *Free Radic Biol Med* 2003;**35**:9–16.
137. Shanely RA, Zergeroglu MA, Lennon SL, Sugiura T, Yimlamai T, Enns D, Belcastro A, Powers SK. Mechanical ventilation-induced diaphragmatic atrophy is associated with oxidative injury and increased proteolytic activity. *Am J Respir Crit Care Med* 2002;**166**:1369–74.
138. Zergeroglu MA, McKenzie MJ, Shanely RA, Van Gammeren D, DeRuisseau KC, Powers SK. Mechanical ventilation-induced oxidative stress in the diaphragm. *J Appl Physiol (1985)* 2003;**95**:1116–24.
139. Levine S, Nguyen T, Taylor N, Friscia ME, Budak MT, Rothenberg P, Zhu J, Sachdeva R, Sonnad S, Kaiser LR, Rubinstein NA, Powers SK, Shrager JB. Rapid disuse atrophy of diaphragm fibers in mechanically ventilated humans. *N Engl J Med* 2008;**358**:1327–35.
140. Kavazis AN, Talbert EE, Smuder AJ, Hudson MB, Nelson WB, Powers SK. Mechanical ventilation induces diaphragmatic mitochondrial dysfunction and increased oxidant production. *Free Radic Biol Med* 2009;**46**:842–50.
141. Whidden MA, McClung JM, Falk DJ, Hudson MB, Smuder AJ, Nelson WB, Powers SK. Xanthine oxidase contributes to mechanical ventilation-induced diaphragmatic oxidative stress and contractile dysfunction. *J Appl Physiol (1985)* 2009;**106**:385–94.
142. Sahin E, Gumuslu S. Stress-dependent induction of protein oxidation, lipid peroxidation and anti-oxidants in peripheral tissues of rats: comparison of three stress models (immobilization, cold and immobilization-cold). *Clin Exp Pharmacol Physiol* 2007;**34**:425–31.
143. Sahin E, Gumuslu S. Immobilization stress in rat tissues: alterations in protein oxidation, lipid peroxidation and antioxidant defense system. *Comp Biochem Physiol C Toxicol Pharmacol* 2007;**144**:342–7.
144. Powers SK, Duarte J, Kavazis AN, Talbert EE. Reactive oxygen species are signalling molecules for skeletal muscle adaptation. *Exp Physiol* 2010;**95**:1–9.
145. Primeau AJ, Adhihetty PJ, Hood DA. Apoptosis in heart and skeletal muscle. *Can J Appl Physiol* 2002;**27**:349–95.
146. Li YP, Chen Y, Li AS, Reid MB. Hydrogen peroxide stimulates ubiquitin-conjugating activity and expression of genes for specific E2 and E3 proteins in skeletal muscle myotubes. *Am J Physiol Cell Physiol* 2003;**285**:C806–12.
147. Himmelfarb J, McMonagle E, McMenamin E. Plasma protein thiol oxidation and carbonyl formation in chronic renal failure. *Kidney Int* 2000;**58**:2571–8.
148. Himmelfarb J, McMonagle E. Albumin is the major plasma protein target of oxidant stress in uremia. *Kidney Int* 2001;**60**:358–63.
149. Agarwal R. Proinflammatory effects of oxidative stress in chronic kidney disease: role of additional angiotensin II blockade. *Am J Physiol Renal Physiol* 2003;**284**:F863–9.

150. Lim PS, Cheng YM, Wei YH. Increase in oxidative damage to lipids and proteins in skeletal muscle of uremic patients. *Free Radic Res* 2002;**36**:295–301.
151. Miyata T, Ueda Y, Saito A, Kurokawa K. 'Carbonyl stress' and dialysis-related amyloidosis. *Nephrol Dial Transplant* 2000;**15**(Suppl. 1):25–8.
152. Miyata T, van Ypersele de Strihou C, Kurokawa K, Baynes JW. Alterations in nonenzymatic biochemistry in uremia: origin and significance of "carbonyl stress" in long-term uremic complications. *Kidney Int* 1999;**55**:389–99.
153. Singh R, Barden A, Mori T, Beilin L. Advanced glycation end-products: a review. *Diabetologia* 2001;**44**:129–46.
154. Halliwell Ba, Gutteridge J. *Free radicals in biology and medicine.* 3rd ed. Oxford University Press; 1999.
155. Cakatay U, Telci A, Salman S, Satman L, Sivas A. Oxidative protein damage in type I diabetic patients with and without complications. *Endocr Res* 2000;**26**:365–79.
156. Dominguez C, Ruiz E, Gussinye M, Carrascosa A. Oxidative stress at onset and in early stages of type 1 diabetes in children and adolescents. *Diabetes Care* 1998;**21**:1736–42.
157. Telci A, Cakatay U, Kayali R, Erdogan C, Orhan Y, Sivas A, Akcay T. Oxidative protein damage in plasma of type 2 diabetic patients. *Horm Metab Res* 2000;**32**:40–3.
158. Telci A, Cakatay U, Salman S, Satman I, Sivas A. Oxidative protein damage in early stage Type 1 diabetic patients. *Diabetes Res Clin Pract* 2000;**50**:213–23.
159. Grattagliano I, Vendemiale G, Boscia F, Micelli-Ferrari T, Cardia L, Altomare E. Oxidative retinal products and ocular damages in diabetic patients. *Free Radic Biol Med* 1998;**25**:369–72.
160. Jakubowski H. Metabolism of homocysteine thiolactone in human cell cultures. Possible mechanism for pathological consequences of elevated homocysteine levels. *J Biol Chem* 1997;**272**:1935–42.
161. Jakubowski H. Protein homocysteinylation: possible mechanism underlying pathological consequences of elevated homocysteine levels. *FASEB J* 1999;**13**:2277–83.
162. Jakubowski H. Calcium-dependent human serum homocysteine thiolactone hydrolase. A protective mechanism against protein N-homocysteinylation. *J Biol Chem* 2000;**275**:3957–62.
163. Jakubowski H, Zhang L, Bardeguez A, Aviv A. Homocysteine thiolactone and protein homocysteinylation in human endothelial cells: implications for atherosclerosis. *Circ Res* 2000;**87**:45–51.
164. Jakubowski H. Homocysteine thiolactone: metabolic origin and protein homocysteinylation in humans. *J Nutr* 2000;**130**:377S–381S.
165. Jakubowski H, Goldman E. Synthesis of homocysteine thiolactone by methionyl-tRNA synthetase in cultured mammalian cells. *FEBS Lett* 1993;**317**:237–40.
166. Zinellu A, Zinellu E, Sotgia S, Formato M, Cherchi GM, Deiana L, Carru C. Factors affecting S-homocysteinylation of LDL apoprotein B. *Clin Chem* 2006;**52**:2054–9.
167. Jalili S, Yousefi R, Papari MM, Moosavi-Movahedi AA. Effect of homocysteine thiolactone on structure and aggregation propensity of bovine pancreatic insulin. *Protein J* 2011;**30**:299–307.
168. Perla-Kajan J, Marczak L, Kajan L, Skowronek P, Twardowski T, Jakubowski H. Modification by homocysteine thiolactone affects redox status of cytochrome C. *Biochemistry* 2007;**46**:6225–31.
169. Zang T, Dai S, Chen D, Lee BW, Liu S, Karger BL, Zhou ZS. Chemical methods for the detection of protein N-homocysteinylation via selective reactions with aldehydes. *Anal Chem* 2009;**81**:9065–71.
170. Yousefi R, Jalili S, Alavi P, Moosavi-Movahedi AA. The enhancing effect of homocysteine thiolactone on insulin fibrillation and cytotoxicity of insulin fibril. *Int J Biol Macromol* 2012;**51**:291–8.
171. Glowacki R, Jakubowski H. Cross-talk between Cys34 and lysine residues in human serum albumin revealed by N-homocysteinylation. *J Biol Chem* 2004;**279**:10864–71.

172. Garel J, Tawfik DS. Mechanism of hydrolysis and aminolysis of homocysteine thiolactone. *Chemistry* 2006;**12**:4144–52.
173. Sharma GS, Kumar T, Singh LR. N-homocysteinylation induces different structural and functional consequences on acidic and basic proteins. *PLoS One* 2014;**9**:e116386.
174. Andreadis A, Brown WM, Kosik KS. Structure and novel exons of the human tau gene. *Biochemistry* 1992;**31**:10626–33.
175. Goedert M, Spillantini MG, Jakes R, Rutherford D, Crowther RA. Multiple isoforms of human microtubule-associated protein tau: sequences and localization in neurofibrillary tangles of Alzheimer's disease. *Neuron* 1989;**3**:519–26.
176. Buee L, Bussiere T, Buee-Scherrer V, Delacourte A, Hof PR. Tau protein isoforms, phosphorylation and role in neurodegenerative disorders. *Brain Res Brain Res Rev* 2000;**33**:95–130.
177. Mandelkow EM, Schweers O, Drewes G, Biernat J, Gustke N, Trinczek B, Mandelkow E. Structure, microtubule interactions, and phosphorylation of tau protein. *Ann N Y Acad Sci* 1996;**777**:96–106.
178. Hong M, Zhukareva V, Vogelsberg-Ragaglia V, Wszolek Z, Reed L, Miller BI, Geschwind DH, Bird TD, McKeel D, Goate A, Morris JC, Wilhelmsen KC, Schellenberg GD, Trojanowski JQ, Lee VM. Mutation-specific functional impairments in distinct tau isoforms of hereditary FTDP-17. *Science* 1998;**282**:1914–7.
179. Karima O, Riazi G, Khodadadi S, Aryapour H, Khalili MA, Yousefi L, Moosavi-Movahedi AA. Altered tubulin assembly dynamics with N-homocysteinylated human 4R/1N tau in vitro. *FEBS Lett* 2012;**586**:3914–9.
180. Trojanowski JQ, Lee VM. The role of tau in Alzheimer's disease. *Med Clin North Am* 2002;**86**:615–27.
181. Prusiner SB. Shattuck lecture—neurodegenerative diseases and prions. *N Engl J Med* 2001;**344**:1516–26.
182. Cisse M, Mucke L. Alzheimer's disease: a prion protein connection. *Nature* 2009;**457**:1090–1.
183. Lauren J, Gimbel DA, Nygaard HB, Gilbert JW, Strittmatter SM. Cellular prion protein mediates impairment of synaptic plasticity by amyloid-beta oligomers. *Nature* 2009;**457**:1128–32.
184. Stroylova YY, Chobert JM, Muronetz VI, Jakubowski H, Haertle T. N-homocysteinylation of ovine prion protein induces amyloid-like transformation. *Arch Biochem Biophys* 2012;**526**:29–37.
185. Lipton SA, Choi YB, Pan ZH, Lei SZ, Chen HS, Sucher NJ, Loscalzo J, Singel DJ, Stamler JS. A redox-based mechanism for the neuroprotective and neurodestructive effects of nitric oxide and related nitroso-compounds. *Nature* 1993;**364**:626–32.
186. Martinez-Ruiz A, Cadenas S, Lamas S. Nitric oxide signaling: classical, less classical, and nonclassical mechanisms. *Free Radic Biol Med* 2011;**51**:17–29.
187. Nakamura T, Tu S, Akhtar MW, Sunico CR, Okamoto S, Lipton SA. Aberrant protein s-nitrosylation in neurodegenerative diseases. *Neuron* 2013;**78**:596–614.
188. Smith BC, Marletta MA. Mechanisms of S-nitrosothiol formation and selectivity in nitric oxide signaling. *Curr Opin Chem Biol* 2012;**16**:498–506.
189. Hess DT, Matsumoto A, Kim SO, Marshall HE, Stamler JS. Protein S-nitrosylation: purview and parameters. *Nat Rev Mol Cell Biol* 2005;**6**:150–66.
190. Stamler JS, Lamas S, Fang FC. Nitrosylation. The prototypic redox-based signaling mechanism. *Cell* 2001;**106**:675–83.
191. Seth D, Stamler JS. The SNO-proteome: causation and classifications. *Curr Opin Chem Biol* 2011;**15**:129–36.
192. Stamler JS, Toone EJ, Lipton SA, Sucher NJ. (S)NO signals: translocation, regulation, and a consensus motif. *Neuron* 1997;**18**:691–6.
193. Doulias PT, Greene JL, Greco TM, Tenopoulou M, Seeholzer SH, Dunbrack RL, Ischiropoulos H. Structural profiling of endogenous S-nitrosocysteine residues reveals

unique features that accommodate diverse mechanisms for protein S-nitrosylation. *Proc Natl Acad Sci U S A* 2010;**107**:16958–63.
194. Benhar M, Forrester MT, Stamler JS. Protein denitrosylation: enzymatic mechanisms and cellular functions. *Nat Rev Mol Cell Biol* 2009;**10**:721–32.
195. Gu Z, Kaul M, Yan B, Kridel SJ, Cui J, Strongin A, Smith JW, Liddington RC, Lipton SA. S-nitrosylation of matrix metalloproteinases: signaling pathway to neuronal cell death. *Science* 2002;**297**:1186–90.
196. Uehara T, Nakamura T, Yao D, Shi ZQ, Gu Z, Ma Y, Masliah E, Nomura Y, Lipton SA. S-nitrosylated protein-disulphide isomerase links protein misfolding to neurodegeneration. *Nature* 2006;**441**:513–7.
197. Lipton SA, Choi YB, Takahashi H, Zhang D, Li W, Godzik A, Bankston LA. Cysteine regulation of protein function—as exemplified by NMDA-receptor modulation. *Trends Neurosci* 2002;**25**:474–80.
198. Ischiropoulos H, Beckman JS. Oxidative stress and nitration in neurodegeneration: cause, effect, or association? *J Clin Invest* 2003;**111**:163–9.
199. Rothstein JD. Current hypotheses for the underlying biology of amyotrophic lateral sclerosis. *Ann Neurol* 2009;**65**(Suppl. 1):S3–9.
200. Schonhoff CM, Matsuoka M, Tummala H, Johnson MA, Estevez AG, Wu R, Kamaid A, Ricart KC, Hashimoto Y, Gaston B, Macdonald TL, Xu Z, Mannick JB. S-nitrosothiol depletion in amyotrophic lateral sclerosis. *Proc Natl Acad Sci U S A* 2006;**103**:2404–9.
201. Cho DH, Nakamura T, Fang J, Cieplak P, Godzik A, Gu Z, Lipton SA. S-nitrosylation of Drp1 mediates beta-amyloid-related mitochondrial fission and neuronal injury. *Science* 2009;**324**:102–5.
202. Savitt JM, Dawson VL, Dawson TM. Diagnosis and treatment of Parkinson disease: molecules to medicine. *J Clin Invest* 2006;**116**:1744–54.
203. Tsang AH, Chung KK. Oxidative and nitrosative stress in Parkinson's disease. *Biochim Biophys Acta* 2009;**1792**:643–50.
204. Thomas B, Beal MF. Parkinson's disease. *Hum Mol Genet* 2007;R183–94. 16 Spec No. 2.
205. Giasson BI, Duda JE, Murray IV, Chen Q, Souza JM, Hurtig HI, Ischiropoulos H, Trojanowski JQ, Lee VM. Oxidative damage linked to neurodegeneration by selective alpha-synuclein nitration in synucleinopathy lesions. *Science* 2000;**290**:985–9.
206. Chung KK, Thomas B, Li X, Pletnikova O, Troncoso JC, Marsh L, Dawson VL, Dawson TM. S-nitrosylation of parkin regulates ubiquitination and compromises parkin's protective function. *Science* 2004;**304**:1328–31.
207. Feany MB, Pallanck LJ. Parkin: a multipurpose neuroprotective agent? *Neuron* 2003;**38**:13–6.
208. Kahle PJ, Haass C. How does parkin ligate ubiquitin to Parkinson's disease? *EMBO Rep* 2004;**5**:681–5.
209. Yao D, Gu Z, Nakamura T, Shi ZQ, Ma Y, Gaston B, Palmer LA, Rockenstein EM, Zhang Z, Masliah E, Uehara T, Lipton SA. Nitrosative stress linked to sporadic Parkinson's disease: S-nitrosylation of parkin regulates its E3 ubiquitin ligase activity. *Proc Natl Acad Sci U S A* 2004;**101**:10810–4.
210. Ito G, Ariga H, Nakagawa Y, Iwatsubo T. Roles of distinct cysteine residues in S-nitrosylation and dimerization of DJ-1. *Biochem Biophys Res Commun* 2006;**339**: 667–72.
211. Chung KK, Zhang Y, Lim KL, Tanaka Y, Huang H, Gao J, Ross CA, Dawson VL, Dawson TM. Parkin ubiquitinates the alpha-synuclein-interacting protein, synphilin-1: implications for Lewy-body formation in Parkinson disease. *Nat Med* 2001;**7**:1144–50.
212. Fang J, Nakamura T, Cho DH, Gu Z, Lipton SA. S-nitrosylation of peroxiredoxin 2 promotes oxidative stress-induced neuronal cell death in Parkinson's disease. *Proc Natl Acad Sci U S A* 2007;**104**:18742–7.
213. Srinivasula SM, Ashwell JD. IAPs: what's in a name? *Mol Cell* 2008;**30**:123–35.
214. Vaux DL, Silke J. IAPs, RINGs and ubiquitylation. *Nat Rev Mol Cell Biol* 2005;**6**:287–97.

215. Wang Z. Protein S-nitrosylation and cancer. *Cancer Lett* 2012;**320**:123–9.
216. Fukumura D, Kashiwagi S, Jain RK. The role of nitric oxide in tumour progression. *Nat Rev Cancer* 2006;**6**:521–34.
217. Jadeski LC, Chakraborty C, Lala PK. Role of nitric oxide in tumour progression with special reference to a murine breast cancer model. *Can J Physiol Pharmacol* 2002;**80**:125–35.
218. Lala PK, Chakraborty C. Role of nitric oxide in carcinogenesis and tumour progression. *Lancet Oncol* 2001;**2**:149–56.
219. Lim KH, Ancrile BB, Kashatus DF, Counter CM. Tumour maintenance is mediated by eNOS. *Nature* 2008;**452**:646–9.
220. Thomsen LL, Miles DW. Role of nitric oxide in tumour progression: lessons from human tumours. *Cancer Metastasis Rev* 1998;**17**:107–18.
221. Bentz BG, Haines 3rd GK, Radosevich JA. Increased protein nitrosylation in head and neck squamous cell carcinogenesis. *Head Neck* 2000;**22**:64–70.
222. Pervin S, Singh R, Chaudhuri G. Nitric oxide-induced cytostasis and cell cycle arrest of a human breast cancer cell line (MDA-MB-231): potential role of cyclin D1. *Proc Natl Acad Sci U S A* 2001;**98**:3583–8.
223. Melino G, Bernassola F, Knight RA, Corasaniti MT, Nistico G, Finazzi-Agro A. S-nitrosylation regulates apoptosis. *Nature* 1997;**388**:432–3.
224. Jaiswal M, LaRusso NF, Nishioka N, Nakabeppu Y, Gores GJ. Human Ogg1, a protein involved in the repair of 8-oxoguanine, is inhibited by nitric oxide. *Cancer Res* 2001;**61**:6388–93.
225. Mantovani A, Allavena P, Sica A, Balkwill F. Cancer-related inflammation. *Nature* 2008;**454**:436–44.
226. Coussens LM, Werb Z. Inflammation and cancer. *Nature* 2002;**420**:860–7.
227. Fitzpatrick B, Mehibel M, Cowen RL, Stratford IJ. iNOS as a therapeutic target for treatment of human tumors. *Nitric Oxide* 2008;**19**:217–24.
228. Li F, Sonveaux P, Rabbani ZN, Liu S, Yan B, Huang Q, Vujaskovic Z, Dewhirst MW, Li CY. Regulation of HIF-1alpha stability through S-nitrosylation. *Mol Cell* 2007;**26**:63–74.
229. Muntane J, la Mata MD. Nitric oxide and cancer. *World J Hepatol* 2010;**2**:337–44.
230. Boo YC, Kim HJ, Song H, Fulton D, Sessa W, Jo H. Coordinated regulation of endothelial nitric oxide synthase activity by phosphorylation and subcellular localization. *Free Radic Biol Med* 2006;**41**:144–53.
231. Iwakiri Y, Satoh A, Chatterjee S, Toomre DK, Chalouni CM, Fulton D, Groszmann RJ, Shah VH, Sessa WC. Nitric oxide synthase generates nitric oxide locally to regulate compartmentalized protein S-nitrosylation and protein trafficking. *Proc Natl Acad Sci U S A* 2006;**103**:19777–82.
232. Oess S, Icking A, Fulton D, Govers R, Muller-Esterl W. Subcellular targeting and trafficking of nitric oxide synthases. *Biochem J* 2006;**396**:401–9.
233. Alderton WK, Cooper CE, Knowles RG. Nitric oxide synthases: structure, function and inhibition. *Biochem J* 2001;**357**:593–615.
234. Vicente Miranda H, El-Agnaf OM, Outeiro TF. Glycation in Parkinson's disease and Alzheimer's disease. *Mov Disord* 2016;**31**:782–90.
235. Maillard L. Action des acides amines sur les sucre: formation des melanoidines par voie methodique. *C R Hebd Seances Acad Sci* 1912;**154**:66–8.
236. Yan SF, Ramasamy R, Schmidt AM. Mechanisms of disease: advanced glycation end-products and their receptor in inflammation and diabetes complications. *Nat Clin Pract Endocrinol Metab* 2008;**4**:285–93.
237. Monnier VM, Elmets CA, Frank KE, Vishwanath V, Yamashita T. Age-related normalization of the browning rate of collagen in diabetic subjects without retinopathy. *J Clin Invest* 1986;**78**:832–5.

238. Baynes JW, Monnier VM. *The Maillard reaction in aging, diabetes, and nutrition*. New York: Alan R. Liss; 1989.
239. Ahmed MU, Thorpe SR, Baynes JW. Identification of carboxymethyllysine as a degradation product of fructoselysine in glycated protein. *J Biol Chem* 1986;**261**:8816–21.
240. Wells-Knecht MC, Thorpe SR, Baynes JW. Pathways of formation of glycoxidation products during glycation of collagen. *Biochemistry* 1995;**34**:15134–41.
241. Thornalley PJ. The glyoxalase system: new developments towards functional characterization of a metabolic pathway fundamental to biological life. *Biochem J* 1990;**269**:1–11.
242. Thornalley PJ, Langborg A, Minhas HS. Formation of glyoxal, methylglyoxal and 3-deoxyglucosone in the glycation of proteins by glucose. *Biochem J* 1999;**344**(Pt 1):109–16.
243. Forbes JM, Soldatos G, Thomas MC. Below the radar: advanced glycation end products that detour "around the side". Is HbA1c not an accurate enough predictor of long term progression and glycaemic control in diabetes? *Clin Biochem Rev* 2005;**26**:123–34.
244. Jang C, Lim JH, Park CW, Cho YJ. Regulator of calcineurin 1 isoform 4 (RCAN1.4) is overexpressed in the glomeruli of diabetic mice. *Kor J Physiol Pharmacol* 2011;**15**:299–305.
245. Chen M, Curtis TM, Stitt AW. Advanced glycation end products and diabetic retinopathy. *Curr Med Chem* 2013;**20**:3234–40.
246. Frank RN. Diabetic retinopathy. *N Engl J Med* 2004;**350**:48–58.
247. Zong H, Ward M, Madden A, Yong PH, Limb GA, Curtis TM, Stitt AW. Hyperglycaemia-induced pro-inflammatory responses by retinal Muller glia are regulated by the receptor for advanced glycation end-products (RAGE). *Diabetologia* 2010;**53**:2656–66.
248. Stitt AW, Curtis TM. Diabetes-related adduct formation and retinopathy. *J Ocul Biol Dis Inform* 2011;**4**:10–8.
249. Zong H, Ward M, Stitt AW. AGEs, RAGE, and diabetic retinopathy. *Curr Diab Rep* 2011;**11**:244–52.
250. Curtis TM, Hamilton R, Yong PH, McVicar CM, Berner A, Pringle R, Uchida K, Nagai R, Brockbank S, Stitt AW. Muller glial dysfunction during diabetic retinopathy in rats is linked to accumulation of advanced glycation end-products and advanced lipoxidation end-products. *Diabetologia* 2011;**54**:690–8.
251. Miller AG, Zhu T, Wilkinson-Berka JL. The renin-angiotensin system and advanced glycation end-products in diabetic retinopathy: impacts and synergies. *Curr Clin Pharmacol* 2013;**8**:285–96.
252. Berner AK, Brouwers O, Pringle R, Klaassen I, Colhoun L, McVicar C, Brockbank S, Curry JW, Miyata T, Brownlee M, Schlingemann RO, Schalkwijk C, Stitt AW. Protection against methylglyoxal-derived AGEs by regulation of glyoxalase 1 prevents retinal neuroglial and vasodegenerative pathology. *Diabetologia* 2012;**55**:845–54.
253. Ai J, Liu Y, Sun JH. Advanced glycation end-products stimulate basic fibroblast growth factor expression in cultured Muller cells. *Mol Med Rep* 2013;**7**:16–20.
254. Nagaraj RH, Linetsky M, Stitt AW. The pathogenic role of Maillard reaction in the aging eye. *Amino Acids* 2012;**42**:1205–20.
255. Choudhuri S, Dutta D, Sen A, Chowdhury IH, Mitra B, Mondal LK, Saha A, Bhadhuri G, Bhattacharya B. Role of N-epsilon-carboxy methyl lysine, advanced glycation end products and reactive oxygen species for the development of nonproliferative and proliferative retinopathy in type 2 diabetes mellitus. *Mol Vis* 2013;**19**:100–13.
256. Vasan S, Foiles P, Founds H. Therapeutic potential of breakers of advanced glycation end product-protein crosslinks. *Arch Biochem Biophys* 2003;**419**:89–96.
257. Neeper M, Schmidt AM, Brett J, Yan SD, Wang F, Pan YC, Elliston K, Stern D, Shaw A. Cloning and expression of a cell surface receptor for advanced glycosylation end products of proteins. *J Biol Chem* 1992;**267**:14998–5004.
258. Bucciarelli LG, Wendt T, Qu W, Lu Y, Lalla E, Rong LL, Goova MT, Moser B, Kislinger T, Lee DC, Kashyap Y, Stern DM, Schmidt AM. RAGE blockade stabilizes established atherosclerosis in diabetic apolipoprotein E-null mice. *Circulation* 2002;**106**:2827–35.

259. Chibber R, Molinatti PA, Rosatto N, Lambourne B, Kohner EM. Toxic action of advanced glycation end products on cultured retinal capillary pericytes and endothelial cells: relevance to diabetic retinopathy. *Diabetologia* 1997;**40**:156–64.
260. Tanaka N, Yonekura H, Yamagishi S, Fujimori H, Yamamoto Y, Yamamoto H. The receptor for advanced glycation end products is induced by the glycation products themselves and tumor necrosis factor-alpha through nuclear factor-kappa B, and by 17beta-estradiol through Sp-1 in human vascular endothelial cells. *J Biol Chem* 2000;**275**:25781–90.
261. Mamputu JC, Renier G. Advanced glycation end-products increase monocyte adhesion to retinal endothelial cells through vascular endothelial growth factor-induced ICAM-1 expression: inhibitory effect of antioxidants. *J Leukoc Biol* 2004;**75**:1062–9.
262. Moore TC, Moore JE, Kaji Y, Frizzell N, Usui T, Poulaki V, Campbell IL, Stitt AW, Gardiner TA, Archer DB, Adamis AP. The role of advanced glycation end products in retinal microvascular leukostasis. *Invest Ophthalmol Vis Sci* 2003;**44**:4457–64.
263. Antonetti DA, Barber AJ, Khin S, Lieth E, Tarbell JM, Gardner TW. Vascular permeability in experimental diabetes is associated with reduced endothelial occludin content: vascular endothelial growth factor decreases occludin in retinal endothelial cells. Penn State Retina Research Group. *Diabetes* 1998;**47**:1953–9.
264. Nakamura N, Hasegawa G, Obayashi H, Yamazaki M, Ogata M, Nakano K, Yoshikawa T, Watanabe A, Kinoshita S, Fujinami A, Ohta M, Imamura Y, Ikeda T. Increased concentration of pentosidine, an advanced glycation end product, and interleukin-6 in the vitreous of patients with proliferative diabetic retinopathy. *Diabetes Res Clin Pract* 2003;**61**:93–101.
265. Joussen AM, Poulaki V, Mitsiades N, Cai WY, Suzuma I, Pak J, Ju ST, Rook SL, Esser P, Mitsiades CS, Kirchhof B, Adamis AP, Aiello LP. Suppression of Fas-FasL-induced endothelial cell apoptosis prevents diabetic blood-retinal barrier breakdown in a model of streptozotocin-induced diabetes. *FASEB J* 2003;**17**:76–8.
266. Podesta F, Romeo G, Liu WH, Krajewski S, Reed JC, Gerhardinger C, Lorenzi M. Bax is increased in the retina of diabetic subjects and is associated with pericyte apoptosis in vivo and in vitro. *Am J Pathol* 2000;**156**:1025–32.
267. Caldwell RB, Bartoli M, Behzadian MA, El-Remessy AE, Al-Shabrawey M, Platt DH, Caldwell RW. Vascular endothelial growth factor and diabetic retinopathy: pathophysiological mechanisms and treatment perspectives. *Diabetes Metab Res Rev* 2003;**19**:442–55.
268. Singh VP, Bali A, Singh N, Jaggi AS. Advanced glycation end products and diabetic complications. *Kor J Physiol Pharmacol* 2014;**18**:1–14.
269. Bell DS. Diabetic cardiomyopathy. *Diabetes Care* 2003;**26**:2949–51.
270. Montagnani M. Diabetic cardiomyopathy: how much does it depend on AGE? *Br J Pharmacol* 2008;**154**:725–6.
271. Bodiga VL, Eda SR, Bodiga S. Advanced glycation end products: role in pathology of diabetic cardiomyopathy. *Heart Fail Rev* 2014;**19**:49–63.
272. Ma H, Li SY, Xu P, Babcock SA, Dolence EK, Brownlee M, Li J, Ren J. Advanced glycation endproduct (AGE) accumulation and AGE receptor (RAGE) up-regulation contribute to the onset of diabetic cardiomyopathy. *J Cell Mol Med* 2009;**13**:1751–64.
273. Bidasee KR, Zhang Y, Shao CH, Wang M, Patel KP, Dincer UD, Besch Jr. HR. Diabetes increases formation of advanced glycation end products on Sarco(endo)plasmic reticulum Ca2+-ATPase. *Diabetes* 2004;**53**:463–73.
274. Ren J, Gintant GA, Miller RE, Davidoff AJ. High extracellular glucose impairs cardiac E-C coupling in a glycosylation-dependent manner. *Am J Physiol* 1997;**273**: H2876–83.
275. Smit AJ, Lutgers HL. The clinical relevance of advanced glycation endproducts (AGE) and recent developments in pharmaceutics to reduce AGE accumulation. *Curr Med Chem* 2004;**11**:2767–84.

276. Brownlee M. Advanced protein glycosylation in diabetes and aging. *Annu Rev Med* 1995;**46**:223–34.
277. Kass DA, Shapiro EP, Kawaguchi M, Capriotti AR, Scuteri A, deGroof RC, Lakatta EG. Improved arterial compliance by a novel advanced glycation end-product crosslink breaker. *Circulation* 2001;**104**:1464–70.
278. Posch K, Simecek S, Wascher TC, Jurgens G, Baumgartner-Parzer S, Kostner GM, Graier WF. Glycated low-density lipoprotein attenuates shear stress-induced nitric oxide synthesis by inhibition of shear stress-activated L-arginine uptake in endothelial cells. *Diabetes* 1999;**48**:1331–7.
279. Brett J, Schmidt AM, Yan SD, Zou YS, Weidman E, Pinsky D, Nowygrod R, Neeper M, Przysiecki C, Shaw A, et al. Survey of the distribution of a newly characterized receptor for advanced glycation end products in tissues. *Am J Pathol* 1993;**143**:1699–712.
280. Striker LJ, Striker GE. Administration of AGEs in vivo induces extracellular matrix gene expression. *Nephrol Dial Transplant* 1996;**11**(Suppl. 5):62–5.
281. Throckmorton DC, Brogden AP, Min B, Rasmussen H, Kashgarian M. PDGF and TGF-beta mediate collagen production by mesangial cells exposed to advanced glycosylation end products. *Kidney Int* 1995;**48**:111–7.
282. Petrova R, Yamamoto Y, Muraki K, Yonekura H, Sakurai S, Watanabe T, Li H, Takeuchi M, Makita Z, Kato I, Takasawa S, Okamoto H, Imaizumi Y, Yamamoto H. Advanced glycation endproduct-induced calcium handling impairment in mouse cardiac myocytes. *J Mol Cell Cardiol* 2002;**34**:1425–31.
283. Vagelatos NT, Eslick GD. Type 2 diabetes as a risk factor for Alzheimer's disease: the confounders, interactions, and neuropathology associated with this relationship. *Epidemiol Rev* 2013;**35**:152–60.
284. Biessels GJ, De Leeuw FE, Lindeboom J, Barkhof F, Scheltens P. Increased cortical atrophy in patients with Alzheimer's disease and type 2 diabetes mellitus. *J Neurol Neurosurg Psychiatry* 2006;**77**:304–7.
285. Moloney AM, Griffin RJ, Timmons S, O'Connor R, Ravid R, O'Neill C. Defects in IGF-1 receptor, insulin receptor and IRS-1/2 in Alzheimer's disease indicate possible resistance to IGF-1 and insulin signalling. *Neurobiol Aging* 2010;**31**:224–43.
286. Trudeau F, Gagnon S, Massicotte G. Hippocampal synaptic plasticity and glutamate receptor regulation: influences of diabetes mellitus. *Eur J Pharmacol* 2004;**490**:177–86.
287. Craft S, Baker LD, Montine TJ, Minoshima S, Watson GS, Claxton A, Arbuckle M, Callaghan M, Tsai E, Plymate SR, Green PS, Leverenz J, Cross D, Gerton B. Intranasal insulin therapy for Alzheimer disease and amnestic mild cognitive impairment: a pilot clinical trial. *Arch Neurol* 2012;**69**:29–38.
288. Carvalho C, Cardoso S, Correia SC, Santos RX, Santos MS, Baldeiras I, Oliveira CR, Moreira PI. Metabolic alterations induced by sucrose intake and Alzheimer's disease promote similar brain mitochondrial abnormalities. *Diabetes* 2012;**61**:1234–42.
289. Vitek MP, Bhattacharya K, Glendening JM, Stopa E, Vlassara H, Bucala R, Manogue K, Cerami A. Advanced glycation end products contribute to amyloidosis in Alzheimer disease. *Proc Natl Acad Sci U S A* 1994;**91**:4766–70.
290. Smith MA, Taneda S, Richey PL, Miyata S, Yan SD, Stern D, Sayre LM, Monnier VM, Perry G. Advanced Maillard reaction end products are associated with Alzheimer disease pathology. *Proc Natl Acad Sci U S A* 1994;**91**:5710–4.
291. Faure P, Troncy L, Lecomte M, Wiernsperger N, Lagarde M, Ruggiero D, Halimi S. Albumin antioxidant capacity is modified by methylglyoxal. *Diabetes Metab* 2005;**31**:169–77.
292. Kimura T, Takamatsu J, Araki N, Goto M, Kondo A, Miyakawa T, Horiuchi S. Are advanced glycation end-products associated with amyloidosis in Alzheimer's disease? *Neuroreport* 1995;**6**:866–8.
293. Ledesma MD, Bonay P, Avila J. Tau protein from Alzheimer's disease patients is glycated at its tubulin-binding domain. *J Neurochem* 1995;**65**:1658–64.

294. Chen L, Thiruchelvam MJ, Madura K, Richfield EK. Proteasome dysfunction in aged human alpha-synuclein transgenic mice. *Neurobiol Dis* 2006;**23**:120–6.
295. Ledesma MD, Bonay P, Colaco C, Avila J. Analysis of microtubule-associated protein tau glycation in paired helical filaments. *J Biol Chem* 1994;**269**:21614–9.
296. Munch G, Mayer S, Michaelis J, Hipkiss AR, Riederer P, Muller R, Neumann A, Schinzel R, Cunningham AM. Influence of advanced glycation end-products and AGE-inhibitors on nucleation-dependent polymerization of beta-amyloid peptide. *Biochim Biophys Acta* 1997;**1360**:17–29.
297. Necula M, Kuret J. Pseudophosphorylation and glycation of tau protein enhance but do not trigger fibrillization in vitro. *J Biol Chem* 2004;**279**:49694–703.
298. Ko SY, Lin YP, Lin YS, Chang SS. Advanced glycation end products enhance amyloid precursor protein expression by inducing reactive oxygen species. *Free Radic Biol Med* 2010;**49**:474–80.
299. Liu BF, Miyata S, Hirota Y, Higo S, Miyazaki H, Fukunaga M, Hamada Y, Ueyama S, Muramoto O, Uriuhara A, Kasuga M. Methylglyoxal induces apoptosis through activation of p38 mitogen-activated protein kinase in rat mesangial cells. *Kidney Int* 2003;**63**:947–57.
300. Reynolds CH, Betts JC, Blackstock WP, Nebreda AR, Anderton BH. Phosphorylation sites on tau identified by nanoelectrospray mass spectrometry: differences in vitro between the mitogen-activated protein kinases ERK2, c-Jun N-terminal kinase and P38, and glycogen synthase kinase-3beta. *J Neurochem* 2000;**74**:1587–95.
301. Zhu X, Rottkamp CA, Boux H, Takeda A, Perry G, Smith MA. Activation of p38 kinase links tau phosphorylation, oxidative stress, and cell cycle-related events in Alzheimer disease. *J Neuropathol Exp Neurol* 2000;**59**:880–8.
302. Luth HJ, Ogunlade V, Kuhla B, Kientsch-Engel R, Stahl P, Webster J, Arendt T, Munch G. Age- and stage-dependent accumulation of advanced glycation end products in intracellular deposits in normal and Alzheimer's disease brains. *Cereb Cortex* 2005;**15**:211–20.
303. Ahmed N, Ahmed U, Thornalley PJ, Hager K, Fleischer G, Munch G. Protein glycation, oxidation and nitration adduct residues and free adducts of cerebrospinal fluid in Alzheimer's disease and link to cognitive impairment. *J Neurochem* 2005;**92**:255–63.
304. Bar KJ, Franke S, Wenda B, Muller S, Kientsch-Engel R, Stein G, Sauer H. Pentosidine and N(epsilon)-(carboxymethyl)-lysine in Alzheimer's disease and vascular dementia. *Neurobiol Aging* 2003;**24**:333–8.
305. Shuvaev VV, Laffont I, Serot JM, Fujii J, Taniguchi N, Siest G. Increased protein glycation in cerebrospinal fluid of Alzheimer's disease. *Neurobiol Aging* 2001;**22**:397–402.
306. Barbeau A, Giguere R, Hardy J. Clinical experience with tolbutamide in Parkinson's disease. *Union Med Can* 1961;**90**:147–51.
307. Lipman IJ, Boykin ME, Flora RE. Glucose intolerance in Parkinson's disease. *J Chronic Dis* 1974;**27**:573–9.
308. Sandyk R. The relationship between diabetes mellitus and Parkinson's disease. *Int J Neurosci* 1993;**69**:125–30.
309. Boyd 3rd AE, Lebovitz HE, Feldman JM. Endocrine function and glucose metabolism in patients with Parkinson's disease and their alternation by L-Dopa. *J Clin Endocrinol Metab* 1971;**33**:829–37.
310. Hu G, Jousilahti P, Bidel S, Antikainen R, Tuomilehto J. Type 2 diabetes and the risk of Parkinson's disease. *Diabetes Care* 2007;**30**:842–7.
311. Xu Q, Park Y, Huang X, Hollenbeck A, Blair A, Schatzkin A, Chen H. Diabetes and risk of Parkinson's disease. *Diabetes Care* 2011;**34**:910–5.
312. Arvanitakis Z, Wilson RS, Bienias JL, Bennett DA. Diabetes and parkinsonian signs in older persons. *Alzheimer Dis Assoc Disord* 2007;**21**:144–9.
313. Cereda E, Barichella M, Cassani E, Caccialanza R, Pezzoli G. Clinical features of Parkinson disease when onset of diabetes came first: a case-control study. *Neurology* 2012;**78**:1507–11.

314. Becker C, Brobert GP, Johansson S, Jick SS, Meier CR. Diabetes in patients with idiopathic Parkinson's disease. *Diabetes Care* 2008;**31**:1808–12.
315. Cereda E, Barichella M, Pedrolli C, Klersy C, Cassani E, Caccialanza R, Pezzoli G. Diabetes and risk of Parkinson's disease: a systematic review and meta-analysis. *Diabetes Care* 2011;**34**:2614–23.
316. Driver JA, Smith A, Buring JE, Gaziano JM, Kurth T, Logroscino G. Prospective cohort study of type 2 diabetes and the risk of Parkinson's disease. *Diabetes Care* 2008;**31**:2003–5.
317. Lu L, Fu DL, Li HQ, Liu AJ, Li JH, Zheng GQ. Diabetes and risk of Parkinson's disease: an updated meta-analysis of case-control studies. *PLoS One* 2014;**9**:e85781.
318. Uchiki T, Weikel KA, Jiao W, Shang F, Caceres A, Pawlak D, Handa JT, Brownlee M, Nagaraj R, Taylor A. Glycation-altered proteolysis as a pathobiologic mechanism that links dietary glycemic index, aging, and age-related disease (in nondiabetics). *Aging Cell* 2012;**11**:1–13.
319. Morris JK, Bomhoff GL, Stanford JA, Geiger PC. Neurodegeneration in an animal model of Parkinson's disease is exacerbated by a high-fat diet. *Am J Physiol Regul Integr Comp Physiol* 2010;**299**:R1082–90.
320. Han Y, Randell E, Vasdev S, Gill V, Gadag V, Newhook LA, Grant M, Hagerty D. Plasma methylglyoxal and glyoxal are elevated and related to early membrane alteration in young, complication-free patients with Type 1 diabetes. *Mol Cell Biochem* 2007;**305**:123–31.
321. McLellan AC, Thornalley PJ, Benn J, Sonksen PH. Glyoxalase system in clinical diabetes mellitus and correlation with diabetic complications. *Clin Sci (Lond)* 1994;**87**:21–9.
322. Dhar I, Dhar A, Wu L, Desai KM. Increased methylglyoxal formation with upregulation of renin angiotensin system in fructose fed Sprague Dawley rats. *PLoS One* 2013;**8**: e74212.
323. Wang X, Desai K, Chang T, Wu L. Vascular methylglyoxal metabolism and the development of hypertension. *J Hypertens* 2005;**23**:1565–73.
324. Wang X, Jia X, Chang T, Desai K, Wu L. Attenuation of hypertension development by scavenging methylglyoxal in fructose-treated rats. *J Hypertens* 2008;**26**:765–72.
325. Li W, Maloney RE, Aw TY. High glucose, glucose fluctuation and carbonyl stress enhance brain microvascular endothelial barrier dysfunction: implications for diabetic cerebral microvasculature. *Redox Biol* 2015;**5**:80–90.
326. Castellani R, Smith MA, Richey PL, Perry G. Glycoxidation and oxidative stress in Parkinson disease and diffuse Lewy body disease. *Brain Res* 1996;**737**:195–200.
327. Dalfo E, Portero-Otin M, Ayala V, Martinez A, Pamplona R, Ferrer I. Evidence of oxidative stress in the neocortex in incidental Lewy body disease. *J Neuropathol Exp Neurol* 2005;**64**:816–30.
328. Munch G, Luth HJ, Wong A, Arendt T, Hirsch E, Ravid R, Riederer P. Crosslinking of alpha-synuclein by advanced glycation endproducts—an early pathophysiological step in Lewy body formation? *J Chem Neuroanat* 2000;**20**:253–7.
329. Choi YG, Lim S. N(varepsilon)-(carboxymethyl)lysine linkage to alpha-synuclein and involvement of advanced glycation end products in alpha-synuclein deposits in an MPTP-intoxicated mouse model. *Biochimie* 2010;**92**:1379–86.
330. Shaikh S, Nicholson LF. Advanced glycation end products induce in vitro cross-linking of alpha-synuclein and accelerate the process of intracellular inclusion body formation. *J Neurosci Res* 2008;**86**:2071–82.
331. Lee D, Park CW, Paik SR, Choi KY. The modification of alpha-synuclein by dicarbonyl compounds inhibits its fibril-forming process. *Biochim Biophys Acta* 2009;**1794**:421–30.
332. Guerrero E, Vasudevaraju P, Hegde ML, Britton GB, Rao KS. Recent advances in alpha-synuclein functions, advanced glycation, and toxicity: implications for Parkinson's disease. *Mol Neurobiol* 2013;**47**:525–36.
333. Outeiro TF, Kazantsev A. Drug targeting of alpha-synuclein oligomerization in synucleinopathies. *Perspect Med Chem* 2008;**2**:41–9.

334. Outeiro TF, Putcha P, Tetzlaff JE, Spoelgen R, Koker M, Carvalho F, Hyman BT, McLean PJ. Formation of toxic oligomeric alpha-synuclein species in living cells. *PLoS One* 2008;**3**:e1867.
335. Winner B, Jappelli R, Maji SK, Desplats PA, Boyer L, Aigner S, Hetzer C, Loher T, Vilar M, Campioni S, Tzitzilonis C, Soragni A, Jessberger S, Mira H, Consiglio A, Pham E, Masliah E, Gage FH, Riek R. In vivo demonstration that alpha-synuclein oligomers are toxic. *Proc Natl Acad Sci U S A* 2011;**108**:4194–9.
336. Kurz A, Rabbani N, Walter M, Bonin M, Thornalley P, Auburger G, Gispert S. Alpha-synuclein deficiency leads to increased glyoxalase I expression and glycation stress. *Cell Mol Life Sci* 2011;**68**:721–33.
337. Thornalley PJ. Glutathione-dependent detoxification of alpha-oxoaldehydes by the glyoxalase system: involvement in disease mechanisms and antiproliferative activity of glyoxalase I inhibitors. *Chem Biol Interact* 1998;**111–112**:137–51.
338. Jenner P. Altered mitochondrial function, iron metabolism and glutathione levels in Parkinson's disease. *Acta Neurol Scand Suppl* 1993;**146**:6–13.
339. Sian J, Dexter DT, Lees AJ, Daniel S, Agid Y, Javoy-Agid F, Jenner P, Marsden CD. Alterations in glutathione levels in Parkinson's disease and other neurodegenerative disorders affecting basal ganglia. *Ann Neurol* 1994;**36**:348–55.
340. Kuhla B, Boeck K, Luth HJ, Schmidt A, Weigle B, Schmitz M, Ogunlade V, Munch G, Arendt T. Age-dependent changes of glyoxalase I expression in human brain. *Neurobiol Aging* 2006;**27**:815–22.
341. Tanaka Y, Engelender S, Igarashi S, Rao RK, Wanner T, Tanzi RE, Sawa A, V, L. D., Dawson, T. M., and Ross, C. A.. Inducible expression of mutant alpha-synuclein decreases proteasome activity and increases sensitivity to mitochondria-dependent apoptosis. *Hum Mol Genet* 2001;**10**:919–26.
342. Im JY, Lee KW, Woo JM, Junn E, Mouradian MM. DJ-1 induces thioredoxin 1 expression through the Nrf2 pathway. *Hum Mol Genet* 2012;**21**:3013–24.
343. Xue M, Rabbani N, Momiji H, Imbasi P, Anwar MM, Kitteringham N, Park BK, Souma T, Moriguchi T, Yamamoto M, Thornalley PJ. Transcriptional control of glyoxalase 1 by Nrf2 provides a stress-responsive defence against dicarbonyl glycation. *Biochem J* 2012;**443**:213–22.
344. Lee JY, Song J, Kwon K, Jang S, Kim C, Baek K, Kim J, Park C. Human DJ-1 and its homologs are novel glyoxalases. *Hum Mol Genet* 2012;**21**:3215–25.
345. Richarme G, Mihoub M, Dairou J, Bui LC, Leger T, Lamouri A. Parkinsonism-associated protein DJ-1/Park7 is a major protein deglycase that repairs methylglyoxal- and glyoxal-glycated cysteine, arginine, and lysine residues. *J Biol Chem* 2015;**290**:1885–97.
346. Beisswenger PJ, Howell SK, Smith K, Szwergold BS. Glyceraldehyde-3-phosphate dehydrogenase activity as an independent modifier of methylglyoxal levels in diabetes. *Biochim Biophys Acta* 2003;**1637**:98–106.
347. Liu L, Xiong N, Zhang P, Chen C, Huang J, Zhang G, Xu X, Shen Y, Lin Z, Wang T. Genetic variants in GAPDH confer susceptibility to sporadic Parkinson's disease in a Chinese Han population. *PLoS One* 2015;**10**:e0135425.
348. Deng Y, Zhang Y, Li Y, Xiao S, Song D, Qing H, Li Q, Rajput AH. Occurrence and distribution of salsolinol-like compound, 1-acetyl-6,7-dihydroxy-1,2,3,4-tetrahydroisoquinoline (ADTIQ) in parkinsonian brains. *J Neural Transm (Vienna)* 2012;**119**:435–41.
349. Song DW, Xin N, Xie BJ, Li YJ, Meng LY, Li HM, Schlappi M, Deng YL. Formation of a salsolinol-like compound, the neurotoxin, 1-acetyl-6,7-dihydroxy-1,2,3,4-tetrahydroisoquinoline, in a cellular model of hyperglycemia and a rat model of diabetes. *Int J Mol Med* 2014;**33**:736–42.
350. Xie B, Lin F, Ullah K, Peng L, Ding W, Dai R, Qing H, Deng Y. A newly discovered neurotoxin ADTIQ associated with hyperglycemia and Parkinson's disease. *Biochem Biophys Res Commun* 2015;**459**:361–6.

CHAPTER

11

Protein Covalent Modification by Homocysteine: Consequences and Clinical Implications

Gurumayum Suraj Sharma, Reshmee Bhattacharya, Laishram Rajendrakumar Singh

Dr. B.R. Ambedkar Center for Biomedical Research, University of Delhi, Delhi, India

1 INTRODUCTION

After proteins are synthesized by ribosomes, they may undergo some modifications to form a mature, functionally active protein product. These kinds of modifications, known as posttranslational modifications (PTMs), generally involve addition/removal of functional groups at specific residues, proteolytic cleavage of regulatory subunits, and also folding processes (induced by glycosylation) in order to generate a mature protein product with correct function/activity.[1] PTMs have also been known to be one of the major strategies to regulate cellular metabolism, homeostasis, and disease.[2] A failure in control of these complex molecular processes could be fatal for cell survival.[3] Several PTMs have been implicated in different human diseases such as heart disease, neurological disorders, the dysmetabolic syndrome, diabetes, cancer, etc.[4] Various forms of post translational modifications include phosphorylation, ubiquitinylation, palmitoylation, myristoylation, neddylation, lipoylation methylation, etc.[5] The enzymes that are involved in these types of modifications include phosphatases, kinases, ligases, and transferases, among others, which can add or remove functional groups, sugars, proteins, or lipids to or from amino acid side chains. Proteases are another category of enzymes that cleave peptide bonds to remove particular sequences or regulatory subunits, therefore, bringing about

Protein Modificomics
https://doi.org/10.1016/B978-0-12-811913-6.00011-4

reversible/irreversible alterations to the target protein. Kinases reversibly add phosphate group(s) to specific amino acid residues in the proteins, which is a prevalent way of catalytic activation/inactivation in various biological processes. Similarly, phosphatases remove phosphate from those groups, thus performing reverse function to kinases.[6, 7] On the other hand, proteolytic cleavage in proteins is a thermodynamically favorable process, and is thus a permanent/irreversible process of removal of peptide sequences or regulatory domains. There is a well-known class of enzymes called zymogens. Zymogens are released as inactive precursors and later on proteolytically cleaved to form active compounds. Enzymes like chymotrypsin and pepsin are secreted as chymotrypsinogen and pepsinogen, respectively, which then undergo proteolysis to form the active enzyme.[8]

Protein modifications involving addition of functional groups most often target specific amino acid residues. For example, in phosphorylation, a process involving addition of phosphate groups during signal cascade, tyrosine, serine, threonine, and arginine residues are specifically targeted.[9] Another PTM that specifically targets lysine and arginine is methylation, a mechanism that is widely known to be involved in epigenetic regulation, since the methylation and demethylation of histones influences the availability of DNA for transcription.[10] Many nonenzymatic processes may also lead to protein structural modifications. Metabolites such as homocysteine (Hcy) and homocysteine thiolactone (HTL), which are formed during cysteine/methionine metabolic pathways,[11] or glyoxals and methylglyoxals,[12] which are produced due to degradation of glucose, may directly interact with specific amino acid residues, causing changes in protein conformation that may ultimately alter protein functionality and induce aggregate formation.[13] While Hcy attacks cysteine residues, HTL,[14] as well as glyoxals and methylglyoxals, associate with lysine residues.[15] Nonenzymatic modification of proteins by sugars and their metabolites are termed protein glycation,[12] while modifications brought about by Hcy and HTL are termed S- and N-homocysteinylation,[11, 13] respectively.

In this chapter, we will place special focus on protein modification via homocysteinylation (S- and N-homocysteinylation) and aim to describe protein structural and functional changes resulting from incorporation of Hcy/HTL to proteins, as well as the cellular and clinical consequences due to these changes.

2 HOMOCYSTEINE AND PROTEIN HOMOCYSTEINYLATION

As mentioned previously, elevated Hcy levels in blood leads to homocystinuria (or hyperhomocysteinemia), a condition that is believed to be connected with many clinical manifestations including cardiovascular complications, bone deformations, thrombosis, and even cancer.[16] In healthy humans, normal levels of Hcy range from 5 to 10 μM. However, mutational defects in the enzymes involved in detoxification pathway of Hcy, cystathionine β-synthase (CBS), and methylene tetrahydrofolate reductase (MTHFR), etc., leads to Hcy accumulation in plasma as well as cells because of the inability to convert Hcy to cystathionine. Elevated levels may range from 15 to 20 μM (mild forms) up to 500 μM (severe forms).[17–19] Besides these genetic factors, deficiency of Vitamin B6, Vitamin B12, and folate have been a major contributing acquired factors for elevated plasma Hcy levels.[20] In addition, administration of anticonvulsants in epileptic patients has been reported to be responsible for rise in Hcy levels in these patients, as most of these drugs administered are known to lower folate levels.[21, 22]

A new mechanism has recently been identified, called "protein homocysteinylation," which is considered to be a prime cause of Hcy toxicity. It is now evident from several studies that Hcy and its thiolactone (HTL) can covalently modify Cys and Lys residues in proteins, respectively. This mechanism can be broadly discussed under two headings, namely "S-homocysteinylation", modification of Cys residue by Hcy, and "N-homocysteinylation," which is the modification of Lys residues by HTL. In the sections that follow, we will discuss protein S- and N-homocysteinylation, consider its major consequences, and emphasize the relevance of these modifications to pathophysiological conditions.

3 PROTEIN S-HOMOCYSTEINYLATION

Hcy forms stable disulfide linkages with the free Cys residues of proteins, in a process termed S-homocysteinylation. Modification of protein Cys residues via S-homocysteinylation results in alterations in the functional and structural properties of protein.[23] Hcy is shown to exhibit highest propensity to generate disulfide bonds with protein -SH groups as compared to free Cys or glutathione. Since the -SH of Cys in proteins participate in functionality of several enzymes and structural and functional properties of proteins, their modification with Hcy would ultimately disrupt cellular function.[24] The binding of Hcy to proteins occur in a biphasic manner. Firstly, the displacement of Cys from the protein occurs rapidly (which is independent of oxygen). During the second phase, the reaction proceeds slowly, and is an oxygen dependent phase, where the -SH oxidation occurs.[25] S-homocysteinylated proteins have been detected in plasma. The major pool of human plasma S-homocysteinylated proteins exists as linked to albumin and γ-globulin. Human serum albumin (HSA) is the predominant plasma protein, which makes 50% or more of the total plasma protein. The Cys-34 of albumin, which does not participate in disulfide bridge, makes up the bulk of free -SH in the protein.[26] Since the pK_a value of this free -SH is very low (~5) in physiological conditions, the protein exists primarily as a thiolate anion, which renders it very highly reactive toward metals, other -SH groups, and disulfides. It has been shown that a third of plasma albumin has Cys-34 disulfide bound with -SH groups (of Hcy or Cys).[27] There has been very limited literature for the actual measurements of structural and functional parameters of proteins upon modification via S-homocysteinylation. A few in vitro and in vivo studies have now revealed that S-homocysteinylation does cause certain impairment in protein function.[28–30]

In a study to predict the likelihood of proteins toward S-homocysteinylation, three factors/parameters including solvent accessibility of cysteine in a protein, the pK_a of free cysteines, and the dihedral strain energy (DSE) of a cysteine disulfide, were measured.[31] It was concluded that several extracellular matrix proteins known to be associated with control of connective tissue organization, inflammation, coagulation, angiogenesis, arthritis, and cell migration, being cysteine-rich and having high energy disulfides could possibly be more susceptible to S-homocysteinylation. This might be a possible mechanism for thrombotic pathogenesis of hyperhomocysteinemia. Metallo-proteins were found to contain substantial amount of cysteine residues. However, in a majority of such cases, cysteine directly linked to the metallic ligand was buried deep inside the protein, thus not available for interaction with Hcy. In contrast, some metallo-proteins contain cysteine residues that are accessible to the solvent and hence they could be potential targets for S-homocysteinylation. For example, metallothionein-2 (MT2) has been experimentally proven to be nonfunctional after being homocysteinylated. The zinc-binding ability of this protein has been known to be altered upon homocysteinylation.[32, 33] Notch, a transmembrane protein, participates in dorso-ventral patterning, neural crest migration and cell fate. It is a well-known fact that Hcy reduces the disulfide bonds, which triggers the release of calcium from the EGF domain, thus, making it vulnerable to proteolysis and activation.[31] A list of protein reported to be modified by Hcy is presented in Table 1. However, studies on the structural and functional aspects of S-homocysteinylation on proteins are still lacking. Understanding the structural and functional consequences of Hcy on cellular proteins and identification of target proteins, would be helpful in better understanding for the mechanisms of Hcy-induced toxicity under homocysteinemic conditions.

TABLE 1 List of Reported S-Homocysteinylated Proteins and Their Consequences

	Proteins	Effects of S-homocysteinylation	References
1.	Serum albumin	Mask -SH responsible for free radicals scavenging	34
2.	Transthyretin	Possible structural alterations and aggregation	35
3.	Annexin II	Structural alterations and loss of function	36
4.	LDL	Reduces endothelial cell vitality and proliferation	37
5.	Fibronectin	Alter thrombosis and interferes with the wound healing process by disrupting the interaction between fibronectin and fibrin	38
6.	Dimethylarginine dimethylaminohydrolase (DDAH)	Inhibition of DDAH enzyme activity, causing ADMA to accumulate and inhibit nitric oxide synthesis	39, 40
7.	Plasminogen	Decreased urokinase-induced plasmin activity	41

4 PROTEIN N-HOMOCYSTEINYLATION

Several studies have now shown that HTL can covalently modify Lys residues in proteins (a process termed as "protein N-homocysteinylation").[14] HTL is a cyclic thioester that is synthesized during an error editing process catalyzed by methionyl-tRNA synthetase (MetRS), which prevents translational incorporation of homocysteine in the growing polypeptide chain. This reaction occurs in two steps. In the first step, MetRS converts Hcy to homocysteinyl adenylate. On the other hand, the -SH group on side chain of Hcy reacts with Hcy activated carboxyl group and HTL is formed. HTL synthesis depends on the Hcy/Met ratio; the higher this ratio, the greater the HTL synthesis. This thiolactone is highly reactive and is known to nonenzymatically target proteins at ε-amino group of lysine residues in a process known as protein N-homocysteinylation.[11] Arginine and histidine residues have also been found to be targeted by HTL, however, lysine residues are more susceptible to N-homocysteinylation by HTL.[42] This linkage is formed by acylation of the free amino group of lysine and activated carboxyl group of HTL. Homocysteinylation by HTL is a very rapid process. The thiolactone has a short half-life and disappears after 1 h when incubated with human serum, which is about 25-fold faster than the rate of its nonenzymatic hydrolysis.[14] Most of the thiolactone is covalently incorporated with the proteins by the end of 3h. Furthermore, modified proteins have been shown to have altered structure and function.

4.1 HTL Induced Covalent Modification Leads to Molten Globule State Formation

The molten globule (MG) is a partially folded state that may induce the advent of protein aggregation. MG states are important protein folding intermediates with a perturbed tertiary interaction and native-like compact secondary structures. These intermediates are characterized by three important aspects, i.e., the loss of tertiary structure, presence of intact secondary structure, and 10%–30% rise in the radius of gyration.[43] A relation between N-homocysteinylation and the formation of molten globule states has been established recently.[44] In a study, the consequence of protein N-homocysteinylation on protein conformation at different time intervals was carried out with three proteins (cytochrome *c*, α-LA, and lysozyme). It was demonstrated that protein N-homocysteinylation leads to MG state formation, which further initiates aggregates formation in α-LA and cytochrome *c*. However, the effect was not true in general, because no structural changes were seen in another protein studied (lysozyme). It was suggested that changes in tertiary structure upon homocysteinylation, must be the limiting step to convert protein's native state to molten globule (MG) state. It was also suggested that the MG state could be more lethal than the HTL-induced aggregates. Similar results have also been reported for protein (hemoglobin) modification by glyoxal,[45] suggesting the MG state formation to be a crucial step for the emergence of aggregate species.

In another study, HTL-induced modification of Cyt c was also shown to result in formation of the MG-like state with perturbed tertiary interactions with significant exposure of the hydrophobic core, as concluded by ANS binding assay. These alterations resulted in heme exposure with concomitant disruption of the heme-Met80 interactions. N-homcysteinylation of Cyt c results in peroxidase activity, which is believed to be absent in native Cyt c. Since Cyt c peroxidase function and cardiolipin (CL) oxidation are prerequisite steps for apoptosis, such modification could be crucial in intrinsic apoptotic pathway. It was concluded that HTL modification leads for the conversion of the native peroxidase-inactive Cyt c to a peroxidase-active, apoptotically competent species.[46] In fact, disruption of heme coordination in Cyt c is believed to be responsible for injury of mitochondria during ischemia,[47] which also reflects a common complication of hyperhomocysteinemia. Furthermore, Cyt c homocysteinylation levels were found to be increased in brains of an ischemic mouse model.

4.2 N-Homocysteinylation-Induced Structural Alteration is Protein Dependent

At the time of writing, there have been only a handful of studies addressing the problem of protein N-homocysteinylation. Nonetheless, most modified proteins tend to be rendered nonfunctional due to structural loss or extensive oligomerization due to cross-linking. In a study carried out by Sharma et al.,[42] it was observed that acidic and basic proteins showed different reactivity toward HTL. Furthermore, the extents of N-homocysteinylation in proteins were directly correlated with the functional activity of the said protein. While basic proteins used in the study (Lysozyme and RNase-A) did not show any significant change in their activities, acidic protein (Carbonic anhydrase) was found to have almost complete loss of enzyme function. In addition to this, conformational assessments gave a hint toward a direct correlation of the reactivity of HTL and the structural consequences. Acidic proteins used in the study (Alpha lactalbumin, Carbonic anhydrase, and Alpha casein) were found to be rendered denatured upon modification via N-homocysteinylation. The findings suggest that acidic proteins could be the prime target for covalent modification by HTL and, thereby, more prone to HTL-induced alterations. In fact, most cytoskeletal proteins are acidic in nature. Thus, there is a high propensity of these proteins undergoing N-homocysteinylation. There are reports of a few cytoskeletal proteins (namely microtubule-associated tau protein, neuronal, and glial intermediate filaments (IF)) being affected by elevated Hcy levels.[48] At present, there is no promising explanation for the differential reactivity of HTL toward acidic and basic proteins. It is argued that targeting specific lysine residues in the native state involved in maintaining the fold type of the protein results in disruption of the native structure, thereby resulting in functional loss. Furthermore, a number of serum proteins have also been reported to have no structural consequences upon N-homocysteinylation.[21] A large portion of cellular and plasma Hcy are linked to hemoglobin and albumin existing in the form of protein N-linked-Hcy.[14] However, HTL-binding leads to subtle/no effect on the structural characteristics of these proteins. Therefore, these proteins are believed to be the major reservoir of HTL/Hcy in human blood.

4.3 Reactivity Toward HTL-Induced Covalent Modification

At the time of writing, there have been countless works done in understanding the reactivities of HTL toward various proteins. Pioneering works by Jakubowski have led to a concept that the number of lysine residues present on the proteins in question is one of the contributing factors that determine the reactivity of HTL. Indeed, there exists a positive association between the rate of protein N-homocysteinylation and the number of lysine residues in the proteins.[14] For proteins varying in size between 104 and 698 amino acid (AA) residues, there was a positive correlation (correlation coefficient of about 0.97) between their lysine content and their reactivities with HTL. There exists no such definite correlation between size of the proteins and their reactivities with HTL. However, in bigger proteins like fibrinogen (3588 AA residues), α2- macroglobulin (5896 AA residues), and the extent of N-homocysteinylation was much less than expected with reference to either their lysine contents or their sizes. One possible explanation for this observation could be that some of their lysine residues may be inaccessible to the solvent.

Several later studies have challenged the fact that the extent of N-homocysteinylation is dependent on the number of lysine residues. In case of cytochrome-c, it has been shown that targeting four lysine residues (Lys-8/Lys-13, Lys-86/Lys-87, Lys-99, and Lys-100) brings about no promising changes in its tertiary structure. However, in case of insulin a single lysine residue modification leads to complete unfolding, ultimately leading to aggregate formation. Furthermore, the reactivities of proteins toward HTL have also been shown to be dependent on the pK_a values of amino acid side chains. The microenvironment where lysine is buried plays a role in determining the pK_a value since the pK_a value of lysine residues in proteins is ~10.6, which is significantly different from that of free lysine in buffer. Amines with a pK_a of 7.7 were shown to react with HTL approximately 6.5 times faster than free lysines in solutions to aggregate formation. Here it must be emphasized that all the cytoskeleton proteins are acidic and therefore their propensity of getting homocysteinylated is too high. As a result, these proteins may lose their structural integrity and hence their neuronal function.[42] N-homocysteinylation of two of the cytoskeletal proteins, microtubule-associated tau protein and neuronal and glial intermediate filaments (IF), clearly confirm the fact that acidic proteins could be ready target for N-homocysteinylation. There exists a close association between the pI of the protein and the rate of homocysteinylation, as mentioned in a study by Sharma et al.[42] α-LA, α-casein and CA (protein with low pI) have been shown to be more prone to homocysteinylation and have higher reactivity toward HTL. While proteins with higher pI values (RNase A and lysozyme) have a lesser tendency of homocysteinylation.

4.4 N-Homocysteinylation Induces Aggregate or Amyloid Formation

Accumulation of toxic protein aggregates or amyloids is one of the well-studied mechanisms of neurodegenerative diseases like Alzheimer's disease, Parkinson's disease, etc. Amyloids are fibrous protein aggregates typically characterized by predominant β-pleated sheet. So far, numerous proteins have been reported to show amyloidogenic transformation upon homocysteinylation. The actual mechanism behind HTL-modified proteins and their tendency to aggregate is still not well understood; however, it is believed that HTL-induced protein modification might induce an increase in exposed hydrophobic patches, which tend to interact through hydrophobic interactions, resulting in the formation of aggregated species.

Elevated levels of homocysteine have been reported in type II diabetes. Since insulin is the drug/hormone that regulates blood glucose level and is implicated in type I and II diabetes, a study was carried out to determine whether structural properties of insulin and its aggregation tendencies in presence of HTL were related.[49] Insulin is a small, helical protein having two polypeptide chains—chain A and B. It has a single lysine residue in its B chain. Modification of this lysine has been shown to result in aggregate formation. In addition, elevated Hcy has been reported in most patients with end-stage renal disease. Both N-linked as well as S-linked Hcy were found to be significantly higher in case of uremic patients as compared to controls, suggesting such modifications to be key players in the progression of associated diseases.

In Aβ, two key stacking interactions play a critical role in aggregation and fibrillation processes: the electrostatic interaction between Lys28 and Glu22/Asp23, and the stacking interaction in aggregation motif of Aβ (16KLVFF20).[50–57] Factors that abolish these interactions could be a potential strategy in inhibiting aggregation process. However, N-homocysteinylation of $A\beta_{1-42}$ results in an increase in nucleation time (lag phase), and concomitant decrease in the aggregation process. The modification leads to stabilization of oligomeric species, thereby attenuating maturation of these oligomers into fibrils.[58, 59] This modified $A\beta_{1-42}$ was reported to be more toxic than the unmodified $A\beta_{1-42}$. It might be possible that HTL reduces β-sheet propensity in Aβ, which decreases oligomer and/or fibril formation, thus leading to new toxic protofibrils generation with lower self-assembly and higher cellular toxicity.[58–61] N-homocysteinylation of human prion proteins has also been shown to occur at a very rapid rate, leading to aggregate formation.[62] Since the advent of AD is also known to be governed by interactions of Aβ with normal or infectious prion proteins, such modification could be a key process in the etiology. In fact, N-homocysteinylation induces significant conformational alterations in normal PrP^C and triggers generation of insoluble multimers and amyloidogenic conversion.[62] Structural investigation revealed that prion proteins, upon modification by HTL, resulted in increased content of β-sheet structure compared to unmodified proteins.[62] Furthermore, the modified protein was shown to be resistant to proteolytic cleavage. Thus, N-Hcy-PrP is likely to stay in the blood for a much longer period because of a longer half-life and an increased resistance to proteolysis.[62] In case of tau protein, it has been shown that in vitro modification of the lysine residues results in an altered tubulin assembly dynamics.[63] Tau is also known to mediate assembly and nucleation of tubulin, and stabilizes microtubule proteins (MTP) against disassembly. Eliminating the interactions between N-homocysteinylated tau and MTP results in release of tau from MTP. Once released, it can undergo self-association and aggregate formation, resulting in production of paired helical filaments,[64, 65] which stabilize to form higher ordered structures, thus, could favor early advent of AD.

In addition, the amyloidogenic transformation of N-homocysteinylated protein has been so far reported for many other proteins including crystalline, BSA, kappa casein, etc.[60–62, 66, 67] Working on HTL-induced modification of BSA, Paoli et al. have revealed the existence of amyloid pore-like circular structures as described by other authors.[68] Such BSA aggregates were also reported to induce cell death and have the ability of acting as a seed, favoring further conversion of the native proteins to other molecular species which highly tend to form aggregates.[60]

In case of proteins like HSA, casein, hemoglobin, cyt c, etc., protein N-homocysteinylation has also been reported to induce amorphous aggregate formation.[14, 61, 69, 70] However, the real molecular insights for HTL-induced protein modification and aggregation are not yet well understood. Considering the fact that HTL induces structural changes in protein, it is speculated that HTL-modified nonnative proteins might have exposed increased hydrophobic group clusters, which have the tendency to interact through hydrophobic interactions, thereby forming aggregated species.[44] Another probable mechanism could be the fact that HTL binding leads to free -SH group generation in the native protein, thereby resulting in formation of dimers, trimers, tetramers, and higher order multimers.[14] Therefore, from the above-mentioned mechanisms, it can be stated that elevated levels of N-homocysteinylated proteins may induce aggregate/amyloid formation, which could contribute to Hcy neurotoxicity and several other human pathologies.

4.5 Functional Consequences of N-Homocysteinylation

Structural changes in proteins due to homocysteinylation may be followed by change in their functions. Several studies have established a direct correlation of homocysteinylation and loss of protein activity or a gain of toxic function. An interesting study by Zinellu et al.[71] on the binding affinities of catechins with human serum albumin revealed that N-homocysteinylation and S-homocysteinylation of HSA considerably reduced its affinity for catechins. Consequently, the serum concentration of catechins, which are more susceptible to oxidation and reduced half-lives, may increase, leading to a considerable decrease in their functions as antioxidants. This result could also be applicable to other xenobiotics, particularly drugs, which bind to HSA. If such interactions of HSA with the drugs get affected by Hcy or HTL induced modifications, it could have important biological and clinically significant implications in their pharmacokinetics and efficacy.

Besides losing their functional activity due to structural alterations, homocysteinylated proteins also become more susceptible to degradation via oxidation. Loss of enzymatic function may not necessarily be due to alterations in the active site, but in most cases, it is outside of the catalytic domains, namely, their exposed lysine residues. A study was conducted by Stroylova et al. on the effect of N-homocysteinylation on two enzymes, lactate dehydrogenase (LDH) and glyceraldehyde-3-phosphate dehydrogenase (GAPDH).[72] It was observed that an increase in the concentration of HTL, lead to an increased inhibitory effect on the enzyme function ultimately leading to a complete loss of enzyme activity. This loss of activity was said to be due to aggregates/oligomers formation, thereby restricting diffusion of substrates. HTL can also cause reduction of the heme moiety of cyt-c.[70] The heme iron of cyt-c, being a transition metal, is known to cause oxidation of thiol group containing molecules or even exposed lysine residues of some proteins, while itself undergoing reduction. Reduction of the Cyt c heme group can cause conversion of ferricytochrome-c to ferrocytochrome-c, which has a tertiary structure, being more stable to unfolding compared to its oxidized form due to stronger bonding between heme iron and Met80 in ferrocytochrome-c. Due to this enhanced stability, ferrocytochrome-c shows a decreased susceptibility toward unfolding or degradation by proteolysis. However, this increased resistance to proteolysis in cyt-c modified by HTL can also be independent of reduced heme moiety as N-homocysteinylation of some important lysine residues Lys8 or -13, Lys86 or -87, Lys99, and Lys100, can also in itself cause changes in cyt-c structure, making them more resistant to degradation by proteolysis. Another example of loss of function due to N-homocysteinylation is microtubular associated protein (MAP) human tau protein.[63] Binding of its lysine residue K280 with microtubular protein (MTP) is critical for tubulin assembly and stabilization. MTP nucleation and assembly have important physiological roles, such as axonal transport.

Incubation of the tau protein with varying concentrations of HTL leads to the homocysteinylation of the K280 residue and this modification was shown to affect interaction of tau and MTP and, hence, a considerable lowering of the tubulin assembly was observed. It was concluded from the study that modification of K280 residue by HTL can cause a disruption of the normal interaction of the two proteins, thus, leading to loss of its normal function. Several other instances have also reported loss of protein function due to homocysteine via S-homocysteinylation. Human fibronectin, for example, was found to lose its ability to specifically bind to fibrin in the blood upon modification with homocysteine.[38] This loss in its ability could cause delay in the blood clotting mechanism and the wound healing process.

4.6 Effect on the Reservoir Proteins

There are a number of serum proteins that, although being modified by N-homocysteinylation, do not show any structural consequences. Out of the total N-Hcy-protein present in the blood, hemoglobin contributes ~75% and albumin ~22%.[11] However, these modified proteins are considered to be the major reservoirs of Hcy in human blood, as Hcy/HTL binding with them brings negligible alterations in the structural properties and, therefore, function as a part of a protective mechanism which detoxifies Hcy and its reactive metabolites. Both the proteins, as already discussed, did not show any change in structure or function even though they have been shown to undergo N-homocysteinylation. The possible explanation given for such an observation was that the proteins serve as reservoir for HTL, contributing significantly in reduction of HTL levels from the human blood. HSA is an interesting example for studying the effects of homocysteinylation. Though the protein does not show a significant structural modification, as has already been mentioned, it has been found to show a decrease in its function. HSA is found to be both N-homocysteinylated and S-homocysteinylated, accounting for the majority (˃90%) of the total serum Hcy. Besides accommodating the major portions of the total Hcy levels in the blood, N-homocysteinylated and S-homocysteinylated forms of albumin get more readily degraded in the liver via proteolysis, an important step in HTL detoxification.[73]

5 HOMOCYSTINURIA/HYPERHOMOCYSTEINEMIA: CONDITION PERSISTING FOR PROTEIN MODIFICATIONS

Homocysteine, an intermediate produced during methionine metabolism pathway, is a nonprotein amino acid containing sulfur, which is formed as an intermediate between trans-sulfuration and remethylation pathway of methionine and cysteine. Four different forms of it exist in the plasma: free thiol (1%), some remain as disulfide bound to plasma proteins (70%–80%), some bind with another molecule of homocysteine, and some bind with other thiols to form dimer (20%–30%).[11] Hcy normal levels in healthy humans are 5–10 μM. However, elevated levels of Hcy ranging from 15 to 20 μM (mild forms) up to 500 μM (severe forms) have been reported to be found in hyperhomocysteinemia/homocystinuria conditions.[11, 14, 17, 74] In homocystinuria, the elevated level of Hcy is thought to be due to the mutational defects in one important enzyme involved in the biosynthesis pathway, cystathionine β-synthase (CBS). Deficiency of CBS results in an inability to convert Hcy to cystathionine, leading to elevated cellular and plasma Hcy levels.[13, 16, 75] Since Hcy is a toxic metabolite, it needs to be removed from the plasma. This removal is carried out by either remethylation pathway, where Hcy gets converted to methionine, or trans-sulfuration, where it converts to cysteine.

While Hcy is formed in all organs of the body, its detoxification occurs mainly in kidney and liver. In vascular tissues and the skin, owing to the fact that they lack enzymes involved in trans-sulfuration, Hcy detoxification occurs via remethylation pathway.[76, 77] In other tissues of the body, including liver and kidney, Hcy removal occurs mainly via remethylation pathways. Hcy conversion to either methionine or cysteine depends on cellular levels of methionine. When there is an excess of methionine concentration in the cells, the trans-sulfuration pathway plays an important role in Hcy removal. The enzyme CBS converts Hcy to cystathionine, which further gets converted to cysteine via cystathionine γ-lyase.[78, 79] In cases where methionine levels are low, Hcy is remethylated to methionine. Methionine synthase (MS) catalyzes the remethylation of homocysteine to methionine, a reaction that links the folate cycle with homocysteine metabolism. Homocysteine remethylation requires vitamin B12 5, 10-methytetrahydrofolate, which is generated by 5, 10-methylene tetrahydrofolate reductase (MTHFR).[79–81] In 1962, the disease of homocystinuria was first discovered when Hcy was identified in the urine.[82] Later the enzyme deficient activity of CBS was identified and the association of homocystinuria with CBS deficiency was well documented.[83, 84]

In contrast to homocystinuria, hyperhomocysteinemia is caused due to other factors not related to CBS deficiency. Other genetic factors including mutations in methylene tetrahydrofolate reductase (MTHFR) or methionine synthase (MS) gene also bring about inability to metabolize toxic Hcy. Besides the genetic factors, food habits, for instance deficient intake of vitamin B6, vitamin B12, and folate deficiency also serve as an acquired factor for elevation of plasma Hcy levels.[85–88] Inborn errors in cobalamin metabolism, which include adenosylcobalamine deficiency, and combined adenosylcobalamine and methylcobalamine deficiencies, also contribute toward Hyperhomocysteinemia. In a study carried out by Mills et al.,[89] it was shown that many women who gave birth to children with neural tube defects have a disturbed homocysteine metabolism, which may be overcome by folic acid and vitamin B12 supplements during conception. Smoking, alcoholism, coffee intake, prolonged use of nonsteroidal antiinflammatory analgesics (NSAIDS), antiepileptic drugs, and birth-control pills also lead to extreme hyperhomocysteinemia.[90] Homocysteine level has also been known to upregulate through various diseases (e.g., cancer) and stresses.[91, 92] Thirteen single nucleotide polymorphisms (SNPs) in genes, including MTHFR and MTR, are identified to be involved in folate metabolism. MTHFR 677C > T and MTHFR 1298A > C SNPs are associated with plasma folate and homocysteine, with 1298A > C polymorphism showing a stronger impact on Hcy metabolism. 677C → A leads to alanine to valine substitution, whereas, 1298A → C results in glutamine to alanine substitution. About 33% lower DNA uracil content was found in MTHFR 677 TT genotype.[93]

In contrast to this, cells have evolved protective mechanisms against Hcy toxicity. The extracellular Hcy is eliminated from the body through urinary excretion in the kidneys. This mode of excretion is catalyzed by the enzyme Hcy-thiolactonase/paraoxonase (PON1) associated with high-density lipoprotein. HTL in the circulation is known to be hydrolyzed by Hcy-thiolactonase, which serves as the extracellular means of HTL clearance. On the other hand, Bleomycin hydrolase (BLH), an enzyme that belongs to a family of cysteine aminopeptidases, serves as the intracellular mode of HTL elimination. BLH hydrolyzes HTL with a catalytic efficiency that is ~100-fold higher than that of the Hcy-thiolactonase.

6 CLINICAL COMPLICATIONS ASSOCIATED WITH ELEVATED Hcy

6.1 Homocystinuria is Associated With Various Birth Defects

One of the most common clinical manifestation associated with homocystinuria are birth defects. In fact, early finding of elevated Hcy levels in human disease came from the discovery of birth defects in CBS deficient patients.[94, 95] Among the birth defects associated with homocystinuria, eye defects and skeletal deformities present the most common form. Major eye defects include ectopia lentis (displacement of eye lens from its normal position), which were found to occur in up to 70% of patients by 8 years of age, increasing up to 95% by age of 40.[96] In addition to this, untreated individuals also develop retinal detachment, optic atrophy, and secondary glaucoma, as their age progress. The mechanism by which the observed biochemical abnormalities produce the clinical characteristics of this disorder has not been clearly explained, but it is believed to be associated with abnormal levels of Hcy and Cys. Deficiency of Cys in lens zonules, which normally have high Cys levels, may affect normal development of zonular environment, thereby predisposing to myopia and even retinal dislocation. Furthermore, disruption of cross-linkage in collagen and elastic tissue by Hcy results in zonule degeneration. The lens dislocation is inevitably bilateral and is commonly inferior, but the position is not diagnostic, as the lens may migrate in any direction.[97] Early diagnosis and treatment may alter these complications significantly. However, the treatment may be only beneficial for betaine- and pyridoxine-responsive patients, and not in the case of nonresponsive patients.[96]

Another birth defect associated with homocystinuria is bone deformities. High arched palate and scoliosis are major defects that result from defective CBS enzyme. Arachnodactyly and joint abnormalities have also been shown to be associated with elevated Hcy levels.[98] There are evidences suggesting Sulfur-containing amino acids affect the collagen structure. It is known that aminothiols compromise the tensile strength of skin and tendon experimentally and probably do so by reacting with the aldehyde groups formed from hydroxylysine residues in the collagen peptide chain.[99]

6.2 Increased Hcy and Megaloblastic Anemia

Megaloblastic anemia is characterized by presence of large immature and dysfunctional RBCs (termed megaloblasts) in bone marrow and also by the presence of hypersegmented neutrophils. The association of elevated Hcy with megaloblastic anemia is not a direct association. In fact, the association of such anemia with elevated Hcy level is rather an indirect one. Classically, megaloblastic anemia has been known to be associated with hyperhomocysteinemia, which results from defects in remethylation pathways and not with CBS deficiency. The development of megaloblastic anemia in hyperhomocysteinemia is via folate deficiency due to excessive consumption of methyltetrahydrofolate in the methylation of Hcy to form methionine. The mechanism by which Vitamin B deficiency leads to megaloblastic anemia remains arguable. There is, however, general agreement that disturbed folate metabolism impairs DNA synthesis. Vitamin B acts as a cofactor in the remethylation pathway of Hcy to methionine, the reaction in which 5-methyltetrahydrofolate is converted to tetrahydrofolate.[100] Nonetheless, megaloblastic anemia has been shown to be intimately associated with mild homocysteinemic conditions, wherein the Hcy levels reach $\sim 50\,\mu M$.

6.3 Hyperhomocysteinemia Results in Cardiovascular Complications

A large volume of evidences have suggested that Hcy plays a crucial role in cardiovascular complications in humans. The hypothesis of Hcy-induced vascular complications was first proposed by McCully.[101]

It was observed that inborn errors of methionine metabolism in children developed advanced arterial lesions. Association between slight elevations of blood Hcy levels and coronary heart disease (CHD) was further shown in 1976.[102] Further studies confirmed that these vascular complications generally arose due to the inborn error of homocystinuria due to CBS.[103] These genetic disorders produce severe hyperhomocysteinemia, with elevation of plasma total Hcy levels >100 μM.[104] When left untreated, the affected individuals have a very high incidence of developing some major vascular events (>50%), including stroke, myocardial infarction, or venous thromboembolism, at very early stage in life (usually before 30 years).[96] Homozygous CBS deficiency in humans results in severe hyperhomocysteinemia, with Hcy concentrations ranging from 100 to 500 μM. However, this is a rare autosomal recessive disease that occurs with a frequency of approximately 1 in 75,000. Severe hyperhomocysteinemic patients usually show neurological abnormalities or premature arteriosclerosis and develop myocardial infarction early, at around the age of 30.[105]

Several studies carried out at cellular levels and animal models have also suggested various biochemical mechanisms for the cardiovascular pathological consequences associated with Hcy. Major potential mechanisms include oxidative stress, inflammation, endothelial dysfunction, thrombosis, and modification of proteins by homocysteinylation.[73, 106] One of the earliest events in development of arteriosclerosis is the impairment of endoplasmic reticulum (ER) induced by increased Hcy levels in various animal models. This ER dysfunction is mainly explained through oxidative inactivation of nitric oxide (NO) via formation of superoxide radicals, which leads to inhibition of antioxidant enzyme glutathionine peroxidase (GPX-1) and depletion of intracellular glutathionine. This happens by uncoupling NOS through the formation of asymmetric dimethylarginine (ADMA), by inhibiting tetrahydrobiopterin (a cofactor of NOS), or by decreasing arginine transport activity.[107] Increased Hcy levels can also induce endothelial cell disruption, an initial step in progression of atherosclerosis, by directly causing toxicity on the endothelium, and inducing ER stress and apoptosis in cultured endothelial cells (EC).[108–111] In addition, Hcy is reported to result in promoting thrombosis, by switching EC to a prothrombotic phenotype with enhanced platelet adhesion.[112]

Hyperhomocysteinemia is also an established risk factor of venous thrombosis. Many hypotheses support the fact that Hyperhomocysteinemia leads to damage of vascular endothelium and clotting cascade. Study was carried out in adult patients between the age of 20 and 70 years, and Hyperhomocysteinemia was found to be a risk for the occurrence of recurrent thrombosis.[113] On the other hand, a work by Falcon et al.[114] suggested that hyperhomocysteinemia may lead to juvenile thrombosis as well.

However, whether Hcy lowering treatment strategies, such as folate and vitamin supplementation, can lower the occurrence of venous thrombosis remains a subject of ongoing research. Hyperhomocysteinemia is not a cause but a marker for venous thromboembolism (VTE). Hyperhomocysteinemia causes endothelial damage to such an extent that it can be compared to the damage caused by hypercholesterolemia and hypertension. The most common reason for this damage could be the excessive increase in oxidative stress. Hcy has different effects on different coagulation factors. It increases tissue factor expression, factor VII activity, and TFPI (tissue factor pathway inhibitor) activity and decreases tPA binding, plasmin generation, and thrombomodulin expression. However, the association of disturbed Hcy metabolism with VTE still remains unclear and the experimentation results are ambiguous.

6.4 Hcy and Renal Failure

Hyperhomocysteinemia is recognized to accompany decreased kidney function, with plasma Hcy showing an inverse relation with the glomerular filtration rate (GFR). Hyperhomocysteinemia occurs prevalently in end-stage renal disease with patients undergoing dialysis.[115] As a matter of fact, creatinine is an independent determinant of tHcy status. In the methionine metabolic pathway, S-adenosylmethionine converts to s-adenosylhomocysteine, the generation of which is coupled to creatine-creatinine synthesis, which explains a direct correlation of creatinine and tHcy levels. Also, it has been found that Hcy remethylation is impaired in dialysis patients, although it is not clear whether this impairment exaggerates other Hcy-related abnormalities in renal failure.[116]

6.5 Hcy and Bone Deformation

Other than targeting different organs of the body like liver and kidney, Hyperhomocysteinemia is also linked with bone deformation. Hcy modulates the bone remodeling process by targeting both osteoclast and osteoblast processes. Elevated Hcy levels can have direct effect on the bone matrix. It (Hcy) is proposed to induce ROS generation that has the ability to activate matrix metalloproteinases (MMPs). MMPs can directly decrease the bone strength by degrading the extracellular bone matrix. In addition, deficiency of folate and vitamin and Hyperhomocysteinemia can alter the bone properties. However, more studies should be carried out to look at the actual mechanism of Hyperhomocysteinemia and bone deformation.

6.6 Elevated Hcy Levels are Associated With Cancer

Recent advances have also proposed that there is a close link between hyperhomocysteinemia and cancer. Firstly, cancer patients were observed to have elevated levels of plasma Hcy and higher incidence of venous thromboembolism (VTE), which is the second most common cause of death in cancer patients. Secondly, several polymorphisms in the enzymes involved in the Hcy detoxification pathways (the trans-sulfuration and remethylation) have close clinical relation with several cancer types.[117–123] Thirdly, folate, which is pivotal for cell proliferation, has an inverse relation with Hcy.

6.6.1 Low Plasma Folate Predisposes to Cancer

Folate is not only involved in nucleotide biosynthesis, but is also required for the conversion of deoxyuridine monophosphate (dUMP) into thymidine monophosphate.[124] In normal conditions, thymidylate synthetase converts dUMP into thymidine monophosphate using 5,10-methylenetetrahydrofolate (derived from folate) as a methyl group donor. If folate is limiting, there is accumulation of dUMP as the methyl donor, 5,10-methylenetetrahydrofolate is restrictive. The condition results in de-oxyribonucleotide pool imbalance, and consequently there is excessive incorporation of uracil into DNA instead of thymine, which, in general, is taken care of by the enzyme uracil DNA glycosylase by removing the mis-incorporated uracil from the DNA strand.[125] Under disturbed folate concentration, the DNA glycosylase fails to cope with the DNA repair mechanism. The condition leads to chromosomal damage, which may result in malignant transformations of cells. Chromosomal aberrations are also associated with inappropriate differentiation and morphology of lineage-specific cells often seen in tumors.[126] The mechanism by which folate deficiency predisposes to cancer is unclear. It has been demonstrated that the excision repair of uracil residues 12 base pairs apart results in double strand breaks; this may increase DNA instability due to relaxation of DNA super coiling and chromosomal remodeling causing an increase in malignant transformations.

A second reason wherein low plasma folate is linked with cancer is via DNA methylation. DNA methylation is an epigenetic modification crucial to normal genome regulation and development. Indeed, it is Hcy that is recycled to methionine with the help of methionine synthase. DNA methylation is carried out with the help of methyl donor, S-adenosyl-l-methionine (SAM), which is obtained from methionine by the action of *S*-adenomethyl synthetase in an ATP dependent reaction.[127] The DNA methylation results as a joint action of three types of DNA methyltransferases (DNMTs): DNMT1, DNMT3a, and DNMT3b on SAM. Since, SAM is generated from 5-methyltetrahydrofolate (5′-MTHF), low folate limits the substrate for methionine synthase, thereby resulting in DNA hypomethylation. DNA hypomethylation leads to decondensation of pericentromeric heterochromatin or the activation of retrotransposon elements.[128]

Global genomic hypomethylation has been found in many types of cancer, including prostate metastatic tumors, chronic lymphocytic tumors, hepatocellular carcinoma, etc.[129] However, in cancer, hypomethylation of transcription regulatory regions seems to be much less frequent than hypermethylation of CpG islands overlapping promoters. During the early stages of tumorigenesis, or in abnormal nonneoplastic tissue, such as hyperplasia, there often seems to be regional hypomethylation of DNA sequences.[123, 125, 130] For example, in a subset of gastrointestinal cancers, age-related hypomethylation appears to precede aneuploidy. DNA hypomethylation has also been shown to result in decondensation of pericentromeric heterochromatin or the activation of retrotransposon elements.[128] Some of the cancer-associated loss of DNA methylation encompasses transcription control regions (e.g., PLAU/uPA) is over expressed and hypomethylated in conjugation with tumor progression in breast cancer (DNA hypomethylation in cancer cells). In addition, CpG island sequences are 1–2 kb that are (C+G)-rich, they may overlap promoters and they are generally unmethylated constitutively in normal tissues. CpG islands also result in faulty activation of the corresponding genes in cancer cells.[131] It has also been argued that hypomethylation of DNA is generally more pronounced with tumor progressions or the degree of malignancy. Such other genes displaying transcriptional upregulation and decreased methylation are PGP9.5, PMOC, claudin4, S100A4, etc. Serum folate levels are lower in smokers as compared to that of nonsmokers. Regional folate deficiency in the bronchial epithelium of smokers, which is believed to be caused by chemical inactivation of the metabolite by cigarette smoke, has been postulated to be involved in carcinogenesis.

6.6.2 *Cancer Patients Have High Plasma Hcy*

As mentioned in the earlier sections, there exists an inverse relation between plasma Hcy and folate. In cancer patients, the level of plasma folate is expected to be low because tumor cells have to pick folate from the blood, as it is required for the de novo purine synthesis. Interestingly, hyperhomocysteinemia has been shown to be associated with several types of cancer. It also appears that the Hcy toxicity leading to cancer is not dependent on the type of organ/tissue or on the type of cancer. In fact, all cancer types in the advanced stage exhibit high plasma Hcy, while there was no significant change in plasma Hcy in early stage cancers. Furthermore, once patients are subjected to surgery or chemotherapy, there is also a sharp increase in the plasma Hcy and hence thromboembolemic events appear. Most chemotherapeutic agents commonly used in the clinics (such as alkylating agents, antimetabolites, methotrexate, hormones, and antagonists) are antifolate drugs,[132] which cause a decrease in plasma concentrations of folic acid. In another development, it has also been shown that older cancer patients are at a higher risk of developing hyperhomocysteinemia than younger patients.[133]

It is still not clear why the levels of Hcy are highly variable in the early and late stages of cancer. However, we speculate that cells in early stages might not secrete Hcy as it helps the proliferation process of cancer cells.[134] On the other hand, advanced stage cancer cells might secrete Hcy because very high concentration of Hcy might also be cytotoxic to the cancer cells. Therefore, it may be important for proliferating cells to maintain an optimum Hcy concentration. This speculation, however, requires further experimental validation.

6.6.3 Cancer Patients Develop Thromboembolism Due to Hcy Toxicity

One major symptom of hyperhomocysteinemia is VTE. VTE is the most frequent complication and second most usual reason for death among cancer patients.[135] Advanced stage cancer patients develop both hyperhomocysteinemia and VTE. Alternatively, in early cancer (wherein hyperhomocysteinemia does not occur), the incidence of VTE is absent. Indeed the advanced stage cancer patients have a greater risk of developing VTE, with a frequency of 5%–15%[136, 137] in comparison to the general population (0.1%). Postchemo cancer patients (that are known to have risk for hyperhomocysteinemia) account for 13% of total VTE patients.[138] In postsurgery patients, the susceptibility toward embolism is increased by three times, and twice toward thrombosis.[139] Use of central venous catheters and hormonal adjuvant therapy (e.g., Tamoxifen), also predisposes a person to VTE[136] due to increase in plasma Hcy. Thus, there is a close link between hyperhomocysteinemia and development of VTE.

The mechanism of cancer-related thrombosis induced by elevated Hcy[140] is complex and not well known. However, it has been considered to be a result of endothelial disturbances due to formation of Hcy-mediated free-radicals.[141] Hcy is pro-oxidant and formation of Hcy-Hcy dimers and Hcy-protein adducts that will help to generate free radicals are well evident.[30] Hcy can also form a more highly reactive compound called HTL. HTL is also known to form covalent adducts with the Lys (or arginine) residues in proteins, thereby resulting in the formation of insoluble toxic protein aggregate or amyloid.[13] Their deposition in blood or heart may therefore impede normal functioning and physiology of the heart. Additionally, modification of homeostatic proteins (via N-homocysteinylation or S-homocysteinylation) has also been reported to impede NO metabolism, which may cause biotoxicity in the endothelial cells.[142] Hcy also inhibits thrombomodulin and Protein C toward Factor V_a[143] and, therefore, the coagulation behavior of blood is enhanced in the presence of Hcy. Furthermore, Hcy limits the secretion of nitric oxide

(NO), leading to increase platelet aggregation and decrease in antithrombic activities of the endothelial cells.[144, 145]

6.7 Elevated Hcy Levels and Its Association With NDs

Neurodegenerative diseases (NDs) are characterized by disturbed skilled movements, cognition, memory, etc. NDs usually target specific regions of the nervous system and lead to death of neuronal cells resulting in various functional consequences. These conditions arise primarily via multifactorial basis, which mainly include (i) oxidative stress and generation of free radicals, (ii) impaired bioenergetics and mitochondrial dysfunction, and (iii) anomalous protein dynamics and defective protein degradation associated with misfolding and aggregation. Alzheimer's (AD), Prion disease, Huntington's (HD), dementia, Parkinson's disease (PD), etc. are commonly occurring neurodegenerative diseases associated with aging. Among these, AD is the prevalent form of irreversible dementia.[146] Defining characteristic patterns of cell death and identifying disease specific markers have also aided in classification of these NDs. For example, AD is mainly characterized by the accumulation of senile Aβ plaques and NFT, accompanied with neuronal cell death, and acetylcholine deficiency. Lewy bodies and depletion of dopamine characterize the etiology in the case of PD. Intracellular inclusions of SOD1 and swollen motor axons are known to be major events in the progression of amyotrophic lateral sclerosis (ALS), and intracellular inclusions of huntingtin protein combined with the loss of γ-aminobutyric acid-containing neurons of neostriatum characterize HD.[147]

Out of hundreds of neurodegenerative conditions, a major share of our attention has been given toward this handful of conditions.[148] Despite significant dissimilarities in clinical manifestations, this diverse group of diseases shares at least a few common features, such as accumulation of proteinaceous deposits in the cerebral tissues as neuronal plaques and inclusions,[146, 149] which are the representative signatures of most of the neurological disorders. These deposits have the potential to trigger cascade of events that can, sooner or later, result in synaptic dysfunction and even neuronal death. In addition to the proteinaceous deposits, oxidative stress and the consequent cellular macromolecular damage, neuroinflammatory processes, and disruption of cellular/axonal transport are also characteristic features of the etiologies of NDs. Furthermore, alterations in acetylcholine esterase (AChE) activity have also been shown to be closely associated with AD pathology.[150] In addition, the role of chemical neurotoxin exposures and environmental factors in NDs has now been well characterized.[151] A list of common neurodegenerative conditions along with their causal proteins and the characteristic aggregate features are presented in Table 2.

TABLE 2 List of Reported N-Homocysteinylated Proteins and Resulting Consequences

	Proteins	Effects of N-homocysteinylation	References
1.	Fibrinogen	Resistant to Fibrinolysis, Protein aggregation	152
2.	Cyt c	Reduction of heme, activation of peroxidase function and protein aggregation	44, 46, 70
3.	BSA	Aggregation and Amyloid formation	60
4.	Alpha S1 Casein	Aggregation	42, 61
5.	Beta casein	Aggregation	61
6.	Kappa casein	Aggregation and amyloid formation	61
7.	Insulin	Aggregation and fibrillation	67
8.	Crystallin	Aggregation and fibrillation	66
9.	Prion	Aggregation and Amyloid-like transformation	62
10.	Aβ	Stabilizes soluble oligomer-intermediate states and increases neurotoxicity	153
11.	Tau	Alters tubulin binding and enhances self-association and aggregation	63
12.	LDLP	Aggregation and spontaneous precipitation	154
13.	Myoglobin	Aggregation	14
14.	Gamma Globulin	Aggregation	14
15.	Transferrin	Aggregation	14
16.	Hemoglobin	Aggregation	14
17.	α-LA	Aggregation	42, 44
18	CA	Aggregation	42
19.	GAPDH	Oligomerization and aggregation	72
20.	LDH	Oligomerization and aggregation	72
21.	Plasminogen	Decreased urokinase-induced plasmin activity	41
22.	Lysyl oxidase	Inhibition of enzyme function	155

Adapted from our earlier published article (Sharma GS, et al. Protein N-homocysteinylation: from cellular toxicity to neurodegeneration. Biochim Biophys Acta Gen Subj *2015;1850(11):2239–2245, BBA General Subjects*

Recently, blood and cellular aggravated Hcy levels were reported to be linked to various NDs.[81, 156, 157] High Hcy levels have been discovered to be associated with central nervous system disorders for the first time in severe CBS-deficient patients.[96, 158] These patients tend to have higher incidence of mental deterioration, seizure, cerebral degeneration, etc.[96, 158–160] Although, in the case of neuronal cells, the existence of CBS and cystathionase has been arguable, Hcy levels in the brain and cerebrospinal fluid (CSF) are reported to be increased in many neurological diseases.[161–164] Furthermore, increased Hcy can lead to severe effects in the brain as Hcy has been reported to cross the blood brain barrier.[11] An interesting study by Seshadri et al. has shown that high levels of Hcy in the blood is a strong indicator for AD.[156, 165] In addition, several case–control and cross-sectional studies have shown the association between Hcy levels and AD.[85] An increase in Hcy concentration in the serum to a level of over 50 μM has been reported to result in dementia, with spontaneous convulsions when it exceeds 300 μM.[166] A schematic representation of the association of elevated Hcy levels with cardiovascular complications and NDs is elaborated in Fig. 1.

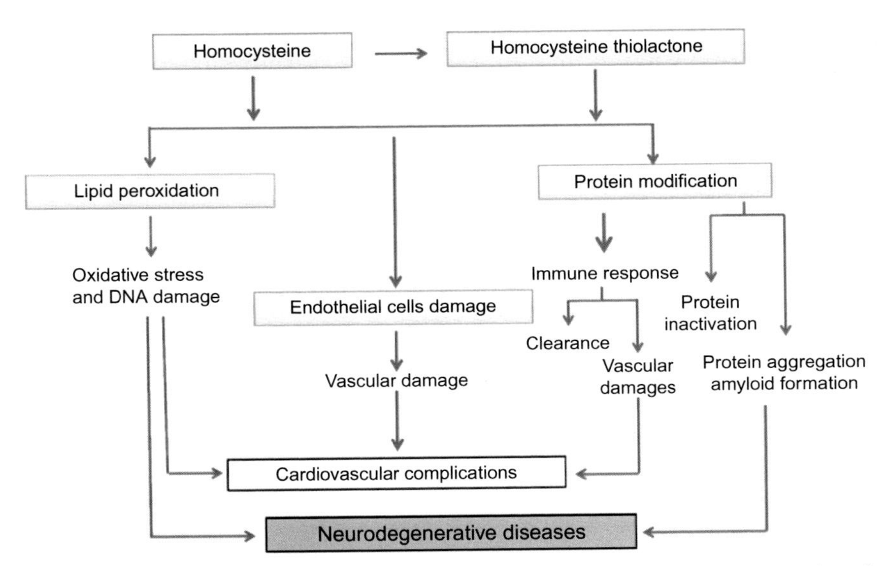

FIG. 1 A schematic representation of the association of elevated Hcy and HTL levels with cardiovascular complications and NDs. *Modified and reproduced from our earlier published article (Sharma GS, et al. Protein N-homocysteinylation: from cellular toxicity to neurodegeneration.* Biochim Biophys Acta Gen Subj *2015;1850(11):2239–2245, BBA General Subjects.*

7 CONCLUSIONS AND FUTURE PROSPECTS

The past two decades have seen a tremendous increase in the interest in Hcy biology. This very interest stems from the fact that elevated plasma and cellular Hcy levels are independent risk factors for several clinical complications. Two major clinical conditions closely linked with elevated Hcy levels include vascular complications and neurodegenerative diseases. Furthermore, defects in Hcy metabolism have also been shown to be linked to several types of cancer. In fact, covalent modification of protein by the metabolite of HTL via N-homocysteinylation has been attributed to be the key player toward the toxicity of elevated Hcy. Besides the close association of HTL-induced protein aggregation and neurodegeneration, HTL-modified proteins are also known to be antigenic in nature and can invoke immune response, which is the central basis of the association of Hcy and vascular injury under Hyperhomocysteinemia conditions. In addition, elevated Hcy level has also been shown to be associated with decreased expression of several antioxidant enzymes including SOD, catalase, and peroxidase.[167] Furthermore, Hcy, due to its highly pro-oxidizing nature, tend to gets oxidized to form homocystine (Hcy-Hcy dimer), which is enhanced in presence of transition metal ions (such as iron).[168] Thus, Hcy could indirectly hamper functionality of the cellular antioxidant proteins, which contain heme moiety. Hence, an elevated Hcy level could also hamper cellular processes without direct interaction with proteins. It seems that S-linked modification of cellular proteins also has significant contribution in the etiology of hyperhomocysteinemia, besides the N-linked modification. A clearer knowledge of these modifications would provide useful insights in understanding the progression of associated diseases. Folic acid, vitamin-B, and betaine supplements are currently the widely used treatments for lowering the Hcy levels. Additionally, a methionine-restricted diet has been shown to provide positive results in lowering Hcy levels. Furthermore, therapeutic agents that can buffer the toxic Hcy/HTL by preventing them from reacting with cellular/serum proteins could prove to be hopeful strategies for treating hyperhomocysteinemic conditions.

References

1. Mann M, Jensen ON. Proteomic analysis of post-translational modifications. *Nat Biotechnol* 2003;**21**(3):255–61.
2. Yan SCB, Grinnell BW, Wold F. Post-translational modifications of proteins: some problems left to solve. *Trends Biochem Sci* 1989;**14**(7):264–8.
3. Prabakaran S, et al. Post-translational modification: nature's escape from genetic imprisonment and the basis for dynamic information encoding. *Wiley Interdiscip Rev Syst Biol Med* 2012;**4**(6):565–83.

4. Karve TM, Cheema AK. Small changes huge impact: the role of protein posttranslational modifications in cellular homeostasis and disease. *J Amino Acids* 2011;**2011**:1–13.
5. Duan G, Walther D. The roles of post-translational modifications in the context of protein interaction networks. *PLoS Comput Biol* 2015;**11**(2).
6. Cheng H-C, et al. Regulation and function of protein kinases and phosphatases. *Enzym Res* 2011:1–3.
7. Majerus PW, Kisseleva MV, Norris FA. The role of phosphatases in inositol signaling reactions. *J Biol Chem* 1999;**274**(16):10669–72.
8. Rinderknecht H. Activation of pancreatic zymogens. *Dig Dis Sci* 1986;**31**(3):314–21.
9. Parker CE, et al. Mass spectrometry for post-translational modifications. *Neuroproteomics* 2010;**2010**. [PMID: 21882444].
10. Bird A. DNA methylation patterns and epigenetic memory. *Genes Dev* 2002;**16**(1):6–21.
11. Jakubowski H. Pathophysiological consequences of homocysteine excess. *J Nutr* 2006;**136**(6):1741S–1749S.
12. Thornalley PJ, Langborg A, Minhas HS. Formation of glyoxal, methylglyoxal and 3-deoxyglucosone in the glycation of proteins by glucose. *Biochem J* 1999;**344**(1):109–16.
13. Sharma GS, et al. Protein N-homocysteinylation: from cellular toxicity to neurodegeneration. *Biochim Biophys Acta Gen Subj* 2015;**1850**(11):2239–45.
14. Jakubowski H. Protein homocysteinylation: possible mechanism underlying pathological consequences of elevated homocysteine levels. *FASEB J* 1999;**13**(15):2277–83.
15. Glomb MA, Nagaraj H. Protein modifications by glyoxal and methylglyoxal during the maillard. In: *The Maillard Reaction in Foods and Medicine, Woodhead Publishing Series in Food Science, Technology and Nutrition*, 2005, 250–255.
16. Schuh S, et al. Homocystinuria and megaloblastic anemia responsive to vitamin B12 therapy. An inborn error of metabolism due to a defect in cobalamin metabolism. *N Engl J Med* 1984;**310**(11):686–90.
17. Dodds L, et al. Effect of homocysteine concentration in early pregnancy on gestational hypertensive disorders and other pregnancy outcomes. *Clin Chem* 2008;**54**(2):326–34.
18. Gellekink H, et al. Genetic determinants of plasma total homocysteine. *Semin Vasc Med* 2005;**5**(2):98–109.
19. Refsum H, et al. Homocysteine and cardiovascular disease. *Annu Rev Med* 1998;**49**:31–62.
20. Obeid R, et al. Vitamin B12 status in the elderly as judged by available biochemical markers. *Clin Chem* 2004;**50**(1):238–41.
21. Diaz-Arrastia R. Homocysteine and neurologic disease. *Arch Neurol* 2000;**57**(10):1422–7.
22. Morrell MJ. Guidelines for the care of women with epilepsy. *Neurology* 1998;**51**(5 Suppl 4):S21–7.
23. Jacobsen DW, et al. Molecular targeting by homocysteine: a mechanism for vascular pathogenesis. *Clin Chem Lab Med* 2005;**43**(10):1076–83.
24. Hultberg B, Andersson A, Isaksson A. Hypomethylation as a cause of homocysteine-induced cell damage in human cell lines. *Toxicology* 2000;**147**(2):69–75.
25. Sengupta S, et al. Albumin thiolate anion is an intermediate in the formation of albumin-S-S-homocysteine. *J Biol Chem* 2001;**276**(32):30111–7.
26. Peters Jr. T. *All about albumin: biochemistry, genetics, and medical applications.* Academic press; 1995.
27. Carter DC, Ho JX. Structure of serum albumin. *Adv Protein Chem* 1994;**45**:153–203.
28. Hong L, Fast W. Inhibition of human Dimethylarginine Dimethylaminohydrolase-1 by S-Nitroso-L-homocysteine and hydrogen peroxide analysis, quantification, and implications for hyperhomocysteinemia. *J Biol Chem* 2007;**282**(48):34684–92.
29. Zinellu A, et al. Factors affecting S-homocysteinylation of LDL apoprotein B. *Clin Chem* 2006;**52**(11):2054–9.
30. Jacobsen DW. Hyperhomocysteinemia and oxidative stress: time for a reality check? *Arterioscler Thromb Vasc Biol* 2000;**20**(5):1182–4.

31. Sundaramoorthy E, et al. Predicting protein homocysteinylation targets based on dihedral strain energy and pKa of cysteines. *Proteins Struct Funct Bioinf* 2008;**71**(3):1475–83.
32. Barbato JC, et al. Targeting of Metallothionein by L-homocysteine. *Arterioscler Thromb Vasc Biol* 2007;**27**(1):49–54.
33. Maret W. Metallothionein redox biology in the cytoprotective and cytotoxic functions of zinc. *Exp Gerontol* 2008;**43**(5):363–9.
34. Refsum H, Helland S, Ueland PM. Radioenzymic determination of homocysteine in plasma and urine. *Clin Chem* 1985;**31**(4):624–8.
35. Sass JrO, et al. S-homocysteinylation of transthyretin is detected in plasma and serum of humans with different types of hyperhomocysteinemia. *Biochem Biophys Res Commun* 2003;**310**(1):242–6.
36. Hajjar KA, et al. Tissue plasminogen activator binding to the annexin II tail domain direct modulation by homocysteine. *J Biol Chem* 1998;**273**(16):9987–93.
37. Zinellu A, et al. Low density lipoprotein S-homocysteinylation is increased in acute myocardial infarction patients. *Clin Biochem* 2012;**45**(4):359–62.
38. Majors AK, et al. Homocysteine binds to human plasma fibronectin and inhibits its interaction with fibrin. *Arterioscler Thromb Vasc Biol* 2002;**22**(8):1354–9.
39. Hong L, Fast W. Inhibition ofHumanDimethylarginine Dimethylaminohydrolase-1 by S-Nitroso-L-homocysteine and hydrogen peroxide: analysis, quantification, and implications for hyperhomocysteinemia. *J Biol Chem* 2007;**282**(48):34684–92.
40. Stühlinger MC, et al. Homocysteine impairs the nitric oxide synthase pathway role of asymmetric Dimethylarginine. *Circulation* 2001;**104**(3):2569–75.
41. Kolodziejczyk-Czepas J, et al. Homocysteine and its thiolactone impair plasmin activity induced by urokinase or streptokinase in vitro. *Int J Biol Macromol* 2012;**50**(3):754–8.
42. Sharma GS, Kumar T, Singh LR. N-homocysteinylation induces different structural and functional consequences on acidic and basic proteins. *PLoS ONE* 2014;**9**(12).
43. Ptitsyn OB. Molten globule and protein folding. *Adv Protein Chem* 1995;**47**:83–229.
44. Kumar T, Sharma GS, Singh LR. Existence of molten globule state in homocysteine-induced protein covalent modifications. *PLoS ONE* 2014;**9**(11).
45. Iram A, et al. Molten globule of hemoglobin proceeds into aggregates and advanced glycated end products. *PLoS One* 2013;**8**(8):e72075.
46. Sharma GS, Singh LR. Conformational status of cytochrome c upon N-homocysteinylation: Implications to cytochrome c release. *Arch Biochem Biophys* 2017;**614**:23–7.
47. Birk AV, et al. Disruption of cytochrome c heme coordination is responsible for mitochondrial injury during ischemia. *Biochim Biophys Acta Bioenerg* 2015;**1847**(10):1075–84.
48. Sontag E, et al. Protein phosphatase 2A methyltransferase links homocysteine metabolism with tau and amyloid precursor protein regulation. *J Neurosci* 2007;**27**(11):2751–9.
49. Jalili S, et al. Effect of homocysteine thiolactone on structure and aggregation propensity of bovine pancreatic insulin. *Protein J* 2011;**30**(5):299–307.
50. Borreguero JM, et al. Folding events in the 21–30 region of amyloid beta-protein (Abeta) studied in silico. *Proc Natl Acad Sci U S A* 2005;**102**(17):6015–20.
51. Esler WP, et al. Point substitution in the central hydrophobic cluster of a human beta-amyloid congener disrupts peptide folding and abolishes plaque competence. *Biochemistry* 1996;**35**(44):13914–21.
52. Karsai A, et al. Effect of lysine-28 side-chain acetylation on the nanomechanical behavior of alzheimer amyloid beta25–35 fibrils. *J Chem Inf Model* 2005;**45**(6):1641–6.
53. Luhrs T, et al. 3D structure of Alzheimer's amyloid-beta(1–42) fibrils. *Proc Natl Acad Sci U S A* 2005;**102**(48):17342–7.
54. Petkova AT, et al. A structural model for Alzheimer's beta -amyloid fibrils based on experimental constraints from solid state NMR. *Proc Natl Acad Sci U S A* 2002;**99**(26):16742–7.

55. Rochet JC, Lansbury Jr. PT. Amyloid fibrillogenesis: themes and variations. *Curr Opin Struct Biol* 2000;**10**(1):60–8.
56. Tjernberg LO, et al. Controlling amyloid beta-peptide fibril formation with protease-stable ligands. *J Biol Chem* 1997;**272**(19):12601–5.
57. Tjernberg LO, et al. Arrest of beta-amyloid fibril formation by a pentapeptide ligand. *J Biol Chem* 1996;**271**(15):8545–8.
58. Bellesia G, Shea JE. Effect of beta-sheet propensity on peptide aggregation. *J Chem Phys* 2009;**130**(14).
59. Pellarin R, Guarnera E, Caflisch A. Pathways and intermediates of amyloid fibril formation. *J Mol Biol* 2007;**374**(4):917–24.
60. Paoli P, et al. Protein N-homocysteinylation induces the formation of toxic amyloid-like protofibrils. *J Mol Biol* 2010;**400**(4):889–907.
61. Stroylova YY, et al. Aggregation and structural changes of alpha(S1)-, beta- and kappa-caseins induced by homocysteinylation. *Biochim Biophys Acta* 2011;**1814**(10):1234–45.
62. Stroylova YY, et al. N-homocysteinylation of ovine prion protein induces amyloid-like transformation. *Arch Biochem Biophys* 2012;**526**(1):29–37.
63. Karima O, et al. Altered tubulin assembly dynamics with N-homocysteinylated human 4R/1N tau in vitro. *FEBS Lett* 2012;**586**(21):3914–9.
64. Buee L, et al. Tau protein isoforms, phosphorylation and role in neurodegenerative disorders. *Brain Res Brain Res Rev* 2000;**33**(1):95–130.
65. Trojanowski JQ, Lee VM. The role of tau in Alzheimer's disease. *Med Clin North Am* 2002;**86**(3):615–27.
66. Khazaei S, Yousefi R, Alavian-Mehr MM. Aggregation and fibrillation of eye lens crystallins by homocysteinylation; implication in the eye pathological disorders. *Protein J* 2012;**31**(8):717–27.
67. Yousefi R, et al. The enhancing effect of homocysteine thiolactone on insulin fibrillation and cytotoxicity of insulin fibril. *Int J Biol Macromol* 2012;**51**(3):291–8.
68. Lashuel HA, et al. Neurodegenerative disease: amyloid pores from pathogenic mutations. *Nature* 2002;**418**(6895):291.
69. Glowacki R, Jakubowski H. Cross-talk between Cys34 and lysine residues in human serum albumin revealed by N-homocysteinylation. *J Biol Chem* 2004;**279**(12):10864–71.
70. Perla-Kajan J, et al. Modification by homocysteine thiolactone affects redox status of cytochrome C. *Biochemistry* 2007;**46**(21):6225–31.
71. Zinellu A, et al. N-and S-homocysteinylation reduce the binding of human serum albumin to catechins. *Eur J Nutr* 2017;**56**(2):785–91.
72. Stroylova YY, et al. Creation of catalytically active particles from enzymes crosslinked with a natural bifunctional agent—homocysteine thiolactone. *Biopolymers* 2014;**101**(9):975–84.
73. Jakubowski H. Molecular basis of homocysteine toxicity in humans. *Cell Mol Life Sci* 2004;**61**(4):470–87.
74. Jakubowski H. Homocysteine thiolactone: metabolic origin and protein homocysteinylation in humans. *J Nutr* 2000;**130**(2):377S–381S.
75. Perla-Kajan J, Twardowski T, Jakubowski H. Mechanisms of homocysteine toxicity in humans. *Amino Acids* 2007;**32**(4):561–72.
76. Stipanuk MH, Ueki I. Dealing with methionine/homocysteine sulfur: cysteine metabolism to taurine and inorganic sulfur. *J Inherit Metab Dis* 2011;**34**(1):17–32.
77. Hoffer LJ. Homocysteine remethylation and trans-sulfuration. *Metabolism* 2004;**53**(11):1480–3.
78. Singh S, et al. Relative contributions of cystathionine Î^2-synthase and Î^3-cystathionase to H2S biogenesis via alternative trans-sulfuration reactions. *J Biol Chem* 2009;**284**(33):22457–66.

79. Finkelstein JD. The metabolism of homocysteine: pathways and regulation. *Eur J Pediatr* 1998;**157**(2):S40–4.
80. Girgis S, et al. 5-Formyltetrahydrofolate regulates homocysteine remethylation in human neuroblastoma. *J Biol Chem* 1997;**272**(8):4729–34.
81. Mattson MP, Shea TB. Folate and homocysteine metabolism in neural plasticity and neurodegenerative disorders. *Trends Neurosci* 2003;**26**(3):137–46.
82. Mudd SH, et al. Homocystinuria associated with decreased methylenetetrahydrofolate reductase activity. *Biochem Biophys Res Commun* 1972;**46**(2):905–12.
83. Kluijtmans LAJ, et al. The molecular basis of cystathionine Î^2-synthase deficiency in Dutch patients with homocystinuria: effect of CBS genotype on biochemical and clinical phenotype and on response to treatment. *Am J Hum Genet* 1999;**65**(1):59–67.
84. Skovby F, Gaustadnes M, Mudd SH. A revisit to the natural history of homocystinuria due to cystathionine Î^2-synthase deficiency. *Mol Genet Metab* 2010;**99**(1):1–3.
85. Ravaglia G, et al. Homocysteine and folate as risk factors for dementia and Alzheimer disease. *Am J Clin Nutr* 2005;**82**(3):636–43.
86. Nelen WLDM, et al. Homocysteine and folate levels as risk factors for recurrent early pregnancy loss. *Obstet Gynecol* 2000;**95**(4):519–24.
87. Clarke R, et al. Folate, vitamin B12, and serum total homocysteine levels in confirmed Alzheimer disease. *Arch Neurol* 1998;**55**(11):1449–55.
88. Schnyder G, et al. Effect of homocysteine-lowering therapy with folic acid, vitamin B12, and vitamin B6 on clinical outcome after percutaneous coronary intervention: the swiss heart study: a randomized controlled trial. *JAMA* 2002;**288**(8):973–9.
89. Mills JL, et al. Homocysteine metabolism in pregnancies complicated by neural-tube defects. *Lancet* 1995;**345**(8943):149–51.
90. Brustolin S, Giugliani R, Félix TM. Genetics of homocysteine metabolism and associated disorders. *Braz J Med Biol Res* 2010;**43**(1):1–7.
91. Wu LL, Wu JT. Hyperhomocysteinemia is a risk factor for cancer and a new potential tumor marker. *Clin Chim Acta* 2002;**322**(1):21–8.
92. Sawai A, et al. Influence of mental stress on the plasma homocysteine level and blood pressure change in young men. *Clin Exp Hypertens* 2008;**30**(3–4):233–41.
93. DeVos L, et al. Associations between single nucleotide polymorphisms in folate uptake and metabolizing genes with blood folate, homocysteine, and DNA uracil concentrations. *Am J Clin Nutr* 2008;**88**(4):1149–58.
94. Carson NAJ, Neill DW. Metabolic abnormalities detected in a survey of mentally backward individuals in Northern Ireland. *Arch Dis Child* 1962;**37**(195):505.
95. Gerritsen T, Vaughn JG, Waisman HA. The identification of homocystine in the urine. *Biochem Biophys Res Commun* 1962;**9**(6):493–6.
96. Mudd SH, et al. The natural history of homocystinuria due to cystathionine beta-synthase deficiency. *Am J Hum Genet* 1985;**37**(1):1–31.
97. Cross HE, Jensen AD. Ocular manifestations in the Marfan syndrome and homocystinuria. *Am J Ophthalmol* 1973;**75**(3):405–20.
98. Brenton DP, et al. Homocystinuria and Marfan's syndrome. *J Bone Joint Surg Br* 1972;**54** (2):277–98.
99. Pinnell SR, Martin GR. The cross-linking of collagen and elastin: enzymatic conversion of lysine in peptide linkage to alpha-aminoadipic-delta-semialdehyde (allysine) by an extract from bone. *Proc Natl Acad Sci* 1968;**61**(2):708–16.
100. Ishida S, et al. Homocystinuria due to cystathionine beta-synthase deficiency associated with megaloblastic anaemia. *J Intern Med* 2001;**250**(5):453–6.
101. McCully KS. Vascular pathology of homocysteinemia: implications for the pathogenesis of arteriosclerosis. *Am J Pathol* 1969;**56**(1):111.
102. Wilcken DE, Wilcken B. The pathogenesis of coronary artery disease. A possible role for methionine metabolism. *J Clin Investig* 1976;**57**(4):1079.

103. Hankey GJ, Eikelboom JW. Homocysteine and vascular disease. *Lancet* 1999;**354**(9176):407–13.
104. Mudd SH, et al. Homocysteine and its disulfide derivatives. *Arterioscler Thromb Vasc Biol* 2000;**20**(7):1704–6.
105. Mudd SH, Levy HL, Skovby F, Beaudet AL, Sly WS, Valle D. Disorders of transsulfuration. In: Scriver CR, editor. *The metabolic and molecular bases of inherited disease*. New York: McGraw-Hill, Inc; 1995. p. 1279–327.
106. Lentz SR. Mechanisms of homocysteine-induced atherothrombosis. *J Thromb Haemost* 2005;**3**(8):1646–54.
107. Wang H, Tan H, Yang F. Mechanisms in homocysteine-induced vascular disease. *Drug Discov Today Dis Mech* 2005;**2**(1):25–31.
108. Eberhardt RT, et al. Endothelial dysfunction in a murine model of mild hyperhomocyst(e) inemia. *J Clin Invest* 2000;**106**(4):483–91.
109. Harker LA, Harlan JM, Ross R. Effect of sulfinpyrazone on homocysteine-induced endothelial injury and arteriosclerosis in baboons. *Circ Res* 1983;**53**(6):731–9.
110. Roybal CN, et al. Homocysteine increases the expression of vascular endothelial growth factor by a mechanism involving endoplasmic reticulum stress and transcription factor ATF4. *J Biol Chem* 2004;**279**(15):14844–52.
111. Zhang C, et al. Homocysteine induces programmed cell death in human vascular endothelial cells through activation of the unfolded protein response. *J Biol Chem* 2001;**276**(38):35867–74.
112. Dardik R, et al. Homocysteine and oxidized low density lipoprotein enhance platelet adhesion to endothelial cells under flow conditions: distinct mechanisms of thrombogenic modulation. *Thromb Haemost* 2000;**83**(2):338–57.
113. Den Heijer M, et al. Hyperhomocysteinemia as a risk factor for deep-vein thrombosis. *N Engl J Med* 1996;**334**(12):759–62.
114. Falcon CR, et al. High prevalence of hyperhomocyst (e) inemia in patients with juvenile venous thrombosis. *Arterioscler Thromb Vasc Biol* 1994;**14**(7):1080–3.
115. Bostom AG, Lathrop L. Hyperhomocysteinemia in end-stage renal disease: prevalence, etiology, and potential relationship to arteriosclerotic outcomes. *Kidney Int* 1997;**52**(1):10–20.
116. Van Guldener C, Stam F, Stehouwer CDA. Homocysteine metabolism in renal failure. *Kidney Int* 2001;**59**:S234–7.
117. Giovannucci E, et al. Alcohol, low-methionine-low-folate diets, and risk of colon cancer in men. *J Natl Cancer Inst* 1995;**87**(4):265–73.
118. Singal R, et al. Polymorphisms in the methylenetetrahydrofolate reductase gene and prostate cancer risk. *Int J Oncol* 2004;**25**(5):1465–71.
119. Robien K, Ulrich CM. 5, 10-methylenetetrahydrofolate reductase polymorphisms and leukemia risk: a HuGE minireview. *Am J Epidemiol* 2003;**157**(7):571–82.
120. Bravata V. Controversial roles of methylenetetrahydrofolate reductase polymorphisms and folate in breast cancer disease. *Int J Food Sci Nutr* 2015;**66**(1):43–9.
121. Ma J, et al. Methylenetetrahydrofolate reductase polymorphism, dietary interactions, and risk of colorectal cancer. *Cancer Res* 1997;**57**(6):1098–102.
122. Krajinovic M, et al. Role of MTHFR genetic polymorphisms in the susceptibility to childhood acute lymphoblastic leukemia. *Blood* 2004;**103**(1):252–7.
123. Kato I, et al. Serum folate, homocysteine and colorectal cancer risk in women: a nested case-control study. *Br J Cancer* 1999;**79**(11 – 12):1917.
124. Montfort WR, et al. Structure, multiple site binding, and segmental accommodation in thymidylate synthase on binding dUMP and an anti-folate. *Biochemistry* 1990;**29**(30):6964–77.
125. Blount BC, et al. Folate deficiency causes uracil misincorporation into human DNA and chromosome breakage: implications for cancer and neuronal damage. *Proc Natl Acad Sci* 1997;**94**(7):3290–5.

126. Drexler HG, et al. *Hematopoietic cell lines*. Orlando: Academic Press; 1994.
127. Crider KS, et al. Folate and DNA methylation: a review of molecular mechanisms and the evidence for folate's role. *Adv Nutr* 2012;**3**(1):21–38.
128. Hall LE, Mitchell SE, O'Neill RJ. Pericentric and centromeric transcription: a perfect balance required. *Chromosom Res* 2012;**20**(5):535–46.
129. Ehrlich M. DNA methylation in cancer: too much, but also too little. *Oncogene* 2002;**21** (35):5400.
130. Locasale JW. Serine, glycine and one-carbon units: cancer metabolism in full circle. *Nat Rev Cancer* 2013;**13**(8):572–83.
131. Sharma S, Kelly TK, Jones PA. Epigenetics in cancer. *Carcinogenesis* 2010;**31**(1):27–36.
132. Stathopoulou A, et al. Molecular detection of cytokeratin-19-positive cells in the peripheral blood of patients with operable breast cancer: evaluation of their prognostic significance. *J Clin Oncol* 2002;**20**(16):3404–12.
133. Refsum H, et al. The Hordaland homocysteine study: a community-based study of homocysteine, its determinants, and associations with disease. *J Nutr* 2006;**136** (6):1731S–1740S.
134. Sun C-F, et al. Serum total homocysteine increases with the rapid proliferation rate of tumor cells and decline upon cell death: a potential new tumor marker. *Clin Chim Acta* 2002;**321**(1):55–62.
135. Rickles FR, Levine M, Edwards RL. Hemostatic alterations in cancer patients. *Cancer Metastasis Rev* 1992;**11**(3):237–48.
136. Gatt A, et al. Hyperhomocysteinemia in women with advanced breast cancer. *Int J Lab Hematol* 2007;**29**(6):421–5.
137. Green KB, Silverstein RL. Hypercoagulability in cancer. *Hematol Oncol Clin North Am* 1996;**10**(2):499–530.
138. Heit JA, et al. Relative impact of risk factors for deep vein thrombosis and pulmonary embolism: a population-based study. *Arch Intern Med* 2002;**162**(11):1245–8.
139. Kakkar AK, et al. Prevention of perioperative venous thromboembolism: outcome after cancer and noncancer surgery. *Br J Surg* 2001;**88**(S1).
140. Zhu H, et al. Homocysteine remethylation enzyme polymorphisms and increased risks for neural tube defects. *Mol Genet Metab* 2003;**78**(3):216–21.
141. Welch GN, Loscalzo J. Homocysteine and atherothrombosis. *N Engl J Med* 1998;**338** (15):1042–50.
142. Morganti M, et al. Atherosclerosis and cancer: common pathways on the vascular endothelium. *Biomed Pharmacother* 2002;**56**(7):317–24.
143. Lentz SR, et al. Vascular dysfunction in monkeys with diet-induced hyperhomocyst (e) inemia. *J Clin Investig* 1996;**98**(1):24.
144. FitzGerald GA. Parsing an enigma: the pharmacodynamics of aspirin resistance. *Lancet* 2003;**361**(9357):542–4.
145. Chandrasekharan NV, et al. COX-3, a cyclooxygenase-1 variant inhibited by acetaminophen and other analgesic/antipyretic drugs: cloning, structure, and expression. *Proc Natl Acad Sci* 2002;**99**(21):13926–31.
146. Ross CA, Poirier MA. Protein aggregation and neurodegenerative disease. *Nat Med* 2004;**10**(Suppl):S10–7.
147. Martin JB. Molecular basis of the neurodegenerative disorders. *N Engl J Med* 1999;**340** (25):1970–80.
148. Przedborski S, Vila M, Jackson-Lewis V. Series introduction: neurodegeneration: what is it and where are we? *J Clin Invest* 2003;**111**(1):3–10.
149. Chowhan RK, et al. Ignored avenues in alpha-synuclein associated proteopathy. *CNS Neurol Disord Drug Targets* 2014;**13**(7):1246–57.
150. Skovronsky DM, Lee VMY, Trojanowski JQ. Neurodegenerative diseases: new concepts of pathogenesis and their therapeutic implications. *Annu Rev Pathol: Mech Dis* 2006;**1**:151–70.

151. Di Monte DA. The environment and Parkinson's disease: is the nigrostriatal system preferentially targeted by neurotoxins? *Lancet Neurol* 2003;**2**(9):531–8.
152. Sauls DL, et al. Modification of fibrinogen by homocysteine thiolactone increases resistance to fibrinolysis: a potential mechanism of the thrombotic tendency in hyperhomocysteinemia. *Biochemistry* 2006;**45**(8):2480–7.
153. Khodadadi S, et al. Effect of N-homocysteinylation on physicochemical and cytotoxic properties of amyloid beta-peptide. *FEBS Lett* 2012;**586**(2):127–31.
154. Naruszewicz M, et al. Thiolation of low-density lipoprotein by homocysteine-thiolactone causes increased aggregation and interaction with cultured macrophages. *Nutr Metab Cardiovasc Dis* 1994;(4):70–7.
155. Liu G, Nellaiappan K, Kagan HM. Irreversible inhibition of Lysyl oxidase by homocysteine Thiolactone and its selenium and oxygen analogues implications for homocystinuria. *J Biol Chem* 1997;**272**(51):32370–7.
156. Seshadri S, et al. Plasma homocysteine as a risk factor for dementia and Alzheimer's disease. *N Engl J Med* 2002;**346**(7):476–83.
157. Obeid R, Herrmann W. Mechanisms of homocysteine neurotoxicity in neurodegenerative diseases with special reference to dementia. *FEBS Lett* 2006;**580**(13):2994–3005.
158. Grieco AJ. Homocystinuria: pathogenetic mechanisms. *Am J Med Sci* 1977;**273**(2):120–32.
159. Sachdev PS, et al. Relationship between plasma homocysteine levels and brain atrophy in healthy elderly individuals. *Neurology* 2002;**58**(10):1539–41.
160. van den Berg M, et al. Hyperhomocysteinaemia; with reference to its neuroradiological aspects. *Neuroradiology* 1995;**37**(5):403–11.
161. Isobe C, et al. Increase of total homocysteine concentration in cerebrospinal fluid in patients with Alzheimer's disease and Parkinson's disease. *Life Sci* 2005;**77**(15):1836–43.
162. Eto K, et al. Brain hydrogen sulfide is severely decreased in Alzheimer's disease. *Biochem Biophys Res Commun* 2002;**293**(5):1485–8.
163. Regland B, et al. CSF-methionine is elevated in psychotic patients. *J Neural Transm* 2004;**111**(5):631–40.
164. Yanai Y, et al. Concentrations of sulfur-containing free amino acids in lumbar cerebrospinal fluid from patients with consciousness disturbances. *Acta Neurol Scand* 1983;**68**(6):386–93.
165. Seshadri S. Elevated plasma homocysteine levels: risk factor or risk marker for the development of dementia and Alzheimer's disease? *J Alzheimers Dis* 2006;**9**(4):393–8.
166. Boldyrev AA. Why homocysteine is a risk factor of neurodegenerative diseases. *Neurochem J* 2007;**1**(1):14–20.
167. Outinen PA, et al. Homocysteine-induced endoplasmic reticulum stress and growth arrest leads to specific changes in gene expression in human vascular endothelial cells. *Blood* 1999;**94**(3):959–67.
168. Škovierová H, et al. The molecular and cellular effect of homocysteine metabolism imbalance on human health. *Int J Mol Sci* 2016;**17**(10):1733.

Further Reading

169. Regland B, et al. Increased concentrations of homocysteine in the cerebrospinal fluid in patients with fibromyalgia and chronic fatigue syndrome. *Scand J Rheumatol* 1997;**26**(4):301–7.

CHAPTER

12

Posttranslational Modifications in Algae: Role in Stress Response and Biopharmaceutical Production

*Parvez Ahmad**, *Fareha Bano*†

*Protein Conformation and Enzymology Lab, Department of Biosciences, Jamia Millia Islamia (A Central University), New Delhi, India

†Department of Biology, College of Science and Arts (Female Branch), Al Ula Campus, Taibah Universty, Al Ula, Madina Province, Saudi Arabia

1 INTRODUCTION

Cells are frequently exposed to various kinds of stresses, which include oxidative, metal, hyperosmotic, DNA damage, heat shock, and hypoxia. While a majority of the cellular defense programs against these stresses are maintained by transcriptional induction of stress genes, a rapid alternative response is provided by posttranslational modification (PTM) mechanisms with the help of preexisting stress proteins.[1] There are various PTMs like phosphorylation, SUMOylation, and ubiquitination, which control introduction of preexisting molecules to ensure the stimulative response to various stresses.[2–4] It is very important, in terms of understanding the functional significance of PTM in biological systems, to identify, characterize, and map the modification on specific amino acid residue. A number of mechanisms like allosteric enzymatic regulations, reversible covalent modifications, protein turnover and localization, protein-protein interactions, and protein structural changes induced by different physical factors are involved in controlling the activities of stress proteins by PTM. Although the protein functions are critically governed

Protein Modificomics
https://doi.org/10.1016/B978-0-12-811913-6.00012-6

by their folding and refolding, the structural and functional diversity in proteins are significantly contributed by amino acids and their side chains modifications. These modifications are responsible for enhancing the complexity of eukaryotic proteome as compared to the coding capacity of the genome. Here, we have tried to summarize few of the major and well-studied PTMs in algae in relation to their importance in various stress conditions and their possible use in therapeutics.

2 POSTTRANSLATIONAL MODIFICATIONS IN ALGAE WITH RESPECT TO STRESS RESPONSE

2.1 Phosphorylation and Stress

A number of well-defined mechanisms accountable for the proper functioning of the proteins in eukaryotes under changing environmental conditions have been observed. Phosphorylation is a major and important reversible PTM, which is mediated by the kinases. The significance of protein phosphorylation can be guessed by the fact that protein kinases alone are encoded by approximately 1%–2% of eukaryotic genes.[4a] For a particular substrate, the level of phosphorylation is balanced by both specific kinase and phosphatase activities. It has been shown that phosphorylation may cause different effects on protein functions, viz. stability, protein-protein interaction, subcellular localization, and catalytic activity.[5] Due to the huge technical advancement, it has now become possible to map various PTM sites in proteins, including the phosphorylation sites. Vectorial proteomics has been developed to map the in vivo phosphorylated sites on the membrane surface of proteins (integral and peripheral) present on thylakoid membranes in plants as well as in *Chlamydomonas reinhardtii*, a green alga. Plants exposed to stresses like intense light or elevated temperature show significant changes in phosphorylation of thylakoid proteins, especially in photosystem II. While in the case of algae, a quicker response was observed when the organism was exposed to various stresses like light exposure, ambient redox condition, and availability of CO_2, resulting in phosphorylation of the thylakoid proteins. Under high light condition, the CP29 linker protein (the minor light-harvesting protein of PSII) of green micro-algae undergoes multiple and differential phosphorylation in a light-dependent manner to regulate the transitions of photosynthetic states and uncouples some of the light harvesting proteins from photosystem II. Like CP29 of algae, TSP9 in plants performs the same function by regulating the dynamic distribution of light harvesting proteins and is phosphorylated in a specific environment/redox-dependent fashion, along with other newly discovered thylakoid surface proteins.[6]

The environment-stimulated-differential-phosphorylation of the proteins involved in photosynthesis, largely in stressed conditions, has been well established with the help of various biochemical techniques. For example, it has been observed that high light stress exposed plants and green micro-algae, and drought-exposed plants, show a significantly increased phosphorylation of polypeptides of PSII.[7–11] Not only this, but it has also been seen that PSII reaction center proteins undergo a dark-sustained phosphorylation in response to combined magnesium and sulfur deficiency,[12] subfreezing temperatures,[13] or high light stress.[14] It has recently been documented that when green micro-algae are exposed to a limiting CO_2 environment, few surface proteins of thylakoid membrane get rapidly phosphorylated.[15]

Techniques like vectorial proteomics are used to detect the exact amino acid undergoing phosphorylation at real molecular level.[16] Such techniques have emerged as a boon for the scientists working in this field. Phosphorylation has been termed as an intrinsically derived hydrophilic process because it takes place only on the domains exposed to the peripheral surfaces of membrane proteins and on the membrane associated peripheral proteins.[11] Moreover, the CP29 protein of micro-algae when exposed to intense light conditions undergoes hyperphosphorylation at seven distinct sites, resulting in dissociation of this protein together with light harvesting complex (LHCII) from PSI and PSII (Fig. 1). Markedly, CP26, a PSII-LHCII linker protein, also undergoes multiple phosphorylation in green alga, at the same time, contributing to the dissociation of LHCII from PSII after being exposed to high light stress.[11] The genetic analysis of *C. reinhardtii* mutants revealed the critical significance of the most abundant Lhcbm1 (LHCII protein) for thermal energy dissipation when exposed to high light stress, which is in agreement with the mechanism described earlier.[17] Furthermore, the light harvesting and the flow

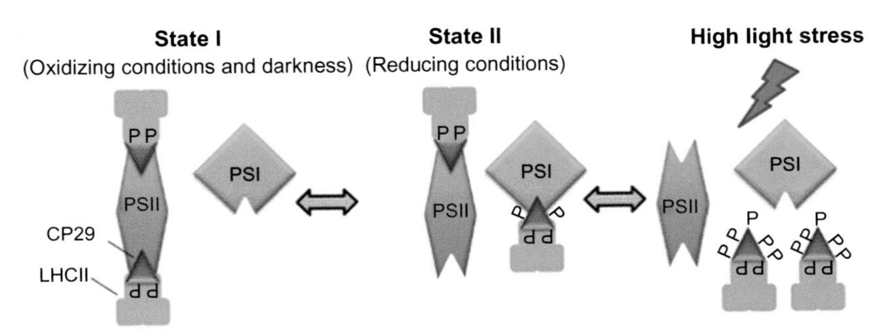

FIG. 1 Phosphorylation in CP29 under various stress conditions. *Scheme of this figure has been taken from Turkina, MV, Kargul, J, Blanco-Rivero, A, Villarejo, A, Barber, J, Vener, AV. Environmentally modulated phosphoproteome of photosynthetic membranes in the green alga Chlamydomonas reinhardtii.* Mol Cell Proteomics *2006a;5:1412–1425.*

of light absorbed may be controlled due to the varied degree of phosphorylation of the LHCII proteins, a light-dependent phenomenon, leading to the forceful re-structuring of LHCII complexes.[18,19]

It has been reported that there are two peripheral proteins, Low CO_2 inducible protein 5 (Lci5) and Unknown Expressed Protein (UEP), which are specifically found on the surface of thylakoid and are phosphorylated when the green algae is acclimatized to low CO_2 concentration, and hence, can be considered as an early cellular response to inadequate level of inorganic carbon in the environment.[15] The phosphorylation of Lci5 protein could occur at multiple sites (Thr^3 and Ser^4) by redox-dependent thylakoid protein kinases like the plant specific TSP9. Generally, this is the basic protein released from membranes in conditions of high ionic strength.[15] In view of the literature, these recently identified peripheral proteins work as per the changing environment by switching on/off mechanisms via surface-induced phosphorylation of the thylakoid membrane and hence, can be considered as potent signal transducing agents and/or mobile regulators. As photosynthetic machinery is very important for sustenance under changing environmental conditions, it has been proved that phosphorylation is a very important mechanism for adaptation and acclimation, and in most of the cases it depends on the environmental conditions. From studies carried out in this direction, it can be concluded that induction of phosphorylation of peripheral thylakoid proteins in both plants and algae by the environment has been promoted to be a unique signaling and regulatory response of the cell toward light and environmental carbon limitation. Further experimentation and characterization studies need to be done in future to learn more about phosphorylation with respect to the changing environmental conditions in algae. Table 1 shows the phosphorylation sites in various proteins of algae.

2.2 Ubiquitination and Stress

Although the involvement of ubiquitin, a highly conserved protein, in higher plants exposed to stress is well known, in the case of algal species, its role is less characterized even though they serve as primary producers. A difference of only a single amino acid has been reported in ubiquitin extension protein sequence of unicellular algae, *C. reinhardtii,* with that of the higher plants,[20] indicating its near similar roles in response to various stresses. Different studies have been done and proved that there are various cellular functions in which ubiquitin is directly involved, viz. response to stress, intracellular protein degradation, the cell cycle, and DNA repair.[21–23] Compared to the other algae, *C. reinhardtii* has been exploited more for studying different biochemical pathways and cellular functions due to its small genome, vulnerability to manipulation of the gene, its haploid nature, and being easy to grow in bulk.[24,25]

TABLE 1 Posttranslational Modifications Sites in Some of the Important Algal Proteins

Protein	Phosphorylation position	Condition	Reference
PsbH	Thr^{3}	–	19a
Lci 5	$Thr^{116, 176, 237}$ $Ser^{136, 137, 196, 197}$	Low CO_2	11, 15[a]
D2 PS II	Thr^{2}	Dark, anaerobic with nitrogen bubbling	11, 15
CP 29 (Lhcb4)	$Thr^{7, 33}$	Dark, aerobic/Dark, anaerobic with nitrogen	11, 15
CP 29 (Lhcb4)	Thr^{17} Ser^{103}	Dark, anaerobic with nitrogen bubbling	11, 15
CP 29 (Lhcb4)	$Thr^{7, 11, 18, 20}$	High Light	11, 15
D1 PS II	Thr^{2}	Moderate/High Light	11, 15
CP 43	Thr^{2}	Moderate/High Light	11, 15
CP 26 (Lhcb5)	Thr^{10}	High Light	11, 15
Lhcbm 1	Thr^{27}	Dark, anaerobic with nitrogen	11, 15
Lhcbm 4	$Thr^{19, 23}$	High Light	11, 15
Lhcbm 6	$Thr^{18, 22}$	High Light	11, 15
Lhcbm 9	$Thr^{19, 23}$	High Light	11, 15
Lhcbm 10	Thr^{26}	Dark, anaerobic with nitrogen	11, 15
Lhcbm 11	$Thr^{19, 23}$	High Light	11, 15
Psbr	Ser^{43}	Dark, anaerobic with nitrogen bubbling	11, 15
UEP A	DVDsEEAR	Moderate/High Light	11, 15
UEP A	VFEsEAGEPEAK	Light	11, 15
UEP B	GEIEEADsDDEAR	Dark, anaerobic with nitrogen/ Light	11, 15
UEP	Ser^{7}	Low CO_2	11, 15[a]

[a] *Proteomics.*

Patrick Vallentine and co-workers have shown that, on exposure of the green alga *C. reinhardtii* to selenium (Se) stress, the role of ubiquitin proteasome pathway (UPP) was found to be both time and dose dependent. When algae *C. reinhardtii* was exposed to moderate selenite stress, an increase in protein ubiquitination, the proteasomal removal of malformed selenoproteins, and proteasome activity were observed. While almost opposite results were obtained in the case of severe Se stress for a prolonged time, i.e., inhibited protein ubiquitination prevented proteasome dependent removal of selenoproteins and decreased proteasome activity. They explained that reactive oxygen species (ROS) accumulation during severe Se stress may be the reason behind the impaired activity of UPP. In addition to that, an inhibition in proteasome activity led to the decreased chlorophyll concentration in cultures challenged with Se.[26]

The function of UPP was fully understood in respect to abiotic stress and only its indirect involvement was explored. The study was carried out to know the role of *C. reinhardtii* UPP when the organism was exposed to selenite stress. Selenium (Se) stress in plants has been shown to be quite unique, with two different toxicity forms: selenate and selenite.[27] In fact, Selenate and selenite are considered as predominant forms of fresh water and soil solution Se. When transported into plants, both the forms function as pro-oxidants by depleting the level of glutathione in cell.[28,29] This results in accumulation of hydrogen peroxide and superoxide[30] in *Arabidopsis*[31] and in green algae *Ulva sp.*[32] Furthermore, the oxidative stress induced by Se can sometimes lead to replacement of sulfur in proteins.[33] The mechanism behind this substitution can be understood as assimilation of inorganic Se into selenocysteine and selenomethionine, causing competition with tRNAcys and tRNAmet, respectively, resulting in substitution of cysteine and methionine in protein molecules.[34] Among various factors, disulfide bond formation by cysteine residues stabilize the native structure of proteins and are also important for protein folding. Replacement of cysteine by selenocysteine leads to the disruption of the native structure of proteins and generation of malformed selenoproteins, leading to misfolding of proteins.[35] Therefore, Se stress can lead to the formation of both oxidized proteins as well as abnormal selenoproteins, both of which are able to provoke a UPP response. It has been shown that *C. reinhardtii* exposed to Se stress can be protected by the UPP, which could remove the malformed selenoproteins and hence, lessens Se toxicity (Fig. 2).

As against the eukaryotic cells, where the participation of ubiquitin as a heat shock and other stress response is very well known, yeast cells are sensitive to the said stresses because of the lack of polyubiquitin (UBI4) gene.[36–38] Among various observations, the one that connects ubiquitin and heat shock in mammalian cells being exposed to stress is the remarkable changes in the ubiquitination patterns of cellular proteins. It has been observed that the quantity of the conjugates of high-molecular-mass increased with a remarkable decrease in the level of ubiquitination in histone uH2A during stress.[39–41] Some aspects of the pattern of conjugated-ubiquitin in *C. reinhardtii* resemble that of the pattern observed

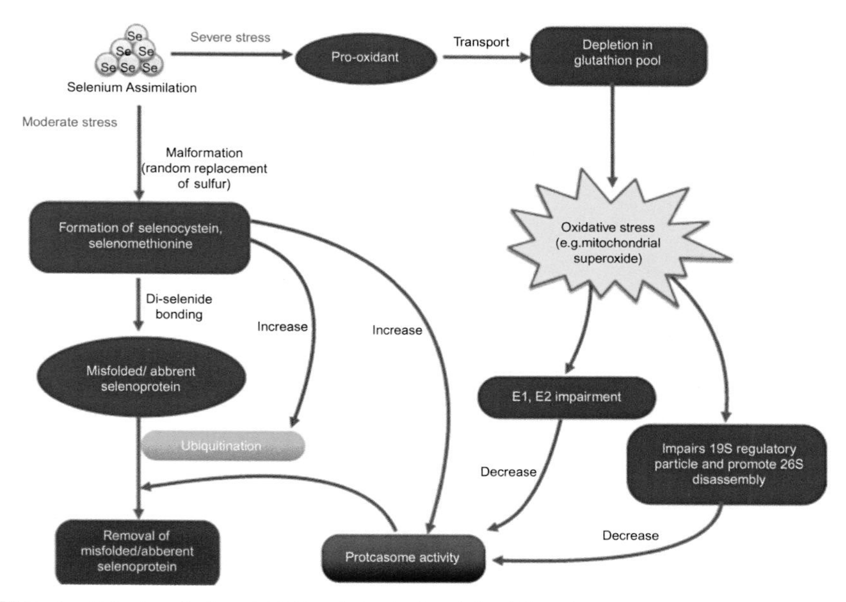

FIG. 2 Selenium induced UPP activity in *C. reinhardtii* under moderate and severe stress conditions. *Scheme of figure is based on the information collected from Vallentine, P, Hung, C-Y, Xie, J, Van Hoewyk, D. The ubiquitin-proteasome pathway protects Chlamydomonas reinhardtii against selenite toxicity, but is impaired as reactive oxygen species accumulate.* AoB Plants *2014;6:plu062. doi: 10.1093/aobpla/plu062.*

in mammalian cells during heat shock stress.[39,40,42] These conjugates, which are accumulated after a moderate heat shock and disappear when the shock is removed, are basically denatured proteins intended for ubiquitin-dependent protease degradation.[38,41] The ubiquitinated histone uH2A is being considered as the most important specific conjugate of mammalian cells whose level is decreased by heat shock and brought back when the condition becomes normal.[39–41] In a study, a similar pattern in the ubiquitination of heat shocked *C. reinhardtii* has been reported, describing depletion in the level of some 28-kDa conjugate polypeptide.[43] Extensive work must be done to determine the identity of the major ubiquitin conjugates in *C. reinhardtii* during various stresses.

2.3 SUMOylation and Stress

Apart from other small biomolecules, sometimes peptides and small proteins can also be involved in protein modifications. Among various small polypeptides that carry out covalent modification of various cellular proteins like ubiquitin, Small Ubiquitin-like Modifier (SUMO) and Neural

Precursor Cell Expressed Developmentally Down-Regulated Protein 8 (Nedd 8) are the most studied ones.[43a,44–46] The method of protein modification by SUMO was discovered in 1996 in eukaryotes and has been marked as an important factor for normal cell development/growth.[44] The process is a reversible regulation.[47,48] The target proteins modification by SUMO is responsible for regulating a number of biological functions like their localization, function, stability, enzyme activity, and a range of other cell processes.[43a,49,50] Like ubiquitination, the modification of proteins by SUMO also involves mainly three enzymes: a SUMO activating enzyme (SAE/SUMO E1), a SUMO conjugating enzyme (SCE/SUMO E2), and a SUMO ligase enzyme (SUMO E3) in their enzymatic pathway mechanism.[44] SUMO and ubiquitin are alike in terms of protein structures, regardless of the limited sequence identity with a difference of the presence of a extended and flexible N-terminal in SUMO.[43a,49] This is not the only difference as functional differences are also found in SUMO as compared to that of ubiquitin.

The degradation of the polyubiquitinated substrates are usually done by proteasome,[51] while SUMOylation often causes alteration in protein functions, protein-proteins interactions, or the target protein subcellular localization.[43a,49] In addition, the primary products translated by SUMO proteins are produced as precursors rather than being synthesized in active forms similar to ubiquitin. Thereafter, to make the active molecules from the primarily synthesized precursors, specific protease enzymes work, which expose C-terminal glycine-glycine motif of molecules proteolytically to enable the active form conjugation by isopeptide linkages with the lysine residues of the target proteins.[43a] The first step of the SUMOylation reaction is the formation of thiolester linkage between the activated SUMO and SAE (E1) cysteine residue. The next step is the shifting of SUMO to the active-site cysteine of the SCE (E2/Ubc9), resulting in the formation of a SUMO-Ubc9 thiolester intermediate.[52–55] While in most of the cases, an efficient and proper shift of SUMO from E2 to a target protein takes place through a specific E3,[50] in some cases, like Ran GTPase-activating protein 1 (RanGAP1), the use of E3 is avoided and SUMO is directly transferred from E2 to the substrate.[56] Remarkably, polymeric SUMO chains can be formed if substrates are further conjugated. The SUMO-specific proteases (essential enzyme for SUMO precursor processing) catalyze the deconjugation of SUMOylated substrates, and also exhibit SUMO isopeptidase and endopeptidase activities. Various SUMO-specific peptidases have been detected in animals, yeast, plants, and other eukaryotes, which are also termed as Ubl-specific proteases.[50, 57–59]

Exposure of cells to various stress conditions results in the formation of distinctly different sized SUMO-protein conjugates. It shows that the variation in the level of SUMOylation occurs as per the signal the cells receive according to the changing environmental conditions. The study carried out by Ying Wang and co-workers has shown that formation of SUMO-protein conjugates does not necessarily correspond to the duration of stress. They observed that when the *C. reinhardtii* was exposed to various stress conditions, in each case, a 30 min of stress exposure resulted in formation of more SUMO-protein conjugates than that of 1 h exposure, showing that the process of SUMOylation could be a quick and momentary process. As soon as the stress signal is gone, the process of deSUMOylation may take place. The whole process of SUMOylation and deSUMOylation is governed by the presence of highly regulated respective enzymes.[60] The process of SUMOylation is very important for cell viability, which can be understood by the fact that if mutation occurs in E2 conjugase enzyme, it results in loss of its functional ability, leading to lethal effect.[61–64]

With respect to the cell viability, the process of SUMOylation is very important but, at the same time, the pattern of SUMOylation changes with differential response towards various stress conditions.[65,66] The protein *C. reinhardtii* ubiquitin conjugating enzyme9, (CrUBC9), is most likely considered as a SUMO E2 conjugase as it shows activities equivalent to the activities of SUMO E2 conjugase in vitro.[60] In addition, the pattern of SUMOylation, detected by immunoblot analysis, markedly changed with the change in various abiotic stresses.[60] The *C. reinhardtii* is a unique organism compared to the others due to the presence of two distinct SUMO E2 conjugases carrying out SUMOylation in two distinct environmental conditions, i.e., CrUBC9 is vital for the SUMOylation in stress environment, while CrUBC3 (*C. reinhardtii* ubiquitin conjugating enzyme3) carries out SUMOylation under normal conditions and is required in cell growth and division.

A recent study, carried out by Knobbe and co-workers, reports accidental isolation of the *C. reinhardtii* mutant5 (mut5) where, in the previously identified SUMO E2 conjugase, gene coding CrUBC9 protein was found to be removed. Although mut5 was just like the viable wild-type cells of *C. reinhardtii* in carrying out the process of SUMOylation normally at 25°C, when the mut5 was exposed to various stress conditions, the process of SUMOylation was hampered for a number of proteins resulting in strikingly decreased or absolute absence of tolerance. When theCrUBC9 gene transformed in mut5 cells, the mutants showed normal SUMOylation pattern in both stress and nonstress conditions showing the necessity of CrUBC9 as an essential SUMO conjugase.[67] In addition to this, the importance of SUMOylation for the survival of *C. reinhardtii* under stress conditions can be understood by the fact that extensive SUMOylation has also been observed in harsh conditions in other eukaryotes.[65,66,68]

2.4 Psychrophilic Microalga and PTM

It has been shown that the Antarctic species of microalgae *Chlamydomonas* sp. ICE-L, a psychrophilic microalga, can survive under high environmental stress conditions such as freezing temperature and high salinity.[69] For molecular level understanding of their survival mechanism, their biological pathways were compared with their closely related microalgae species, *Volvox carteri* and *C. reinhardtii*. When ICE-L transcripome was studied, it was found that the sequences were frequently homologous with that of plant or bacterial proteins involved in the posttranscriptional and posttranslational modification, and signal-transduction KEGG pathways, showing the probable enhanced survival reason of ICE-L, while those homologies were absent in nonpsychrophilic green algae. The presence of these complex pathways might be the reason behind enhanced stress tolerance capacity of ICE-L. In fact, it has been observed that on exposure of ICE-L to freezing stress, all the genes associated with PTM, lipid metabolism, and nitrogen metabolisms responded.[69]

The frozen polar sea ice, constituting of a system composed of brine channels (high salinity, low temperature, limited gas exchange, low light, and highly toxic conditions) can be considered as one of the most dangerous environmental conditions for living creatures.[70] There are a number of psychrophilic organisms like microalgae and bacteria inhabiting and adapted well to such adverse and fluctuating environment conditions.[71] It has recently been proved that ubiquitin 26S proteasome system (UPS) plays an important role in various abiotic stresses.[72] This may be a reason behind the abundance of the proteins implicated in the ubiquitin-dependent proteolytic pathway more in ICE-L than that present in *C. reinhardtii* and *V. carteri*. Moreover, it has been observed that under freezing conditions in ICE-L, the genes are up-regulated encoding proteins, which are involved in translation, transportation, and sorting processes. The role of PTMs in algae acclimatizing themselves during the starting stages of freezing stress has also been observed and can be considered as one of the most important process of cold acclimatization.

The comparative study between biological pathways of ICE-L with microalgal species *C. reinhardtii* and *V. carteri* showed that in the case of psychrophilic species, ICE-L, the number of functional proteins involved in the posttranscription, posttranslation and signal-transduction pathways were much higher than those of moderate temperature species. To gain mechanistic insights for the adaptation of sea algae in such extreme conditions, a deeper investigation of these pathways is warranted.[69]

3 POSTTRANSLATIONAL MODIFICATION OF THERAPEUTIC PROTEINS IN MICROALGAE

In recent years, microalgae have gained interest as a treasure house of bioactive compounds and as a scaled down industrial facility for recombinant protein generation. The bioactive compounds, viz. pigments, PUFA, and polysaccharides, which exhibit multiple applications in the dietary, nutraceutical, and pharmaceutical fields, are currently generated from nontransgenic microalgae. The transgenic microalgae are also utilized as bioreactors for the creation of helpful proteins for human welfare and demonstrate incredible potential to improve the generation process as well as reduce the generation cost. Until now, various recombinant proteins have been used as therapeutics, like monoclonal antibodies, hormones, and pharmaceutical proteins created from the genome of nucleus or chloroplast of transgenic *C. reinhardtii*.[73]

The larger part of biopharmaceutical proteins endorsed or in clinical trials harbor some degree of PTMs, which can significantly influence protein properties identified with their curative applications.[74] The relationships between structure and function of the therapeutic proteins for many PTMs in animal as well as in plant systems have been studied but remain incomplete, especially, in the case of algae. Structural-functional characterization of the protein of interest will thus aid to engineer it as desired, therapeutically in micro algae. It will be interesting to focus on the involvement of PTMs in the manufacturing of recombinant therapeutic proteins from transgenic microalgae and advantages of microalgae as therapeutic protein factories over the other expression systems.

3.1 Algae as an Emerging Recombinant Protein Production Platform

Established expression systems (bacteria and yeast) for the generation of recombinant proteins have been explored and engineered with high effectiveness but there are still disadvantages associated with expression of complex eukaryotic proteins, which require PTMs or the assembly of numerous protein units. Mammalian frameworks like Chinese hamster ovary (CHO) cell lines have been utilized for more than 60% of recombinant protein pharmaceuticals because of their capacity to perform complex PTMs including human-like glycosylation, while the production is exceptionally costly and bears the danger of human pathogenic contaminations.[75,76] Among various production systems, there is growing interest in the utilization of microalgae as bioreactors for substantial scale up generation of biopharmaceuticals.[77,78] Green algae, for example, *C. reinhardtii*, are generally regarded as safe organisms by the FDA because of the fact that microalgae do not harbor human, restoring trust that natural consumable antibodies can be delivered in a photosynthetic organism.[79]

Similar to CHO cells, microalgae exhibits high growth rates, simple handling, the ability to execute PTMs like N-glycosylation in a precise and balanced way[80–83] that possess important roles in folding, half-life, and activity of biopharmaceuticals.[84] Moreover, microalgae contain a good amount of proteins (50%–70% of f.w.), showing a substantially higher rate than other organisms.[85] Specifically, the micro algae *C. reinhardtii* has been proved as a model framework and has completely sequenced (nuclear, mitochondrial, and plastidial) genomes,[86] with the simple and speedy production of stable transgenic or transplastomic lines.[87,88] The nuclear genome expression depends upon the position effects and gene silencing,[89,90] yet the expression from the chloroplast genome is well proven.[91,92] The microalgae chloroplast does not possess the machinery for performing complex PTMs, like glycosylation, as in bacteria, however, the *C. reinhardtii* chloroplast can well arrange disulfide bonds and play out a few other types of PTMs like phosphorylation—an advantage over bacteria. Moreover, it contains a low level of protease and additionally, a few molecular chaperones supporting protein folding,[91] proving it a better candidate for recombinant therapeutics production. Although eukaryotic microalgae share common mechanisms with plants but show dissimilarity with higher plants in containing around 100 genomes copies/chloroplasts, *C. reinhardtii* possesses around 80 genome copies only in one chloroplast. This is an advantage of *C. reinhardtii* that it expresses all the genome copies in the same recombinant form (homoplasmy) due to solitary chloroplast.

While plant expressed recombinant proteins are different at some level because of their formation in various chloroplasts, *C. reinhardtii* has been utilized for the production of recombinants vaccines,[93–95] completely functional antibodies,[96,97] therapeutics,[98] and different proteins of biotechnological significance,[99] with undetectable levels to 5% of Total Soluble Protein (TSP) yield.[87] Although real advances achieved in the production of therapeutic proteins from transgenic microalgae like *C. reinhardtii* reflects that, biotechnologically, microalgae are not a very much focused on group. The major curb in the protein expression at commercial level from microalgae is the unavailability of standard methods of genetic transformation.[99a] Despite this, the cost to deliver recombinant antibody per gram in mammalian cell culture is US $150 g^{-1}$, in transgenic plants is US $0.05 g^{-1}$ and in microalgae it is around US $0.002 g^{-1}$, which is impressively lower than others.[96] A comparison among different systems for therapeutic protein production has been illustrated in Fig. 3.

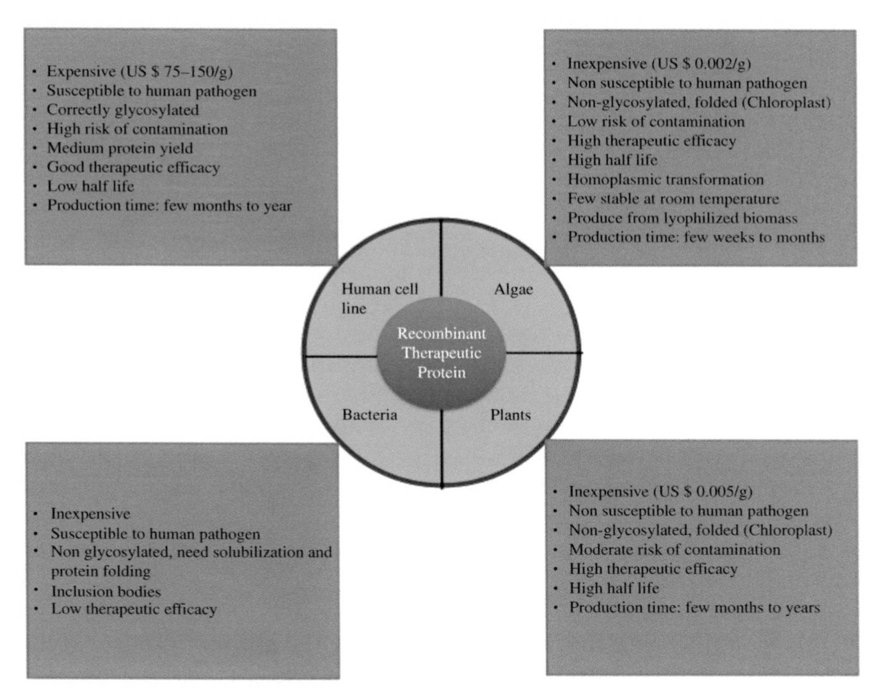

FIG. 3 Comparison among different systems for production of therapeutic proteins.

3.2 Posttranslational Modification in Perspective of Expression of Therapeutic Proteins in Algae

In spite of showing an extensive scope of PTMs, a small set of these PTMs are related to therapeutic proteins currently used. One of the PTMs, glycosylation, is most commonly found in normal and biopharmaceutical proteins and roughly half of the human proteins are glycosylated.[100] The importance of glycosylation has been proved already with the fact that defects in the glycosylation pathway in a number of genes can cause severe congenital disorders.[101] The final products of most of the recombinant proteins, which are being produced in artificial generation systems like NS0 myloma cells and CHO cells, are subjected to thorough downstream procedures to make them usable. Utilization of prokaryotic generation systems like *Escherichia coli*, may bring about an aglycosylated protein recovered as an inclusion body that needs in vitro solubilization and refolding to make it usable. In contrast, sugar side chains with high mannose content are added when expressed in yeasts,[102] also mammalian cells expressing the glycosylation patterns of the proteins are different in insect cell lines.[103] Plants are reported to glycosylate proteins, but they consistently add α1,3-fucose and β1,3-xylose sugars, which show immune or allergic response in humans.[104] The chloroplast-expressed proteins are nonglycosylated, like bacterial expression, due to the absence of machinery that establishes complex protein PTMs. In mammalian systems, most of the expressed proteins, including antibodies, are generally glycosylated during synthesis and then transported. The carbohydrate moieties attached to proteins during glycosylation do affect the antibody's assembly, serum half-life, and its functions.[105] Hence, it is essential to figure out the functioning of nonglycosylated proteins transported to chloroplasts.

Nonglycosylated antibodies give different degrees of actions on the activities of various antibodies and activate complementary systems with less ability, in addition to the decreased Fc-mediated binding in activating antibody-dependent cell-mediated cytotoxicity (ADCC).[105] Antigen binding is not necessary in fixation of complement and activation of ADCC; and activation of ADCC is not mandatory for the function of those therapeutic antibodies whose function is to sequester molecules or blocking binding sites. Also, in terms of retaining its biological activity and clearance rate, the nonglycosylated antibodies exhibit similar patterns to glycosylated ones without evoking unwanted immune responses.[105a] Therefore, the nonglycosylated antibodies produced in microalgal chloroplast may be considered as more advantageous than the glycosylated ones for various therapeutic applications. Disulfide bond formation, an important PTM, is critically required for the functioning of antibodies and other eukaryotic proteins. In prokaryotic organism like *E. Coli*, disulfide bond formation does not take place in proteins that accumulate in cytoplasm.[106] Therefore, it is obvious that bacterial systems are not ideal for production of proteins requiring disulfide bond formation. Human somatotropin produced in the chloroplast of tobacco was the first plastid-expressed recombinant protein having disulfide bond.[107] From then onwards, numerous antibodies and proteins have been expressed in chloroplasts of *C. reinhardtii* and higher plants with expected disulfide bonds.[96,108–111] As we have mentioned earlier, *C. reinhardtii* has multiple genome copies in a solitary chloroplast that express all genome of the chloroplast into recombinant protein (homoplasmy), which can be considered as an advantage over plants. Accordingly, chloroplast could be established as an amazing platform for the outflow of proteins that require disulfide bonding for the proper structuring. The proteins responsible for transducing the light activation signals responsible for regulation of chloroplast translation could be the same proteins accountable for the formation of disulfide bonds, as among them there is a presence of a protein disulfide isomerase enzyme.[112]

It has been also proposed, especially for antibodies, that proteolytic degradation of recombinant proteins can be an important aspect in protein aggregation[112a] and downstream processing by following protein heterogeneity and a decline in protein yield. The chloroplast in microalgae and higher plants shares few common proteases (Clp, Deg, and FtsH) that are found in bacteria also, and at least one ortholog of these proteases found in nuclear genome of *C. reinhardtii* is targeted to the chloroplast. Nonetheless, these are the extent of the presence of commonly occurring proteases in the eukaryotic cells' cytoplasm. Subsequently, the chloroplast expression system could conceivably be a safe platform for proper folded and assembled proteins that are prone to proteolysis and Eventually for improper folding and assemblage.

A wide range of molecular chaperones, containing prokaryotic and eukaryotic orthologs, have been distinguished in the chloroplasts of *C. reinhardtii*. In fact, *C. reinhardtii* has been observed to be able to produce varieties of proteins through chloroplast expression (Table 2) because of the internal condition of chloroplast stroma which is equipped for the proper folding and assembly of heterologous proteins.[91]

3.3 Posttranslational Modification in Algae Derived Therapeutic Proteins

As compared to other systems, very small numbers of therapeutic proteins are produced in algal chloroplasts. Mayfield and co-workers in 2003 were able to produce a monoclonal antibody (mAbs) directed against herpes simplex virus D (HSV8) glycoprotein in a transplastomic strain of *C. reinhardtii*, which efficiently binds with herpes virus protein D in vitro.[96] The antibody was expressed from the coding sequence of a large single chain (lsc) of the whole IgA heavy chain protein that is bonded by a flexible linker peptide to the variable region of the light chain. The lsc antibodies assembled as dimers consisting of disulfide bonds in the chloroplast.

Another *C. reinhardtii* chloroplast expressed therapeutic was a mAbs collected into a functional tetrameric structure of two heavy and two light-chain subunits. The mAbs was against a protein from *Bacillus anthracis*. After comparison with antibody 83K7C that was expressed in CHO cells, it showed a protection against anthrax toxicity in vitro as well as in vivo. It showed that the nonglycosylated antibody generated in chloroplast have the same PA83 binding properties similar to that of the mammalian system expressed antibody.[97] For at least one year, the strain was able to produce a stable antibody.

The way the chloroplasts do not glycosylate proteins may be viewed as a disadvantage, but this quality is being used positively for expression of some proteins. A prototype of malaria vaccine (*Plasmodium falciparum* surface protein 25 (Pfs25) and 28 (Pfs28) antigens) was expressed in a

TABLE 2 Posttranslational Modification in Therapeutic Proteins in Microalgae

Therapeutic protein	Organism	Type of PTM	Site of expression	Application	Reference
HBsAg mAbs	*P. tricornutum*	N-glycosylation	ER	Human antihepatitis-B monoclonal antibody	113
Full-length IgG1 human monoclonal antibody (mAb) 83K7C	*C. reinhardtii*	Di-sulfide bridge	Chloroplast	*Bacillus anthracis* anthrax protective antigen 83 (PA83)	97
E7GGG variants vaccine	*C. reinhardtii*	Phosphorylation	Chloroplast	Human papillomavirus 16 E7 vaccine	114
Human monoclonal IgG1 antibody CL4mAb	*P. tricornutum*	Glycosylation	ER	Hepatitis B virus surface antigen (HBsAg)	115
Human monoclonal IgG1 antibody CL4mAb	*P. tricornutum*	Di-sulfide bridge	ER	Hepatitis B virus surface antigen (HBsAg)	116
Pfs25 elicit antibody	*C. reinhardtii*	Di-sulfide bridge	Chloroplast	Antimalarial	94
Antibody hsv8-lsc	*C. reinhardtii*	Di-sulfide bridge	Chloroplast	Herpes simplex virus glycoprotein D	96
Pfs48/45	*C. reinhardtii*	Di-sulfide bridge	Chloroplast	Antimalarial	117
Pfs25-CTB	*C. reinhardtii*	Di-sulfide bridge	Chloroplast	Antimalarial oral vaccine	110
Hemagglutinins	*Schizochytrium* sp.	Glycosylation	Nucleus	Influenza B	118

transplastomic strain of *C. reinhardtii* by Gregory and his group in 2013,[110] which showed that the algae derived prototypes Pfs25 and Pfs28 accumulated as 0.5% and 0.2% of TSP, respectively. The immunization was then done with the pure algal derived antigen, resulting in the production of highly specific antibodies against the native proteins of sexual stage lysates of *P. falciparum*. The results showed a reduction in the formation of *P. falciparum* sporozoites (accountable for the transmission of parasite from mosquito to human) with only Pfs25 effective for eliciting antibodies with transmission blocking activity. When the same protein was expressed in other eukaryotic systems, it became glycosylated, while the native Pfs25 and Pfs28 from the parasite are nonglycosylated. However, microalgal chloroplast generated antibody showed no glycosylation with correctly mimicking protein folding, disulfide bridge formation, and produced parasite transmitting blocking antibodies by stopping the sporozoite formation in the mosquito under in vitro conditions. All this demonstrates the suitability of the chloroplast of *C. reinhardtii* for expression of correctly mimicking proteins.

Likewise, the E7 protein of human papilloma virus shows phosphorylation when expressed in *C. reinhardtii* and did not necessitate fusion to attain consistent expression. The E7 protein when produced in algal chloroplast was found to be soluble in nature, whereas the plant-produced E7 was found in insoluble fraction in different solubilization buffers. While affinity purified protein generated an excellent titer of antibodies, an equally efficient prevention of the tumor development and promotion of mouse survival was evident when crude algal extract was used.[114]

A recombinant human antihepatitis-B mAbs was expressed in *Phaeodactylum tricornutum*[113] and it was observed that the light and heavy chains along with C-terminal end were conserved in the expressed mAbs. A high dissimilarity with mAbs expressed in CHO cells was observed. In addition, elimination of the signal peptide was seen to be like that of the mammalian cells. Furthermore, high mannose-type *N*-glycans were recognized in the antibody produced in algae, which could prompt a low-level half-life modifications of the effector functions interceded by the Fc region.[119] Hempel and co-workers in 2011 also reported human IgG mAbs expression against the hepatitis-B surface antigen in *P. Tricornutum*.[115] The recombinant antibody was accumulated in the ER because of the presence of a retention signal fused with the antibody. By using this process, a complex glycosylation was avoided that occurs in the golgi apparatus. The algae-made antibody showed positive results in ELISA assays with the specific target proteins and gave an 8.7% TSP yield. The same antibody was produced by Hempel and Maier[116] in *P. tricornutum* in the absence of ER retention signal and due to that, the recombinant antibody was secreted in the culture medium. This approach would pave a path for simple way of purification of recombinant antibody, as endogenous proteins from micro algae *P. tricornutum* are not secreted, and thus it was easier to separate antibody as compared to purifying it from the cellular extract.[116] The gross yield (2250 ng mL^{-1}) for the production of same antibody was higher than the antibody procured with the ER retention method. The *P. tricornutum* generated antibody was found to be stable for a minimum of 2 days in the culture medium and the antibody production was restored after changing the media.

The benefit of already applied processes for PUFA production for animals and human health has been utilized for the production of hemagglutinin proteins of influenza virus in micro algae *Schizochytrium* sp. The micro algae clones were transformed with the full-length reading frames (two H1, one H5, one influenza B hemagglutinins) at nuclear level under the control of the promoter (EF-1) and terminator (PFA3). The resulting clones were quite efficient to secrete full-length membrane-bound proteins with hemagglutination activities with a range of 16–512 HAU/50 μL of the supernatant, and its immunogenic properties were examined in mice against A/Puerto Rico/8/34 (H1N1) influenza virus.[118] It is important to note that algae-made proteins were used for studying glycosylation patterns but none of the glycoforms showed the presence of fucose or xylose or sialic acid residues. However, the presence of high mannose structures, like $(GlcNAc)_2$ $(Man)_{5–6}$, were detected. As such, glycans have been proved to be immunogenic in mammals; their presence has opened up several questions for the researchers. Table 2 summarizes all the aforementioned updates in this area of research.

4 CONCLUSION AND FUTURE PROSPECTS

Similar to that of plants and animals, micro algae like *C. reinhardtii* undergo a variety of PTMs to cope up with environmental stress conditions. Algae, being advantageous over other conventional expression systems, offer a better system to be exploited for the production of therapeutic proteins. Algae are very important eukaryotic organisms, which adapt themselves quite effectively during various stress conditions and can also serve as an efficient system for the manufacturing of therapeutic proteins. Evaluating them in detail seems to be the current need and may be quite fruitful for human welfare. In fact, microalgae like *C. reinhardtii*- and *P. tricornutum*-based systems for therapeutic proteins production are quite effective and feasible, and also serve as an easy way for the production and purification with the advantage of cost effectiveness and good yield. Furthermore, it will be quite promising to use the whole algal biomass for the production of oral vaccines as well. Although a continuous study in this field has shown the importance of PTMs in algae, a detailed study is yet to be established using various approaches and techniques.

As activities of preexisting proteins linked to the stress responses are directly regulated by various PTMs, a detailed study is very important to know what exactly happens at molecular level when cells are being exposed to different environmental stresses. More importantly, microalgae in the context of PTM related to stress response and therapeutic protein production systems has not been so far explored in detail. Compared to the other eukaryotic systems, in some cases, the algae adopt quite effective and rapid methods during stress conditions for their survival. Keeping this in view, it will be quite promising to promote algae as an efficient and alternate system for not only survival but also for therapeutic protein production system. In addition to this, it seems quite important to compare the algae PTM events with that of the other prokaryotic/eukaryotic organisms to establish an ecological correlation among them.

Although a number of methods exploiting PTMs in algae for therapeutic protein production have been developed, there is still a large scope for exploring much better large-scale production methods. A few of the therapeutic proteins, being produced from *C. reinhardtii,* offer good quality and cost effectiveness but they have yet to be approved for commercial application.

References

1. Zhang Q, Bhattacharya S, Pi J, Clewell RA, Carmichael PL, Andersen ME. Adaptive posttranslational control in cellular stress response pathways and its relationship to toxicity testing and safety assessment. *Toxicol Sci* 2015;**147**(2):302–16.
2. Schaber J, Baltanas R, Bush A, Klipp E, Colman-Lerner A. Modelling reveals novel roles of two parallel signaling pathways and homeostatic feedbacks in yeast. *Mol Syst Biol* 2012;**8**:622.
3. Winter J, Jakob U. Beyond transcription—new mechanisms for the regulation of molecular chaperones. *Crit Rev Biochem Mol Biol* 2004;**39**:297–317.
4. Zhang Q, Pi J, Woods CG, Andersen ME. A systems biology perspective on Nrf2-mediated antioxidant response. *Toxicol Appl Pharmacol* 2010;**244**:84–97.

4a. Lehti-Shiu MD, Shiu SH. Diversity, classification and function of the plant protein kinase superfamily. *Philos Trans R Soc Lond B Biol Sci* 2012;**367**(1602):2619–39.

5. Whitmarsh AJ, Davis RJ. Regulation of transcription factor function by phosphorylation. *Cell Mol Life Sci* 2000;**57**:1172–83.
6. Vener AV. Environmentally modulated phosphorylation and dynamics of proteins in photosynthetic membranes. *Biochim Biophys Acta* 2007;**1767**:449–57.
7. Baena-Gonzalez E, Barbato R, Aro E-M. Role of phosphorylation in repair cycle and oligomeric structure of photosystem two. *Planta* 1999;**208**:196–204.
8. Ebbert V, Godde D. Phosphorylation of PS II polypeptides inhibits D1 protein-degradation and increases PS II stability. *Photosynth Res* 1996;**50**:257–69.
9. Giardi MT. Phosphorylation and disassembly of the photosystem II core as an early stage of photoinhibition. *Planta* 1993;**190**:107–13.
10. Giardi MT, Cona A, Geiken B, Kucera T, Masojidek J, Mattoo AK. Long-term drought stress induces structural and functional reorganization of photosystem II. *Planta* 1996;**199**:118–25.

11. Turkina MV, Kargul J, Blanco-Rivero A, Villarejo A, Barber J, Vener AV. Environmentally modulated phosphoproteome of photosynthetic membranes in the green alga *Chlamydomonas reinhardtii*. *Mol Cell Proteomics* 2006;**5**:1412–25.
12. Dannehl H, Herbik A, Godde D. Stress-induced degradation of the photosynthetic apparatus is accompanied by changes in thylakoid proteinturnover and phosphorylation. *Physiol Plant* 1995;**93**:179–86.
13. Adams WW, Demmig-Adams B, Rosenstiel TN, Ebbert V. Dependence of photosynthesis and energy dissipation activity upon growth form and light environment during the winter. *Photosynth Res* 2001;**67**:51–62.
14. Ebbert V, Demmig-Adams B, Adams WW, Mueh KE, Staehelin LA. Correlation between persistent forms of zeaxanthin-dependent energy dissipation and thylakoid protein phosphorylation. *Photosynth Res* 2001;**67**:63–78.
15. Turkina MV, Blanco-Rivero A, Vainonen JP, Vener AV, Villarejo A. CO(2) limitation induces specific redox-dependent protein phosphorylation in *Chlamydomonas reinhardtii*. *Proteomics* 2006;**6**:2693–704.
16. Vener AV, Stralfors P. Vectorial proteomics. *IUBMB Life* 2005;**57**:433–40.
17. Elrad D, Niyogi KK, Grossman AR. A major light-harvesting polypeptide of photosystem II functions in thermal dissipation. *Plant Cell* 2002;**14**:1801–16.
18. Horton P, Wentworth M, Ruban A. Control of the light harvesting function of chloroplast membranes: the LHCII-aggregation model for nonphotochemical quenching. *FEBS Lett* 2005;**579**:4201–6.
19. Pascal AA, Liu Z, Broess K, van Oort B, van Amerongen H, Wang C, Horton P, Robert B, Chang W, Ruban A. Molecular basis of photoprotection and control of photosynthetic light-harvesting. *Nature* 2005;**436**:134–7.

19a. Dedner N, Meyer HE, Ashton C, Wildner GF. N-terminal sequence analysis of the 8 kDa protein in *Chlamydomonas reinhardii* localization of the phosphothreonine. *FEBS Lett* 1988;**236**:77–82.

20. Callis J, Pollmann L, Wettern M, Shanklin J, Vierstra RD. Sequence of a cDNA from Chlamydomonas reinhardii encoding a ubiquitin 52 amino acid extension protein. *Nucleic Acids Res* 1989;**17**:8377.
21. Hershko A. Ubiquitin-mediated protein degradation. *J Biol Chem* 1988;**263**:15237–40.
22. Rechsteiner M. *Ubiquitin*. New York: Plenum Press; 1988.
23. Schlesinger M, Hershko A. *The ubiquitin system*. Cold Spring Harbor NY: Cold Spring Harbor Laboratory; 1988.
24. Rochaix J-D, Goldschmidt-Clermont M, Merchant S. *The molecular biology of chloroplasts and mitochondria of Chlamydomonas*. New York: Kluwer Academic/Plenum Publishers; 1998.
25. Weeks DP. *Chlamydomonas*: an increasingly powerful model plant cell system. *Plant Cell* 1992;**4**:871–8.
26. Vallentine P, Hung C-Y, Xie J, Van Hoewyk D. The ubiquitin–proteasome pathway protects *Chlamydomonas reinhardtii* against selenite toxicity, but is impaired as reactive oxygen species accumulate. *AoB Plants* 2014;**6**:plu062. https://doi.org/10.1093/aobpla/plu062.
27. Hoewyk DV. A tale of two toxicities: malformed selenoproteins and oxidative stress both contribute to selenium stress in plants. *Ann Bot* 2013;**112**:965–72.
28. Grant K, Carey NM, Mendoza M, Schulze J, Pilon M, Pilon-Smits EA, Van HD. Adenosine 5′-phosphosulfate reductase (APR2) mutation in Arabidopsis implicates glutathione deficiency in selenate toxicity. *Biochem J* 2011;**438**:325–35.
29. Hoewyk DV, Takahashi H, Inoue E, Hess A, Tamaoki M, Pilon-Smits EA. Transcriptome analyses give insights into selenium-stress responses and selenium tolerance mechanisms in *Arabidopsis*. *Physiol Plant* 2008;**132**:236–53.
30. Mroczek-Zdyrska M, Wojcik M. The influence of selenium on root growth and oxidative stress induced by lead in *Vicia faba* L. minor plants. *Biol Trace Elem Res* 2012;**147**:320–8.

31. Lehotai N, Kolbert Z, Peto A, Feigl G, Ördög A, Kumar D, Tari I, Erdei L. Selenite-induced hormonal and signalling mechanisms during root growth of Arabidopsis thaliana L. *J Exp Bot* 2012;**63**:5677–87.
32. Schiavon M, Moro I, Pilon-Smits EA, Matozzo V, Malagoli M, Dalla Vecchia F. Accumulation of selenium in *Ulva sp.* and effects on morphology, ultrastructure and antioxidant enzymes and metabolites. *Aquat Toxicol* 2012;**122**:222–31.
33. Elizabeth AH, Pilon-Smits, Quinn CF. Selenium metabolism in plants. In: Hell R, Mendel R-R, editors. *Cell biology of metals and nutrients, plant cell monographs*. Springer-Verlag Berlin Heidelberg; 2010. p. 225–41.
34. Zhu Y-G, Pilon-Smits EA, Zhao F-J, Williams PN, Meharg AA. Selenium in higher plants: understanding mechanisms for biofortification and phytoremediation. *Trends Plant Sci* 2009;**14**:436–42.
35. Stadtman TC. Selenium biochemistry. *Annu Rev Biochem* 1990;**59**:111–27.
36. Finley D, Ozkaynak E, Varshavsky A. The yeast polyubiquitin gene is essential for resistance to high temperatures, starvation, and other stresses. *Cell* 1987;**48**(6): 1035–46.
37. Goff SA, Voellmy R, Goldberg AL. Protein breakdown and the heat-shock response. In: Rechsteiner M, editor. *Ubiquitin*. New York: Plenum Press; 1988. p. 207–38.
38. Kulka RG. In: Grisolia S, Knecht E, editors. *Current trends in intracellular protein degradation*. Berlin: Springer; 1990.
39. Bond U, Agell N, Haas AL, Redman K, Schlesinger MJ. Ubiquitin in stressed chicken embryo fibroblasts. *J Biol Chem* 1988;**263**:2384–8.
40. Carlson N, Rogers S, Rechsteiner M. Microinjection of ubiquitin: changes in protein degradation in HeLa cells subjected to heat-shock. *J Cell Biol* 1987;**104**:547–55.
41. Parag HA, Raboy B, Kulka RG. Effect of heat shock on protein degradation in mammalian cells: involvement of the ubiquitin system. *EMBO J* 1987;**6**:55–61.
42. Finley D, Ciechanover A, Varshavsky A. Thermolability of ubiquitin-activating enzyme from the mammalian cell cycle mutant ts85. *Cell* 1984;**37**(1):43–55.
43. Shimogawara K, Muto S. Heat shock induced change in protein ubiquitination in *Chlamydomonas*. *Plant Cell Physiol* 1989;**30**:9–16.
43a. Johnson ES. Protein modification by SUMO. *Annu Rev Biochem* 2004;**73**:355–82.
44. Geiss-Friedlander R, Melchior F. Concepts in sumoylation: a decade on. *Nat Rev Mol Cell Biol* 2007;**8**:947–56.
45. Kerscher O, Felberbaum R, Hochstrasser M. Modification of proteins by ubiquitin and ubiquitin-like proteins. *Annu Rev Cell Dev Biol* 2006;**22**:159–80.
46. Palancade B, Doye V. Sumoylating and desumoylating enzymes at nuclear pores: underpinning their expected duties? *Trends Cell Biol* 2008;**18**:174–83.
47. Hay RT. SUMO-specific proteases: a twist in the tail. *Trends Cell Biol* 2007;**17**:370–6.
48. Mukhopadhyay D, Dasso M. Modification in reverse: the SUMO proteases. *Trends Biochem Sci* 2007;**32**:286–95.
49. Dohmen RJ. SUMO protein modification. *Biochim Biophys Acta* 2004;**1695**:113–31.
50. Hochstrasser M. Evolution and function of ubiquitin-like protein-conjugation systems. *Nat Cell Biol* 2000;**2**:E153–7.
51. Smalle J, Vierstra RD. The ubiquitin 26S proteasome proteolytic pathway. *Annu Rev Plant Biol* 2004;**55**:555–90.
52. Desterro JMP, Thomson J, Hay RT. Ubch9 conjugates SUMO but not ubiquitin. *FEBS Lett* 1997;**417**:297–300.
53. Johnson ES, Blobel G. Ubc9p is the conjugating enzyme for the ubiquitin-like protein Smt3p. *J Biol Chem* 1997;**272**:26799–802.
54. Sampson DA, Wang M, Matunis MJ. The small ubiquitin-like modifier-1 (SUMO-1) consensus sequence mediates Ubc9 binding and is essential for SUMO-1 modification. *J Biol Chem* 2001;**276**:21664–9.

55. Schwarz SE, Matuschewski K, Liakopoulos D, Scheffner M. The ubiquitin-like proteins SMT3 and SUMO-1 are conjugated by the UBC9 E2 enzyme. *Proc Natl Acad Sci U S A* 1999;**95**:560–4.
56. Matunis MJ, Coutavas E, Blobel G. A novel ubiquitin-like modification modulates the partitioning of the Ran-GTPase-activating protein RanGAP1 between the cytosol and the nuclear pore complex. *J Biol Chem* 1996;**135**:1457–70.
57. Colby T, Matthai A, Boeckelmann A, Stuible HP. SUMO-conjugating and SUMO-deconjugating enzymes from Arabidopsis. *Plant Physiol* 2006;**142**:318–32.
58. Li SJ, Hochstrasser M. A new protease required for cell-cycle progression in yeast. *Nature* 1999;**398**:246–51.
59. Suzuki T, Ichiyama A, Saitoh H, Kawakami T, Omata M, Chung CH, Kimura M, Shimbara N, Tanaka K. A new 30-kDa ubiquitin-related SUMO-1 hydrolase from bovine brain. *J Biol Chem* 1999;**274**:31131–4.
60. Wang Y, Ladunga I, Miller AR, Horken KM, Plucinak T, Weeks DP, Bailey CP. The small ubiquitin-like modifier (SUMO) and SUMO-conjugating system of *Chlamydomonas reinhardtii*. *Genetics* 2008;**179**:177–92.
61. Hayashi T, Seki M, Maeda D, Wang W, Kawabe Y, Seki T, Saitoh H, Fukagawa T, Yagi H, Enomoto T. Ubc9 is essential for viability of higher eukaryotic cells. *Exp Cell Res* 2002;**280**:212–21.
62. Nacerddine K, Lehembre F, Bhaumik M, Artus J, Cohen-Tannoudji M, Babinet C, Pandolfi PP, Dejean A. The SUMO pathway is essential for nuclear integrity and chromosome segregation in mice. *Dev Cell* 2005;**9**:769–79.
63. Saracco SA, Miller MJ, Kurepa J, Vierstra RD. Genetic analysis of SUMOylation in Arabidopsis: conjugation of SUMO1 and SUMO2 to nuclear proteins is essential. *Plant Physiol* 2007;**145**:119–34.
64. Seufert W, Futcher B, Jentsch S. Role of a ubiquitin-conjugating enzyme in degradation of S- and M-phase cyclins. *Nature* 1995;**373**:78–81.
65. Saitoh H, Hinchey J. Functional heterogeneity of small ubiquitin-related protein modifiers SUMO-1 versus SUMO-2/3. *J Biol Chem* 2000;**275**:6252–8.
66. Šramko M, Markus J, Kabát J, Wolff L, Bies J. Stress-induced inactivation of the c-Myb transcription factor through conjugation of SUMO-2/3 proteins. *J Biol Chem* 2006;**281**:40065–75.
67. Knobbe AR, Horken KM, Plucinak TM, Balassa E, Cerutti H, Weeks DP. SUMOylation by a stress-specific small ubiquitin-like modifier E2 Conjugase is essential for survival of *Chlamydomonas reinhardtii* under stress conditions. *Plant Physiol* 2015;**167**:753–65.
68. Kurepa J, Walker JM, Smalle J, Gosink MM, Davis SJ, Durham TL, Sung DY, Vierstra RD. The small ubiquitin-like modifier (SUMO) protein modification system in Arabidopsis. Accumulation of SUMO1 and −2 conjugates is increased by stress. *J Biol Chem* 2003;**278**:6862–72.
69. Liu C, Wang X, Wang X, Sun C. Acclimation of Antarctic Chlamydomonas to the sea-ice environment: a transcriptomic analysis. *Extremophiles* 2016;**20**:437–50.
70. Thomas DN, Dieckmann GS. Antarctic Sea ice: a habitat for extremophiles. *Science* 2002;**295**:641–4.
71. Arrigo KR, Mock T, Lizotte MP. Primary production in sea ice. In: Thomas D, Dieckmann G, editors. *Sea ice: an introduction to its physics, chemistry, biology and geology*. USA: Wiley-Blackwell, Malden, MA; 2010. p. 143–83.
72. Lyzenga WJ, Stone SL. Abiotic stress tolerance mediated by protein ubiquitination. *J Exp Bot* 2012;**63**:599–616.
73. Gong Y, Hu H, Gao Y, Xu X, Gao H. Microalgae as platforms for production of recombinant proteins and valuable compounds: progress and prospects. *J Ind Microbiol Biotechnol* 2011;**38**:1879–90.

74. Walsh G, Jefferis R. Post-translational modifications in the context of therapeutic proteins. *Nat Biotechnol* 2006;**24**:1241–52.
75. Farid SS. Established bioprocesses for producing antibodies as a basis for future planning. *Adv Biochem Eng Biotechnol* 2006;**101**:1–42.
76. Jayapal KP, Wlaschin KF, Hu W, Yap MG. Recombinant protein therapeutics from CHO cells–20 years and counting. *Chem Eng Prog* 2007;**103**(10):40–7.
77. Mathieu-Rivet E, Kiefer-Meyer M-C, Vanier G, Ovide C, Burel C, Lerouge P, Bardor M. Protein N-glycosylation in eukaryotic microalgae and its impact on the production of nuclear expressed biopharmaceuticals. *Plant Physiol* 2014;**5**:359.
78. Rasala BA, Mayfield S. Photosynthetic biomanufacturing in green algae; production of recombinant proteins for industrial, nutritional, and medical uses. *Photosynth Res* 2014;1–13.
79. Specht EA, Mayfield SP. Algae-based oral recombinant vaccines. *Front Microbiol* 2014;**5**:60.
80. Baïet B, Burel C, Saint-Jean B, Louvet R, Menu-Bouaouiche L, Kiefer-Meyer M-C, Mathieu-Rivet E, Lefebvre T, Castel H, Carlier A, Cadoret J-P, Lerouge P, Bardor M. N-glycans of *Phaeodactylum tricornutum* diatom and functional characterization of its N-acetylglucosaminyltransferase I enzyme. *J Biol Chem* 2011;**286**:6152–64.
81. Levy-Ontman O, Arad SM, Harvey DJ, Parsons TB, Fairbanks A, Tekoah Y. Unique N-glycan moieties of the 66-kDa cell wall glycoprotein from the red microalga *Porphyridium sp. J Biol Chem* 2011;**286**:21340–52.
82. Levy-Ontman O, Fisher M, Shotland Y, Weinstein Y, Tekoah Y, Arad SM. Genes involved in the endoplasmic reticulum N-glycosylation pathway of the red microalga *Porphyridium sp.*: a bioinformatic study. *Int J Mol Sci* 2014;**15**:2305–26.
83. Mathieu-Rivet E, Scholz M, Arias C, Dardelle F, Schulze S, Mauff FL, Teo G, Hochmal AK, Blanco-Rivero A, Loutelier-Bourhis C, Kiefer-Meyer M-C, Fufezan C, Burel C, Lerouge P, Martinez F, Bardor M, Hippler M. Exploring the N-glycosylation pathway in *Chlamydomonas reinhardtii* unravels novel complex structures. *Mol Cell Proteomics* 2013;**12**:3160–83.
84. Lingg N, Zhang P, Song Z, Bardor M. The sweet tooth of biopharmaceuticals: importance of recombinant protein glycosylation analysis. *Biotechnol J* 2012;**7**:1462–72.
85. Passwater R, Soloman N, editors. Algae: the next generation of superfoods. *Experts Optimal Health J* 1997;**1**:2–10.
86. Merchant SS, Prochnik SE, Vallon O, Harris EH, Karpowicz SJ, Witman GB, Terry A, Salamov A, Fritz-Laylin LK, Maréchal-Drouard L, Marshall WF, Qu LH, Nelson DR, Sanderfoot AA, Spalding MH, Kapitonov VV, Ren Q, Ferris P, Lindquist E, Shapiro H, Lucas SM, Grimwood J, Schmutz J, Cardol P, Cerutti H, Chanfreau G, Chen CL, Cognat V, Croft MT, Dent R, Dutcher S, Fernández E, Fukuzawa H, González-Ballester D, González-Halphen D, Hallmann A, Hanikenne M, Hippler M, Inwood W, Jabbari K, Kalanon M, Kuras R, Lefebvre PA, Lemaire SD, Lobanov AV, Lohr M, Manuell A, Meier I, Mets L, Mittag M, Mittelmeier T, Moroney JV, Moseley J, Napoli C, Nedelcu AM, Niyogi K, Novoselov SV, Paulsen IT, Pazour G, Purton S, Ral JP, Riaño-Pachón DM, Riekhof W, Rymarquis L, Schroda M, Stern D, Umen J, Willows R, Wilson N, Zimmer SL, Allmer J, Balk J, Bisova K, Chen CJ, Elias M, Gendler K, Hauser C, Lamb MR, Ledford H, Long JC, Minagawa J, Page MD, Pan J, Pootakham W, Roje S, Rose A, Stahlberg E, Terauchi AM, Yang P, Ball S, Bowler C, Dieckmann CL, Gladyshev VN, Green P, Jorgensen R, Mayfield S, Mueller-Roeber B, Rajamani S, Sayre RT, Brokstein P, Dubchak I, Goodstein D, Hornick L, Huang YW, Jhaveri J, Luo Y, Martínez D, Ngau WC, Otillar B, Poliakov A, Porter A, Szajkowski L, Werner G, Zhou K, Grigoriev IV, Rokhsar DS, Grossman AR. The *Chlamydomonas* genome reveals the evolution of key animal and plant functions. *Science* 2007;**318**:245–50.

87. Specht E, Miyake-Stoner S, Mayfield S. Micro-algae come of age as a platform for recombinant protein production. *Biotechnol Lett* 2010;**32**(10):1373–83.
88. Walker TL, Purton S, Becker DK, Collet C. Microalgae as bioreactors. *Plant Cell Rep* 2005;**24**:629–41.
89. Debuchy R, Purton S, Rochaix JD. The argininosuccinate lyase gene of *Chlamydomonas-reinhardtii*: an important tool for nuclear transformation and for correlating the genetic and molecular maps of the ARG7 locus. *EMBO J* 1989;**8**:2803–9.
90. Schroda M. RNA silencing in *Chlamydomonas*: mechanisms and tools. *Curr Genet* 2006; **49**(2):69–84.
91. Mayfield SP, Manuell AL, Chen S, Wu J, Tran M, Siefker D, Muto M, Marin-Navarro J. Chlamydomonas reinhardtii chloroplasts as protein factories. *Curr Opin Biotechnol* 2007;**18**:126–33.
92. Rasala BA, Mayfield SP. The microalga *Chlamydomonasreinhardtii* as a platform for the production of human protein therapeutics. *Bioengineered Bugs* 2011;**2**(1):50–4.
93. Dreesen IA, Charpin-El, Hamri G, Fussenegger M. Heat-stable oral alga-based vaccine protects mice from *Staphylococcus aureus* infection. *J Biotechnol* 2010;**145**:273–80.
94. Gregory JA, Li F, Tomosada LM, Cox CJ, Topol AB, Vinetz JM, Mayfield S. Algae-produced pfs25 elicits antibodies that inhibit malaria transmission. *PLoS ONE* 2012;**7**(5).
95. Sun M, Qian K, Su N, Chang H, Liu J, Shen G. Foot-and- mouth disease virus VP1 protein fused with cholera toxin B subunit expressed in *Chlamydomonasreinhardtii* chloroplast. *Biotechnol Lett* 2003;**25**:1087–92.
96. Mayfield SP, Franklin SE, Lerner RA. Expression and assembly of a fully active antibody in algae. *Proc Natl Acad Sci U S A* 2003;**100**:438–42.
97. Tran M, Zhou B, Pettersson PL, Gonzalez MJ, Mayfield SP. Synthesis and assembly of a full-length human monoclonal antibody in algal chloroplasts. *Biotechnol Bioeng* 2009;**104**:663–73.
98. Rasala BA, Muto M, Lee PA, Jager M, Cardoso RM, Behnke CA, Kirk P, Hokanson CA, Crea R, Mendez M, Mayfield SP. Production of therapeutic proteins in algae, analysis of expression of seven human proteins in the chloroplast of *Chlamydomonas reinhardtii*. *Plant Biotechnol J* 2010;**8**:719–33.
99. Li SS, Tsai HJ. Transgenic microalgae as a non-antibiotic bactericide producer to defend against bacterial pathogen infection in the fish digestive tract. *Fish Shellfish Immunol* 2009;**26**:316–25.
99a. Surzycki R, Greenham K, Kitayama K, Dibal F, Wagner R, Rochaix JD, Ajam T, Surzycki S. Factors effecting expression of vaccines in microalgae. *Biologicals* 2009;**37**:133–8.
100. Wong CH. Protein glycosylation: new challenges and opportunities. *J Org Chem* 2005;**70**:4219–25.
101. Freeze HH. Genetic defects in the human glycome. *Nat Rev Genet* 2006;**7**:537–51.
102. Gemmill TR, Trimble RB. Overview of N- and O-linked oligosaccharide structures found in various yeast species. *Biochim Biophys Acta* 1999;**1426**:227–37.
103. Jarvis DL, Kawar ZS, Hollister JR. Engineering N-glycosylation pathways in the baculovirus-insect cell system. *Curr Opin Biotechnol* 1998;**9**:528–33.
104. Gomord W, Sourrouille C, Fitchette AC, Bardor M, Pagny S, Lerouge P, Faye L. Production and glycosylation of plant-made pharmaceuticals: the antibodies as a challenge. *Plant Biotechnol J* 2004;**2**:83–100.
105. Wright A, Morrison SL. Effect of glycosylation on antibody function: implications for genetic engineering. *Trends Biotechnol* 1997;**15**:26–32.
105a. Simmons LC, Reilly D, Klimowski L, Raju TS, Meng G, Sims P, Hong K, Sheild RL, Damico LA, Rancatore P, Yansura DG. Expression of full-length immunoglobulins in *Escherichia coli*: rapid and efficient production of aglycosylated antibodies. *J Immunol Methods* 2002;**263**:133–47.

106. Derman AI, Prinz WA, Belin D, Beckwith J. Mutations that allow disulfide bond formation in the cytoplasm of *Escherichia coli*. *Science* 1993;**262**:1744–7.
107. Staub JM, Garcia B, Graves J, Hajdukiewicz PT, Hunter P, Nehra N, Paradkar V, Schlittler M, Carroll JA, Spatola L, Ward D, Ye G, Russell DA. High-yield production of a human therapeutic protein in tobacco chloroplasts. *Nat Biotechnol* 2000;**18**:333–8.
108. Arlen PA, Falconer R, Cherukumilli S, Cole A, Cole AM, Oishi KK, Daniell H. Field productionand functional evaluation of chloroplast-derived interferon-alpha2b. *Plant Biotechnol J* 2007;**5**:511–25.
109. Boyhan D, Daniell H. Low-cost production of proinsulin in tobacco and lettuce chloroplastsfor injectable or oral delivery of functional insulin and C-peptide. *Plant Biotechnol J* 2011;**9**:585–98.
110. Gregory JA, Topol AB, Doerner DZ, Mayfield S. Alga-produced cholera toxin-Pfs25 fusion proteins as oral vaccines. *Appl Environ Microbiol* 2013;**79**:3917–25.
111. Lee SB, Li B, Jin S, Daniell H. Expression and characterization of antimicrobial peptides Retrocyclin-101 and Protegrin-1 in chloroplasts to control viral and bacterial infections. *Plant Biotechnol J* 2010;**9**:100–15.
112. Kim J, Mayfield SP. Protein disulfide isomerase as a regulator of chloroplast translational activation. *Science* 1997;**278**:1954–7.
112a. Doran PM. Foreign protein degradation and instability in plants and plant tissuecultures. *Trends Biotechnol* 2006;**24**:426–32.
113. Vanier G, Hempel F, Chan P, Rodamer M, Vaudry D, Maier UG, Lerouge P, Bardor M. Biochemical characterization of human anti-hepatitis B monoclonal antibody produced in the microalgae *Phaeodactylum tricornutum*. *PLoS ONE* 2015;**10**(10).
114. Demurtas OC, Massa S, Ferrante P, Venuti A, Franconi R, Giuliano G. A *Chlamydomonas* derived human papillomavirus 16 E7 vaccine induces specific tumor protection. *PLoS ONE* 2013;**8**(4).
115. Hempel F, Lau J, Klingl A, Maier UG. Algae as protein factories: expression of a human antibody and the respective antigen in the diatom *Phaeodactylum tricornutum*. *PLoS ONE* 2011;**6**(12).
116. Hempel F, Maier UG. An engineered diatom acting like a plasma cell secreting human IgG antibodies with high efficiency. *Microb Cell Factories* 2012;**11**:126.
117. Jones CS, Luong T, Hannon M, Tran M, Gregory JA, Shen Z, Briggs SP, Mayfield SP. Heterologous expression of the C-terminal antigenic domain of the malaria vaccine candidate Pfs48/45 in the green algae *Chlamydomonasreiinhardtii*. *Appl Microbiol Biotechnol* 2013;**97**:1987–95.
118. Bayne AC, Boltz D, Owen C, Betz Y, Maia G, Azadi P, Archer-Hartmann S, Zirkle R, Lippmeier JC. Vaccination against influenza with recombinant hemagglutinin expressed by *Schizochytrium sp.* confers protective immunity. *PLoS One* 2013;**8**.
119. Loos A, Steinkellner H. IgG-Fc glycoengineering in non-mammalian expression hosts. *Arch Biochem Biophys* 2012;**526**(2):167–73.

Further Reading

120. Mazzucotelli E, Mastrangelo AM, Crosatti C, Guerra D, Stanca AM, Cattivelli L. Abiotic stress response in plants: when post-transcriptional and post-translational regulations control transcription. *Plant Sci* 2008;**174**:420–31.
121. Monroy AF, Sangwan V, Rajinder S. Low temperature signal transduction during cold acclimation: protein phosphatase 2A as an early target for cold-inactivation. *Plant J* 1998;**13**:653–60.
122. Wong KH, Todd RB, Oakley BR, Oakley CE, Hynes MJ, Davis MA. Sumoylation in *Aspergillus nidulans*: sumO inactivation, overexpression and live-cell imaging. *Fungal Genet Biol* 2008;**45**:728–37.

CHAPTER

13

Protein Glycosylation: An Important Tool for Diagnosis or Early Detection of Diseases

Humayra Bashir, Barqul Afaq Wani*[†], *Bashir A. Ganai*, Shabir Ahmad Mir*[‡]

*Centre of Research for Development, University of Kashmir, Srinagar, India

[†]Govt. Medical College, Srinagar, India

[‡]College of Applied Medical Science, Majmah University, Al-Majmaah, Saudi Arabia

1 INTRODUCTION

Proteins are the principal constituents of living cells, which perform numerous cellular functions essential to health. The study of all proteins expressed under stressful conditions of environment and aging provide insights into the health of the organism. The changes induced during stressful conditions are expressed within the dynamic alteration of the proteins expressed by the integrated genome pool. Due to stressful conditions, proteins fail to function properly, hence causing diseases. Determining the protein pools thus expressed in a given cell, tissue, or whole body under particular conditions, their relative quantities, locations, and the type/extent of their modifications may provide great insights in understanding vital biological processes related to disease conditions. Protein forms, which emerge as a result of posttranslational modifications (PTMs), can have markedly different properties due to a single PTM.[1] These modifications of proteins (phosphorylations, methylations, acetylations, etc.) are expected to establish suitable biomarkers potentially useful for diagnosis and early detection of diseases.

Protein Modificomics
https://doi.org/10.1016/B978-0-12-811913-6.00013-8

Protein glycosylation as a PTM is critical for regulation of the biological activities and biophysical properties of diverse proteins. Among all the PTMs, glycosylation is expected to be structurally diverse and one of the most abundant modifications, and accounts for about 80%–90% of all nucleoplasmic and extracellular proteins.[2] Although protein glycosylation is a rigorously controlled modification process in healthy state, these may be dysregulated in various chronic metabolic diseases as well as cancers,[3, 4] diabetes,[5] inflammation,[6] and neurodegenerative diseases.[7, 8] The molecular signaling that commands the stoichiometry and organization of sugar branching is not well understood. However, the glycan patterns observed in cancerous cells might be different in distinct cell types, e.g., core-fucosylation.[9, 9a] Therefore, analysis of glycoproteins in these pathological conditions may provide suitable key biomarkers with lasting outcomes in the long run of disease prognosis.

Recent technological advances in the development of protein biomarkers has opened new avenues for early, noninvasive diagnosis of significant diseases such as identification of B-type natriuretic peptides and cardiac troponin, for the detection of congestive heart failure and myocardial infarction, respectively.[10, 11] Although, a plethora of technical approaches have been developed during the last several decades to describe glycosylation changes associated with various clinical conditions, there is still a great demand for robust and proficient tools for glycoprotein analysis. Mass spectrometry, evolving as a primary technology, has potentially contributed to the identification of billions of proteins evolved due to combination of specific PTMs on intact proteins.[12] More recently, quantitative MS studies have been applied to determine the relative quantities of glycoproteins/peptides using isotope labels of known masses or label-free approaches. Improvement of simplistic and efficient approaches for the assignment of site-specific analysis would enable investigation of the important role of protein glycosylation in many cellular processes. Recent advancements in glycoproteomics have opened up innovative platforms for protein biomarker discovery, allowing early diagnosis of disease subtypes and stages at molecular level. This chapter summarizes current methods applied for the characterization of glycoproteins to support the discovery and development of biomarkers that may be helpful for early detection and diagnosis of diseases.

2 PROTEIN POSTTRANSLATIONAL MODIFICATIONS AND QUEST FOR CLINICAL BIOMARKERS

Disease prediction is going to be an important aspect for disease management in coming years. In near future, medical research will focus more toward preventing disease rather than treatment of already diagnosed disease. This important aspect in the development of advanced health care will help in the discovery and development of predictive biomarkers for predictive diagnostics and strengthen the platform for personalized treatment of patients.

Hence, with the development of targeted therapeutics and individualized treatment programs, biomarkers for disease classification and treatment response are becoming increasingly important. Biomarkers epitomize substitute amounts of interrelated biological processes representing one-to-one relationships of a truly correlated process. These associations are, however, infrequent in biology, leading to the extensive standardized testing process essential for the approval by regulatory agencies.[12a] Once a protein is synthesized, it undergoes a number of events until its final degradation. All these events are strictly controlled by specific groups of enzymes, which are present in specific concentrations and perform defined activities. It is expected that more than 50% of proteins are glycosylated and a number of proteins encoded by a single gene can be produced by addition of different glycan structures. A single PTM can starkly alter the properties of a protein. The autosomal dominant spongiform encephalopathy characterized by cerebral atrophy, hypometabolism, and early onset dementia has been associated with the loss of N-linked glycosylation in prion protein (PRNP) due to amino acid substitution T183A.[13] Another example is Kennedy's disease, where the loss of acetylation sites in androgen receptor (AR) due to amino acid substitution K630A, or both K632A and K633A, have been correlated with a significant slowdown of ligand-dependent nuclear translocation.[14] Moreover, rapid growth of databases containing PTMs, disease-related polymorphic genes, and mutations has facilitated the interpretation of novel PTMs in diseases.[15]

The nontemplate-based synthesis of glycoproteins reflecting expression of protein-coding genes, enzymatic glycosylation machinery activity, activated monosaccharide substrate availability for glycan synthesis, and several downstream effectors makes them a potentially rich source of information.[2] Amalgamation of glycobiology into conventional medicine has been strengthened by the fast pace of the discovery of protein glycosylation related disorders. Therefore, glycoproteins have promising potential to serve as markers for a number of clinical conditions. With the advancement of hi-tech next-generation sequencing of exomes and genomes, a new glycosylation disorder has been reported, on average, every 17 days.[16]

3 PROTEIN GLYCOSYLATION AND RELATED DISORDERS

Glycosylation covers a wide class of modifications through the addition of complex carbohydrate moiety (glycans) to specific amino acid residues. There are at least 10 distinct biosynthetic pathways in which protein glycosylation occur. Most of them add a single carbohydrate moiety to the protein while others involve the addition of hundreds to multiple protein and lipid anchors.[17, 18] The attachment of glycans to hydroxyl group of serine or threonine residue (O-glycosylation) and asparagine typically within Asn-X-Ser/Thr consensus sequence (N-glycosylation) are the best-studied types of glycosylation. O-linked glycosylation (mucin-type) occur in the golgi apparatus and is initiated by a group glycosyltransferases that catalyze the stepwise addition of monosaccharides to serine or threonine residues. Based on the type of monosaccharides directly attached to amino group, O-glycosylations can be further divided into several subclasses (Fig. 1). Recently, two novel types, O-Fuc and O-Man, were reported to play a role in the regulation of important proteins, for example, notch and dystroglycan.[19, 20]

N-linked glycosylation takes place in the endoplasmic reticulum with the help of oligosaccharyltransferases, which initiate the transfer of a lipid-linked glycan to a protein. Specific glycosidases then act for trimming several monosaccharides and finally the chain is extended in the golgi compartments with the help of a different set of glycosyltransferases. Both these glycosylations help in maintenance of protein conformation and activity, protein intracellular trafficking, and protection of proteins from proteolytic degradation. Moreover, N-Glycosylation is associated with protein folding, processing, and secretion from endoplasmic reticulum and golgi apparatus. Despite being strictly regulated in humans, protein glycosylation processes may sometimes result in various pathologies leading to certain types of cancers,[21–23] neurodegenerative diseases,[7, 8, 24] inflammation,[6] atherosclerosis,[25] diabetes,[5] and blood-related diseases.[26] Thus, investigation of differential expression of glycoproteins might be of great importance in understanding the mechanism of several cancers, such as gastric carcinoma, breast, ovarian, and colon cancers.[27–29]

Freeze et al. reported that >100 glycosylation related genetic disorders occur in humans.[16] Disease-related alterations in protein glycosylation

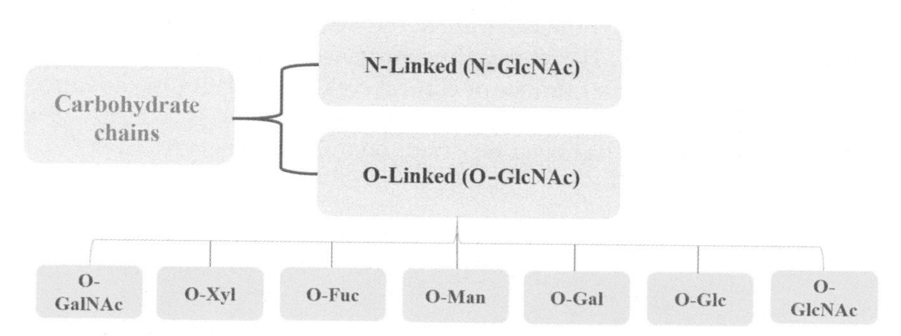

FIG. 1 Classification of carbohydrate chains attached to glycoproteins.

may result from loss or gain of glycosylation, which leads to changes in glycoprotein abundance, occupancy of glycosylation site (macroheterogeneity or "glycosylation efficiency"), and the distribution and type of glycans present on particular sites.[30] Most of the studies stress glucose and glycogen metabolism, since they play an essential role in energy breakdown. However, the synthesis and quantitative contributions of nutritive oligosaccharides, their synthesis, and recovery pathways are least known and probably differ among diverse cells and tissues. Addition of insufficient sugar moieties may be toxic subjected to the type of cell, glycosylation pattern, and the consequent gene expression. This complexity in glycosylation pattern makes it difficult to predict which cell or organ is at risk from developing defects. Therefore, understanding the mechanism by which aberrant glycosylation cause disease will improve the development of targeted therapeutics. To elucidate the importance of glycosylation and developing the next generation therapeutic agents, we need to describe the structure of glycoproteins. Even partial structural information is very helpful in understanding structure/function relationships and in the identification of alterations in the biosynthetic pathway of glycosylations. Almost all known glycosylation biosynthetic pathways are reported to be involved in disease progression.[17] Congenial disorders of glycosylation are associated with aberrant glycoprotein distribution and the resulting aberrant glycoforms serve as candidate markers for the disease.

3.1 Congenital Disorders of Glycosylation

A newly defined group of diseases characterized by aberrant glycosylation are termed congenital disorders of glycosylation (CDG)—previously termed carbohydrate-deficient glycoprotein syndrome.[31–35] These are inborn errors of glycan metabolism and almost 50 inborn errors due to congenital defects in N-glycosylation have been described.[36] These are of two types, based on the enzymatic effect involved: Type I, called CDG—I, is characterized by defectively synthesized dolichol lipid-linked oligosaccharide (LLO) chain, which is transferred to proteins during passage through the endoplasmic reticulum. Type II involves defective processing of oligosaccharides transferred to the proteins. The most common one is CDG Ia, now known as PMM2-CDG, with an estimated frequency of 1 in 20,000 newborns and is caused by defect in phosphomannomutase activity.[37, 38] CDG Ib has been attributed to the defective phosphomannose isomerase activity. A newly described CDG, known as CDG Ic, results from a defect in the glucosyltransferase I enzyme.[39] CDG type II involves defects in processing of N-glycans on the glycosylated proteins either in the ER or golgi. Additional types of N-glycosylation disorders include mucolipidosis II (MLII), also known as I-Cell disease, which is caused by deficiency of golgi GlcNAc phosphotransferase.[39a] Defects related to O-glycosylation include congenital muscular dystrophies (CMDs) and dystroglycanopathies, which are caused by disruptions in dystrophin-glycoprotein complex.[40, 40a] It has been shown that mutations in some glycotransferases, such as protein O-mannosyltransferase 1 and 2, Fukutin related protein, Dolichyl-phosphate mannosyltransferase 2,3, etc. produce altered glycosylation in α-dystroglycan (α-DG), which is an important part of dystrophin-glycoprotein complex.[41, 42]

3.2 Age Related Glycosylation Disorders

Age dependent changes due to alteration in protein glycosylation have been described in a cohort of studies, and have received much attention in recent years. Glycosylation of several proteins, e.g., Amyloid Precursor Protein (APP), Tau protein, β-site APP cleaving 1 (BACE-1), and Transferrin, have been reported in Alzheimer's disease (AD)—a commonly found type of cognitive disease in elderly people.[43, 44, 44a, 44b] With the use of mass-spectrometric methods, several more proteins with altered glycosylation patterns were identified in this disease. These include α1-antitrypsin (cerebrospinal fluid), synaptic proteins (β-Synuclein), chaperones (HSP90 and GP96), metabolic proteins (glucose dehydrogenase and enolase), and GFAP (Cerebrum).[45]

Type II diabetes mellitus (T2DM) is a major health problem, which is characterized by insulin resistance, hyperglycemia, and defects in insulin secretion. Elevated levels of biantennary N-glycan containing a α1,6-fucose with a dissecting GlcNAc were identified in T2DM.[46] O-GlcNAc modifications occur due to hyperglycemic condition in diabetes and are a major risk factor for AD. O-GlcNAc also widely modifies Tau and hampers aggregation.[47] Increased glycosylation of HbA, IgG, IgA, and IgM have been observed in diabetes mellitus.[47a] Age related changes in glycosylation of IgG and IgA were recently detected using advanced high throughput techniques like hydrophilic interaction liquid chromatography (HILIC),[48] matrix-assisted laser desorption ionization time of flight (MALDI-TOF/MS),[49] and capillary gel electrophoresis with laser-induced fluorescence detection (CGE-LIF).[50] Another newly reported age related syndrome called progeroid syndrome is characterized by defects in galactosyltansferase.[51] Borelli et al. observed a reduction in the level of N-glycosylation in plasma proteins in the case of Down's syndrome, which is the most recurrent chromosomal abnormality in humans.[52] Thus, age related glycans are established and vital markers for aging and age related disorders, which may help in improving treatment and management of health.

3.3 Glycosylation in Cancer: Diagnostic and Therapeutic Implications

Cancer is considered to be the second most important cause of deaths worldwide and accounted for 8.8 million deaths in 2015 (WHO). Cancer mortality needs to be ruled out through early diagnosis, prognosis, and therapeutic intervention. Glycosylation has long been found to be involved in several processes of tumor growth and progression.[53] Cancer cells exhibit changes in glycosylation pattern on several proteins like transmembrane receptors, and adhesion molecules, including signaling and inflammatory molecules. Altered synthesis of truncated glycans structures, expression of terminal sialylated glycans, overexpression of complex branched chain N-glycans, and altered fucosylation pattern have been reported in most cancers.[54] Therefore, glycosylations serve as important biomarkers of cancer development and progression and provide a platform of specific targets for therapeutic intervention.

One of the frequently observed glycan modification in tumors is sialylation, e.g., sialyl-Tn antigen is expressed by most carcinomas. Saldova et al. reported that increased sialylation intensely affects the glycan structures produced by cells in ovarian cancer development[55] and a significant increase of alpha 2-3 sialylation of N-glycome in serum has been observed in prostate cancers. It has been shown that increase in modulation of sialylated structures favors cell detachment and enhances communication with cell-extracellular matrix (ECM). The recent advent of glycoproteomics has permitted the analysis of glycosylation sites in cancers.[56] An important irregular glycosylated modification namely altered core fucosylation has been identified in various malignancies, e.g., aggressive prostate cancer, and is being detected by fucosylated prostate specific antigen (PSA) in serum.[57] Sialylated biantennary glycans showed 40% enhanced core fucosylation in serum from pancreatic cancer, suggesting a subgroup of glycoforms associated with tumors.[58] Increased expression and activity of α-1,6-fucosyltransferase (Fut8) have been reported in colorectal cancer,[59] hepatocellular carcinoma,[60] and ovarian serous adenocarcinoma.[61] Core fucosylation of E-cadherin was reported in highly metastatic lung cancers, suggesting its role as a promising biomarker for lung cancer patients. Recently, gastric cancer malignancy was reported to be associated with decreased core-fucosylation.[9] Moreover, the function of EGFR was reported to be associated with the function of core-fucosylation.[61a] Higher expression levels of fucosyltransferase FUT8 were reported in metastatic melanoma than primary tumors, elucidating higher expression of α-1,6 core fucosylation.[62]

A systematic analysis by Agrawal et al. established a link of pro-metastatic behavior of the disease with α-1,6 fucosylation of N-glycans present on neural cell adhesion molecule L1 (L1CAM).[62] In addition, alteration of α-1,2 fucosylation by the ATF2 transcription factor suppressed metastasis.[63] Other biomarkers predictive of therapeutic benefit include the fucosylated AFP (AFP-L3), a hepatocellular carcinoma marker in clinical practice,[64] sialyl-Lewis X (selectin ligand), and sialyl-Lewis A (CA19.9) are widely used serological cancer biomarkers. Their increased levels of these biomarkers correlate with poor patient survival. A potential diagnostic and prognostic marker for lung and pancreatic cancers is *N*-acetylgalactosaminyltransferase (GALNT) 3.[65, 65a] GALNT6 is an indicator of breast cancer development and progression[66, 67] and death-receptor. O-Glycosylation of melanoma cells and nonsmall-cell lung carcinoma and pancreatic carcinoma are regulated by GALNT14, and therefore, may serve as a prognostic biomarker for tumor necrosis factor/Apo2L associated apoptosis-inducing cancer therapy.[68]

4 TECHNOLOGICAL APPROACHES IN ANALYSIS OF PROTEIN GLYCOSYLATION

Due to some technical difficulties, the analysis of glycoproteins has remained severely understudied despite tremendous advances in proteome technologies. In order to unravel the varied functions of glycans and glycoproteins for efficient next gen glycoprotein therapy, a perfect analysis of the structural features of glycans is required. So far, chromatography and mass spectrometry are the most reliable tools for characterization of the glycoproteome (complete set of glycoproteins encoded by a particular genome). In recent decades, a growing interest in the identification and quantification of glycosylations in the development of diseases has been observed. Thus, a realization of the use and development of advanced and efficient technologies in glycoproteomics has been progressing.

Glycoproteome analysis typically involves the identification of glycosylated protein/peptide structure, heterogeneity of the attached glycan, and site of glycosylation. Although current advances in mass-spectrometric technologies provide effective means for glycopeptide quantification, owing to high structural diversity their structural/function relationship is challenging to characterize. In order to overcome the analytical challenges, many strategies have been developed during recent decades. The typical workflows (Fig. 2) for glycoproteomics (a subset of proteomics focused on glycoproteins) are initiated with the identification of the target protein/proteins, which is commonly done using gel electrophoresis followed by antibody recognition (Western blotting) or by staining with commercially available sugar detection stains.[68a] Currently, more sensitive chromogenic gel staining is being utilized, followed by fluorescent detection at 530nm. Some fluorescent stains are extremely sensitive and can detect <18ng of glycoproteins.[69, 69a] These methods however are

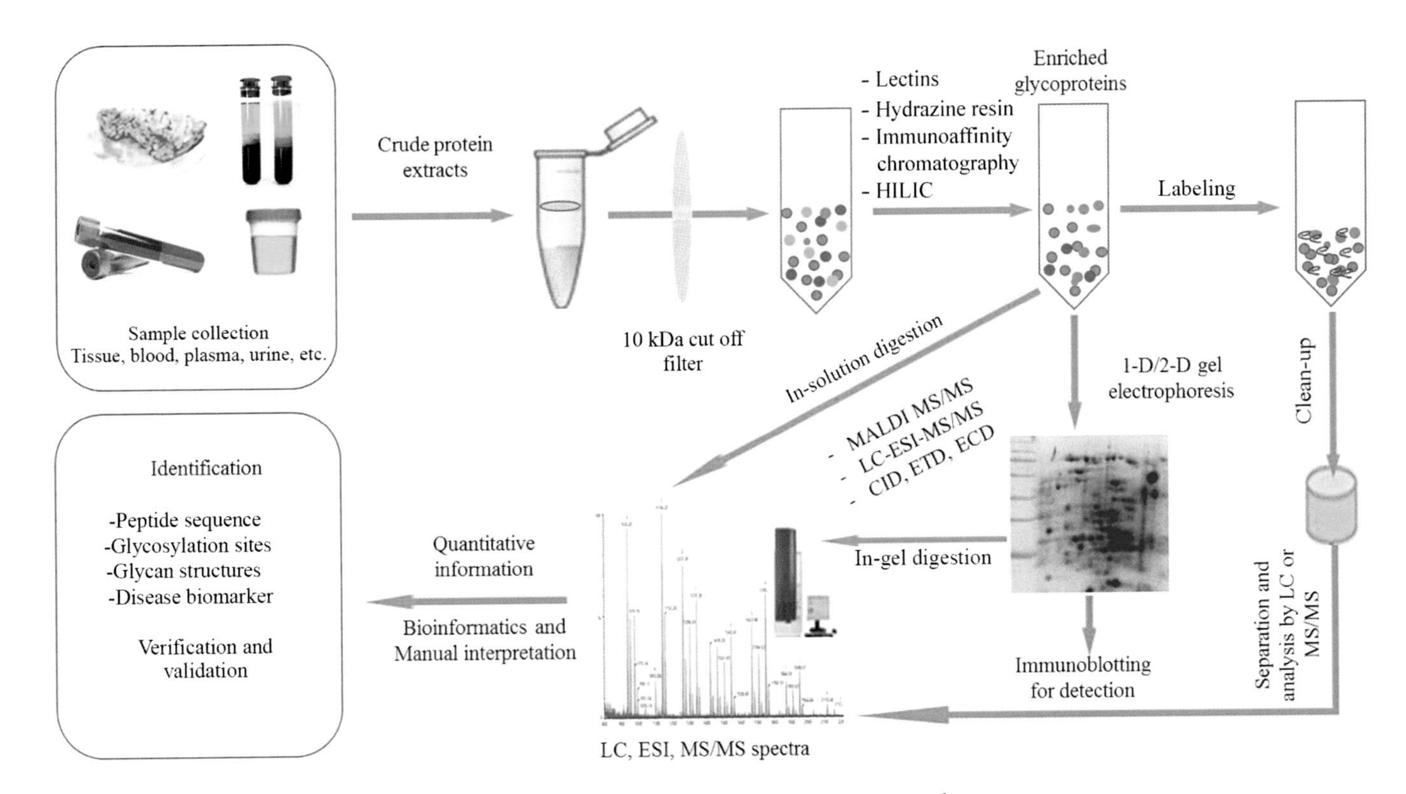

FIG. 2 Workflow of glycoproteome targeting, separation, and MS-based quantification approaches.

not sufficient alone, as they do not distinguish between N- and O-linked glycans. Therefore, suggesting the need for advanced techniques that can detect and characterize specific types of glycans.

Mass spectrometry has turned out to be an established analytical tool for characterization of protein glycosylation patterns and glycan analysis.[70, 70a] However, prior to MS, several enrichment/prefractionation methods are utilized while preparing the sample[71] as discussed below:

4.1 Preparation and Enrichment of Glycoproteins

Lectin affinity chromatography has been a method of choice for large-scale identification and enrichment of glycoproteins, thus, reducing sample complexity prior to identification by MS. Several plant-based lectins are available with varied affinities for different sugars. For example, Concanavalin A (Con A) and Wheat Germ Agglutinin have affinity for mannose and *N*-acetylglucosamine, respectively, and sialic acid, Aleuria Aurantia Lectin (AAL) specifically binds to fucose, and Jaclin lectin shows specificity to galactosyl moieties attached to *N*-acetylgalactosamine (GalNAc) via β (1,3) linkage.[72, 73] Two dimensional gel electrophoresis following affinity purification has become powerful tool in glycoproteomics. Pro Q-Emerald, a specific gel stain for glycoprotein, has recently been released commercially and generates a bright green-fluorescent signal upon reaction with acid-oxidized carbohydrate groups.[69] More recent glycoprotein enrichment methods are based on immunoaffinity[74]—covalent interactions using hydrazine and hydrophilic interation liquid chromatography (HILIC).[75, 75a] Hydrazine enrichment, in which glycans are covalently attached to solid phase hydrazine, has been used in several studies including liver carcinoma,[76] breast cancer,[77] and prostate cancers.[78] Immuno-affinity enrichment methods provide site-specific occupancy of glycan units; however, it proves more expensive than other enrichment techniques. Haptoglobin samples from lung cancer have been analyzed using immunoprecipitation methods.[74] Compared to other enrichment techniques, HILIC has gained much popularity due to its efficient removal of competing peptides from digested biological samples, thus, increasing glycopeptide identification.[79]

Nanotechnology, the newly developed approach, has provided new enrichment methods, e.g., magnetic nanoparticles (MNPs), which are being used as an attractive enrichment practice for improved categorization of specific sets of glycoproteins.[79a] Boronic acid chemistry is another covalent separation technique in which boronic acid is used to capture glycoproteins through covalent bond with 1-2, 1-3 cis-diol groups.[80] Another enrichment technique called electrostatic repulsion hydrophilic interaction chromatography (ERLIC) utilizes the advantage of electrostatic charge exhibited by glycan units to remove the basic, hydrophilic peptides, whereas glycans are attached to functional groups through hydrophilic interaction.[80a] Since most of these enrichment techniques mentioned above are limited by their natural or technical disadvantages, a combination of these techniques may be utilized for better enrichment of glycopeptides.[81]

Once enriched, glycoprotein samples are prepared; the next step is the release of glycan moiety. N-Glycans have been extensively studied and are released enzymatically with the help of PNGase F (Peptide N-glycanase F), which cleaves the amide linkage between Asn residues and oligosaccharide chain.[82] A number of *endo*-glycopeptidases have been reported showing varied activities for releasing N-linked glycans.[83] However, at the time of writing, no endoglycanase have been reported that could release O-linked glycans, which are therefore released by chemical methods such as β-eliminations.[84]

4.2 Mass Spectrometry Techniques for Glycoproteome Analysis

Over the years, mass-spectrometry based techniques have played an essential role in identifying the site of glycosylation, glycan structure, and composition, and are commonly performed by matrix-assisted laser desorption/ionization (MALDI) or electrospray ionization (ESI). MALDI-TOF (time of flight) is widely used for direct analysis of the released glycans.[84, 84a, 84b] MALDI-TOF offers a dynamic range and has the ability to tolerate salts and other contaminants added during processing; thus, it is widely used for peptide mapping. Likewise, liquid chromatography followed either by MALDI-Mass Spectrometry (LC-MALDI-MS) or ESI-MS (LC-ESI-MS) has been a powerful method for rapid and efficient analysis of glycoproteins/glycopeptides.[84c, 84d] LC-ESI-MS is commonly used in peptide mapping of N- and O-glycosylation sites in various glycosylated proteins, for example, tissue plasminogen activator and erythropoietin.[85–87] Various combinations like ESI-MS with HILIC column reduce the ion suppression due to salts and other contaminants.[87a]

Mass-spectrometry based techniques have also been successful in capturing of O-glycosylated peptides/proteins. Two novel methods, viz. higher energy collision dissociation (HCD) and collision ionization dissociation (CID), together with electron transfer dissociation, have been employed for O-glycosylation site identification.[87b, 88] ESI-CID-MS/MS[89] and ESI-ETD-MS/MS[89a] methods have also been adopted for facilitating site-specific information of O-GlcNAc proteins. ESI-CID-MS/MS has gained vast recognition for its more-or-less ideal demonstration of identifying and characterizing numerous types of PTMs. Site assignment of O-GlcNAc proteins has also been achieved efficiently using ESI-ETD-MS/MS.[89a, 89b] Moreover, CID or HCD is easily coupled with ETD (i.e., CID/ ETD-MS/MS or HCD/ETD-MS/MS) for consistent identification and comprehensive characterization of peptides with O-GlcNAc. Both these combinations, CID/ ETD-MS/MS[89a, 89c] and HCD/ETD-MS/MS,[89d] have shown increased confidence for identification and location of site in O-GlcNAcylated peptide.[88] In addition to this, coupling of pulsed Q dissociation (PQD) with ETD has also been achieved as a two-stage tandem MS approach for analysis of O-GlcNAc peptides.[89e]

With the growing field of glycoproteomics, revolutionary advances in analytical approaches have been achieved which vary from lab to lab and researcher to researcher. However, new avenues are being explored to achieve linkage specific information at glycopeptide levels.[90] Another variant of the technology universally known as peptide MRM (multiple reaction monitoring) has achieved great appreciation for quantitative measurements of protein biomarkers. Peptide MRM can join the high sensitivity and specificity of MS for the target protein with dynamic range and impressive quantitative precision. Multiple reaction monitoring based analysis have been used for a variety of biological samples and diseases like esophagus diseases,[91] liver diseases,[92] and immunoglobulin subclasses.[92a, 92b] Stable isotope labelling with amino acids in cell culture (SILAC)[85] and isobaric tags for relative and absolute quantitation (iTRAQ)[93, 93a] are two important quantitative methods used in glycobiology. However, as has been reviewed recently by Moh et al.[94], quantitative information regarding site micro-heterogeneity may be more helpful in glycobiology.

For better understanding of disease pathogenesis and development of targeted therapeutics, a clear understanding of the actual mechanism contributing to modification of glycoproteins in a particular disease is essential. Thus, the maturing field of glycoproteomics technologies is expected to yield a wealth of information in the wide spectrum of biological systems.[94a]

5 CONCLUSIONS AND FUTURE PROSPECTUS

Owing to its involvement in the pathological development of various diseases including cancer, diabetes, and neurodegenerative disorders, glycosylation has become one of the promising protein modifications for development of new biomarkers. In fact, due to the rapid increase in discovery of glycosylation disorders, identification, site mapping, and quantification of glycoproteins is a prerequisite to decipher their functions. Glycoproteins have become gradually more important both in the medical sciences and technology—predominantly, biomedicine and biotechnology. Further, a comprehensive investigation of the actual mechanism(s) causing altered glycoproteomes in particular disease conditions is very important for biomarker identification and targeted therapeutics. Glycoproteomics is still limited due to extreme structural heterogeneity and the often low abundance of several glycoforms. These complexities of protein glycosylation pose a daunting analytical challenge in the development of therapeutic glycoproteins. Recent advances and maturation of workflows in MS/MS methods hold great promise for glycoproteomics to develop into a large-scale, handy tool for the entire research community. These advances provide an opportunity for biomolecular scientists and, more specifically, glycoscientists, who would like to investigate the fascinating information stored in the spatiotemporally modulated glycoproteome. Moreover, with the advent of new and improvised protein chip technologies, novel glycobiomarkers can be identified that may be accurate and reliable for disease management and prognosis. Use of multiple glycomarkers and combination of existing and novel molecular markers would be an emerging trend in theranostic tests for disease management. Glyco-oncoproteomics would be a promising approach for unveiling complex molecular events of tumorigenesis.

References

1. Tran JC, Zamdborg L, Ahlf DR, Lee JE, Catherman AD, Durbin KR, Tipton JD, Vellaichamy A, Kellie JF, Li M, Wu C, Sweet SM, Early BP, Siuti N, LeDuc RD, Compton PD, Thomas PM, Kelleher NL. Mapping intact protein isoforms in discovery mode using top-down proteomics. *Nature* 2011;**480**:254–8.
2. Varki A, Freeze HH, Gagneux P. Evolution of glycan diversity. In: Varki A, Cummings RD, Esko JD, Freeze HH, Stanley P, Bertozzi CR, Hart GW, Etzler ME, editors. *Essentials of glycobiology*. 2nd ed. Cold Spring Harbor, NY: Cold Spring Harbor Laboratory Press; 2009.
3. Christiansen MN, Chik J, Lee L, Anugraham M, Abrahams JL, Packer NH. Cell surface protein glycosylation in cancer. *Proteomics* 2013;**14**:525–46.
4. Ma Z, Vosseller K. O-GlcNAc in cancer biology. *Amino Acids* 2013;**45**(4):719–33.
5. Ma J, Hart GW. Protein O-GlcNAcylation in diabetes and diabetic complications. *Expert Rev Proteomics* 2013;**10**:365–80.
6. Scott DW, Patel RP. Endothelial heterogeneity and adhesion molecules N-glycosylation: implications in leukocyte trafficking in inflammation. *Glycobiology* 2013;**23**:622–33.
7. Hwang H, Zhang J, Chung KA, Leverenz JB, Zabetian CP, Peskind ER, Jankovic J, Su Z, Hancock AM, Pan C, Montine TJ, Pan S, Nutt J, Albin R, Gearing M, Beyer RP, Shi M, Zhang J. Glycoproteomics in neurodegenerative diseases. *Mass Spectrom Rev* 2010;**29**:79–125.
8. Zhang J, Keene CD, Pan C, Montine KS, Montine TJ. Proteomics of human neurodegenerative diseases. *J Neuropathol Exp Neurol* 2008;**67**:923–32.
9. Zhao YP, Xu XY, Fang M, Wang H, You Q, Yi CH, Gao CF. Decreased core-fucosylation contributes to malignancy in gastric cancer. *PLoS ONE* 2014;**9**(4):e94536.

9a. Varki A, Kannagi R, Toole B, et al. Glycosylation changes in cancer. In: Varki A, Cummings RD, Esko JD, et al., editors. *Essentials of glycobiology* [Internet]. 3rd ed. Cold Spring Harbor, NY: Cold Spring Harbor Laboratory Press; 2017. [chapter 47].

10. Catalona WJ, Partin AW, Slawin KM, Brawer MK, Flanigan RC, Patel A, Richie JP, deKernion JB, Walsh PC, Scardino PT, Lange PH, Subong EN, Parson RE, Gasior GH, Loveland KG, Southwick PC. Use of the percentage of free prostate specific

antigen to enhance differentiation of prostate cancer from benign prostatic disease: a prospective multicenter clinical trial. *JAMA* 1998;**279**:1542–7.

11. Danesh J, Wheeler JG, Hirschfield GM, Eda S, Eiriksdottir G, Rumley A, Lowe GD, Pepys MB, Gudnason V. C-reactive protein and other circulating markers of inflammation in the prediction of coronary heart disease. *N Engl J Med* 2004;**350**:1387–97.
12. Jensen ON. Modification-specific proteomics: characterization of post-translational modifications by mass spectrometry. *Curr Opin Chem Biol* 2004;**8**(1):33–41.

12a. Gutman S, Kessler LG. The US Food and Drug Administration perspective on cancer biomarker development. *Nat Rev Cancer* 2006;**6**:565–71.

13. Grasbon-Frodl E, et al. Loss of glycosylation associated with the T183A mutation in human prion disease. *Acta Neuropathol* 2004;**108**:476–84.
14. Thomas M, et al. Androgen receptor acetylation site mutations cause trafficking defects, misfolding, and aggregation similar to expanded glutamine tracts. *J Biol Chem* 2004;**279**:8389–95.
15. Wang ZJ, Moult J. SNPs, protein structure, and disease. *Hum Mutat* 2001;**17**:263–70.
16. Freeze HH, Chong JX, Bamshad MJ, Ng BG. Solving glycosylation disorders: fundamental approaches reveal complicated pathways. *Am J Hum Genet* 2014;**94**(2):161–75.
17. Freeze HH. Understanding human glycosylation disorders: biochemistry leads the charge. *J Biol Chem* 2013;**288**:6936–45.
18. Rillahan CD, Paulson JC. Glycan microarrays for decoding the glycome. *Annu Rev Biochem* 2011;**80**(1):797–823.
19. Muntoni F, Torelli S, Wells DJ, Brown SC. Muscular dystrophies due to glycosylation defects: diagnosis and therapeutic strategies. *Curr Opin Neurol* 2011;**24**(5):437–42.
20. Rana NA, Haltiwanger RS. Fringe benefits: functional and structural impacts of O-glycosylation on the extracellular domain of Notch receptors. *Curr Opin Struct Biol* 2011;**21**(5):583–9.
21. Ho WL, Hsu WM, Huang MC, Kadomatsu K, Nakagawara A. Protein glycosylation in cancers and its potential therapeutic applications in neuroblastoma. *J Hematol Oncol* 2016;**9**(1):100.
22. Banh A, Zhang J, Cao H, Bouley DM, Kwok S, Kong C, et al. Tumor galectin-1 mediates tumor growth and metastasis through regulation of T-cell apoptosis. *Cancer Res* 2011;**71**(13):4423–31.
23. Kazuno S, Furukawa JI, Shinohara Y, Murayama K, Fujime M, Ueno T, et al. Glycosylation status of serum immunoglobulin G in patients with prostate diseases. *Cancer Med* 2016;**5**:1137–46.
24. Schedin-Weiss S, Winblad B, Tjernberg LO. The role of protein glycosylation in Alzheimer disease. *FEBS J* 2014;**281**:46–62.
25. Xu Y-X, Ashline D, Liu L, Tassa C, Shaw SY, Ravid K, Layne MD, Reinhold V, Robbins PW. The glycosylation-dependent interaction of perlecan core protein with LDL: implications for atherosclerosis. *J Lipid Res* 2015;**56**(2):266–76.
26. Brooks AR, Sim D, Gritzan U, et al. Glycoengineered factor IX variants with improved pharmacokinetics and subcutaneous efficacy. *J Thromb Haemost* 2013;**11**(9):1699–706.
27. Brockhausen I. Mucin-type O-glycans in human colon and breast cancer: glycodynamics and functions. *EMBO Rep* 2006;**7**:599–604.
28. Gomes J, et al. Expression of UDP-N-acetyl-D-galactosamine: polypeptide N-acetylgalactosaminyltransferase-6 in gastric mucosa, intestinal metaplasia, and gastric carcinoma. *J Histochem Cytochem* 2009;**57**:79–86.
29. Ricardo S, et al. Detection of glyco-mucin profiles improves specificity of MUC16 and MUC1 biomarkers in ovarian serous tumours. *Mol Oncol* 2015;**9**:503–12.
30. Carlsson MC, Cederfur C, Schaar V, Balog CI, Lepur A, Touret F, Salomonsson E, Deelder AM, Ferno M, Olsson H, Wuhrer M, Leffler H. Galectin-1-binding glycoforms of haptoglobin with altered intracellular trafficking, and increase in metastatic breast cancer patients. *PLoS One* 2011;**6**:e26560.

31. Freeze HH. Update and perspectives on congenital disorders of glycosylation. *Glycobiology* 2001;**11**(12):129R–143R.
32. Jaeken J, Hennet T, Matthijs G, Freeze HH. CDG nomenclature: time for a change! *Biochim Biophys Acta* 2009;**1792**:825–6.
33. Kim S, Westphal V, Srikrishna G, Mehta DP, Peterson S, Filiano J, Karnes PS, Patterson MC, Freeze HH. Dolichol phosphate mannose synthase (DPM1) mutations define congenital disorder of glycosylation Ie (CDG-Ie). *J Clin Invest* 2000;**105**:191–8.
34. Matthijs G, Schollen E, Bjursell C, Erlandson A, Freeze H, Imtiaz F, Kjaergaard S, Martinsson T, Schwartz M, Seta N, Vuillaumier-Barrot S, Westphal V, Winchester B. Mutations in PMM2 that cause congenital disorders of glycosylation, type Ia (CDG-Ia). *Hum Mutat* 2000;**16**:386–94.
35. Nakayama Y, Nakamura N, Tsuji D, Itoh I, Kurosaka A. Genetic diseases associated with protein glycosylation disorders in mammals. In: Puiu M, editor. *Genetic disorders*. InTech; 2013. p. 243–69. https://doi.org/10.5772/54097.
36. Scott NE, Marzook NB, Cain JA, Solis N, Thaysen-Andersen M, Djordjevic SP, Packer NH, Larsen MR, Cordwell SJ. Comparative proteomics and glycoproteomics reveal increased N-linked glycosylation and relaxed sequon specificity in *Campylobacter jejuni* NCTC11168 O. *J Proteome Res* 2014;**13**:5136–50.
37. Fletcher JM, Matthijs G, Jaeken J, Van Schaftingen E, Nelson PV. Carbohydrate-deficient glycoprotein syndrome: beyond the screen. *J Inherit Metab Dis* 2000;**23**:396–8.
38. Jaeken J, Matthijs G. Congenital disorders of glycosylation. *Annu Rev Genomics Hum Genet* 2001;**2**:129–51.
39. Morava E, Tiemes V, Thiel C, Seta N, de Lonlay P, de Klerk H, Mulder M, Rubio-Gozalbo E, Visser G, van Hasselt P, Horovitz DD, de Souza CF, Schwartz IV, Green A, Al-Owain M, Uziel G, Sigaudy S, Chabrol B, van Spronsen FJ, Steinert M, Komini E, Wurm D, Bevot A, Ayadi A, Huijben K, Dercksen M, Witters P, Jaeken J, Matthijs G, Lefeber DJ, Wevers RA. ALG6-CDG: a recognizable phenotype with epilepsy, proximal muscle weakness, ataxia and behavioral and limb anomalies. *J Inherit Metab Dis* 2016;**39**:713–23.
39a. Tiede S, Storch S, Lubke T, Henrissat B, Bargal R, Raas-Rothschild A, Braulke T. Mucolipidosis II is caused by mutations in GNPTA encoding the alpha/beta GlcNAc-1-phosphotransferase. *Nat Med* 2005;**11**(10):1109–12.
40. Durbeej M, Campbell KP. Muscular dystrophies involving the dystrophin-glycoprotein complex: an overview of current mouse models. *Curr Opin Genet Dev* 2002;**12**:349–61.
40a. Lapidos KA, Kakkar R, McNally EM. The dystrophin glycoprotein complex: signalling strength and integrity for the sarcolemma. *Circ Res* 2004;**94**:1023–31.
41. Barone R, Aiello C, Race V, Morava E, Foulquier F, Riemersma M, Passarelli C, Concolino D, Carella M, Santorelli F. DPM2-CDG: a muscular dystrophy-dystroglycanopathy syndrome with severe epilepsy. *Ann Neurol* 2012;**72**:550–8.
42. Stevens E, Carss KJ, Cirak S, Foley AR, Torelli S, Willer T, Tambunan DE, Yau S, Brodd L, Sewry CA, Feng L, Haliloglu G, Orhan D, Dobyns WB, Enns GM, Manning M, Krause A, Salih MA, Walsh CA, Hurles M, Campbell KP, Manzini MC, Stemple D, Lin YY, Muntoni F. Mutations in B3GALNT2 cause congenital muscular dystrophy and hypoglycosylation of alpha-dystroglycan. *Am J Hum Genet* 2013;**92**(3):354–65.
43. Perdivara I, Petrovich R, Allinquant B, Deterding LJ, Tomer KB, Przybylski M. Elucidation of O-glycosylation structures of the beta-amyloid precursor protein by liquid chromatography-mass spectrometry using electron transfer dissociation and collision induced dissociation. *J Proteome Res* 2009;**8**:631–42.
44. Van Rensburg SJ, Berman P, Potocnik F, MacGregor P, Hon D, de Villiers N. 5- and 6-glycosylation of transferrin in patients with Alzheimer's disease. *Metab Brain Dis* 2004;**19**:89–96.

44a. Sato N, Imaizumi K, Manabe T, et al. Increased production of beta-amyloid and vulnerability to endoplasmic reticulum stress by an aberrant spliced form of presenilin 2. *J Biol Chem* 2001;**276**(3):2108–14.
44b. Kizuka Y, Kitazume S, Fujinawa R, Saito T, Iwata N, Saido TC, Nakano M, Yamaguchi Y, Hashimoto Y, Staufenbiel M, Hatsuta H, Murayama S, Manya H, Endo T, Taniguchi N. An aberrant sugar modification of BACE1 blocks its lysosomal targeting in Alzheimer's disease. *EMBO Mol Med* 2015;**7**:175–89.
45. Miura Y, Endo T. Glycomics and glycoproteomics focused on aging and age-related diseases-glycans as a potential biomarker for physiological alterations. *Biochim Biophys Acta* 2016; S0304-4165(16)00022-2v.
46. Itoh N, Sakaue S, Nakagawa H, et al. Analysis of N-glycan in serum glycoproteins from db/db mice and humans with type 2 diabetes. *Am J Physiol Endocrinol Metab* 2007;**293**(4):E1069–77.
47. Yuzwa SA, Shan X, Macauley MS, Clark T, Skorobogatko Y, Vosseller K, Vocadlo DJ. Increasing O-GlcNAc slows neurodegeneration and stabilizes tau against aggregation. *Nat Chem Biol* 2012;**8**:393–9.
47a. Kalia K, Sharma S, Mistry K. Non-enzymatic glycosylation of immunoglobulins in diabetic nephropathy. *Clin Chim Acta Int J Clin Chem* 2004;**347**(1–2):169–76.
48. Pucic M, Knezevic A, Vidic J, et al. High throughput isolation and glycosylation analysis of IgG-variability and heritability of the IgG glycome in three isolated human populations. *Mol Cell Proteomics* 2011;**10**(10). M111.010090.
49. Bakovic MP, Selman MH, Hoffmann M, Rudan I, Campbell H, Deelder AM, Lauc G, Wuhrer M. High-throughput IgG Fc N-glycosylation profiling by mass spectrometry of glycopeptides. *J Proteome Res* 2013;**12**:821–31.
50. Ruhaak LR, Koeleman CA, Uh HW, Stam JC, van Heemst D, Maier AB, Houwing-Duistermaat JJ, Hensbergen PJ, Slagboom PE, Deelder AM, Wuhrer M. Targeted biomarker discovery by high throughput glycosylation profiling of human plasma alpha1-antitrypsin and immunoglobulin A. *PLoS One* 2013;**8**:e73082.
51. Okajima T, Fukumoto S, Furukawa K, Urano T. Molecular basis for the progeroid variant of Ehlers-Danlos syndrome. Identification and characterization of two mutations in galactosyltransferase I gene. *J Biol Chem* 1999;**2**(74):28841–4.
52. Borelli V, et al. Plasma N-glycome signature of down syndrome. *J Proteome Res* 2015;**14**:4232–45.
53. Stowell SR, Ju T, Cummings RD. Protein glycosylation in cancer. *Annu Rev Pathol* 2015;**10**:473–510.
54. Pinho SS, Reis CA. *Nat Rev Cancer* 2015;**15**:540–55.
55. Saldova R, Fan Y, Fitzpatrick JM, Watson RW, Rudd PM. Core fucosylation and alpha2-3 sialylation in serum N-glycome is significantly increased in prostate cancer comparing to benign prostate hyperplasia. *Glycobiology* 2011;**21**:195–205.
56. Campos D, Freitas D, Gomes J, Magalhães A, Steentoft C, Gomes C, Vester-Christensen MB, Ferreira JA, Afonso LP, Santos LL, et al. Probing the O-glycoproteome of gastric cancer cell lines for biomarker discovery. *Mol Cell Proteomics* 2015;**14**:1616–29.
57. Li QK, Gabrielson E, Zhang H. Application of glycoproteomics for the discovery of biomarkers in lung cancer. *Proteomics Clin Appl* 2012;**6**:244–56.
58. Barrabés S, Pagès-Pons L, Radcliffe CM, Tabarés G, Fort E. Glycosylation of serum ribonuclease 1 indicates a major endothelial origin and reveals an increase in core fucosylation in pancreatic cancer. *Glycobiology* 2007;**17**:388–400.
59. Muinelo-Romay L, Vázquez-Martín C, Villar-Portela S, Cuevas E, Gil-Martín E, et al. Expression and enzyme activity of alpha(1,6) fucosyltransferase in human colorectal cancer. *Int J Cancer* 2008;**123**:641–6.

60. Comunale MA, Lowman M, Long RE, Krakover J, Philip R, et al. Proteomic analysis of serum associated fucosylated glycoproteins in the development of primary hepatocellular carcinoma. *J Proteome Res* 2006;**5**:308–15.
61. Takahashi T, Ikeda Y, Miyoshi E, Yaginuma Y, Ishikawa M, et al. Alpha1, 6 fucosyltransferase is highly and specifically expressed in human ovarian serous adenocarcinomas. *Int J Cancer* 2000;**88**:914–9.
61a. Wang X, Gu J, Ihara H, Miyoshi E, Honke K, Taniguchi N. Core fucosylation regulates epidermal growth factor receptor-mediated intracellular signaling. *J Biol Chem* 2006;**281**:2572–7.
62. Agrawal P, Fontanals-Cirera B, Sokolova E, Jacob S, Vaiana CA, Argibay D, Davalos V, McDermott M, Nayak S, Darvishian F, et al. *Cancer Cell* 2017;**31**:804–19.
63. Lau E, Feng Y, Claps G, Fukuda MN, Perlina A, Donn D, Jilaveanu L, Kluger H, Freeze HH, Ronai ZA. The transcription factor ATF2 promotes melanoma metastasis by suppressing protein fucosylation. *Sci Signal* 2015;**8**:ra124.
64. Moriwaki K, Miyoshi E. Fucosylation and gastrointestinal cancer. *World J Hepatol* 2010;**2**:151–61.
65. Dosaka-Akita H, Kinoshita I, Yamazaki K, Izumi H, Itoh T, Katoh H, et al. N-acetylgalactosaminyl transferase-3 is a potential new marker for non-small cell lung cancers. *Br J Cancer* 2002;**87**(7):751–5.
65a. Yamamoto S, Nakamori S, Tsujie M, Takahashi Y, Nagano H, Dono K, Umeshita K, Sakon M, Tomita Y, Hoshida Y, Aozasa K, Kohno K, Monden M. Expression of uridine diphosphate N-acetyl-alpha-D-galactosamine: polypeptide N-acetylgalactosaminyl transferase 3 in adenocarcinoma of the pancreas. *Pathobiology* 2004;**71**(1):12–8.
66. Park JH, Katagiri T, Chung S, Kijima K, Nakamura Y. Polypeptide N-acetylgalactosaminyltransferase 6 disrupts mammary acinar morphogenesis through O-glycosylation of fibronectin. *Neoplasia* 2011;**13**(4):320–6.
67. Park JH, Nishidate T, Kijima K, Ohashi T, Takegawa K, Fujikane T, et al. Critical roles of mucin 1 glycosylation by transactivated polypeptide N-acetylgalactosaminyltransferase 6 in mammary carcinogenesis. *Cancer Res* 2010;**70**(7):2759–69.
68. Wagner KW, Punnoose EA, Januario T, Lawrence DA, Pitti RM, Lancaster K, et al. Death-receptor O-glycosylation controls tumor-cell sensitivity to the proapoptotic ligand Apo2L/TRAIL. *Nat Med* 2007;**13**(9):1070–7.
68a. Bardor M, Cabrera G, Stadlmann J, Lerouge P, Cremata JA, Gomord V, Fitchette AC. N-glycosylation of plant recombinant pharmaceuticals. *Methods Mol Biol* 2009;**483**:239–64.
69. Hart C, Schulenberg B, Steinberg TH, Leung WY, Patton WF. Detection of glycoproteins in polyacrylamide gels and on electroblots using Pro-Q Emerald 488 dye, a fluorescent periodate Schiff-base stain. *Electrophoresis* 2003;**24**:588–98.
69a. Steinberg TH. Chapter 31: Protein gel staining methods. An introduction and overview. *Methods Enzymol* 2009;**463**:541–63.
70. Song W, Henquet MG, Mentink RA, van Dijk AJ, Cordewener JH, Bosch D, America AH, van der Krol AR. N-glycoproteomics in plants: perspectives and challenges. *J Proteomics* 2011;**74**:1463–74.
70a. Yates JR, Ruse CI, Nakorchevsky A. Proteomics by mass spectrometry: approaches, advances, and applications. *Annu Rev Biomed Eng* 2009;**11**:49–79.
71. Rudiger H, Gabius HJ. Plant lectins: occurrence, biochemistry, functions and applications. *Glycoconj J* 2001;**18**:589–613.
72. Butterfield DA, Owen JB. Lectin-affinity chromatography brain glycoproteomics and Alzheimer disease: insights into protein alterations consistent with the pathology and progression of this dementing disorder. *Proteomics Clin Appl* 2011;**5**:50–6.
73. Owen JB, Di Domenico F, Sultana R, Perluigi M, Cini C, Pierce WM, Butterfield DA. Proteomics-determined differences in the concanavalin-A-fractionated proteome of

hippocampus and inferior parietal lobule in subjects with Alzheimer's disease and mild cognitive impairment: implications for progression of AD. *J Proteome Res* 2009;**8**:471–82.

74. Wang D, Hincapie M, Rejtar T, Karger BL. Ultrasensitive characterization of site-specific glycosylation of affinity-purified haptoglobin from lung cancer patient plasma using 10 μm i.d. porous layer open tubular liquid chromatography-linear ion trap collision-induced dissociation/electron transfer dissociation mass spectrometry. *Anal Chem* 2011;**83**:2029–37.
75. Ruhaak LR, Huhn C, Waterreus WJ, de Boer AR, Neususs C, Hokke CH, Deelder AM, Wuhrer M. Hydrophilic interaction chromatography-based high-throughput sample preparation method for N-glycan analysis from total human plasma glycoproteins. *Anal Chem* 2008;**80**:6119–26.
75a. Walker SH, Budhathoki-Uprety J, Novak BM, Muddiman DC. Stable-isotope labeled hydrophobic hydrazide reagents for the relative quantification of N-linked glycans by electrospray ionization mass spectrometry. *Anal Chem* 2011;**83**:6738–45.
76. Chen R, Seebun D, Ye M, Zou H, Figeys D. Site-specific characterization of cell membrane N-glycosylation with integrated hydrophilic interaction chromatography solid phase extraction and LCMS/MS. *J Proteomics* 2014;**103**:194–203.
77. Whelan SA, He J, Lu M, Souda P, Saxton RE, Faull KF, Whitelegge JP, Chang HR. Mass spectrometry (LC–MS/MS) identified proteomic biosignatures of breast cancer in proximal fluid. *J Proteome Res* 2012;**11**:5034–45.
78. Chen CY, Jan YH, Juan YH, Yang CJ, Huang MS, et al. Fucosyltransferase 8 as a functional regulator of nonsmall cell lung cancer. *Proc Natl Acad Sci U S A* 2013;**110**:630–5.
79. Zhang L, Luo S, Zhang B. Glycan analysis of therapeutic glycoproteins. *MAbs* 2016;**8**:205–15.
79a. Cova M, Oliveira Silva R, Ferreira JA, Ferreira R, Amado F, Daniel-da-Silva AL, et al. Glycoprotein enrichment method using a selective magnetic nanoprobe platform (MNP) functionalized with lectins. *Methods Mol Biol (Clifton, NJ)* 2015;**1243**:83–100.
80. Ongay S, Boichenko A, Govorukhina N, Bischoff R. Glycopeptide enrichment and separation for protein glycosylation analysis. *J Sep Sci* 2012;**35**:2341–72.
80a. Zacharias LG, Hartmann AK, Song E, Zhao J, Zhu R, Mirzaei P, Mechref Y. HILIC and ERLIC enrichment of glycopeptides derived from breast and brain cancer cells. *J Proteome Res* 2016;**15**(10):3624–34.
81. Li X, Jiang J, Zhao X, Wang J, Han H, Zhao Y, Qian X. N-glycoproteome analysis of the secretome of human metastatic hepatocellular carcinoma cell lines combining hydrazide chemistry, HILIC enrichment and mass spectrometry. *PLoS One* 2013;**8**(12):e81921.
82. Zhang W, Wang H, Zhang L, Yao J, Yang P. Large-scale assignment of N-glycosylation sites using complementary enzymatic deglycosylation. *Talanta* 2011;**85**:499–505.
83. Goodfellow JJ, Baruah K, Yamamoto K, Bonomelli C, Krishna B, Harvey DJ, Crispin M, Scanlan CN, Davis BG. An endoglycosidase with alternative glycan specificity allows broadened glycoprotein remodelling. *J Am Chem Soc* 2012;**134**:8030–3.
84. Wada Y, Dell A, Haslam SM, Tissot B, Canis K, Azadi P, Backstrom M, Costello CE, Hansson GC, Hiki Y, et al. Comparison of methods for profiling O-glycosylation: human proteome organisation human disease glycomics/proteome initiative multi-institutional study of IgA1. *Mol Cell Proteomics* 2010;**9**:719–27.
84a. Stephens E, Sugars J, Maslen SL, Williams DH, Packman LC, Ellar DJ. The N-linked oligosaccharides of aminopeptidase N from Manduca sexta. Site localization and identification of novel N-glycan structures. *Eur J Biochem* 2004;**271**:4241–58.
84b. Liu H, Zhang N, Wan D, Cui M, Liu Z, Liu S. Mass spectrometry-based analysis of glycoproteins and its clinical applications in cancer biomarker discovery. *Clin Proteomics* 2014;**11**(1):14.
84c. Lochnit G, Geyer R. An optimized protocol for nano-LC-MALDI-TOF-MS coupling for the analysis of proteolytic digests of glycoproteins. *Biomed Chromatogr* 2004;**18**:841–8.

84d. Klapoetke S. N-glycosylation characterization by liquid chromatography with mass spectrometry. *Methods Mol Biol* 2014;**1131**:513–24.
85. Parker BL, Thaysen-Andersen M, Fazakerley DJ, Holliday M, Packer NH, James DE. Terminal galactosylation and sialylation switching on membrane glycoproteins upon TNF-alpha-induced insulin resistance in adipocytes. *Mol Cell Proteomics* 2016;**15**:141–53.
86. Parker BL, Thaysen-Andersen M, Solis N, Scott NE, Larsen MR, Graham ME, Packer NH, Cordwell SJ. Site-specific glycan-peptide analysis for determination of N-glycoproteome heterogeneity. *J Proteome Res* 2013;**12**:5791–800.
87. Thaysen-Andersen M, Packer NH. Advances in LC-MS/MS-based glycoproteomics: getting closer to system-wide site-specific mapping of the N and O-glycoproteome. *Biochim Biophys Acta* 2014;**1844**(9):1437–52.
87a. Remoroza C, Cord-Landwehr S, Leijdekkers A, Moerschbacher B, Schols H, Gruppen H. Combined HILIC-ELSD/ESI-MSn enables the separation, identification and quantification of sugar beet pectin derived oligomers. *Carbohydr Polym* 2012;**90**:41–8.
87b. Syka JE, Coon JJ, Schroeder MJ, Shabanowitz J, Hunt DF. Peptide and protein sequence analysis by electron transfer dissociation mass spectrometry. *Proc Natl Acad Sci USA* 2004;**101**:9528–33.
88. Scott NE, Cordwell SJ. Enrichment and identification of bacterial glycopeptides by mass spectrometry. *Methods Mol Biol* 2015;**1295**:355–68.
89. Zhao P, Zhang W, Wang S, et al. HAb18G/CD147 promotes cell motility by regulating annexin II-activated RhoA and Rac1 signaling pathways in hepatocellular carcinoma cells. *Hepatology* 2011;**54**:2012–24.
89a. Berk JM, Maitra S, Dawdy AW, Shabanowitz J, Hunt DF, Wilson KL. O-GlcNAc regulates emerin binding to BAF in a chromatin- and lamin B-enriched 'niche'. *J Biol Chem* 2013;**11**:30192–209.
89b. Myers SA, Daou S, Affar el B, Burlingame A. Electron transfer dissociation (ETD): the mass spectrometric breakthrough essential for O-GlcNAc protein site assignments-a study of the O-GlcNAcylated protein host cell factor C1. *Proteomics* 2013;**13**:982–91.
89c. Isono T. OGlcNAc-specific antibody CTD110.6 cross-reacts with N-GlcNAc$_2$-modified proteins induced under glucose deprivation. *PLoS ONE* 2011;**11**(4):e18959.
89d. Zhao P, Viner R, Teo CF, Boons GJ, Horn D, Wells L. Combining high-energy C-trap dissociation and electron transfer dissociation for protein O-GlcNAc modification site assignment. *J Proteome Res* 2011;**11**:4088–104.
89e. Hahne H, Kuster B. A novel two-stage tandem mass spectrometry approach and scoring scheme for the identification of O-GlcNAc modified peptides. *J Am Soc Mass Spectrom* 2011;**22**:931–42.
90. de Haan N, Reiding KR, Haberger M, Reusch D, Falck D, Wuhrer M. Linkage-specific sialic acid derivatization for MALDITOF-MS profiling of IgG glycopeptides. *Anal Chem* 2015;**87**:8284–91.
91. Song E, Mayampurath A, Yu CY, et al. Glycoproteomics: identifying the glycosylation of prostate specific antigen at normal and high isoelectric points by LCMS/MS. *J Proteome Res* 2014;**13**(12):5570–80.
92. Ahn JM, Sung HJ, Yoon YH, et al. Integrated glycoproteomics demonstrates fucosylated serum paraoxonase 1 alterations in small cell lung cancer. *Mol Cell Proteomics* 2014;**13**(1):30–48.
92a. Hong Q, Lebrilla CB, Miyamoto S, Ruhaak LR. Absolute quantitation of immunoglobulin G and its glycoforms using multiple reaction monitoring. *Anal Chem* 2013;**85**(18):8585–93.
92b. Yuan W, Sanda M, Wu J, Koomen J, Goldman R. Quantitative analysis of immunoglobulin subclasses and subclass specific glycosylation by LC-MS-MRM in liver disease. *J Proteome* 2015;**116**:24–33.

93. Shah P, Wang X, Yang W, Toghi Eshghi S, Sun S, Hoti N, Chen L, Yang S, Pasay J, Rubin A, Zhang H. Integrated proteomic and glycoproteomic analyses of prostate cancer cells reveals glycoprotein alteration in protein abundance and glycosylation. *Mol Cell Proteomics* 2015;**14**:2753–63.
93a. Ahn YH, Kim JY, Yoo JS. Quantitative mass spectrometric analysis of glycoproteins combined with enrichment methods. *Mass Spectrom Rev* 2015;**34**:148–65.
94. Moh ES, Lin CH, Thaysen-Andersen M, et al. Site-specific N-glycosylation of recombinant pentameric and hexameric human IgM. *J Am Soc Mass Spectrom* 2016;**27**(7):1143–55.
94a. Thaysen-Andersen M, Packer NH, Schulz BL. Maturing glycoproteomics technologies provide unique structural insights into the N-glycoproteome and its regulation in health and disease. *Mol Cell Proteomics MCP* 2016;**15**(6):1773–90.

Further Reading

95. Aebi M. N-linked protein glycosylation in the ER. *Biochim Biophys Acta* 2013;**1833**:2430–7.
96. Brockhausen I, Yang J, Lehotay M, Ogata S, Itzkowitz S. Pathways of mucin O-glycosylation in normal and malignant rat colonic epithelial cells reveal a mechanism for cancer-associated sialyl-Tn antigen expression. *Biol Chem* 2001;**382**:219–32.
97. Brooks M. Breast cancer screening and biomarkers. *Methods Mol Biol* 2009;**472**:307–21.
98. Bunkenborg J, Pilch BJ, Podtelejnikov AV, Wisniewski, and J. R.. Screening for N-glycosylated proteins by liquid chromatography mass spectrometry. *Proteomics* 2004;**4**:454–65.
99. Chen G. *Characterization of protein therapeutics using mass spectrometry.* Springer; 2013. ISBN 978-1-4419-7862-2.
100. Cheng C, Ru P, Geng F, et al. Glucose-mediated N-glycosylation of SCAP is essential for SREBP-1 activation and tumor growth. *Cancer Cell* 2015;**28**:569–81.
101. Dall'Olio F, Malagolini N, Trinchera M, Chiricolo M. Mechanisms of cancer-associated glycosylation changes. *Front Biosci (Landmark Ed)* 2012;**17**:670–99.
102. Dennis JW, Laferte S, Waghorne C, et al. Beta 1-6 branching of Asn-linked oligosaccharides is directly associated with metastasis. *Science* 1987;**236**:582–5.
103. Ferreira JA, et al. Overexpression of tumour-associated carbohydrate antigen sialyl-Tn in advanced bladder tumours. *Mol Oncol* 2013;**7**:719–31.
104. Freeze HH, Eklund EA, Ng BG, Patterson MC. Neurology of inherited glycosylation disorders. *Lancet Neurol* 2012;**11**:453–66.
105. Gillette MA, Carr SA. Protein biomarker discovery and validation: the long and uncertain path to clinical utility. *Nat Biotechnol* 2006;**24**:971–83.
106. Goldstein IJ, Hollerman CE, Smith EE. Protein-carbohydrate interaction. II. Inhibition studies on the interaction of concanavalin A with polysaccharides. *Biochemistry* 1965;**4**:876–83.
107. Guo HB, et al. Specific posttranslational modification regulates early events in mammary carcinoma formation. *Proc Natl Acad Sci U S A* 2010;**107**:21116–21.
108. Guo H, Nagy T, Pierce M. Post-translational glycoprotein modifications regulate colon cancer stem cells and colon adenoma progression in $Apc^{min/+}$ mice through altered Wnt receptor signaling. *J Biol Chem* 2014;**289**:31534–49.
109. Jansen BC, Reiding KR, Bondt A, et al. MassyTools: a high-throughput targeted data processing tool for relative quantitation and quality control developed for glycomic and glycoproteomic MALDI-MS. *J Proteome Res* 2015;**14**(12):5088–98.
110. Kellokumpu S, Sormunen R, Kellokumpu I. Abnormal glycosylation and altered Golgi structure in colorectal cancer: dependence on intra-Golgi pH. *FEBS Lett* 2002;**516**:217–24.

111. Kim YJ, Varki A. Perspectives on the significance of altered glycosylation of glycoproteins in cancer. *Glycoconj J* 1997;**14**:569–76.
112. Kizuka Y, Kitazume S, Fujinawa R, Saito T, Iwata N, Saido TC, Nakano M, Yamaguchi Y, Hashimoto Y, Staufenbiel M, Hatsuta H, Murayama S, Manya H, Endo T, Taniguchi N. An aberrant sugar modification of BACE1 blocks its lysosomal targeting in Alzheimer's disease. *EMBO Mol Med* 2015;**7**:175–89.
113. Ma J, Hart GW. O-GlcNAc profiling: from proteins to proteomes. *Clin Proteomics* 2014;**11**:8.
114. Ma Z, Vosseller K. Cancer metabolism and elevated O-GlcNAc in oncogenic signaling. *J Biol Chem* 2014;**289**:34457–65.
115. Magalhães A, Duarte HO, Reis CA. Aberrant glycosylation in cancer: a novel molecular mechanism controlling metastasis. *Cancer Cell* 2017;**31**:733–5.
116. Mann M, Jensen ON. Proteomic analysis of post-translational modifications. *Nat Biotechnol* 2003;**21**:255–61.
117. Moh ES, Thaysen-Andersen M, Packer NH. Relative versus absolute quantitation in disease glycomics. *Proteomics Clin Appl* 2015;**9**:368–82.
118. Reticker-Flynn NE, Bhatia SN. Aberrant glycosylation promotes lung cancer metastasis through adhesion to galectins in the metastatic niche. *Cancer Discov* 2015;**5**(2):168–81.
119. Ruhaak LR, et al. Plasma protein N-glycan profiles are associated with calendar age, familial longevity and health. *J Proteome Res* 2011;**10**:1667–74.
120. Ruhaak LR, Uh H-W, Deelder AM, Dolhain REJM, Wuhrer M. Total plasma N-glycome changes during pregnancy. *J Proteome Res* 2014;**13**:1657–68.
121. Song E, Hu Y, Hussein A, Yu CY, Tang H, Mechref Y. Characterization of the glycosylation site of human PSA prompted by missense mutation using LC-MS/MS. *J Proteome Res* 2015;**14**:2872–83.
122. Song E, Pyreddy S, Mechref Y. Quantification of glycopeptides by multiple reaction monitoring liquid chromatography/tandem mass spectrometry. *Rapid Commun Mass Spectrom* 2012;**26**:1941–54.
123. Takeuchi H, Haltiwanger RS. Significance of glycosylation in Notch signaling. *Biochem Biophys Res Commun* 2014;**453**:235–42.
124. Walsh CT. *Posttranslational modification of proteins: Expanding nature's inventory.* Englewood, CO: Roberts and Company Publishers; 2006.
125. Wu R, Haas W, Dephoure N, Huttlin EL, Zhai B, Sowa ME, Gygi SP. A large-scale method to measure absolute protein phosphorylation stoichiometries. *Nat Methods* 2011;**8**:677–83.
126. Zhao YP, Ruan CP, Wang H, Hu ZQ, Fang M, et al. Identification and assessment of new biomarkers for colorectal cancer with serum N-glycan profiling. *Cancer* 2012;**118**:639–50.

Index

Note: Page numbers followed by *f* indicate figures, *t* indicate tables, and *s* indicate schemes.

H

R

S